MTP International Review of Science

Volume 8

Macromolecular Science

Edited by **C. E. H. Bawn, F.R.S.**
University of Liverpool

Butterworths · London
University Park Press · Baltimore

THE BUTTERWORTH GROUP

ENGLAND
Butterworth & Co (Publishers) Ltd
London: 88 Kingsway, WC2B 6AB

AUSTRALIA
Butterworth & Co (Australia) Ltd
Sydney: 586 Pacific Highway 2067
Melbourne: 343 Little Collins Street, 3000
Brisbane: 240 Queen Street, 4000

NEW ZEALAND
Butterworth & Co (New Zealand) Ltd
Wellington: 26–28 Waring Taylor Street, 1

SOUTH AFRICA
Butterworth & Co (South Africa) (Pty) Ltd
Durban: 152–154 Gale Street

ISBN 0 408 70269 9

UNIVERSITY PARK PRESS

U.S.A. and CANADA
University Park Press Inc
Chamber of Commerce Building
Baltimore, Maryland, 21202

Library of Congress Cataloging in Publication Data

Bawn, Cecil Edwin Henry
Macromolecular science.

(Physical chemistry, series one, v. 8) (MTP international review of science)
Includes bibliographies.
1. Polymers and polymerization. I. Title.
QD453.2.P58 vol. 8 [QD381] 541′.3′08s [547′.84]
ISBN O-8391-1022-7 72-2331

MTP MEDICAL AND TECHNICAL PUBLISHING CO. LTD.
Seacourt Tower
West Way
Oxford, OX2 OJW
and
BUTTERWORTH & CO. (PUBLISHERS) LTD.

Filmset by Photoprint Plates Ltd., Rayleigh, Essex
Printed in England by Redwood Press Ltd., Trowbridge, Wilts
and bound by R. J. Acford Ltd., Chichester, Sussex

MTP International Review of Science

Macromolecular Science

MTP International Review of Science

Publisher's Note

The MTP International Review of Science is an important new venture in scientific publishing, which we present in association with MTP Medical and Technical Publishing Co. Ltd. and University Park Press, Baltimore. The basic concept of the Review is to provide regular authoritative reviews of entire disciplines. We are starting with chemistry because the problems of literature survey are probably more acute in this subject than in any other. As a matter of policy, the authorship of the MTP Review of Chemistry is international and distinguished; the subject coverage is extensive, systematic and critical; and most important of all, new issues of the Review will be published every two years.

In the MTP Review of Chemistry (Series One), Inorganic, Physical and Organic Chemistry are comprehensively reviewed in 33 text volumes and 3 index volumes, details of which are shown opposite. In general, the reviews cover the period 1967 to 1971. In 1974, it is planned to issue the MTP Review of Chemistry (Series Two), consisting of a similar set of volumes covering the period 1971 to 1973. Series Three is planned for 1976, and so on.

The MTP Review of Chemistry has been conceived within a carefully organised editorial framework. The over-all plan was drawn up, and the volume editors were appointed, by three consultant editors. In turn, each volume editor planned the coverage of his field and appointed authors to write on subjects which were within the area of their own research experience. No geographical restriction was imposed. Hence, the 300 or so contributions to the MTP Review of Chemistry come from many countries of the world and provide an authoritative account of progress in chemistry.

To facilitate rapid production, individual volumes do not have an index. Instead, each chapter has been prefaced with a detailed list of contents, and an index to the 13 volumes of the MTP Review of Physical Chemistry (Series One) will appear, as a separate volume, after publication of the final volume. Similar arrangements will apply to the MTP Review of Organic Chemistry (Series One) and to subsequent series.

Butterworth & Co. (Publishers) Ltd.

Physical Chemistry
Series One
Consultant Editor
A. D. Buckingham
Department of Chemistry
University of Cambridge

Volume titles and Editors

1 **THEORETICAL CHEMISTRY**
Professor W. Byers Brown, *University of Manchester*

2 **MOLECULAR STRUCTURE AND PROPERTIES**
Professor G. Allen, *University of Manchester*

3 **SPECTROSCOPY**
Dr. D. A. Ramsay, F.R.S.C., *National Research Council of Canada*

4 **MAGNETIC RESONANCE**
Professor C. A. McDowell, *University of British Columbia*

5 **MASS SPECTROMETRY**
Professor A. Maccoll, *University College, University of London*

6 **ELECTROCHEMISTRY**
Professor J. O'M Bockris, *University of Pennsylvania*

7 **SURFACE CHEMISTRY AND COLLOIDS**
Professor M. Kerker, *Clarkson College of Technology, New York*

8 **MACROMOLECULAR SCIENCE**
Professor C. E. H. Bawn, F.R.S., *University of Liverpool*

9 **CHEMICAL KINETICS**
Professor J. C. Polanyi, F.R.S., *University of Toronto*

10 **THERMOCHEMISTRY AND THERMODYNAMICS**
Dr. H. A. Skinner, *University of Manchester*

11 **CHEMICAL CRYSTALLOGRAPHY**
Professor J. Monteath Robertson, F.R.S., *University of Glasgow*

12 **ANALYTICAL CHEMISTRY – PART 1**
Professor T. S. West, *Imperial College, University of London*

13 **ANALYTICAL CHEMISTRY – PART 2**
Professor T. S. West, *Imperial College, University of London*

INDEX VOLUME

Inorganic Chemistry
Series One
Consultant Editor
H. J. Emeléus, F.R.S.
Department of Chemistry
University of Cambridge

Volume titles and Editors

1 **MAIN GROUP ELEMENTS—HYDROGEN AND GROUPS I–IV**
Professor M. F. Lappert, *University of Sussex*

2 **MAIN GROUP ELEMENTS—GROUPS V AND VI**
Professor C. C. Addison, F.R.S. and Dr. D. B. Sowerby, *University of Nottingham*

3 **MAIN GROUP ELEMENTS—GROUP VII AND NOBLE GASES**
Professor Viktor Gutmann, *Technical University of Vienna*

4 **ORGANOMETALLIC DERIVATIVES OF THE MAIN GROUP ELEMENTS**
Dr. B. J. Aylett, *Westfield College, University of London*

5 **TRANSITION METALS—PART 1**
Professor D. W. A. Sharp, *University of Glasgow*

6 **TRANSITION METALS—PART 2**
Dr. M. J. Mays, *University of Cambridge*

7 **LANTHANIDES AND ACTINIDES**
Professor K. W. Bagnall, *University of Manchester*

8 **RADIOCHEMISTRY**
Dr. A. G. Maddock, *University of Cambridge*

9 **REACTION MECHANISMS IN INORGANIC CHEMISTRY**
Professor M. L. Tobe, *University College, University of London*

10 **SOLID STATE CHEMISTRY**
Dr. L. E. J. Roberts, *Atomic Energy Research Establishment, Harwell*

INDEX VOLUME

Organic Chemistry
Series One
Consultant Editor
D. H. Hey, F.R.S.
Department of Chemistry
King's College, University of London

Volume titles and Editors

1 **STRUCTURE DETERMINATION IN ORGANIC CHEMISTRY**
Professor W. D. Ollis, F.R.S., *University of Sheffield*

2 **ALIPHATIC COMPOUNDS**
Professor N. B. Chapman, *Hull University*

3 **AROMATIC COMPOUNDS**
Professor H. Zollinger, *Swiss Federal Institute of Technology*

4 **HETEROCYCLIC COMPOUNDS**
Dr. K. Schofield, *University of Exeter*

5 **ALICYCLIC COMPOUNDS**
Professor W. Parker, *University of Stirling*

6 **AMINO ACIDS, PEPTIDES AND RELATED COMPOUNDS**
Professor D. H. Hey, F.R.S. and Dr. D. I. John, *King's College, University of London*

7 **CARBOHYDRATES**
Professor G. O. Aspinall, *University of Trent, Ontario*

8 **STEROIDS**
Dr. W. D. Johns, *G. D. Searle & Co., Chicago*

9 **ALKALOIDS**
Professor K. F. Wiesner, F.R.S., *University of New Brunswick*

10 **FREE RADICAL REACTIONS**
Professor W. A. Waters, F.R.S., *University of Oxford*

INDEX VOLUME

Physical Chemistry Series One

Consultant Editor
A. D. Buckingham

Consultant Editor's Note

The MTP International Review of Science is designed to provide a comprehensive, critical and continuing survey of progress in research. The difficult problem of keeping up with advances on a reasonably broad front makes the idea of the Review especially appealing, and I was grateful to be given the opportunity of helping to plan it.

This particular 13-volume section is concerned with Physical Chemistry, Chemical Crystallography and Analytical Chemistry. The subdivision of Physical Chemistry adopted is not completely conventional, but it has been designed to reflect current research trends and it is hoped that it will appeal to the reader. Each volume has been edited by a distinguished chemist and has been written by a team of authoritative scientists. Each author has assessed and interpreted research progress in a specialised topic in terms of his own experience. I believe that their efforts have produced very useful and timely accounts of progress in these branches of chemistry, and that the volumes will make a valuable contribution towards the solution of our problem of keeping abreast of progress in research.

It is my pleasure to thank all those who have collaborated in making this venture possible – the volume editors, the chapter authors and the publishers.

Cambridge — A. D. Buckingham

Preface

In most highly developed countries a large fraction of the chemical industry is now concerned with the manufacture and processing of polymers and at the same time the effort devoted to teaching and research in polymer science is expanding at an ever-increasing rate. This rapid growth of macromolecular science and technology has resulted in a phenomenal expansion of publications on the subject. Macromolecular science, which embraces chemistry, physics, material sciences, molecular biology, and technology and engineering, is now so highly developed that it is vital for the scientist and technologist wishing to keep abreast of advances in the subject to have access to up-todate review articles not only describing current research and progress but also giving informative surveys of the more established aspects of the subject.

This volume, the first of a biennial series, provides a source of authoritative, timely reviews of specialised topics in macromolecular science. Clearly, in a single volume not all of the more interesting and attractive areas of progress could be covered and inevitably some selection was necessary. The eight articles present a coverage and critical assessment of work carried out in the past five years in the very broad areas of the synthesis and properties of polymers and the relationships between polymer properties and molecular structure. An article entitled the 'Chemical Structure of Polymers' is published in a companion volume. It is our intention in subsequent volumes of this biennial series to review systematically developments in macromolecular science and also to include additional articles on topics in which significant advances have been made.

Liverpool C. E. H. Bawn

Contents

1
Anionic Polymerisation and Block Copolymers

M. MORTON
The University of Akron, Ohio

1.1 INTRODUCTION

1.1.1 Scope of review

The term 'anionic polymerisation' has become considerably more diffuse ever since the revolution in polymerisation chemistry brought on by the wide use of organometallic catalysts, especially the coordinated organometallic types (Ziegler–Natta). It is generally conceded that the anionic mechanism prevails whenever the growing chain-end has an actual or formal negative charge and therefore involves a positive counter-ion. However, many of the organometallic 'counter-ions' involved in these systems are so complex that they obscure the 'anionic' or 'cationic' nature of the monomer 'insertion' reaction. Furthermore, many of these systems involve heterogeneous reactions at the catalyst surface which further complicate the problem of defining the mechanism.

In view of these considerations, it has become convenient to discuss the anionic mechanism of polymerisation only in the context of homogeneous reactions initiated by simple organometallic compounds, mainly of the alkali metal type. One of the reasons for this limitation has been the relative 'simplicity' of such reactions, whose mechanism has been greatly elucidated during the past 15 years. The monomers involved in these alkali metal polymerisations have been of two types: conjugated unsaturated compounds and heterocyclics. Even with the imposition of these limitations on our definition, the field of 'anionic' polymerisation has grown rapidly during recent years, so much so that a comprehensive review of all the research publications of the past 2 or 3 years would be rather voluminous. Hence it was thought best to restrict this review to two aspects of this field: (i) recent advances in understanding of the polymerisation mechanism, and (ii) recent developments in block co-polymerisation by anionic mechanism.

1.1.2 Status of anionic polymerisation prior to current review

In order to put the more recent developments into proper perspective, it would be helpful to review *briefly* the known features of the mechanism of anionic polymerisation as they were understood prior to the recent period under review, i.e. about 1966 to 1967.

1.1.2.1 Synthetic aspects

The specific capabilities of the anionic mechanism in the synthesis of polymers are, of course, based on the special features of this mechanism, and these are described below.

(*a*) *'Living' polymers*—The term 'living' polymerisation was invoked[1] to describe the *absence of termination reactions* in anionic systems, e.g. those involving organoalkali initiators such as sodium naphthalene. In such systems, *each* polymer chain can continue to grow until all the monomer is depleted. It is this feature which leads to many other unusual features of these polymerisations.

(*b*) *Molecular weight distribution*—Non-terminating polymer chains, if initiated more or less simultaneously, will tend to attain similar molecular weights. Such a narrow molecular weight distribution has been characterised[2] as belonging to the Poisson distribution function, defined by the relation

$$x_w/x_n = 1+1/x_n$$

where x_w = weight-average number of units per chain
x_n = number-average number of units per chain

In theory, therefore, it is possible, even at relatively low molecular weight, to attain a distribution described by $1.01 > x_w/x_n > 1$. Experimentally, values of $x_w/x_n < 1.05$ have been attained by many investigators and have been recently reviewed[3].

(c) *Functional end-group polymers*—Also because of the 'living' nature of the polymer chain-ends, it is possible to form various functional end groups by simply adding a suitable reagent at the conclusion of the polymerisation. Thus organoalkali chain-ends, when treated by carbon dioxide, form carboxylate end groups; while reactions with cyclic oxides or carbonyl compounds lead to hydroxyl end groups. The various reactions used for the formation of functional end-group polymers have been recently reviewed[3]. α,ω-Difunctional polymers are especially interesting as substrates for chain extension reactions in liquid polymer technology (castable melts).

(*d*) *Block copolymers*—Possibly the most unique synthetic feature of the anionic 'living' polymers lies in the opportunities they offer for block copolymerisation of two or more monomers. To the extent that it is possible to keep the system termination-free by rigorous exclusion of impurities or side reactions, it is possible to prepare 'pure' block copolymers of controlled and uniform block size. This method has raised the synthesis of precisely defined block copolymers to an entirely new dimension. The various block copolymers that have been thus synthesised have been recently reviewed[3].

(*e*) *Stereospecificity*—Although homogeneous anionic polymerisations, such as those initiated by organoalkalies, do not show any great tendency to form stereospecific chain structures, there are a number of such cases. Thus a moderate degree of tacticity (both iso and syndio) can be found in polymethylmethacrylate[4] polymerised in various solvents. The outstanding example of anionic stereospecific polymerisation is, of course, that of a high poly(*cis*-1,4-polyisoprene) (natural rubber) by use of lithium initiators in non-polar solvents[5]. The various instances of stereospecific polymerisations initiated by organoalkali reagents have been listed in a recent review[3]. It is important, in this connection, to note the profound influence of both the counter-ion and the solvent on the microstructure of the polymer chain.

(*f*) *Effect of environment on copolymerisation*—Just as in the case of chain microstructure, the nature of the counter-ion and solvent may have a remarkable effect on copolymerisation ratios of monomer pairs. The classic example can be seen in the copolymerisation of styrene and butadiene by organolithium, where polar solvents favour styrene while non-polar solvents reverse the preference almost entirely to the butadiene. These various environmental influences on anionic copolymerisation kinetics have been discussed in a recent review[6].

1.1.2.2 Mechanistic aspects

As might be expected, the mechanism of anionic polymerisation is also subject to solvent effects, similar to the case of other ionic processes. The most profound differences have been found, of course, between polar and non-polar media, the former being characterised by the occurrence of ionisation while the latter lead to association phenomena.

(a) *Ions and ion pairs in polar media*—It has been found that many polar solvents can exert *two* distinct effects on these ionic species: (a) a *general* solvent effect due to the dielectric constant, and (b) a *specific* solvent effect, e.g. by solvation of the ion-pair or the cation (the latter commonly occurs by coordination with electron donors, e.g. ethers, amines). In this way, it has been suggested[7] that the 'anionic' growing chain can have several identities in equilibrium with each other and *each possessing its individual reactivity* towards the monomer. This can be depicted as follows:

$$\underset{\text{covalent}}{A{:}X} \rightleftharpoons \underset{\text{contact ion pair}}{A^-X^+} \rightleftharpoons \underset{\text{solvent separated ion pair}}{A^-|S|X^+} \rightleftharpoons \underset{\text{free ions}}{A^- + X^+} \quad (1.1)$$

where A^- represents the anionic chain-end, while X^+ is the counter-ion. Considerable evidence, mainly from kinetic studies, has been found for the existence of these ion-pairs as well as of the free anions. As expected, the latter are generally the most reactive species, exhibiting propagation rate constants of the order of 10^4–10^5 M^{-1} s^{-1} as compared to 10^2–10^3 M^{-1} s^{-1} for the ion pairs. The structure and reactivity of these anionic propagating species have been recently reviewed[8].

(b) *Association phenomena in non-polar media*—The non-polar solvents, i.e. hydrocarbons, are generally poor solvents for organometallic compounds of highly electropositive metals, e.g. organoalkali compounds. The one outstanding exception is the organolithium compounds, which are generally soluble in hydrocarbons as well as polar solvents, even when the organic moiety attached to the lithium is as small as an ethyl group. This has been ascribed to the small size and low electropositivity of the lithium atom.

However, because of the non-polar environment in such solvents as hydrocarbons, organolithium species are generally strongly associated in dimers and higher aggregates. As a result, the mechanism and kinetics of both the initiation and propagation steps in these polymerisations have been complex and difficult to elucidate. Attempts to ascribe reactivity solely to a small proportion of monomeric organolithium, in equilibrium with the associated species, thus

$$\underset{\text{Unreactive}}{(\text{RLi})_x} \rightleftharpoons \underset{\text{Reactive}}{x\text{RLi}} \xrightarrow{\text{M}} \quad (1.2)$$

have only been partly successful in explaining the low order of these reaction kinetics[9–11]. Thus, for example, the one-half order of the propagation rate for styrene could satisfactorily be ascribed to the observed dimeric state of association of the 'living' chain-ends. However, a similar explanation could not be proposed to account for the much lower reaction order found in diene polymerisation, since there was no evidence of correspondingly higher states of association for these chain-ends.

It should be remembered, too, that the above anomalies in kinetics of diene polymerisation parallel the previously-mentioned anomalies in their copolymerisation with styrene.

1.2 RECENT ADVANCES IN ANIONIC POLYMERISATION

1.2.1 Mechanistic aspects

1.2.1.1 Ion and ion pair processes in polar solvents

During the current review period considerable research results have been published on anionic polymerisations of various monomers in polar solvents. However, very little, *if any*, additional light has been thrown on the basic mechanism, which had already been largely elucidated in prior years. The concept of simultaneous propagation by ion pairs and free ions[7] seems to have stood the test of time so far, and enabled the determination of the propagation rate constants for both types of growing chains by means of the well-known equation[7]

$$R_p/[\text{M}]\,[\text{R}^-\text{X}^+]^{\frac{1}{2}} = k_p^{\pm}[\text{R}^-\text{X}^+]^{\frac{1}{2}} + k_p^- K_e^{\frac{1}{2}} \quad (1.3)$$

where R_p = propagation rate, $[\text{M}]$ = monomer concentration, $[\text{R}^-\text{X}^+]$ = concentration of ion pair chain-ends ($\sim$total concentration of active chain-ends when dissociation into free ions is small), $k_p^{\pm}$ = propagation rate constant for ion pairs, k_p^- = propagation rate constant for free ions and K_e = ionisation constant of ion pairs.

Of course, the determination of k_p^- requires knowledge of the value of the dissociation constant, K_e, and this has been independently measured in the usual way. It is interesting to note that values of K_e are typically about 10^{-7} M for styrene in tetrahydrofuran, with various alkali metal counter-ions.

A fairly recent publication[12] contains a compilation of data on $k_p^{\pm}$, k_p^- and K_e for various monomers in different solvents and with the five alkali metal cations (Li to Cs). It would be redundant to reproduce these data here, although they are not extensive. They help to confirm that the free carbanions, e.g. for styrene, are very much more reactive ($k_p^- \sim 10^5$ M^{-1} s^{-1}) than the ion pairs ($k_p^{\pm} \sim 10^2$ in THF with various cations), so that most of the chain growth process is performed by the free carbanions, despite their low concentration.

In this connection, it would be well to remember that the participation of ions and ion pairs in the propagation process is a *hypothesis* which has been advanced to explain the observed kinetics, and should be used with caution. Thus the existence of various kinds of ion pairs, (e.g. contact vs. solvent-separated) has also been used to explain various anomalies in these kinetics. Such explanation can only be considered as speculative, unless there is real and independent evidence for the existence of the various proposed species.

1.2.1.2 Kinetics and association phenomena in non-polar solvents, with special reference to diene polymerisation by organolithium

(a) *Chain-end association and propagation kinetics*—Recent data have provided an unambiguous answer to the question concerning the state of association of the growing chain in the organolithium polymerisation of isoprene and butadiene. This question had been the subject of much speculation, suggesting higher states of association than two for these systems, based on the low kinetic orders found for the propagation rate of these two monomers, in accordance with equation (1.3). Thus, because the propagation rate of isoprene had been found to be about $\frac{1}{4}$ order with respect to chain-ends in various hydrocarbon solvents, such as hexane[13], heptane[14, 15], cyclohexane[13, 16] and benzene[13], the assumption was made[16] that x was equal to 4 in equation (1.2). Similarly, since butadiene showed a kinetic order varying from $\frac{1}{4}$ in hexane[17] to $\frac{1}{6}$ in heptane[15], cyclohexane[18] and benzene[17], it was proposed that polybutadienyllithium chain-ends were associated in hexamers[10, 18].

The above proposals regarding the state of association of dienyllithium chain-ends were advanced simply to explain the low kinetic orders, without any experimental evidence for the actual state of association. The latter had, however, actually been determined by Morton and his associates[19], by means of viscometric measurements of the relative molecular weights of the associated, 'living' chains. The method involved a determination of the viscosity of a polymerised solution (5–20% concentration) before and after termination of the chain-ends. The relative molecular weight of the associated ('living') and unassociated ('killed') polymer chains can then be deduced from the established relation between viscosity (η) and molecular weight for such solutions, as given by[20]

$$\eta = KM^{3.4} \tag{1.4}$$

where K is a constant which includes the concentration of the solution. This relation applies only to linear chains and is therefore valid where *two* chain-ends are associated, so that $M_{\text{assoc.}} = 2\,M_{\text{unassoc.}}$. The high exponent of 3.4 makes this method very sensitive, since a doubling in the molecular weight results in a 10.5-fold increase in viscosity. By means of this method, it had previously been established[19, 21] that polystyryllithium and polyisoprenyllithium chains were associated as *dimers*, while polybutadienyllithium was also, although not as definitely, thus associated.

In view of the later proposals[16, 18] assigning a higher state of association to the isoprene and butadiene chain-ends, further measurements of the actual state of association of these polymers were made and recently reported[22]. In this light scattering measurements were made on 'living' and terminated polyisoprene (in cyclohexane), as an absolute molecular weight method; and the results of these experiments are shown in Table 1.1 below

Table 1.1 Light scattering molecular weight of 'living' and terminated polyisoprenelithium in n-hexane

(From Morton[22], by courtesy of the American Chemical Society)

Sample	*Weight-average mol. wt.* ($M_w \times 10^{-5}$ g mol^{-1})	
	Living	*Terminated*
1	3.0	1.5
2a	2.6	1.4
2b	2.6	1.3

Note: 2a and 2b are duplicates of same polymer solution.

These results again establish unequivocally the dimeric state of association of polyisoprenyllithium chain-ends. Of course, all such measurements depend in the final analysis on the maintenance of a high state of purity in the system and the avoidance of any fortuitous termination of 'living' chain-ends, but this can be assured by rigorous, high-vacuum techniques[23].

In addition to the above experiments, further confirmation of the state of association of both isoprene and butadiene chain-ends was obtained[22] by the above viscometric method, utilising a special 'capping' technique applied to polystyryllithium chain-ends. In this work, a completely polymerised solution of polystyryllithium was first prepared, under conditions such that equation (1.4) was applicable, and to this was then added a very small proportion of isoprene or butadiene, sufficient to convert all of the styryl chain-ends to dienyl and to form 30–40 units of the diene on each chain-end. Since it is known that the reaction between styryllithium and these dienes is very rapid[24] in these solvents, this technique can guarantee the complete conversion of styryl chain-ends (which is also demonstrated by the discharge of the styryllithium colour). Furthermore, the addition of so few units of the diene has an almost negligible effect on the chain length of about 2500 units. The results of this work are shown in Table 1.2.

It can be seen at once that there is no dramatic rise in viscosity of the solutions upon the addition of the isoprene or butadiene, such as would be expected if the state of association changed from dimeric for the styrene to tetrameric or hexameric for the dienes. Furthermore, the viscosities of the 'living' and terminated isoprenyl and butadienyl chain-ends again indicate an association number (N) of 2. This is especially interesting in the case of the butadiene chain-ends, since previous results on polybutadiene[19] had shown a variable and higher association number, between 2 and 3 (presumably due to some cross-linking of this polymer which would lead to more than one lithium chain-end per chain). In the case of the above 'capping' experiments, there was little opportunity for cross-linking, because of the small number of butadiene units per chain, so that an accurate value of 2 was obtained for the association number.

The above results, therefore, help to prove quite conclusively that all of the above propagating species (styrene, isoprene and butadiene) are associated in pairs, and that this does not necessarily bear any relation to the kinetic order of the reaction. While it may still be permissible to apply equation (1.2) in the case of styrene, it is obvious that no such scheme is valid for isoprene or butadiene. It must, therefore, be concluded that the propagation reaction of these dienes does not involve a simple association–dissociation equilibrium, even if such a mechanism is valid for styrene. Thus the detailed mechanism of propagation of these dienes requires further elucidation, and this will be discussed in a further section of this review.

Table 1.2 Viscometric data of 'capped' polystyryllithium in benzene at 30°C
(From Morton *et al.*[22], by courtesy of the American Chemical Society)

[RLi] $\times 10^3$	*Monomer conc. [M]*	*Flow time/s* 'Living polymer' *Polystyrene*	'Capped' polymer	*Terminated polymer*	*'Capping' diene*	*Polydiene block mol. wt.** $g\ mol^{-1} \times 10^{-3}$	*N*
0.60	1.5	—	398.8	38.4	Butadiene	2.0	1.99
0.63	1.5	—	268.9	26.2	Butadiene	2.0	1.98
2.40	3.2	1899	1974	187.5	Isoprene	2.0	1.98
2.40	3.0	1567	1591	158.4	Isoprene†	2.0	1.96

*Based on stoichiometry
†Cyclohexane solvent.

(b) *Cross-association of chain-ends with initiator*—At this point, however, it might be of interest to discuss some recent advances in understanding of other association phenomena in organolithium polymerisations in non-polar solvents, which would be expected to have an important bearing on the kinetics and mechanism of both the initiation and propagation reactions. These involve the 'cross-association' of the propagating chain-ends with the organolithium initiator itself. It has been known for some time that such alkyllithium initiators are themselves highly associated in non-polar solvents. Thus the hexameric associate of n-butyllithium has been invoked to explain the $\frac{1}{6}$ order found for the initiation rate of styrene in benzene[25], using a relation of the type shown by equation (1.2). However, more recent considerations have shown that there are two very powerful objections to this simple scheme, i.e. the nature of the associated alkyllithium species, and the possibility of cross-association between the initiator and the propagating chains.

Thus, with regard to the equilibrium between the n-butyllithium hexamers and monomeric species, Brown[26] recently pointed out that any assumption of such an equilibrium may not be justified, because of the excessive energy (>100 kcal mol^{-1}) which such a dissociation would demand. A more likely possibility, then, would be the formation of an intermediate complex between the associated alkyllithium initiator and the unsaturated monomer, followed by some rearrangement reactions whose nature cannot be suggested at this time. Such a complex sequence of steps can then be responsible for the very low order found in the kinetic studies.

However, aside from the above considerations involving the interaction of initiator and monomer, there is also the other factor involving cross-association of initiator and propagating chains. Although this would be expected to lead to a complex mixture of associated species, difficult to elucidate, it has recently been found possible to measure such cross-association[27] without too much difficulty. This was done by determining the effect of added initiator (ethyl lithium) on the state of association of poly-isoprenyllithium using the sensitive viscometric technique described by equation (1.4). In this way, by using various amounts of chain-ends and initiator, it was possible to calculate the cross-association equilibrium which would best fit the data, which are shown in Table 1.3. It can be seen that

Table 1.3 Cross-association of ethyllithium with polyisoprenyllithium
(From Morton *et al.*[27], by courtesy of the American Chemical Society)

$[PLi] \times 10^3$ /mol l^{-1}	$\frac{[C_2H_5Li]}{[PLi]}$	N	K_e $x = 5$	$x = 4$	$x = 3$	$x = 1$*
1.35	1.80	1.627	—	104	6.6	133
	3.55	1.415	1210	109	6.6	133
2.81	9.47	1.195	108	39	6.3	27
5.00	6.23	1.333	54	32	4.5	15

$$K_e = \frac{[PLi{\cdot}x\ EtLi]^2}{[(PLi)_2]\,[(EtLi)_6]^{\frac{x}{3}}} \qquad {}^*K_e = \frac{[PLi{\cdot}EtLi]^2}{[(PLi)_2]\,[EtLi]^2}$$

the interaction between the ethyllithium and polyisoprenyllithium can best be expressed by equation (1.5)

$$(C_2H_5Li)_6 + (PLi)_2 \rightleftharpoons 2(PLi{\cdot}3\,C_2H_5Li) \qquad (1.5)$$

with the equilibrium favouring the formation of the cross-complex. It should be noted, of course, that this relation is derived on the basis of the previously known association of ethyllithium as hexamers[28] and the chain-ends as dimers[22].

The relation illustrated by equation (1.5) means that, as soon as each molecule of ethyllithium reacts with isoprene to form an isoprenyllithium, the latter can tie up three molecules of the ethyllithium into a cross-complex, which thus comprises a new initiator species. Nothing is known at present about the reactivity of this cross-complex with regard to the monomer, except, of course, that all of the ethyllithium eventually is consumed. However, it is obvious that the formation of such a cross-complex between the residual initiator and the propagating chains can seriously complicate any measurement of initiation kinetics. The converse need not apply to the propagation kinetics since it is possible to work in systems containing no residual initiator.

1.2.1.3 Structure of propagating chain-end in organolithium polymerisation of dienes

The anomalous kinetics of the propagation reaction of dienes, as well as of their copolymerisation with styrene, as discussed above, have thus far not been satisfactorily explained. Neither has, for that matter, the mode of entry of the diene monomers into the polymer chain, i.e., *cis*- and *trans*-1,4 v. 1,2 or 3,4 etc. It is perhaps not surprising, therefore, that there have been some very recent attempts to study the structure of the 'living' chain-end of these polydienes in an effort to find reasons for some, if not all, of these phenomena. Thus several recent publications have described the application of n.m.r. techniques to the study of the structure of low molecular weight (oligomeric) polybutadienyllithium[29, 30], polyisoprenyllithium[29–32] and poly(2,3-dimethylbutadienyl)lithium[29, 30]. These studies throw considerable light on the 'living' chain-end structure, even though these chains were very short, ranging in size from 1 to 20 units, in order to provide a sufficiently high chain-end concentration for accurate analysis.

Before discussing these latest studies of chain-end structures, some comments may be in order about a parallel n.m.r. study of monomer–chain-end complex formation. As previously discussed (Section 1.2.1.2), there is a strong presumption that the mechanism of polymerisation of dienes by organolithium compounds may involve an intermediate complex formation between monomer and growing chain-end. To check the possibility of actually detecting the presence of such a complex, proton magnetic resonance spectra were obtained[29, 30] of butadiene and isoprene both in presence and absence of a high concentration of butadienyl and isoprenyllithium (*c.* 0.2 M). No differences could be detected in the monomer spectra, indicating that a monomer–lithium complex, if present, was quite fugitive, resulting in an undetectable ($<1\%$) steady state concentration.

The findings concerning the chain-end structures were, in general, quite similar by both groups of workers[29, 32], although there were some differences which could have an important bearing on mechanistic interpretations. The approach used in this work was to use high concentrations of initiator in order to develop a high concentration (0.2–1.0 M) of organolithium chain-ends, and to identify the chain-end n.m.r. peaks by observing any changes caused by termination. One interesting approach used by one group[29] was to 'pseudo-terminate' the active chain-ends by adding a 'transparent' monomer (butadiene-d_6 was the only one available for this purpose). This accomplished the objective of 'removing' the lithium atom from the chain-end, i.e. converting the latter into an in-chain unit, without the possible complexities introduced by the usual termination by hydrolysis by highly polar compounds (water, alcohol, etc.). The 'transparent' butadiene-d_6 was also useful in noting the effect of chain length on chain-end structure, without 'swamping' the n.m.r. spectrum with a high concentration of in-chain units.

The proton magnetic resonance spectra[29] for the butadiene, isoprene and 2,3-dimethylbutadiene chain-ends in benzene-d_6 are shown in Figures 1.1 and 1.2. The peaks corresponding to the chain-end protons are clearly recognisable by comparing the 'living' and pseudo-terminated species, the notation for the carbon atoms of the chain-end unit being as follows:

$$\underset{\delta}{-\overset{|}{\underset{|}{C}}}-\underset{\gamma}{\overset{|}{\underset{|}{C}}}=\underset{\beta}{\overset{|}{\underset{|}{C}}}-\underset{\alpha}{\overset{|}{\underset{|}{C}}}\quad Li$$

As indicated, these oligomers were all prepared by the reaction of ethyl-

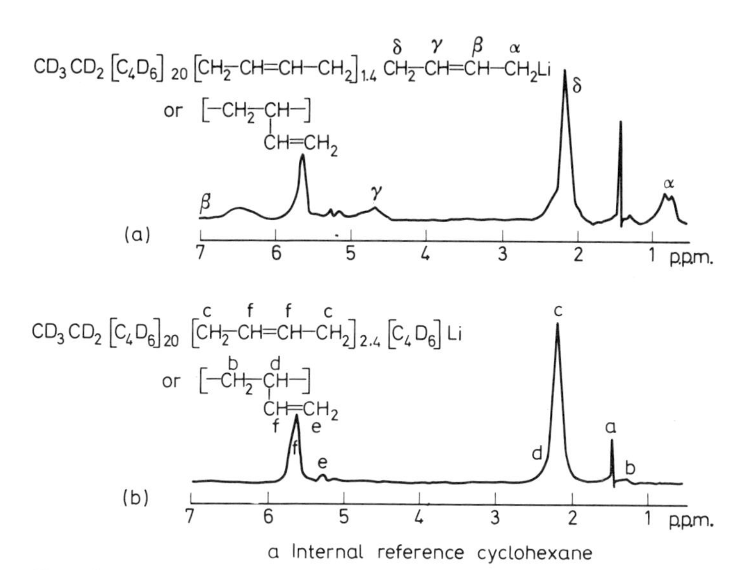

Figure 1.1 100 MHz n.m.r. spectra of polybutadienyllithium in benzene-d_6 at 23 °C; (a) 'living', (b) pseudo-terminated
(From Morton *et al.*[29] by permission of J. Wiley)

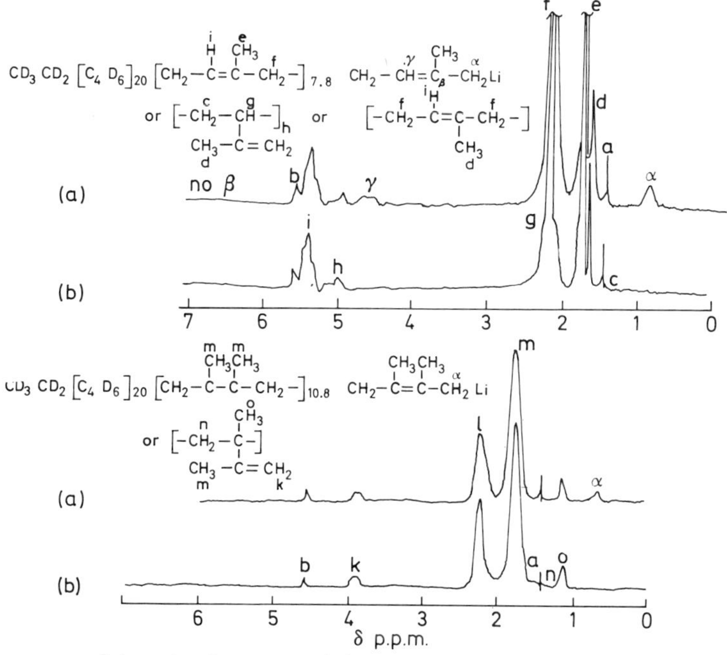

a Internal reference cyclohexane

b Remnant non-deuteration in perdeuterobutadiene units

Figure 1.2 100 MHz n.m.r. spectra of polyisoprenyllithium and poly-2,3-dimethylbutadienyllithium in benzene-d_6 at 23 °C; (a) 'living', (b) pseudo-terminated
(From Morton *et al.*[29], by permission of J. Wiley)

lithium-d_5 with the monomer. Since this initiator does not react too rapidly with the monomer, there was a considerable amount of residual initiator (transparent) present in the case of the lower oligomers of isoprene and 2,3-dimethylbutadiene. This was later remedied by using the much more reactive initiator, s-butyllithium, to make very low oligomers of isoprene[30, 32]. As for the butadiene, the chains were grown to a length of 20 or more units (using butadiene-d_6), in order to avoid the anomalous chain microstructures reported for low oligobutadienes[33].

An examination of these n.m.r. spectra, as exemplified by Figures 1.1 and 1.2, led the authors[29] to the following conclusions regarding the structure of these dienyllithium chain-ends in hydrocarbon media:

(i) The butadienyllithium chain-end unit (Figure 1.1) is 100% 1,4 with no 1,2 structures observable, although the in-chain units show a 9% 1,2 content, as expected. The lithium is *σ-bonded* to the α-carbon, as indicated by the *two equivalent* α-protons, hence there is no evidence of a π-allyl type of delocalised bonding involving the γ-carbon.

(ii)The isoprenyllithium *chain-ends* (Figure 1.2) show an exclusive 4,1 structure with no 4,3 addition, despite the 10% 3,4 *in-chain* units which are

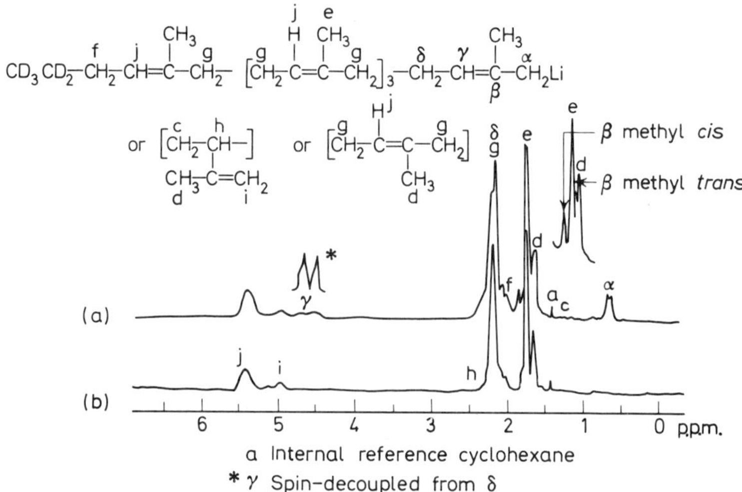

Figure 1.3 100 MHz n.m.r. spectra of polyisoprenyllithium (low mol. wt.) in benzene-d_6 at 23 °C, showing *cis*- and *trans*-1,4 chain-ends; (a) 'living', (b) pseudo-terminated
(From Morton *et al.*[29], by permission of J. Wiley)

present as expected. In this case, too, the lithium is σ-bonded to the α-carbon (two equivalent α-protons, no β-protons and no perturbation of the γ-protons which would be expected for π-allyl structures).

(iii) The dimethylbutadienyllithium also shows exclusive 1,4 structures in its chain-end units, despite the presence of 15–20% 1,2 in-chain units; there are, of course, no β- or γ-protons in this case, but the α-protons again show only a localised carbon–lithium bond.

All of these spectra, therefore, indicate exclusive 1,4 (or 4,1) chain-ends with a covalent carbon–lithium bond, yet these must somehow account for the observed 1,2 (or 3,4) content of the overall polymer chain. One obvious explanation is that there *is* a small proportion of 1,2 chain-ends, too small to be detected (e.g. $<1\%$), and that such chain-ends are much more reactive than the 1,4 chain-ends, leading to a noticeable amount of in-chain 1,2 units. Such a 'kinetic' hypothesis, however, is vitiated by the observation that the 1,2 content of the in-chain units of the 'live' polybutadiene (Figure 1.1) *is the same* (9 %) as the 1,2 content of the in-chain units of the 'pseudo-terminated' polymer despite the fact that the chain-ends of the former have become the in-chain units of the latter. This type of behaviour can then only be explained by postulating the presence of an *undetectable* amount of chain-ends which can lead to 1,2 units and are *in equilibrium* with the 1,4 chain-ends. Such chain-ends can possibly be of the 1,2 σ-bonded type or of the π-allyl type, involving the γ-carbon. This question is further elucidated by the spectra of the oligoisoprenyllithium.

Figure 1.3 shows the 100 MHz spectra of an oligoisoprene, both 'living' and 'pseudo-terminated', having a much shorter chain-length (4 in-chain units), since it was initiated directly by the ethyllithium-d_5 rather than by 20-unit long butadienyllithium-d_6. This results in a much higher concentration of chain-ends and a concomitant better resolution of the spectra. Thus the *cis* and *trans* forms of the γ-hydrogen can be resolved, as well as the *cis* and *trans* β-methyl protons. Assuming that the *cis* form lies downfield of the *trans*, the peak areas indicate an approximate 2:1 ratio of *cis*:*trans*, which is close to (but somewhat higher than) that found for the in-chain units.

This resolution of the *cis*- and *trans*-4,1 chain-ends of the isoprenyllithium makes it possible to draw some conclusions about the mechanism leading to the formation of the 3,4 in-chain units, as discussed above. Thus, using butadienyllithium as a model, it is possible to depict the two types of equilibria, i.e. σ-bonded 1,4 v. 1,2 chain-ends, as shown in Figure 1.4, and σ-

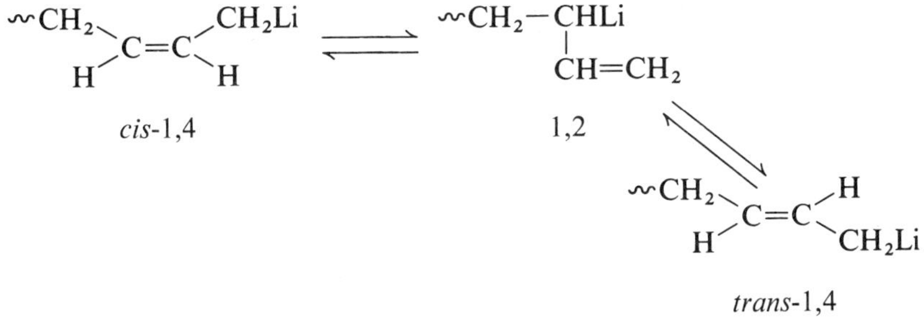

~CH₂(H)C=C(H)CH₂Li ⇌ ~CH₂—CHLi(CH=CH₂) ⇌ ~CH₂(H)C=C(H)CH₂Li

cis-1,4   1,2   *trans*-1,4

Figure 1.4 Equilibrium between 1,4 and 1,2 chain-ends of polybutadienyllithium, showing *cis–trans* isomerisation

bonded 1,4 v. π-allyl, as shown in Figure 1.5. It can be seen at once that the former, i.e. 1,4 v. 1,2 tautomerism, must lead to an isomerisation equilibrium between the *cis*-1,4 and *trans*-1,4 forms, whereas the σ–π equilibrium does not affect the configuration around the 2,3 bond. Fortunately, experimental evidence is available[29,30] to distinguish between these two possibilities. Thus if a polyisoprenyllithium solution, such as depicted in Figure 1.3, is kept at 50 °C for several days, it undergoes a slow destruction of chain-ends. Careful observation of such solutions has shown that the *cis* and *trans* forms of the chain-end do not decompose at the same rate, the *trans* generally decomposing more rapidly. Such an imbalance offers strong evidence of the *absence* of a σ-bonded 1,4 v. 1,2 equilibrium, which would permit isomerisation. Additional data on this point have become available very recently from more recent work[34], and these are shown in Table 1.4 for isoprenyllithium initiated by s-butyl- and isoprenyllithium. It can be noted that the

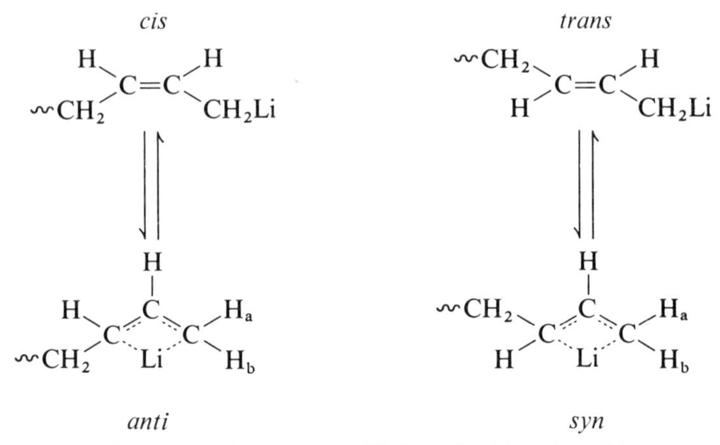

Figure 1.5 σ–π Bond structure equilibrium of polybutadienyllithium chain-ends, showing no *cis–trans* isomerisation
(From Morton *et al.*[29], by permission of J. Wiley)

cis–trans ratio seems to level off eventually, but this always appears to happen only when the decomposition has led to some precipitation and n.m.r. readings become dubious. It is also interesting to note that the chain-ends initiated by the isopropyllithium show a *decrease* in *cis–trans* ratio on ageing, unlike the increase shown by the oligomer initiated by the s-butyllithium. Apparently the relative rates of decomposition of the chain-ends is

Table 1.4 Ageing of oligoisoprenyl-lithium in d_6-benzene at 50 °C
(From Morton *et al.*[34], by courtesy of the American Chemical Society)

Initiator	$s\text{-}C_4H_9Li$		$i\text{-}C_3H_7Li$
chain length	1.23	2.70	1.32
Time/days	*Cis–trans ratio* (γ–H)		
0	1.6	1.7	1.8
3	1.9	2.4	1.3
5	2.3	2.3	—
10	2.3	2.5*	—
11	—	—	1.1†

*45 % of chain-ends destroyed
†26 % of chain-ends destroyed.

influenced by the initiator, which is, of course, always present in excess and is presumably cross-associated with the chain-ends.

In view of the strong evidence against isomerisation of the 2,3-bond in these oligomers, one can conclude that the σ-π type equilibrium depicted in Figure 1.5 represents the more likely structure of the chain-ends. Thus, for butadiene in hydrocarbon media, the 1,2 in-chain units can be postulated as occurring when an incoming monomer adds to the γ-position of the π-allyl form of the chain-end, the latter being present in undetectably small proportion but being much more reactive than the σ-bonded form with which it is in rapid equilibrium. Unfortunately, attempts to 'freeze out' the π-allyl form at low temperatures were unsuccessful due to substantial line broadening caused by the high viscosity. It is interesting to note, however, that excellent confirmation of undetectably small amounts of π-allyl structures in these systems has recently been found[30] in the polymerisation of *trans*-penta-1,3-diene. This is evidenced by the fact that the 20% 1,2-structure noted in the in-chain units contains *c.* 25–35% cis *configuration for the* 3,4 *bond instead of the* 100% trans *of the original monomer.* Since it can be shown that the monomer itself does not isomerise under these conditions, this phenomenon is most easily explained by a σ–π equilibrium of the chain-end, as depicted in Figure 1.6. Incidentally, the 70:30 *trans:cis*

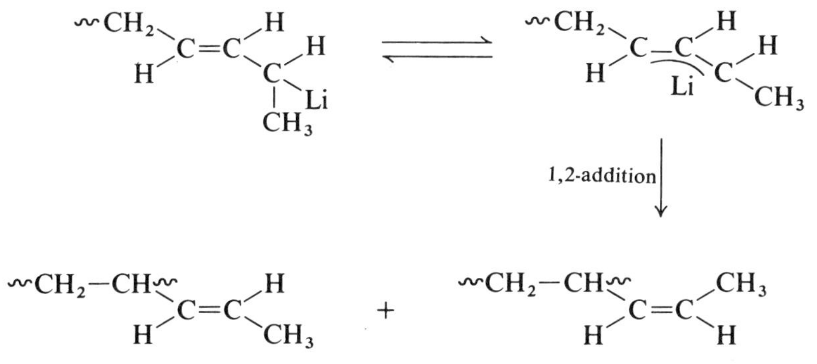

1,2-addition

trans-1,2 + *cis*-1,2

Figure 1.6 σ–π Bond structure equilibrium of poly-l-*trans*-3-pentadienyllithium chain-ends, showing isomerisation of 3,4 bonds

ratio shown by the 3,4 bonds corresponds with the thermodynamic equilibrium ratios usually observed for this type of interconversion.

The effect of ethers, and other polar solvents which can coordinate with, or solvate, carbon–lithium bonds, is to alter drastically the character of such bonds, shifting them, as might be expected, toward the delocalised π-bonding. This can be easily seen in the spectra shown in Figure 1.7 for polybutadienyllithium in tetrahydrofuran (THF). At low temperatures, the α-protons appear as two overlapping doublets, corresponding to the *two distinguishable* hydrogen atoms of the π-allyl chain-end (see Figure 1.5). These, however, tend to 'coalesce' into one doublet at higher temperature, presumably due to a sufficiently rapid σ–π equilibrium. The γ-H also is markedly affected in the π-allyl form, appearing sharply upfield at 3.3 δ instead of at 4.7 δ as was the case for the σ-bonded chain-end (see Figures 1.1 and 1.2). Hence there is convincing evidence that the predominant form of chain-end in THF is π-allylic in structure, and this, of course, correlates with the very high predominance (80–90%) of 1,2 chain microstructure observed for this polymer.

It appears reasonable, therefore, to propose a mechanism[29] as illustrated in Figure 1.8 for the propagation process for butadiene, where the σ-bonded

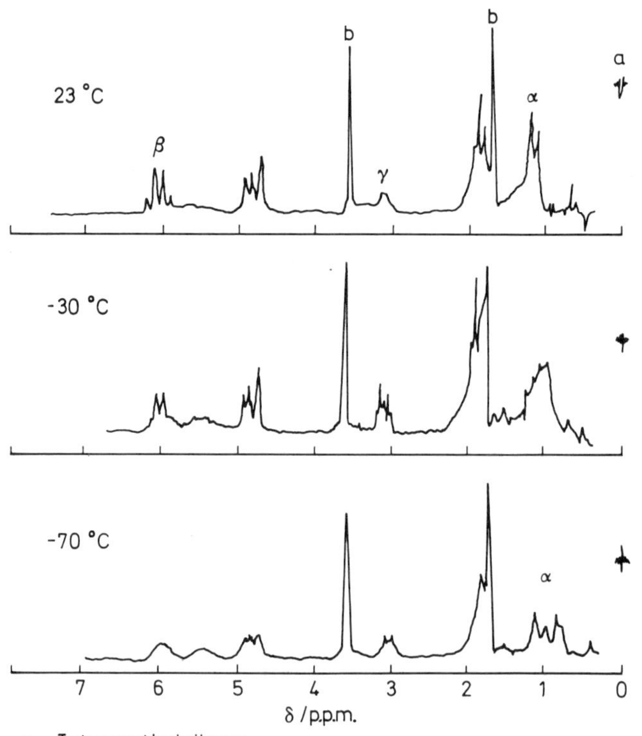

Figure 1.7 100 MHz n.m.r. spectra of polybutadienyllithium in tetrahydrofuran-d_8
(From Morton *et al.*[34], by permission of J. Wiley)

chain-end, which predominates in non-polar media, undergoes a concerted 1,4 reaction with the diene, while the π-bonded chain-end, which prevails in polar solvents, leads mainly to 1,2 addition, as shown. However, the observed fact that polyisoprene chain-ends undergo *noticeable* cis–trans *isomerisation in THF*, unlike the case in hydrocarbon solvents, makes it necessary to postulate an additional equilibrium between the π-allyl structures and a *σ-bonded* 1,2 *chain end*, as shown.

Tetrahydrofuran, being a highly solvating solvent, strongly favours the formation of π-allylic chain-ends. It is interesting to note the effect of a 'weaker' solvent, e.g. diethyl ether and this is shown in the spectra in Figure 1.9, which depict the region of the γ-protons of polybutadienyl-lithium. The spectrum at $-35\,°C$ shows only the γ-H peak (3.3 δ) of the π-bonded chain-end, but, as the temperature is decreased still further, a new peak appears downfield at 4.7 δ, which corresponds very well with that observed for the σ-bonded chain-ends (Figures 1.1 and 1.2). Hence it appears that, at sufficiently low temperatures, it is possible, in this case, to 'freeze out' the covalent bond structure.

The question now arises, how can the chain-end structures observed in these studies account for the basic features of these organolithium polymerisations? The effect of polar solvents on the chain microstructure can apparently be satisfactorily explained by the observed π–σ chain-end structures. Furthermore, the concerted reaction between the diene and the σ-bonded chain-end, which prevails in hydrocarbons, can also account for the anomalous copolymerisation behaviour with styrene, since the dienes could obviously participate much more easily in this 4-centre reaction. Hence both of these phenomena are largely elucidated on this basis. However, these chain-end

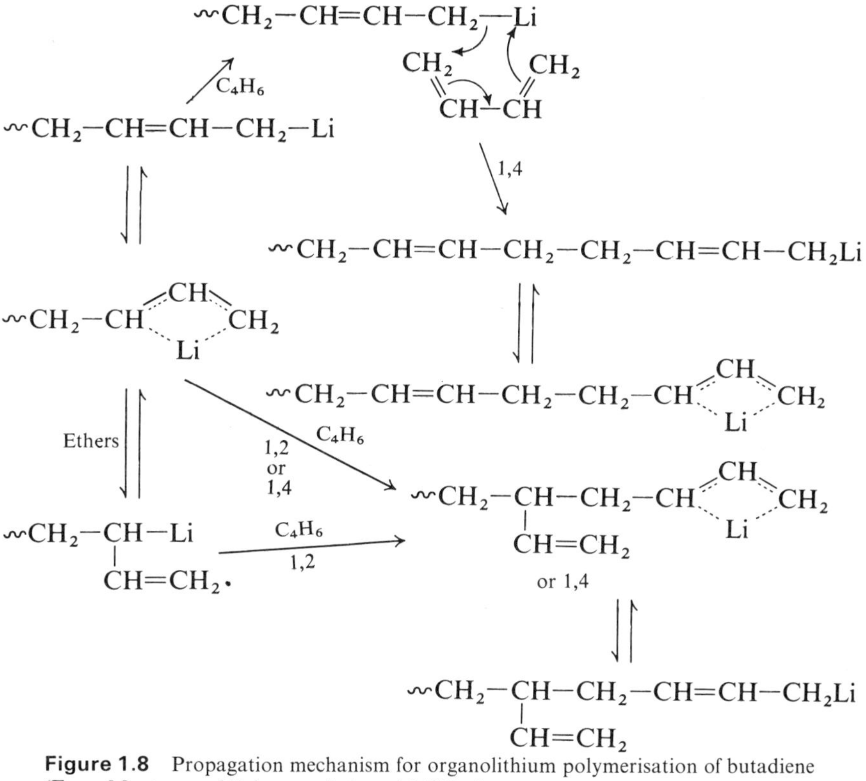

~CH_2—CH=CH—CH_2—Li
CH_2 CH_2
CH—CH
C_4H_6
~CH_2—CH=CH—CH_2—Li
1,4
~CH_2—CH=CH—CH_2—CH_2—CH=CH—CH_2Li
~CH_2—CH(CH)CH_2 Li
~CH_2—CH=CH—CH_2—CH_2—CH(CH)CH_2 Li
Ethers
C_4H_6
1,2 or 1,4
~CH_2—CH(CH=CH_2)—CH_2—CH(CH)CH_2 Li
~CH_2—CH(CH=CH_2)—Li
C_4H_6
1,2
or 1,4
~CH_2—CH(CH=CH_2)—CH_2—CH=CH—CH_2Li

Figure 1.8 Propagation mechanism for organolithium polymerisation of butadiene (From Morton *et al.*[29], by permission of J. Wiley)

structures throw little light on the other features of this mechanism, i.e., the high *cis*-polyisoprene and the low kinetic order of the diene propagation rates obtained in hydrocarbons. It has already been noted that the active chain-ends of the isoprene are largely *cis* in structure, hence the stereochemistry must already be decided at the time of monomer insertion by the concerted reaction shown in Figure 1.8. Thus isoprene strongly prefers a *cis*-addition while butadiene apparently does not, as judged by the known chain microstructure. It would be tempting to ascribe this preference to a predominant cisoid conformation for the isoprene monomer, if that were the case. However, the most recent evidence[35] favours a high transoid conformation for both isoprene and butadiene, hence this approach cannot provide an adequate explanation. It can only be stated at this time, therefore, that a concerted *cis*-addition is favoured for the incoming monomer unit, presumably on steric grounds.

As for the low kinetic order of diene propagation rates in non-polar solvents, no obvious mechanistic answer is provided by the above studies.

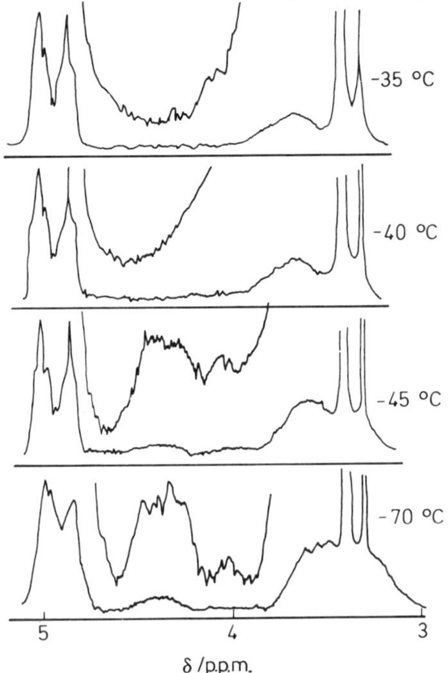

Figure 1.9 100 MHz n.m.r. spectra of polybutadienyllithium in diethyl ether-d_{10}
(From Morton *et al.*[29], by permission of J. Wiley)

It should be noted that these n.m.r. studies had to be done at relatively high concentrations of chain-ends, for the sake of analytical accuracy, and it is unfortunately not possible to study the *effect of concentration on chain-end structure*, especially down to the very low concentrations used for polymerisation rate studies. It may very well be that, since the propagation step in non-polar solvents appears to involve a concerted 4-centre reaction between monomer and chain-end, the *reactivity* of the highly polar carbon–lithium σ-bond could itself be a function of its *concentration*, decreasing as the concentration increased. However, there is no evidence available at present for such a phenomenon, so that the kinetic order must yet await an acceptable explanation.

One further item of recent evidence concerning the structure of polydienyllithium chain-ends in non-polar solvents is worthy of note. This concerns recent work[34, 80] on the active chain-end structure of oligomers of *trans*-penta-1,3-diene and hexa-2,4-diene, initiated by s-butyllithium in benzene-d_6.

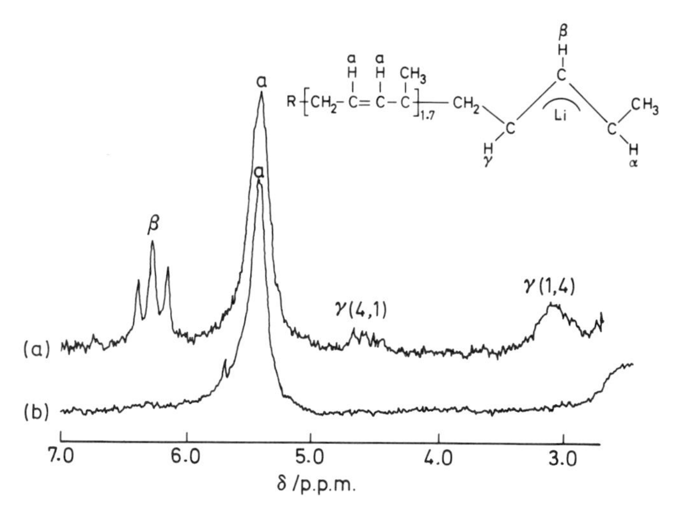

Figure 1.10 100 MHz spectra of poly-1-*trans*-3-pentadienyllithium in benzene-d_6 at 23 °C; (a) 'living', (b) pseudo-terminated (From Morton and Falvo[34], by courtesy of the American Chemical Society)

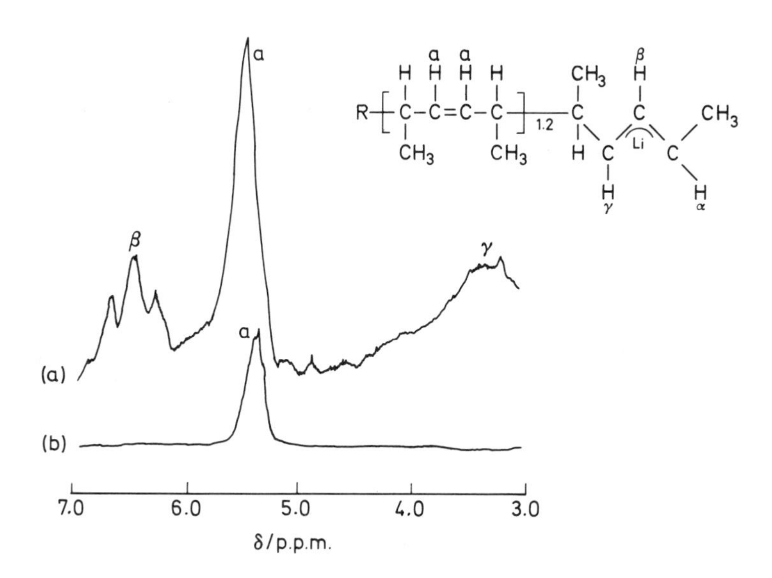

Figure 1.11 100 MHz spectra of poly-2,4-hexadienyllithium in benzene-d_6 at 23 °C; (a) 'living', (b) pseudo-terminated

Their spectra are shown in Figures 1.10 and 1.11 respectively, which depict the β and γ proton regions. In the case of the pentadiene[80], γ-protons corresponding to both 1,4 and 4,1 chain-ends are visible (Figure 1.10), with the 1,4 predominating despite the fact that such a chain-end involves a *secondary carbanion*, which should be much less stable than the primary carbanion resulting from 4,1 addition. In this connection, it is interesting to note that the 1,4 chain-ends are only noticeable in the *very low oligomers* of the penta-1,3-diene, e.g. mainly telomers, and that they disappear rapidly, with further monomer addition, in favour of the more stable 4,1 chain-ends.

In the case of the hexa-2,4-diene[34], there is only one mode of conjugated addition possible, i.e. 2,5, and the γ-protons of these chain-ends are clearly visible in Figure 1.11. These 2,5 chain-ends of the hexadiene also involve *secondary* carbanions, as in the case of the pentadiene 1,4 chain-ends, and it should be noted that the γ-proton peaks of *both* of these types of chains occur at δ 3.3, *corresponding to the γ-protons of the π-allylic chain-ends found for polyisoprenyllithium in THF* (Figure 1.7). Hence, these secondary carbanions appear to be π-bonded to the lithium even in non-polar media. Presumably, this is due to their greater instability, compared to the primary carbanions, so that a delocalisation of the bonding electrons offers a stabilising mechanism. In this connection it is interesting to note that whereas the π-bonded 1,4 chain-ends of the pentadiene apparently lead, as expected, to some 1,2 in-chain units, this is not the case for the π-bonded hexadiene chain-ends, which lead exclusively to 2,5 in-chain units. Apparently, the presence of the methyl group at the 2 position causes sufficient steric hindrance to deactivate the 3 position.

A very intriguing question arises as a result of the observed π-bonded chain-end structure of the hexa-2,4-diene. Since this type of delocalised bond

Table 1.5 Association of hexa-2,4-dienyllithium in benzene at 25 °C[36]

Base polymer	[RLi] (× 10^3)	*Flow time/s*			*N**	N_C†
		'Living' polymer before capping	*'Living' polymer after capping*	*Terminated polymer*		
Polystyryl Li	3.0	3690	1030	368	1.97	1.36
Polyisoprenyl Li	1.4	6350	1870	610	1.99	1.39

*Association number of base polymer.
†Association number of polymer capped with hexa-2,4-diene.

has heretofore only been observed in the presence of solvating ethers, such as THF, where the chain-ends are known *not to be associated*, would such π-bonded chain-ends be associated in a hydrocarbon solvent? A negative answer might be expected, since the association phenomenon is presumably a stabilisation mechanism for localised carbon–lithium bonds, and has been shown to be unnecessary when such bonds are delocalised with the aid of a solvating solvent (THF). Fortunately, some very interesting data on this score have recently become available[36] which indicate the correctness of this line of reasoning. These data were obtained using the same viscometric technique previously used to determine the state of association of styryl-, isoprenyl- and butadienyllithium[22]. In this case, hexa-2,4-diene was used to 'cap' previously prepared polyisoprenyllithium and polystyryllithium and the changes in viscosity noted, as before[22]. The results are shown in Table 1.5 and prove unequivocally that the secondary carbanions of hexadienyllithium are *much less strongly associated* (over 60% dissociated) than the primary carbanions of the other dienes. Hence it appears that this low state of association is the result of the *hybridised state* of the carbon–lithium bond and/or the *steric restrictions* imposed by the secondary carbanions. In this connection, it should be noted that, where n-butyllithium has been shown to be associated as hexamers[37], s-butyllithium exists as tetramers[38]. The hexa-2,4-dienyl chain-ends studied in this case represent the first instance where a polymer-lithium species having *secondary* allylic chain-ends has been subjected to such association studies.

1.2.2 Anionic polymerisation of cyclic sulphides

The anionic ring opening polymerisation of heterocyclics has been known for many years, in fact, it was the polymerisation of ethylene oxide by sodium alkoxides which was originally treated by Flory[2] as a special case of a non-terminating chain growth reaction leading to the Poisson distribution of chain-lengths. In more recent years, it was shown that such cyclic oxides as ethylene oxide[39] or siloxanes[40] could be polymerised by carbanionic initiators like sodium or potassium naphthalene. In all of these cases, the propagating anion resides on the hetero atom which is, of course, more electronegative than the carbon atom. Thus organolithium compounds are known to attack epoxides and higher oxides as follows[41, 42]:

$$RLi + \underset{\text{(O bridging the two } CH_2\text{)}}{CH_2\!-\!CH_2} \rightarrow R\!-\!CH_2\!-\!CH_2\!-\!OLi \rightarrow \ldots \qquad (1.6)$$

It should be noted, therefore, that the reverse reaction, e.g. the formation of a carbanion by the reaction of an alkoxide with an unsaturated monomer is rarely, if ever, found. The sulphur analogues of the cyclic oxides have also been found to behave in a similar manner, but with certain differences which have been only recently discovered and elucidated.

1.2.2.1 *Polymerisation of episulphides*

The anionic polymerisation of episulphides such as methylthiacyclopropane (propylene sulphide) involving the sodium counter-ion has been investigated by Sigwalt and his associates[43–46]. This work established that the sodium thiolate chain-end, under proper precautions, could be prevented from undergoing any termination and thus having a 'living' character. In this way, polythioethers of predicted molecular weight, up to 3×10^5 g mol^{-1} were prepared. More recently, the polymerisation of this monomer, and its analogues, by organolithium initiators was comprehensively investigated[47, 48]. The reaction between organolithium compounds and cyclic sulphides had previously been shown[49, 50] to be much more complex than the analogous reaction of the oxides, resulting in fast elimination reactions to yield olefins, lithium thiolates and other compounds. However, it has only recently been demonstrated[47] that, at low temperatures (−78 °C) in THF, the reaction between propylene sulphide and ethyllithium yields *exclusively* propylene and lithium ethanethiolate, thus

$$C_2H_5Li + CH_3\overset{}{CH}\underset{\diagdown S \diagup}{\text{—}}CH_2 \xrightarrow[\text{THF}]{-78\,^{\circ}\text{C}} CH_3CH{=}CH_2 + C_2H_5SLi \quad (1.7)$$

Any excess propylene sulphide is then polymerised in the expected manner by the *thiolate anion.*

The above reactions indicate that organolithium can react with cyclic sulphides quite differently than with cyclic oxides, in that the carbanionic moiety may attack the *sulphur atom* rather than the carbon atom. Nucleophilic reagents are known to coordinate with bivalent sulphur, and this is usually followed by elimination reactions, such as the one indicated by equation (1.7) above. In terms of polymer synthesis, the net result of equation (1.7) is that the initiation step involves the *loss* of one monomer unit; otherwise the chain propagation proceeds, as expected. In fact, the anionic polymerisation of propylene sulphide is easily initiated by lithium *thiolates* and yields a 'living' thiolate anion[48] which is stable at room temperature.

1.2.2.2 Polymerisation of thietanes

In the same study[47, 48] the reaction of ethyllithium with the four-membered cyclic sulphides (thietanes) was examined, and further confirmation was obtained of the nucleophilic attack on the *sulphur atom*, as follows

$$C_2H_5Li + \begin{array}{cc} CH_2 & \text{—}S \\ | & | \\ CH_2 & \text{—}CH_2 \end{array} \xrightarrow[\text{THF}]{-78\,^{\circ}\text{C}} C_2H_5\text{—}S\text{—}CH_2\text{—}CH_2\text{—}CH_2Li \rightarrow \ldots \quad (1.8)$$

leading to a *carbanionic* chain-end. The presence of the latter was unequivocally established by its n.m.r. spectrum[47]. The polythiabutane formed by reaction (1.8) is a crystalline polymer and hence insoluble, leading to a precipitation during the polymerisation process. However, 2-methylthiacyclobutane, when treated in the same way, was found to lead to a homogeneous, 'living' polymerisation, as indicated by the stoichiometry of the reaction and the resulting molecular weights. Furthermore, the 'living' polythiabutane, with its *carbanionic* chain-end, was shown to be capable of initiating the polymerisation of an *unsaturated monomer*, such as styrene, thus yielding a block copolymer. Hence these thietanes can not only be polymerised by organolithium species (including polymer-lithium) but can also initiate a *carbanionic* polymerisation as depicted in Figure 1.12. This makes them very versatile in the synthesis of block copolymers, as compared to the 'unidirectional' limitation of the other types of heterocyclic monomers.

Several other characteristics of these cyclic sulphides are of interest. Thus

$$C_2H_5Li + CH_3{-}CH\underset{S}{\overset{CH_2}{\diamond}}CH_2 \rightarrow \cdots \rightarrow C_2H_5{+}\left[S{-}\underset{\displaystyle CH_3}{\underset{|}{CH}}{-}CH_2{-}CH_2\right]_x^- Li^+$$

$$C_2H_5{+}\left[S{-}\underset{\displaystyle CH_3}{\underset{|}{CH}}{-}CH_2{-}CH_2\right]_x^- Li^+ + CH_2{=}\underset{\displaystyle C_6H_5}{\underset{|}{CH}} \rightarrow \cdots \rightarrow C_2H_5{+}\left[S{-}\underset{\displaystyle CH_3}{\underset{|}{CH}}{-}CH_2{-}CH_2\right]_x\left[CH_2{-}\underset{\displaystyle C_6H_5}{\underset{|}{CH}}\right]_y^- Li^+$$

Figure 1.12 Block copolymerisation of styrene and 2-methylthiacyclobutane

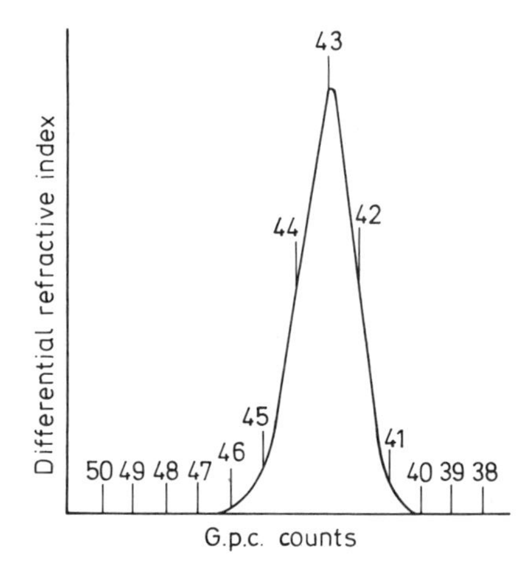

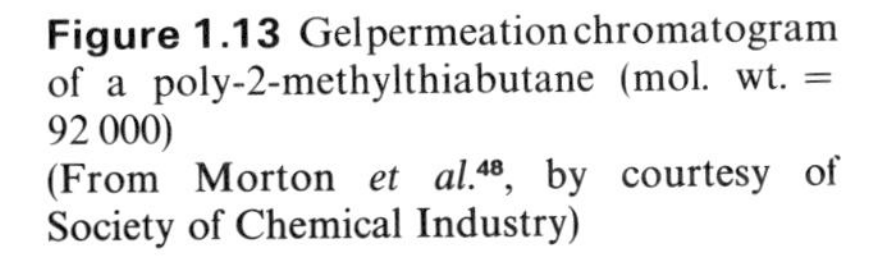
Figure 1.13 Gelpermeation chromatogram of a poly-2-methylthiabutane (mol. wt. = 92 000)
(From Morton *et al.*[48], by courtesy of Society of Chemical Industry)

the thietanes discussed above are *not* polymerised by lithium thiolates*, unlike their three-membered analogues. This is also apparently true for the higher cyclics, thiacyclopentane and thiacycloheptane, which are not attacked by organolithium initiators either. Presumably the ring strain is not sufficiently great in these cases to favour ring opening.

Table 1.6 shows data on the molecular weights of some of the cyclic sulphide polymers described above. The measured number-average molecular

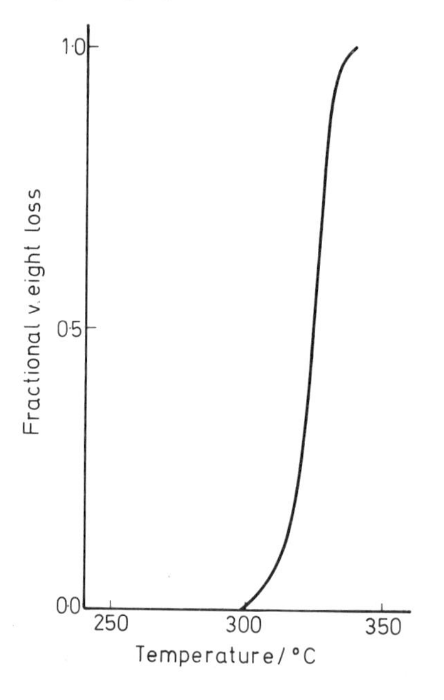

Figure 1.14 Thermogravimetric analysis curve of poly-2-methylthiabutane (*in vacuo*, heating rate 1.5 °C min^{-1})
(From Morton *et al.*[48], by courtesy of the Society of Chemical Industry)

weights (M_n) agree very well with the stoichiometrically calculated values (M_s) from the known amounts of monomer and initiator used. Furthermore, all of these polymers had the characteristic narrow molecular weight distribution of 'living' chains, as attested to by the typical gel permeation chromatograph of a poly-2-methylthiabutane shown in Figure 1.13.

The 'living' character of these polymerisations makes these monomers excellent candidates for block copolymerisations, and these are discussed in a later section of this review. It should be mentioned at this point that these monomers can lead to polymers of unusually interesting properties. All of these polythioethers have a glass temperature (T_g) of about -50 °C (high flexibility of thioether linkage) so that they exhibit excellent rubbery properties at ambient temperatures, unless crystallisation occurs. Thus poly(ethylene sulphide) is crystalline at room temperature ($T_m \sim 205$ °C) and so is the polythiabutane ($T_m \sim 56$ °C), while their methyl-substituted analogues—polypropylene sulphide and poly-2-methylthiabutane, respectively—do not

*A very recent publication[51] reports the polymerisation of thietane by the sodium, potassium or tetraethylammonium salts of thiolates in highly polar solvents, at room temperature, with some evidence that some of these yield 'living' polymers.

crystallise and hence are amorphous elastomers. In addition, the chemical structure of these polythioethers imparts very good thermal stability, as indicated in the thermogram shown in Figure 1.14 for poly-2-methylthiabutane.

Table 1.6 Molecular weights of polymers from cyclic sulphides
(From Morton *et al.*[48], by courtesy of Society of Chemical Industry)

Monomer	*Initiator*	*Mol. wt.*/g mol^{-1} × 10^{-4}	
		M_s	M_n
Propylene sulphide	C_2H_5Li	45.5	47.0
	n-C_4H_9Li	9.0	9.1
	C_2H_5SLi	6.4	6.7
	Li—R—Li*	9.1	9.2
Cyclohexene sulphide†	C_2H_5SLi	2.9	3.0
2-Methylthiacyclobutane	C_2H_5Li	8.5	8.8
	n-C_4H_9Li	1.8	2.0

*1,4-dilithio-1,1,4,4-tetraphenylbutane[52].
†M_s calculated on maximum attainable conversion of 34%.

1.2.3 Block copolymers of the AB and ABA type

Although the concept and possible utility of block and graft polymers have been known for some time[53], it has been difficult to pinpoint or predict the special physical properties which might be expected for these materials. It was recognised that both block and graft copolymers are unique in that two (or more) homopolymers, which are thermodynamically incompatible*, are chemically bonded to each other thus limiting the extent of phase separation which might ordinarily be expected in mixtures of homopolymers. Their properties as 'surfactants' which could assist the dispersion of polymer blends were, of course, recognised at an early stage and have been widely utilised, probably representing the most important application to date.

Unlike graft polymers, block polymers have until recently been utilised to a very minor extent, undoubtedly due, at least in part, to the difficulties involved in their synthesis. This is understandable since block polymers can only be formed by initiation of a 'block' polymerisation by a *chain-end* of an already formed chain, while 'grafting' can occur at any number of sites along a polymer chain. It should be noted at this point that the linking together of the chain-ends of two macromolecules is not a feasible method for the formation of block polymers, by virtue of the *incompatibility of polymers* which makes it virtually impossible to obtain a *molecular dispersion* of two different polymers. Hence the only path open for such synthesis is by the growth of a block initiated by a polymer chain-end, since molecular mixtures of one monomer with another polymer are generally possible. However, until the advent of homogeneous anionic polymerisation systems, such block polymerisation was beset with the expected 'side' reactions of chain termina-

*The 'incompatibility' of homopolymers is a characteristic of macromolecules, since a normally insignificant positive free energy of mixing even two slightly dissimilar polymers is great enough to overcome the very small combinatorial entropy of mixing these macromolecules.

tion and transfer, which invariably led to uncontrolled mixtures of blocks and homopolymers. Needless to say, the properties of such mixtures have mostly been of dubious value and very difficult to control.

As previously mentioned, anionic polymerisations of the 'living' type were the first systems capable of producing, with reasonable precautions, virtually 100% of block polymer structures, uncontaminated by products from side reactions, and possessing a very narrow distribution of each block length. Hence it is not surprising that this type of 'pure' block polymer has been able to demonstrate special physical properties hitherto obscured by the presence of polymeric 'impurities'.

A recent publication[3] contains a review of the monomers which have been used for the preparation of homopolymers and block copolymers by homogeneous anionic polymerisation. In view of the importance of dienes for elastomeric polymers, it is not surprising that various butadiene–styrene and isoprene–styrene block copolymers of the AB type have been synthesised and developed commercially[54]. As might be expected, these generally consist of a long polydiene chain attached to a shorter polystyrene block, so that the latter aggregates to form a dispersed polystyrene phase within the elastomeric polydiene. Such an elastomer containing a finely dispersed polystyrene 'filler' usually exhibits good extrusion properties, i.e. good flow at elevated temperatures and low elastic recovery ('die swell'). Such rubbers can thus be easily processed and vulcanised in the conventional manner with or without the presence of reinforcing fillers such as carbon black. Their only real point of superiority over the usual butadiene–styrene copolymers thus lies in their superior processing characteristics.

When, however, a block copolymer of styrene and butadiene is synthesised in the form of an ABA type, where A represents the polystyrene, then an entirely different and unique phenomenon occurs. These linear polymers show similar behaviour to that of a cross-linked network, presumably due to the fact that the polystyrene end blocks aggregate into 'domains' which, at ambient temperatures, act as glassy network junctions for the elastic polydiene chains. These materials have, therefore, been classified as 'thermoplastic elastomers', in view of their ability to become fluid and mouldable at elevated temperatures, unlike the normal chemically-crosslinked rubber vulcanisates. Aside from their behaviour as rubber vulcanisates, these polymers show unusually high strength, often superior to filler-reinforced elastomers, and this is now known to be due to their precise molecular architecture with regard to block sizes and their distribution.

1.3 ABA BLOCK COPOLYMERS

As indicated above, it was the ABA block copolymers, in which the A blocks comprised a glassy polymer while the B block was elastomeric, which aroused the greatest interest in recent years because of their unusual properties as 'thermoplastic elastomers'. They have therefore been the subject of intensive investigations, especially during the last few years, with regard to their synthesis, molecular structure, morphological features and mechanical properties. The main representatives thus far of these elastomeric block copolymers have been the styrene–diene–styrene type, and it is this type of polymer with which this review will mainly be concerned, with a few exceptions.

1.3.1 Synthesis of ABA block copolymers

1.3.1.1 The problem

It is now known that the desired molecular architecture for these styrene–diene triblock polymers requires polystyrene end blocks in the 10 000–20 000 molecular weight range, with a polydiene centre block of *c.* 50 000–100 000 molecular weight[55]. It has also been established[56] that these blocks should be as monodisperse as possible, to enhance phase separation, hence a consideration of the kinetics of the initiation and propagation of *each block* is of utmost importance. The main factors that must be considered in this synthesis, therefore, are as follows: (a) Influence of solvent on chain microstructure of polydiene. (b) Effect of initiator and solvent on the initiation and propagation rates. (c) Effect of solvent on copolymerisation behaviour of styrene and dienes.

The chain microstructure of the polydiene block is of paramount importance, since a high poly(*cis*-buta-1,4-diene) or polyisoprene is a good elastomer (low T_g), while high *trans*-1,4 or 1,2 or 3,4 content result in poor rubbery properties. A mixed *cis* and *trans* structure in the intermediate range is still satisfactory, provided the side vinyl content is reasonably low ($<20\%$). Studies on the effect of counter-ion and solvent on the microstructure of polyisoprene[16, 57, 58] and polybutadiene[58–60] have shown that the presence of even small concentrations ($\sim 10^{-3}$ M) of highly solvating solvents, such as tetrahydrofuran and other aliphatic ethers, is sufficient to cause the formation of a high proportion of side vinyl groups in the chain of both of the above

Table 1.7 Relative reactivities of alkyllithiums as initiators

For styrene:	s-BuLi > i-PrLi > i-BuLi > n-BuLi or EtLi > t-BuLi
For dienes:	s-BuLi > i-PrLi > t-BuLi > i-BuLi > n-BuLi or EtLi

polydienes. On the other hand, hydrocarbon media, or weakly solvating solvents, such as substituted amines or aromatic ethers, lead[57] to high 1,4 content, i.e. good elastomers.

The effect of solvating solvents on the rates of initiation and propagation in organolithium systems is also very striking. Thus solvents such as tetrahydrofuran make the initiation step virtually instantaneous while also accelerating the propagation step, leading to the attainment of very narrow molecular weight distributions. Even weakly solvating solvents, such as anisole, which are known[57] not to interfere with 1,4 addition in the propagation reaction, have been shown[55] to have a markedly accelerating effect on the initiation step, compared to the rate in hydrocarbon media. In the latter case, the initiation rate may be much too slow for the attainment of the Poisson distribution of chain-lengths, but this depends on the initiator used. Table 1.7 shows the relative reactivities of various alkyllithiums in the initiation of styrene and dienes. To put this comparison in perspective, it should be stated that the use of s-butyllithium is adequate to insure virtually instantaneous initiation and the attainment of a very narrow MWD, while the use of n-butyl- or ethyllithium can lead to a much broader MWD. In fact, if the monomer–initiator ratio falls below a certain value (e.g. < 100 for isoprene at 20 °C) all the monomer is depleted by polymerisation before all the initiator is consumed. It can be concluded, therefore, from the foregoing, that the two possible routes for the preparation of a polymer having a narrow MWD involve either the use of an initiator such as s-butyllithium in a hydrocarbon solvent, or alternately n-butyllithium or ethyllithium in a hydrocarbon solvent containing a small proportion of a weakly solvating solvent, e.g. anisole[55].

Since the formation of block copolymers involves the sequential polymerisation of several monomers, it is important also to consider the relative rates of the 'cross-over' reactions whereby one monomer block is initiated by the chain-end of another. Here again the solvent plays a critical role, as might be expected. Table 1.8 shows this effect in terms of the copolymerisation behaviour of styrene and the two dienes. The great preference for the dienes

Table 1.8 Alkyllithium copolymerisation of styrene and dienes
(From Morton[6], by courtesy of Interscience)

Monomer 1	*Monomer* 2	*Solvent*	r_1	r_2
Styrene	Butadiene	Toluene	0.1	12.5
Styrene	Butadiene	Tetrahydrofurran	8	0.2
Styrene	Isoprene	Toluene	0.25	9.5
Styrene	Isoprene	Triethylamine	0.8	1.0
Styrene	Isoprene	Tetrahydrofurran	9	0.1
Butadiene	Isoprene	n-Hexane	3.4	0.5

in hydrocarbon media is immediately obvious, so that, under these conditions, a 'blocky' copolymer would be obtained, the diene polymerising first, with only a small proportion of styrene involved, and this would be followed, after depletion of the diene, by the formation of a virtual block of polystyrene. In terms of sequential polymerisations required for block copolymerisation, this means that the initiation of a polydiene block by styryllithium would be very fast while the converse would be very slow, and highly undesirable.

1.3.1.2 Possible synthetic routes for styrene–diene–styrene block copolymers

Four possible routes for the synthesis of these triblock copolymers have been outlined[61]. These are listed below: (i) Difunctional initiators, e.g. sodium naphthalene, leading to a 'two-stage' process (i.e. polymerisation of B followed by polymerisation of A). (ii) Three-stage process, using monofunctional initiators, e.g. alkyllithium. (iii) Two-stage process, using monofunctional initiators to synthesise AB diblocks and subsequent *coupling* to ABA. (iv) Two-stage process, with alkyllithium initiators, involving formation of an initial styrene block followed by *copolymerisation* of the styrene and diene, in which the latter is preferentially polymerised.

Each of the above methods has advantages and disadvantages which have a direct bearing on the quality of the final polymer. These will be considered, with special reference to the SDS polymers, in connection with the following parameters: (i) initiation and termination problems, (ii) microstructure of polydiene.

(a) *Difunctional initiators*—These initiators, e.g. sodium naphthalene, dilithium compounds, etc., are, in principle, best for the synthesis of ABA type block polymers, on two counts. In the first place, they involve only a two-stage process, i.e. the sequential addition of two monomers, B followed by A, thus minimising any termination of blocks by adventitious impurities present in the monomers. It should be noted that termination of a B block *at one end only* will lead to an AB polymer while the statistical termination of some of the B blocks at *both ends* will lead only to B homopolymer. It will be seen later that presence of the latter does not have a profound effect on the behaviour of the ABA polymer while the AB diblock polymer does have a very great effect. This may pose a serious problem, since termination at one end only represents the statistically predominant mode.

The main problem in the use of such difunctional initiators for SDS block polymers is due to the fact that they are generally soluble *only in ethers*, or similar solvents, which lead to a *low* 1,4 *structure* in the polydiene, i.e. a non-rubbery polymer. It is possible, of course, to use metallic lithium itself as an initiator, in hydrocarbon solvents, and thus achieve a high 1,4 structure, but the extremely *slow initiation rate* of such heterogeneous systems makes it impossible to control the chain length within the prescribed limits for these block polymers.

In this connection, some new developments[52] in homogeneous polymerisation with dilithium initiators have made it possible to circumvent these difficulties, and to use this system to generate ABA block polymers of high purity. The soluble dilithium initiators were prepared in high concentration by the reaction of lithium with 1,1-diphenylethylene in presence of a small proportion of aromatic ether, and these were used to synthesise an ABA polymer from isoprene (B) and α-methylstyrene (A). This method is particularly important in the case of a monomer like α-methylstyrene, which has a low ceiling temperature and therefore requires low temperatures to avoid the presence of residual monomer. In this particular case, the isoprene was first polymerised by means of the dilithium initiator, leading to a high 1,4 structure ($\sim 90\%$), after which a substantial amount of dimethoxyethane was added to permit the polymerisation of the α-methylstyrene to virtual completion at -78 °C.

(b) *Three-stage process with monofunctional initiators*—These have the disadvantage of involving *three* sequential monomer additions with the attendant dangers of termination by impurities. Furthermore, attention must be paid to the relative rates of initiation and propagation of each block. Actually, it is the initiation of the first block, styrene, that poses the only problem, since this is a short block and requires a very fast initiation rate in order to assure a narrow molecular weight distribution. Such fast initiation can be accomplished either by the use of s-butyllithium or by means of primary alkyllithiums together with small proportions of an aromatic ether, like anisole, which markedly accelerates the initiation rate *without disrupting* the high 1,4 chain structure[55]. The subsequent initiation of the diene block by the styryllithium is very rapid[24], too, either in presence or absence of the aromatic ether, while the initiation of the last styrene block can be made very fast in presence of any added ether, either aromatic or aliphatic.

The efficacy of such a three-stage process in producing a 'pure' block

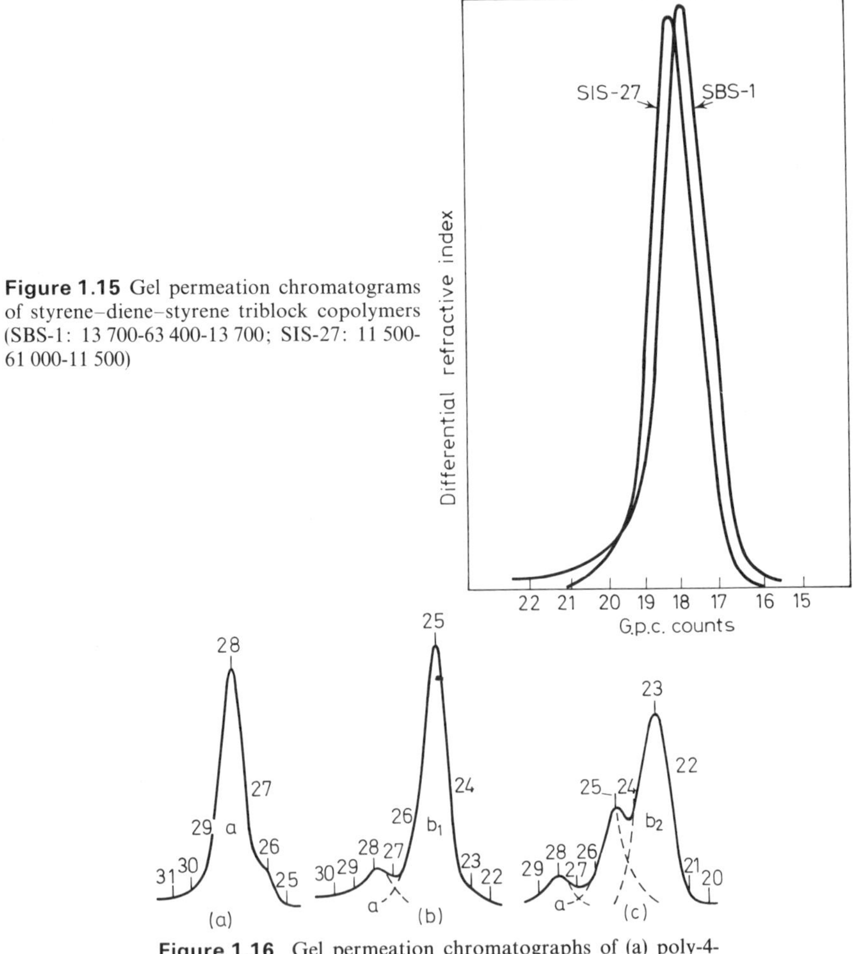

Figure 1.15 Gel permeation chromatograms of styrene–diene–styrene triblock copolymers (SBS-1: 13 700-63 400-13 700; SIS-27: 11 500-61 000-11 500)

Figure 1.16 Gel permeation chromatographs of (a) poly-4-vinylbiphenyl (A block), (b) poly-4-vinylbiphenyl-polyisoprene (AB block), (c) coupled AB block polymer
(From Heller *et al.*[63], by permission of J. Wiley)

polymer is indicated by the gel permeation chromatographs shown in Figure 1.15 for SBS and SIS polymers[55] initiated by ethyllithium in benzene containing a small proportion of anisole. The sharpness of the single peaks, the absence of any shoulders, and the fact that the M_n of these polymers was found to be within 5% of the stoichiometric value attests to the absence of any fortuitous termination under the rigorous high-vacuum purification techniques used.

(c) *Monofunctional initiation and coupling*[54, 62] – This is a two-stage process, analogous to that of difunctional initiation and, similarly, is useful for 'unidirectional' block polymerisation where, for example, A can initiate B but not vice versa. It also has the advantage of a two-stage process in minimising termination. However, it has one very great disadvantage in that it demands a high precision in the stoichiometry of the coupling reaction and in its efficiency. Furthermore, any deficiency in the coupling reaction can lead to the formation of excess diblocks (SD) which have been shown[56] to be very deleterious in very small amounts.

Figure 1.16 shows a g.p.c. trace of a vinylbiphenyl–isoprene–vinylbiphenyl polymer[63] prepared by coupling a 4-vinylbiphenyl–isoprene lithium diblock by means of phosgene. The presence of a small proportion (~5%) of poly-4-vinylbiphenyl as well as of a substantial amount (26%) of the uncoupled diblock is obvious.

(d) *Two-stage process of block polymerisation and copolymerisation of styrene and dienes* – This method is based on the known behaviour of styrene in copolymerisation with dienes in lithium–hydrocarbon systems[6], i.e. the dienes are preferentially polymerised first so that most of the styrene ends up as a homopolymer block at the end. It has the usual advantage of a two-stage system in minimising the effects of termination. Thus, even if the monomers contain terminating impurities, these may not be particularly deleterious. The first addition of styrene may lead to destruction of some of the organolithium initiator but this will merely increase the polystyrene block length to some extent. The second addition, of the proper mixture of styrene and the diene, may result in termination of the first styrene block, i.e. formation of free polystyrene, which is known[56] not to be too important in affecting the polymer properties. Thus no free diblocks should form, if the termination reactions *occurs rapidly relative to propagation.*

The marked disadvantage of this method is due to the fact that *some proportion of styrene* becomes copolymerised with the diene, and this *decreases the incompatibility* of the two phases, leading to more phase blending and hence poorer tensile properties. This is especially noteworthy in SIS polymers, since isoprene does not exclude the styrene as strongly as does butadiene. Hence the 'isoprene block' contains a substantial amount of styrene, and this is reflected in the poorer physical properties of these polymers[64].

1.3.2 Morphology of styrene–diene triblock polymers

The molecular weight limits imposed on the styrene–diene–styrene block copolymers have already been described above. Superposed on these limits is the requirement that the total polystyrene content should be between 20% and 40% by weight. The effect of these variables on the physical properties is described in a later section, but suffice it to say that these limits on block size and proportions are imposed by the need to attain the correct morphology. Thus the lower limit of molecular weight is governed by minimum polystyrene chain length required for formation of heterogeneous domains, while the upper limit is set by viscosity considerations which can seriously hamper good separation of these domains.

A schematic representation of the chain structure and morphology of a styrene–butadiene–styrene thermoplastic elastomer is shown in Figure 1.17. The suggested morphology has actually been demonstrated by direct observation of the polystyrene domains by means of electron microscopy[55, 65–68]. These included both replica specimens and stained thin films, prepared from

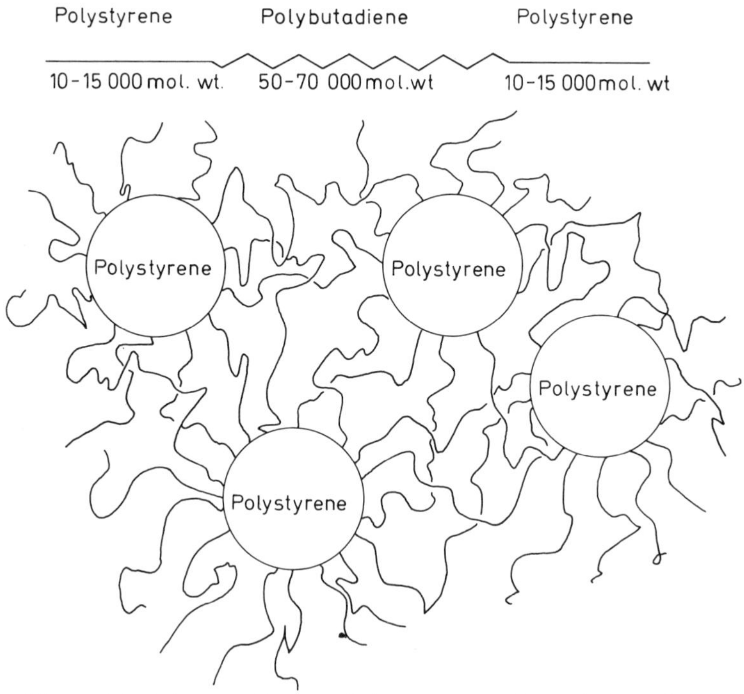

Figure 1.17 Schematic diagram of the structure and morphology of styrene–butadiene–styrene triblock copolymers
(From Morton *et al.*[56], by courtesy of Rubber and Technical Press Ltd.)

moulded or solvent-cast films, and revealed the presence of ordered spherical domains of polystyrene 200–400 Å in diameter. Small angle x-ray scattering studies[69] have also indicated spherical domains arranged regularly as a 'macrolattice'. In this particular case, an SBS polymer of molecular weights 21 100–63 400–21 100 resulted in polystyrene domains 356 Å in diameter, and the orthorhombic macrolattice had unit cell dimensions of 676, 676 and 566 Å. This is diagrammed in Figure 1.18.

The spherical polystyrene domains described above were, of course, observed in 'unperturbed' films, mainly formed by solvent casting. As may be expected, these domains can assume different shapes under shearing forces, e.g. when the material is extruded. Thus an ordered array of cylindrical domains has been observed[70] in these block copolymers, when subjected to an uniaxial shearing force, and this, of course, introduces an anisotropy in elastic properties of the material. It has also been shown, from electron microscope observations[65, 66] that the polystyrene domains undergo a substantial inelastic deformation when the polymer is subjected to large strains. Such deformations are recoverable by thermal treatment, and their significance to the mechanical properties of the material is discussed in a later section.

It should be noted that the size and shape of the polystyrene domains is a direct outcome of the molecular architecture. Thus the driving force for domain formation is the thermodynamic incompatibility of the two different blocks, but the actual aggregation of the polystyrene blocks into domains is hindered by the limitations imposed by the statistical end-to-end distance of the central polydiene block to which they are attached. This accounts for the

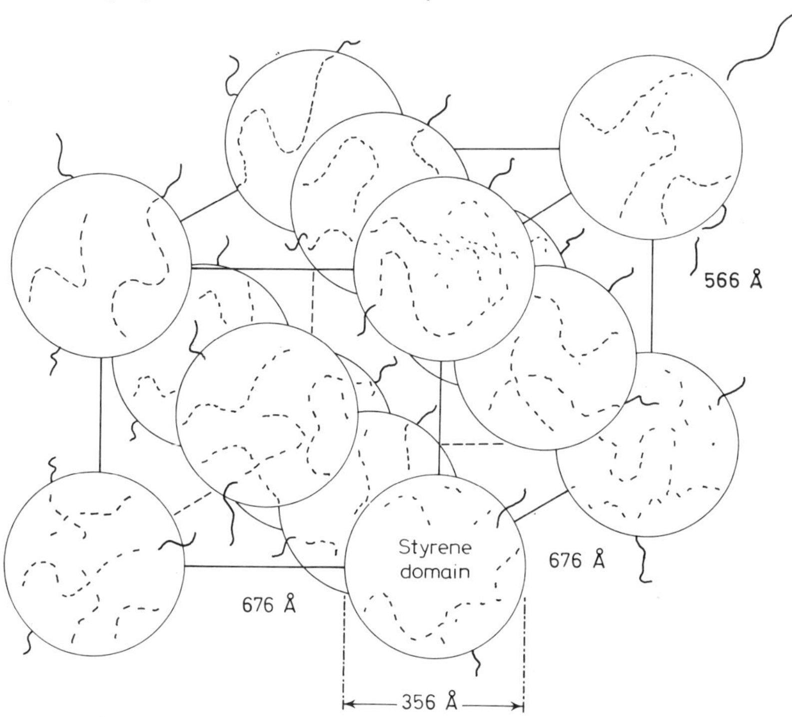

Figure 1.18 Scale model of the 'macrolattice' of a styrene–butadiene–styrene triblock copolymer (21 100-63 400-21 100), from x-ray scattering[69]
(From McIntyre and Campos-Lopez[69], by courtesy of the American Chemical Society)

very small and remarkably uniform size of these domains. Other considerations, such as the surface energy between the two phases, have also been invoked[71] in the theoretical treatment of this morphology. Finally, it should be remembered that all of these thermodynamic considerations must be subject to the time-dependent, or kinetic, factors which would govern the phenomenon of phase separation of macromolecular species.

1.3.3 Mechanical properties of thermoplastic elastomers (ABA block copolymers)

The styrene–diene thermoplastic elastomers have aroused great interest principally because of their physical properties, which can be enumerated as follows: (i) A high degree of elasticity, similar to that of rubber vulcanisates. (ii) Retention of thermoplastic properties, unlike the thermoset character of rubber vulcanisates. (iii) Excellent tensile strength, capable of surpassing that of either gum or filled vulcanisates of natural rubber.

The first two of the above characteristics are an obvious result of the morphology, as described above. It is the exceptionally high strength of these elastomers which is perhaps the most unexpected phenomenon. It should be borne in mind that the polybutadiene or polyisoprene in these block copolymers has the character of an *amorphous* elastomer, incapable of undergoing the strain-induced crystallisation which gives natural rubber its very high tensile strength. Hence these elastomers depend on some type of *reinforcement* (e.g. by a finely divided filler) for the development of high strength, and one must look to their morphology for an answer.

In principle, these elastomers consist of a 'physical' network of elastic chains, held together by glassy domains, which can act both as network junctures and as a finely divided reinforcing filler (their size is somewhat smaller than that of the finest carbon black and they are present in reasonable proportion – *c.* 30% by volume – to provide reinforcement). An additional characteristic of these polymers is, of course, the monodispersity of the elastic chains between junctions (an outcome of the anionic polymerisation mechanism). If these chains are considered to represent the 'molecular weight between cross-links' (M_c) parameter of elastic networks, then they can introduce a marked deviation from the usual, randomly cross-linked rubber vulcanisate. These aspects must be considered in any mechanism that may be advanced to account for the behaviour of these elastomers.

It should be stated, at the outset, that the strength of these elastomers is very sensitive to certain variations in the molecular architecture, as will be seen later, and even to variations in the processing history. Hence such measurements of strengths may be completely unreliable and worthless unless there is unequivocal evidence that the desired molecular architecture has in fact been accomplished. Many of the reported results may therefore be invalid, unless accompanied by such evidence about the macromolecular structures involved. The best evidence of this type is, of course, an accurate measurement of the molecular-weight distribution (MWD) as given, for instance, by gel permeation chromatograms, such as those in Figure 1.15, indicating the presence of a single species of very narrow MWD.

1.3.3.1 Effect of processing variables

Since the integrity of the 'network' in these block copolymers is dependent upon a phase separation phenomenon, it can be readily imagined that their strength could be quite sensitive to the 'processing' variables. Depending upon what kind of processing is involved, it is obvious that some correlation must be obtained between the conditions of the processing and the final properties. For example, in discussing the tensile properties of various modifications of these elastomers, it is necessary to know under what conditions the tensile test specimens were prepared as well as the reproducibility of the test data.

An interesting illustration of the effect of these variables on the integrity of these polymers is available from some recent data[56] on the effect of press moulding versus solvent casting in the preparation of tensile test films. The tensile strengths of two SIS block copolymers are shown in Table 1.9 where a comparison is made between moulded versus solvent-cast films. The SIS-1 (30% styrene) polymer had the structure 13 700–63 400–13 700 in molecular weight units, while SIS-5 (40% styrene) was structured as 21 100–63 400–

Table 1.9 Effect of moulding versus film casting on tensile strength
(From Morton *et al.*[56], by courtesy of Rubber and Technical Press Ltd.)

Polymer	*Tensile strength*/kg cm^{-2}		
	Cast film (0.03 cm)	*Moulded sample**	
		(0.03 cm)	(0.13 cm)
SIS-1	320	250†	290
SIS-5	340	270†	290

*Moulded 10 min. at 140 °C.
†Moulded from 0.03 cm cast film.

21 100. Both polymers had the required narrow MWD as indicated by gel permeation chromatography. The solvent used for casting was a mixture of 90% tetrahydrofuran and 10% of methyl ethyl ketone which had been shown[66] to be the best solvent for this use. Care also had to be taken to remove the last traces of the solvent under vacuum for several days at room temperature.

It can be seen at once from Table 1.9 that solvent casting yields films of

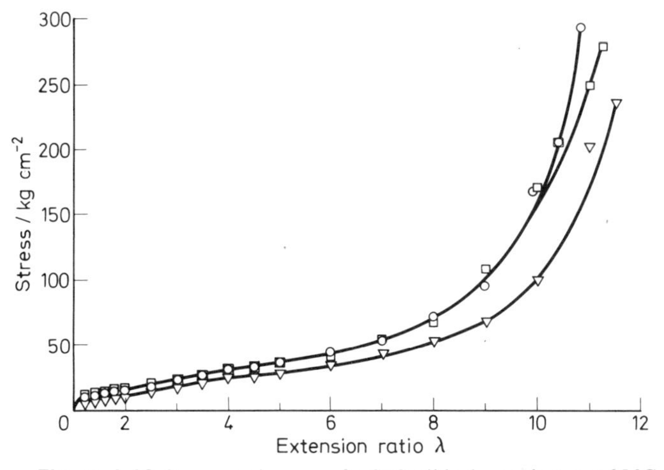

Figure 1.19 Stress–strain curves for SBS triblock copolymer at 25 °C (13 700-63 400-13 700) on films cast from different solvents. O tetrahydrofuran–butanone, □ benzene–heptane, ▽ CCl_4
(From Morton *et al.*[56], by courtesy of Rubber and Technical Press Ltd.)

higher strength and that this is not due to their thinner dimension (thinner films often show higher tensile strengths, per unit cross-section, due to the lower statistical *probability of flaws*). In fact, the remoulding of a cast film resulted in a noticeable decrease in strength. This is an excellent example of the effect of conditions which may influence phase separation in these block copolymers. Presumably solvent casting permits a much better phase separation than is possible in the viscous melt, leading to 'purer' polystyrene domains, which contribute to the integrity of the network. It is interesting to note that even when such a phase separation is attained in the solvent-cast films, it can be disrupted again by subjecting such films to press moulding.

The role of the polystyrene domains in holding the network together has been further demonstrated by other experiments with solvent casting of films. For this purpose, three different solvent systems were used, based on recent morphological studies[66] with electron microscopy which showed that specific solvents for *one* of the two phases led to better phase separation. Thus a 90 : 10 volume ratio mixture of tetrahydrofuran–butanone was found to be a specific solvent for the polystyrene, while a 90 : 10 benzene–heptane mixture was a specific solvent for the polydiene. Carbon tetrachloride is, of course, a mutual solvent.

The stress–strain relations of the films cast from the above three solvents are shown in Figure 1.19, where it can be seen that the two specific solvents (THF–MEK and benzene–heptane) show very similar curves and almost the same tensile strength, while the mutual solvent (CCl_4) shows a lower 'modulus' at any given elongation, as well as a noticeably lower strength. These data agree very well with the electron microscopy studies[66] mentioned above which showed better phase separation with the use of the two specific solvents. Again it is evident that phase blending reduces the efficacy of the glassy polystyrene in reinforcing and maintaining the integrity of the network under stress. This also confirms recent studies of the time–temperature response[72] of these polymers which were taken to indicate that it is the polystyrene phase which primarily supports the tensile stresses imposed on these materials.

1.3.3.2 Effect of molecular architecture

It might be expected that the molecular structure of these block polymers would have a controlling influence on their mechanical properties. It is therefore of prime interest to examine the effect of such variables as block size and proportion of polystyrene on the physical behaviour of these materials. Considerable information is now available on the influence of these parameters on the stress–strain behaviour and tensile strength of these elastomers.

As a working hypothesis, these elastomers can be considered as elastic networks in which the polystyrene domains act *both* as network junctions and as a finely divided rigid filler. The length of the elastic centre block can then be considered as the 'molecular weight between cross-links' (M_c) parameter, as in rubber vulcanisates, and its size should influence the stress–strain properties. Furthermore, since these elastic centre-blocks can be made almost monodisperse in length, it should be possible to observe the effect of such a 'uniform' network on these properties.

Figure 1.20 shows the stress–strain curves obtained on a series of SBS block copolymers in which both the block size and polystyrene content were varied. These polymers were *moulded* (10 min at 140 °C) and microdumbbell specimens tested at a draw rate of 2 in min^{-1}. It is apparent at once that the polystyrene content affects the stress–strain properties in the expected way, i.e. a higher content raises the stress at any given strain. However, there seems to be no effect, *per se*, of the block size on the stress levels, i.e. the centre-block length apparently does *not* represent the M_c value of the network. A re-examination of the situation indicates that this conclusion was, in fact,

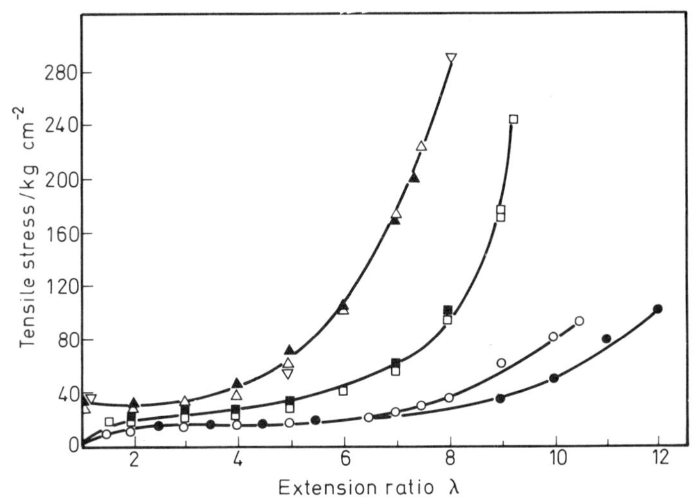

Figure 1.20 Tensile properties of styrene–butadiene–styrene block copolymers at 25 °C.

	% Styrene	*Mol. wt.* ($\times 10^{-3}$)	
○	20	8.4–63.4–8.4	
●	20	13.7–104–13.7	
□	30	13.7–63.4–13.7	
■	30	21.2–97.9–21.2	
△	40	21.1–63.4–21.1	
▲	40	13.7–41.2–13.7	
▽	40	13.7–41.2–13.7	(Annealed)

(From Morton *et al.*[55], by permission of J. Wiley)

predictable, because of the *large size* of this elastic chain (40 000–100 000 mol. wt.). Since polybutadiene is considered as having a 'molecular weight between *entanglements*' of *c*. 6000, it is obvious that each centre-block would contain a substantial number of such entanglements, which would, in effect, act as *network junctions*. This has also been pointed out and demonstrated by means of solvent-swelling experiments[73].

The other interesting feature observed in Figure 1.20 is the effect on tensile strength, which appears to be a monotonic function of the polystyrene content. In this regard, the results on the 40% styrene polymer are especially interesting, since a substantially higher strength was obtained by 'annealing' these specimens, which was *not* the case for the polymers of lower styrene content. The term 'annealing' here refers to a slow cooling rate (1 °C min^{-1}) of the moulded specimen, as opposed to the usual rapid cooling of 20 °C min^{-1}. It is apparent here that, at high volume fractions of polystyrene, more care must be exerted in moulding this material in order to permit adequate phase separation in the highly viscous medium. Another feature of these high styrene block copolymers is the 'yield point' which occurs at the first draw of these specimens (and which is not present on subsequent stretching cycles). It appears that, at 40% styrene content, the polystyrene domains may reach or exceed the critical volume fraction for packing and may in fact form a continuous interpenetrating phase with the polybutadiene. The first draw apparently destroys the inter-domain contacts, resulting in the expected dispersion of the polystyrene within the elastic matrix.

The effect of polystyrene content on the tensile strength (Figure 1.20) can be taken as evidence for the 'filler reinforcement' hypothesis previously

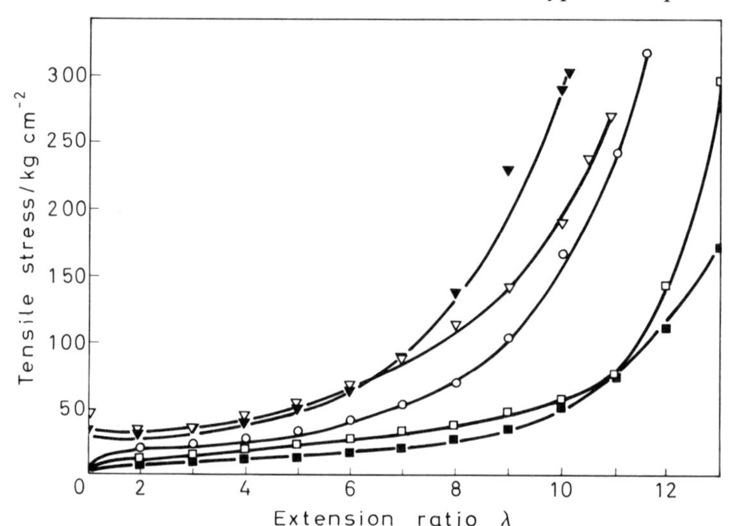

Figure 1.21 Tensile properties of styrene–isoprene–styrene block copolymers at 25 °C.

	% Styrene	*Mol. wt.* ($\times 10^{-3}$)
■	20	8.4–63.4–8.4
□	20	13.7–109.4–13.7
○	30	13.7–63.4–13.7
▽	40	13.7–41.1–13.7
▼	40	21.1–63.4–21.1

(From Morton *et al.*[56], by courtesy of Rubber and Technical Press Ltd.)

described. However, as will be seen later, it may not be valid, in general, to relate the strength of these elastomers directly to the polystyrene content. This is illustrated quite clearly in Figure 1.21, which shows analogous data for a series of SIS block copolymers. Here again, the *stress* levels appear to be directly related to the polystyrene content and largely independent of block size. However, in this case, the tensile strength appears to be *independent* of polystyrene content, with the exception of the one specimen having the lowest content (20%) as well as the lowest polystyrene block size (mol. wt. = 8400). (These SIS polymers exhibit higher tensile strengths than the SBS polymers in Figure 1.20, since these specimens were solution cast rather than moulded.)

These results on the styrene–isoprene–styrene polymers help greatly in elucidating the role of the polystyrene in the tensile strength. Thus it appears that, unlike the case of the usual fillers (e.g. carbon black), it is not simply the amount of 'filler' that is important in this case but its behaviour. Since all the SIS polymers show very similar high tensile strengths, with the exception of the polymer having a very short polystyrene block, it may be suspected that the latter polymer is 'defective' in not being able to provide a good phase separation of polystyrene domains. In other words, the short polystyrene blocks are too *compatible* with the polyisoprene. These considerations of *compatibility* can very well also be invoked to explain the varying tensile strengths found for the SBS polymers (Figure 1.20). Available experimental evidence indicates that polybutadiene is basically more compatible than polyisoprene with polystyrene. Phase separation of polymers is governed not only by their basic incompatibility but by the molecular weight and relative concentrations.

Since, at any given molecular weight, the polystyrene would be more compatible with the polybutadiene than with the polyisoprene, it can thus be suggested that the increase in tensile strength in the SBS series of Figure 1.20 is due to the better phase separation which occurs as both the proportions and block size of the polystyrene increase. In the SIS series (Figure 1.21) this 'critical' phase separation has already been reached even at 20% styrene content (provided the mol. wt. of the polystyrene block is not too low), due to the greater incompatibility in this system, so that no further increase in tensile strength is found with increase in styrene content and block size.

The effect of polystyrene block size on the tensile strength of SIS block polymers is further demonstrated in Table 1.10. The dramatic drop in strength as the polystyrene block is reduced from 13 700 to 8400 to 7000, at constant

Table 1.10 Tensile strength of SIS block polymers
(From Morton *et al.*[56], by courtesy of Rubber and Technical Press Ltd.)

Polymer	*Wt.% styrene*	*Mol. wt.* ($\times 10^{-3}$)	*Tensile strength* /kg cm^{-2}	*Stress at* $\lambda = 4$ /kg cm^{-2}
SIS-5	40	21.1–63.4–21.1	310	37
SIS-2	40	13.7–41.1–13.7	306	43
SIS-1	30	13.7–63.4–13.7	321	24
SIS-3	20	13.7–109.4–13.7	270	18
SIS-4	20	8.4–63.4–8.4	160	11
SIS-9	19	7.0–60.0–7.0	22	13
SIS-10	11	5.0–80.0–5.0	~0	~0

polystyrene content, may be ascribed to the approach to the 'critical block size', where no phase separation would be expected. This apparently occurs at *c.* 6000 mol. wt. polystyrene[71, 74]. It should be pointed out that, although the loss of integrity of the SIS-9 and SIS-10 polymers of Table 1.10 is ascribed to the failure of phase separation, this cannot be an unequivocal conclusion, in the absence of actual morphological evidence, since the rapid decrease in T_g which occurs in low molecular weight polystyrene may be at least a contributing factor to this loss of strength.

All of the above data indicate that the modulus of these elastomers appears to depend solely on the polystyrene content, analogous to filler reinforcement, but that the tensile strength is sensitive to the extent of phase separation of the polystyrene, phase mixing apparently softens the latter domains thus lessening their ability to withstand high stresses. This seems to corroborate the proposed mechanism[55, 72] which ascribes the strength of the elastomers to the integrity of the polystyrene domains, which yield rather than rupture, at high stresses, thus absorbing the elastic energy which would otherwise contribute to rupture by crack propagation. This absorption of energy by a yielding of the polystyrene shows up as a high hysteresis loss[75] in the stress–strain curve, as shown in Figure 1.22. The high hysteresis loss as well as the high 'set' of the block polymers is easily observed. Presumably this reflects the yielding and distortion of the polystyrene domains, which thus act as 'energy sinks', delaying the onset of rupture. This distortion of the poly-

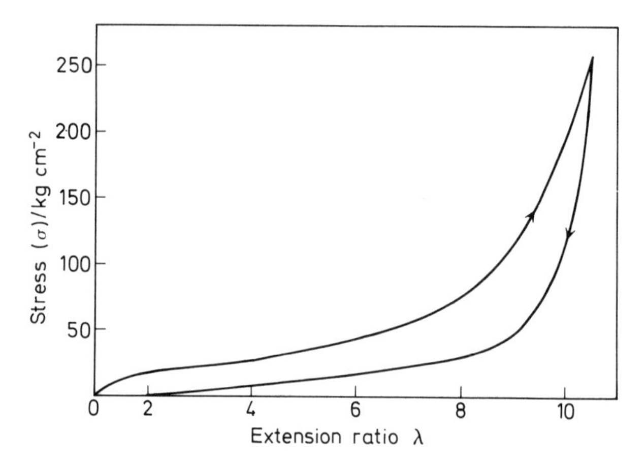

Figure 1.22 Tensile hysteresis of a styrene–isoprene–styrene block copolymer (13 700-63 400-13 700) at 25 °C

styrene has actually been observed by electron microscopy[65, 66]. It has also been corroborated by swelling experiments with selective solvents[75], in which it was shown that tetrahydrofurfuryl alcohol (THFA), a selective solvent for polystyrene, is capable of completely restoring the SIS polymer to its original state, while n-decane, a selective solvent for the polyisoprene, is not.

The predominant role of the polystyrene domains is further confirmed by the effect of temperature on tensile strength of these polymers. Figure 1.23 compares this effect for the SBS and SIS polymers[75]. In accordance with the viscoelastic theory of tensile strength of elastomers[76], the polyisoprene rubber (SIS), with a T_g of -65 °C, should exhibit a considerably higher tensile strength (at any given temperature) than the polybutadiene (SBS), which has the much lower T_g of -95 °C. Yet both the SBS and SIS polymers are virtually on the same curve in Figure 1.23. Hence it must be concluded that it is the thermal response of the polystyrene which controls the tensile strength.

These findings raise questions about the behaviour of an ABA block polymer of this same type in which the end blocks have a higher T_g than polystyrene. Such a polymer has in fact been synthesised[52] in the form of a block copolymer of α-methylstyrene–isoprene–α–methylstyrene, using a dilithium initiator and a two-stage process of polymerisation. Its stress–strain properties are compared in Figure 1.24 with those of an SIS polymer prepared by

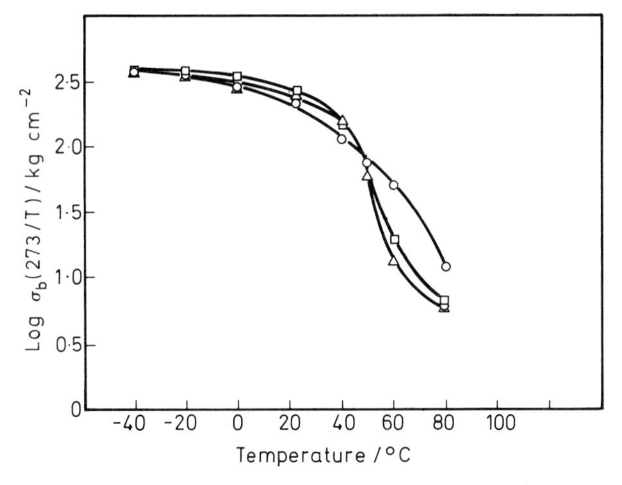

Figure 1.23 Effect of temperature on tensile strength of styrene–diene–styrene block copolymers (30% styrene).

	Centre block	*Mol. wt.* ($\times 10^{-3}$)
○	butadiene	21.2–97.9–21.2
△	butadiene	13.7–63.4–13.7
□	isoprene	13.7–63.4–13.7

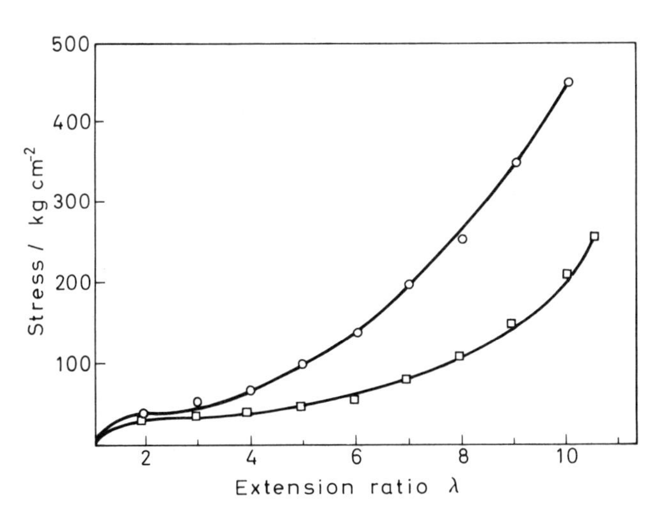

Figure 1.24 Tensile properties of an α-methylstyrene-isoprene-α-methylstyrene block copolymer at 25 °C.

	End block	*Mol. wt.* ($\times 10^{-3}$)
○	α-methylstyrene	21.0–85.0–21.0
□	styrene	20.0–87.0–20.0

(From Morton *et al.*[56], by courtesy of Rubber and Technical Press Ltd.)

the same method. It can be seen at once that the α-methylstyrene end blocks having the higher T_g impart a substantially higher tensile strength to this elastomer, presumably by being able to withstand higher stresses before yielding (also resulting in very high 'set', e.g. ~100%). Similarly, Figure 1.25

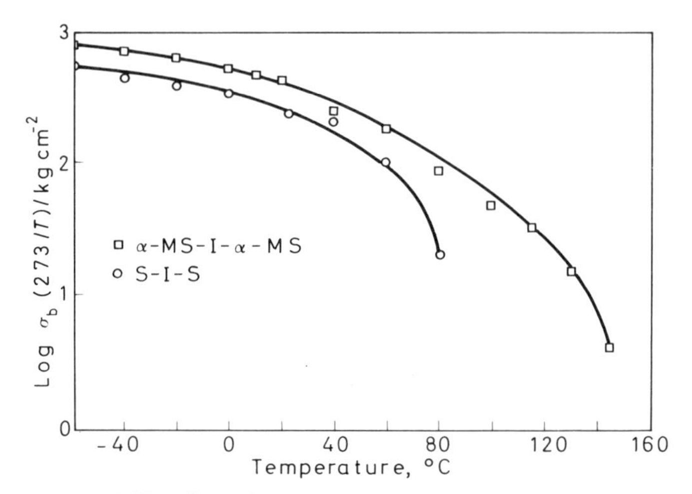

Figure 1.25 Effect of temperature on tensile strength of styrene–isoprene–styrene versus α-methylstyrene–isoprene–α-methylstyrene triblock copolymers.
(From Morton *et al.*[56], by courtesy of Rubber and Technical Press Ltd.)

shows the effect of temperature on the tensile strength of these two block polymers, demonstrating quite unequivocally that the end blocks of higher T_g can of course maintain their integrity to higher temperatures.

In this connection, it might also be mentioned that the introduction of cross-links into the rubber phase of these block polymers has a deleterious effect on the tensile strength, as demonstrated in Table 1.11, which shows

Table 1.11 Effect of dicup cross-linking of an SIS polymer[81]

	*Swelling vol. ratio**	*Stress at 300% elong.* /kg cm^{-2}	*Tensile strength* /kg cm^{-2}	*Strain at break* (λ)/kg cm^{-2}
Before vulcanisation	9.8	50	225	11
After vulcanisation	5.4	53	160	10

*In *n*-decane 48 h.

data on an SIS polymer cross-linked by dicumyl peroxide[81]. The latter is known to cross-link polyisoprene stoichiometrically[77] without any noticeable chain scission. The introduction of cross-links between the elastic chains apparently hinders the efficient transmission and distribution of the applied stress to the polystyrene domains by preventing the chain slippage which can occur in the absence of any cross bonds.

1.3.3.3 *Effect of polymeric impurities*

Since these block polymers are synthesised by the *sequential* polymerisation of monomers by organolithium initiators, premature termination by adventitious impurities (e.g. H_2O, O_2) is always a possibility. Hence the final block polymer may contain either *free polystyrene* or *styrene–diene diblocks* (and sometimes even free polydiene, depending on the mode of synthesis). It is, therefore, important to know what the presence of such polymeric impurities might do to the properties. This has been demonstrated by preparing synthetic mixtures of well characterised triblock copolymers with known amounts of polystyrene or styrene–diene diblocks, by use of solution blending. The results of some of these experiments[56] are shown in Tables 1.12 and 1.13.

Table 1.12 Effect of added homopolymers
(From Morton *et al.*[56], by courtesy of Rubber and Technical Press Ltd.)

Polymer added	*Wt.%*	*Total styrene content/%*	*Stress at* $\lambda = 4$/kg cm^{-2}	*Tensile strength*/kg cm^{-2}
		SIS-3: 13 700–109 400–13 700		
None	—	20.0	17.5	296
PS-15 000	5	22.8	24.0	306
PS-15 000	10	25.5	25.4	302
PS-15 000	20	30.2	33.5	288
PI-84 000	5	19.3	14.2	263
		SBS-20: 13 700–109 400–13 700		
None	—	20.0	—	127
PS-12 500	14.9	30.0	—	342
		SBS-17: 13 700–63 000–13 700		
None	—	30.0	—	313

From Table 1.12 it is obvious that the added polystyrene (Mol. wt. = 15 000) does not affect the tensile strength of the SIS polymer, although it raises the modulus. This indicates that the free polystyrene must enter the domains of the block polystyrene, since the formation of a free polystyrene phase would lead to coarse heterogeneities and a disastrous decrease in strength. Furthermore, the clarity of the films also confirmed the absence of any separation of the polystyrene. In the case of the SBS polymer, the added

Table 1.13 Effect of added diblock polymer

SBS-5 21 100–63 400–21 100
SB diblock 21 100–63 400
(From Morton *et al.*[56], by courtesy of Rubber and Technical Press Ltd.)

% Added diblock	*Stress at* $\lambda = 4$/kg cm^{-2}	*Tensile strength* /kg cm^{-2}	*% Set at break*
0	46.3	319	50
1	50.3	308	50
2	47.5	264	50
5	48.3	244	50
67	14.6	49	200

polystyrene also must have entered the polystyrene domains, since the tensile strength increased substantially, paralleling the effect of an increase in block polystyrene from 20% to 30%. It can be concluded, therefore, that, in the synthesis of these polymers, any adventitious termination of the 'living' polystyrene should not be deleterious to the strength, since the free polystyrene thus formed is apparently capable of being incorporated into the polystyrene domains. The same can be said of any small amounts of free polyisoprene (PI-84 000) which apparently act as a minor diluent in the rubber.

On the other hand, the data on the effect of styrene–butadiene diblocks shown in Table 1.13 indicate that the presence of even a small proportion can have a marked effect in decreasing the tensile strength. This presumably results from the introduction of 'flaws' in the network, in the form of free polybutadiene chain-ends, which prevent the transmission of the applied stress to the polystyrene domains.

These results on the effect of polymeric 'impurities' raise questions about the possibility of blending with other homopolymers which may be sufficiently 'compatible' with the polystyrene end-blocks in order to modify the properties of the material. Some interesting results have been reported[56] in this connection on the effect of adding poly-α-methylstyrene to an SBS block copolymer, and these are shown in Table 1.14. It appears from these data that the poly-α-methylstyrene is indeed capable of being incorporated within

Table 1.14 Effect of added poly-α-methylstyrene to SBS polymers
(From Morton *et al.*[56], by courtesy of Rubber and Technical Press Ltd.)

Polymer added	*Wt.%*	*Total filler/%*	*Temp./°C*	*Tensile strength/ kg cm^{-2}*
		SBS-20: 13 700–109 400–13 700		
None	—	20	22	127
PS (12 500)	14.9	30	22	342
PαMS (13 500)	14.9	30	22	373
		SBS-17: 13 700–63 000–13 700		
None	—	30	22	311
None	—	30	48	125
None	—	30	59	~15
PS	16.7	40	48	267
PαMS	16.7	40	48	290
PS	16.7	40	59	70
PαMS	16.7	40	59	276
PS	16·7	40	65	23
PαMS	16.7	40	65	215
PαMS	16.7	40	70	35

the polystyrene domains, as witnessed by the noticeable rise in tensile strength, and especially by the increased resistance of the blended material to higher temperatures. It is probable that this does not represent a true molecular blend of the two polymers, since the poly-α-methylstyrene is not really compatible with polystyrene. Instead, this rather lends support to the proposed existence of a 'mixed' phase at the interface of the polystyrene domains. The existence of such a third phase has been proposed on the basis of evidence from thermal analysis and dynamic behaviour[78].

1.3.3.4 New non-diene elastomers

Although the behaviour of the elastomeric ABA block copolymers has been ascribed to their general molecular architecture and morphology, the question still arises whether these properties are in some way dependent on the presence of a *polydiene* centre block. There have been very few, if any, other examples of a successful synthesis of an elastomeric ABA block copolymer by anionic polymerisation. One such case has been recently reported[48, 79] in which the centre-block consists of poly(propylene sulphide) while the end blocks are poly-α-methylstyrene. As mentioned previously in Section 1.2.2.1, propylene sulphide may be polymerised by the action of a carbanionic initiator but is itself too weak a base to initiate polymerisation of unsaturated monomers. Hence, for this particular synthesis, the α-methylstyrene was first polymerised by use of ethyllithium, followed by polymerisation of the propylene sulphide to form an AB block. These blocks were then coupled,

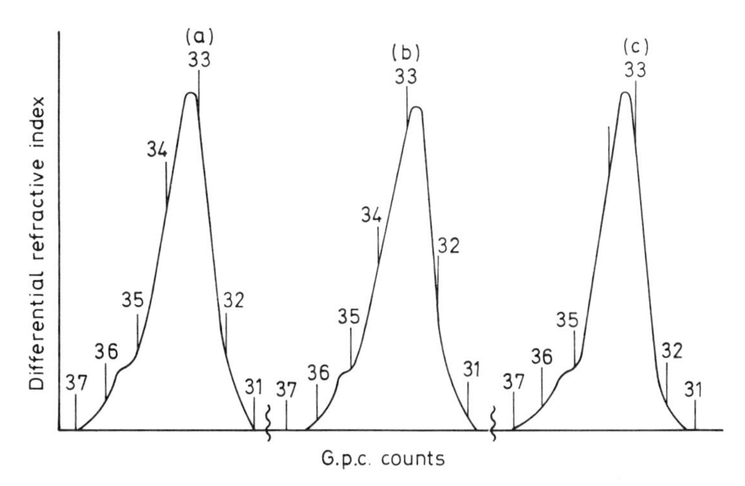

Figure 1.26 Gel permeation chromatograms of α-methylstyrene–(propylene sulphide)–α-methylstyrene block copolymers.

	% α-methylstyrene	*Mol. wt.* ($\times 10^{-3}$)
(a)	20	8–66–8
(b)	30	13–62–13
(c)	40	17–55–17

(From Morton *et al.*[48], by courtesy of the Society of Chemical Industry)

by use of phosgene, to form an ABA block copolymer having the following structure:

(α-methylstyrene)$_x$(propylene sulphide)$_y$—C—(propylene sulphide)$_y$ (α-methylstyrene)$_x$

The gel permeation chromatogram of the three block copolymers thus formed are shown in Figure 1.26. The sharpness of the peaks attests to the narrow MWD present in these polymers, while the small shoulder on the low-molecular-weight end presumably represents about 5% of uncoupled AB blocks. The stress–strain properties of these three polymers are shown in

Figure 1.27. Their remarkable similarity to the styrene–diene triblock copolymers is immediately apparent. However, their strength, while still very high, is substantially below the exceptional values shown by the analogous α-methylstyrene–isoprene triblocks (see Figure 1.24). This may, of course, be easily ascribed to the observed proportion of uncoupled AB material present.

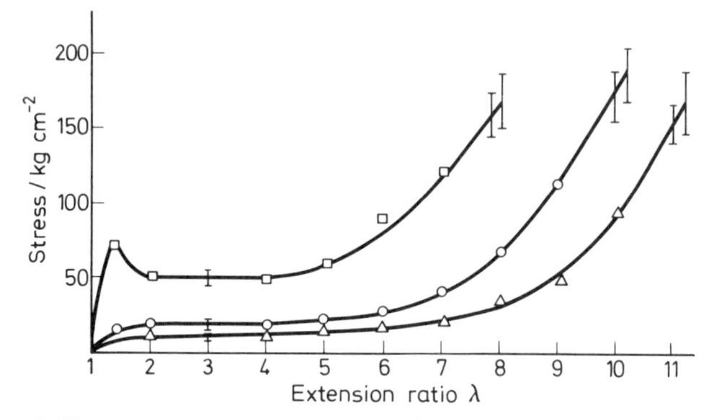

Figure 1.27 Tensile properties of α-methylstyrene–(propylene sulphide)–α-methylstyrene block copolymers at 25 °C.

	% *α-methylstyrene*	*Mol. wt.* ($\times 10^{-3}$)
△	20	8–66–8
○	30	13–62–13
□	40	17–55–17

(From Morton *et al.*[48], by courtesy of Society of Chemical Industry)

However, without detailed knowledge about the degree of phase separation in these materials, it is not possible to draw definite conclusions about these differences.

References

1. Szwarc, M., Levy, M. and Milkovich, R. (1956). *J. Amer. Chem. Soc.*, **78,** 2656
2. Flory, P. J. (1940). *J. Amer. Chem. Soc.*, **62,** 1561
3. Morton, M. and Fetters, L. J. (1967). *Macromolecular Rev.*, Vol. 2, 71. (New York: John Wiley and Sons)
4. Fox, T. G., Garrett, B. S., Goode, W. E., Gratch, S., Kincaid, J. F., Spell, A. and Stroupe, J. D. (1958). *J. Amer. Chem. Soc.*, **80,** 1768
5. Stavely, F. W. and co-workers (1956). *Ind. Eng. Chem.*, **48,** 778
6. Morton, M. (1964). Chap. in *Copolymerisation*, ed. by G. Ham, p. 421. (New York: Interscience Publishers)
7. Bhattacharyya, D. N., Lee, C. L., Smid, J. and Szwarc, M. (1964). *Polymer*, **5,** 54; (1965). *J. Phys. Chem.*, **69,** 612
8. Szwarc, M. (1968). *Carbanions, Living Polymers, and Electron-Transfer Processes.* (New York: Interscience Publishers)
9. Morton, M., Fetters, L. J. and Bostick, E. E. (1963). *J. Polymer Sci.*, **C1,** 311
10. Bywater, S. (1965). *Advan. Polymer Sci.*, **4,** 66
11. Szwarc, M. (1968). *Carbanions, Living Polymers, and Electron-Transfer Processes,* 487. (New York: Interscience Publishers)
12. Ibid., p. 405
13. Morton, M., Pett, R. A. and Fellers, J. F. (1966). Preprint 2.1.10, IUPAC Macromolecular Symposium, Tokyo
14. Spirin, Yu. L., Gantmakher, A. R. and Medvedev, S. S. (1962). *Dokl. Akcad. Nauk SSSR,* **146,** 368
15. Sinn, H. and Patat, F. (1963). *Angew. Chem.*, **75,** 805

16. Worsfold, D. J. and Bywater, S. (1964). *Can. J. Chem.*, **42,** 2884
17. Morton, M. and Pett, R. A., unpublished results
18. Johnson, A. F. and Worsfold, D. J. (1965). *J. Polymer Sci.*, **A3,** 449
19. Morton, M., Bostick, E. E. and Livigni, R. (1961). *Rubber Plastics Age*, **42,** 397
20. Fox, T. G., Gratch, B. S. and Loshaek, S. (1956). *Rheology*, ed. by F. R. Eirich, p. 443 (New York: Academic Press)
21. Morton, M. and Fetters, L. J. (1964). *J. Polymer Sci.*, **A2,** 3311
22. Morton, M., Fetters, L. J., Pett, R. A. and Meier, J. F. (1970). *Macromolecules*, **3,** 327
23. Fetters, L. J. (1966). *J. Research Nat. Bur. Stand.*, **70A,** 421
24. Morton, M. and Ells, F. R. (1962). *J. Polymer Sci.*, **61,** 25
25. Worsfold, D. J. and Bywater, S. (1960). *Can. J. Chem.*, **38,** 1891
26. Brown, T. L. (1966). *J. Organometal. Chem.*, **5,** 191
27. Morton, M., Pett, R. A. and Fetters, L. J. (1970). *Macromolecules*, **3,** 333
28. Brown, T. L. and Rogers, M. T. (1957). *J. Amer. Chem. Soc.*, **79,** 1859; also Brown, T. L., Gerteis, R. L., Bafus, D. A. and Ladd, J. A. (1964). *J. Amer. Chem. Soc.*, **86,** 2134
29. Morton, M., Sanderson, R. D. and Sakata, R. (1971). *J. Polymer Sci.*, **B9,** 61
30. Morton, M., Sanderson, R. D. and Sakata, R., *Macromolecules*, in the press
31. Schue, F., Worsfold, D. J. and Bywater, S. (1969). *J. Polymer Sci.*, **B7,** 821
32. Schue, F., Worsfold, D. J. and Bywater, S. (1970). *Macromolecules*, **3,** 509
33. Makowski, H. S. and Lynn, M. (1966). *J. Macromol. Chem.*, **1,** 443; (1968). *J. Macromol. Sci. Chem.*, **A2,** 683
34. Morton, M. and Falvo, L. A. (1971). *Polymer Preprints*, **12,** 398 (1971)
35. Hsu, S. L., Kemp, M. K., Pochan, J. M., Benson, R. C. and Flygare, W. H. (1969). *J. Chem. Phys.*, **50,** 1482
36. Fetters, L. J., private communication
37. Margerison, D. and Newport, J. P. (1963). *Trans. Faraday Soc.*, **59,** 2058
38. Bywater, S. and Worsfold, D. J. (1967). *J. Organometal. Chem.*, **9,** 1
39. Richards, D. H. and Szwarc, M. (1959). *Trans. Faraday Soc.*, **55,** 1644
40. Morton, M., Rembaum, A. and Bostick, E. E. (1958). *J. Polymer Sci.*, **32,** 530
41. Searles, S. (1951). *J. Amer. Chem. Sci.*, **73,** 124
42. Steiner, E. C., Pelletier, R. R. and Trucks, R. O. (1964). *J. Amer. Chem. Soc.*, **86,** 4678
43. Boileau, S. and Sigwalt, P. (1961). *Compt. Rend. Hebd. Seanc. Acad. Sci., Paris*, **252,** 882
44. Gourdenne, A. and Sigwalt, P. (1967). *Europ. Polymer J.*, **3,** 481
45. Gourdenne, A., Boileau, S., Fontanille, M. and Sigwalt, P. (1969). *Polym. Prepr.*, **10,** 826
46. Boileau, S., Champetier, G. and Sigwalt, P. (1963). *Makromolek. Chem.*, **69,** 180
47. Morton, M. and Kammereck, R. F. (1970). *J. Amer. Chem. Soc.*, **92,** 3217
48. Morton, M., Kammereck, R. F. and Fetters, L. J. (1971). *Br. Polymer, J.*, **3,** 120
49. Bordwell, F. G., Andersen, H. M. and Pitt, B. M. (1954). *J. Amer. Chem. Soc.*, **76,** 1082
50. Trost, B. M. and Ziman, S. (1969). *Chem. Commun.*, 181
51. Machon, J. P. and Nicco, A. (1971). *Europ. Polymer J.*, **7,** 353
52. Fetters, L. J. and Morton, M. (1969). *Macromolecules*, **2,** 453
53. Mark, H. (1953). *Textile Res. J.*, **23,** 294
54. Zelinski, R. and Childers, C. W. (1968). *Rubber Chem. Technol.*, **41,** 161
55. Morton, M., McGrath, J. E. and Juliano, P. C. (1969). *J. Polymer Sci.*, **C26,** 99
56. Morton, M., Fetters, L. J., Schwab, F. C., Strauss, C. R. and Kammereck, R. F. (1969). *4th Synthetic Rubber Symposium*, p. 70. (London: Rubber and Technical Press)
57. Tobolsky, A. V. and Rogers, C. E. (1959). *J. Polymer Sci.*, **40,** 73
58. Forman, L. E. (1969). Chap. in *Polymer Chemistry of Synthetic Elastomers*, Part 2, ed. by J. P. Kennedy and E. G. M. Tornqvist, p. 491. (New York: Interscience Publishers)
59. Kuntz, I. and Gerber, A. (1960). *J. Polymer Sci.*, **42,** 299
60. Rembaum, A. R., Morrow, R. C. and Tobolsky, A. V. (1962). *J. Polymer Sci.*, **61,** 155
61. Morton, M. (1969). *Polymer Preprints*, **10,** 512; (1970). *Block Polymers*, ed. by S. L. Aggarwal, p. 1. (New York-London: Plenum Press)
62. Milkovich, R. (1964). *South African Pat.* 642,271
63. Heller, J. G., Schimsheimer, Pasternak, R. A., Kingsley, C. B. and Moacanin, J. (1969). *J. Polymer Sci.*, **A1, 7,** 73
64. Cunningham, R. E. and Trieber, M. R. (1968). *J. Appl. Polymer Sci.*, **12,** 23
65. Hendus, H., Illers, K. H. and Ropte, E. (1967). *Kolloid-Z. Z. Polymere*, **216–217,** 110
66. Beecher, J. F., Marker, L., Bradford, R. D. and Aggarwal, S. L. (1969). *J. Polymer Sci.*, **C26,** 117

67. Matsuo, M. (1968). *Japanese Plastics,* **2,** 6
68. Lewis, P. R. and Price, C. (1969). *Nature (London),* **223,** 494
69. McIntyre, D. and Campos-Lopez, E. (1970). *Macromolecules,* **3,** 322
70. Keller, A., Pedemonte, E. and Willmouth, F. M. (1970). *Nature (London),* **225,** 538
71. Meier, D. J. (1967). *J. Phys. Chem.,* **71,** 1861; (1969). *J. Polymer Sci.,* **C26,** 81
72. Smith, T. L. and Dickie, R. A. (1969). *J. Polymer Sci.,* **C26,** 163
73. Holden, G., Bishop, E. T. and Legge, N. R. (1969). *J. Polymer Sci.,* **C26,** 37
74. Fedors, R. F. (1969). *J. Polymer Sci.,* **C26,** 189
75. Goh, S., University of Akron, private communication
76. Halpin, J. C. (1965). *Rubber Reviews,* **38,** 1007
77. Scott, K. W. (1962). *J. Polymer Sci.,* **58,** 517
78. Chen, M. and Kaelble, D. H. (1970). *J. Polymer Sci.,* **88,** 149
79. Morton, M., Kammereck, R. F. and Fetters, L. J. (1971). *Macromolecules,* **4,** 11
80. Falvo, L. A., private communication
81. Schwab, F. C. (1970). *Ph.D. Dissertation.* (University of Akron)

2
Cationic Polymerisation

J. P. KENNEDY
The University of Akron, Ohio

2.1 INTRODUCTION TO CATIONIC POLYMERISATIONS

Why cationic polymerisations? What distinguishes this field of polymer science from that of radical or anionic or other ionic polymerisations say, coordination polymerisations? Why is this field of interest to polymer chemists at all?

Table 2.1 Cationically polymerisable (or oligomerisable) compounds

Hydrocarbons (and a few substituted derivatives)

Aliphatic olefins

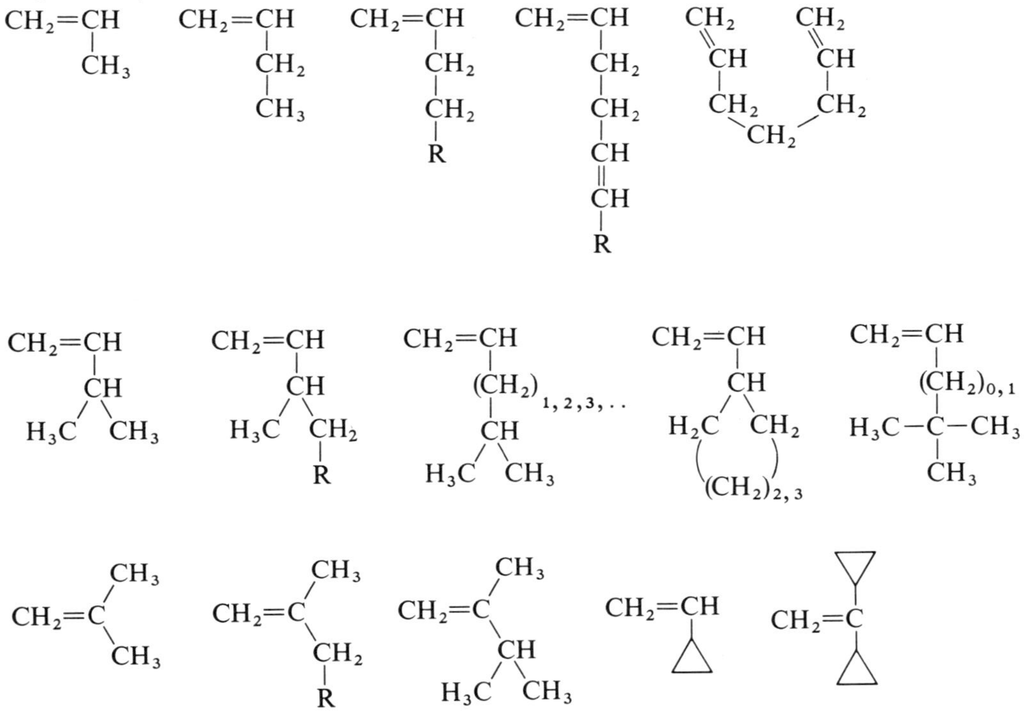

$CH_2{=}CH{-}CH_3$

$CH_2{=}CH{-}CH_2{-}CH_3$

$CH_2{=}CH{-}CH_2{-}CH_2{-}R$

$CH_2{=}CH{-}CH_2{-}CH_2{-}CH{=}CH{-}R$

$CH_2{=}CH{-}CH_2{-}CH_2{-}CH_2{-}CH{=}CH_2$

$CH_2{=}CH{-}CH(CH_3)_2$

$CH_2{=}CH{-}CH(CH_3){-}CH_2{-}R$

$CH_2{=}CH{-}(CH_2)_{1,2,3,\ldots}{-}CH(CH_3)_2$

$CH_2{=}CH{-}CH{<}(CH_2{-}(CH_2)_{2,3}{-}CH_2)$ (cycloalkyl ring)

$CH_2{=}CH{-}(CH_2)_{0,1}{-}C(CH_3)_3$

$CH_2{=}C(CH_3)_2$

$CH_2{=}C(CH_3){-}CH_2{-}R$

$CH_2{=}C(CH_3){-}CH(CH_3)_2$

$CH_2{=}CH{-}C_3H_5$ (cyclopropyl)

$CH_2{=}C(C_3H_5)_2$ (dicyclopropyl)

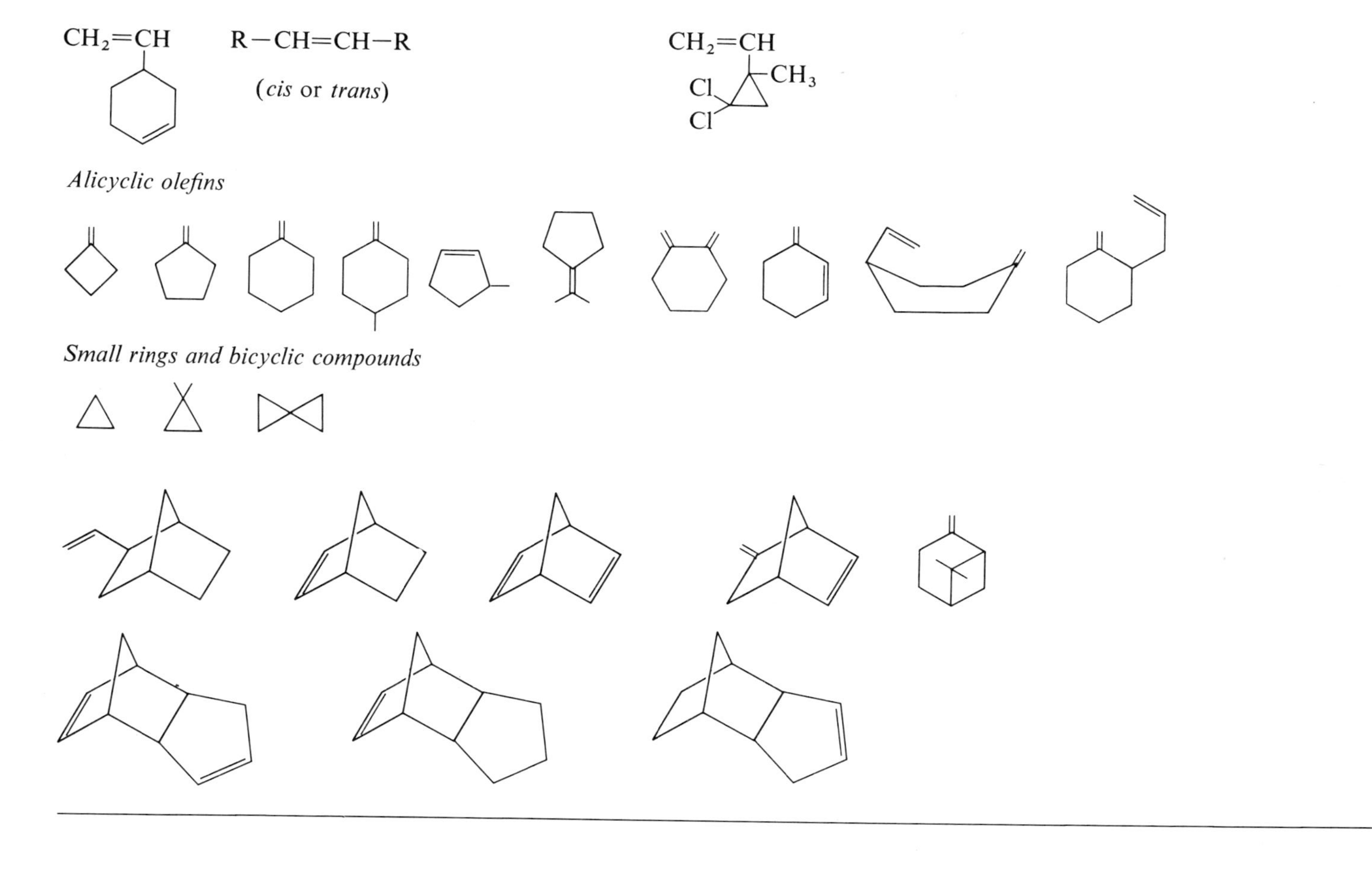

$CH_2=CH$
$R-CH=CH-R$
(*cis* or *trans*)
$CH_2=CH$
CH_3
Cl
Cl
Alicyclic olefins
Small rings and bicyclic compounds

Table 2.1 (continued)

Aromatic olefins (and derivatives)

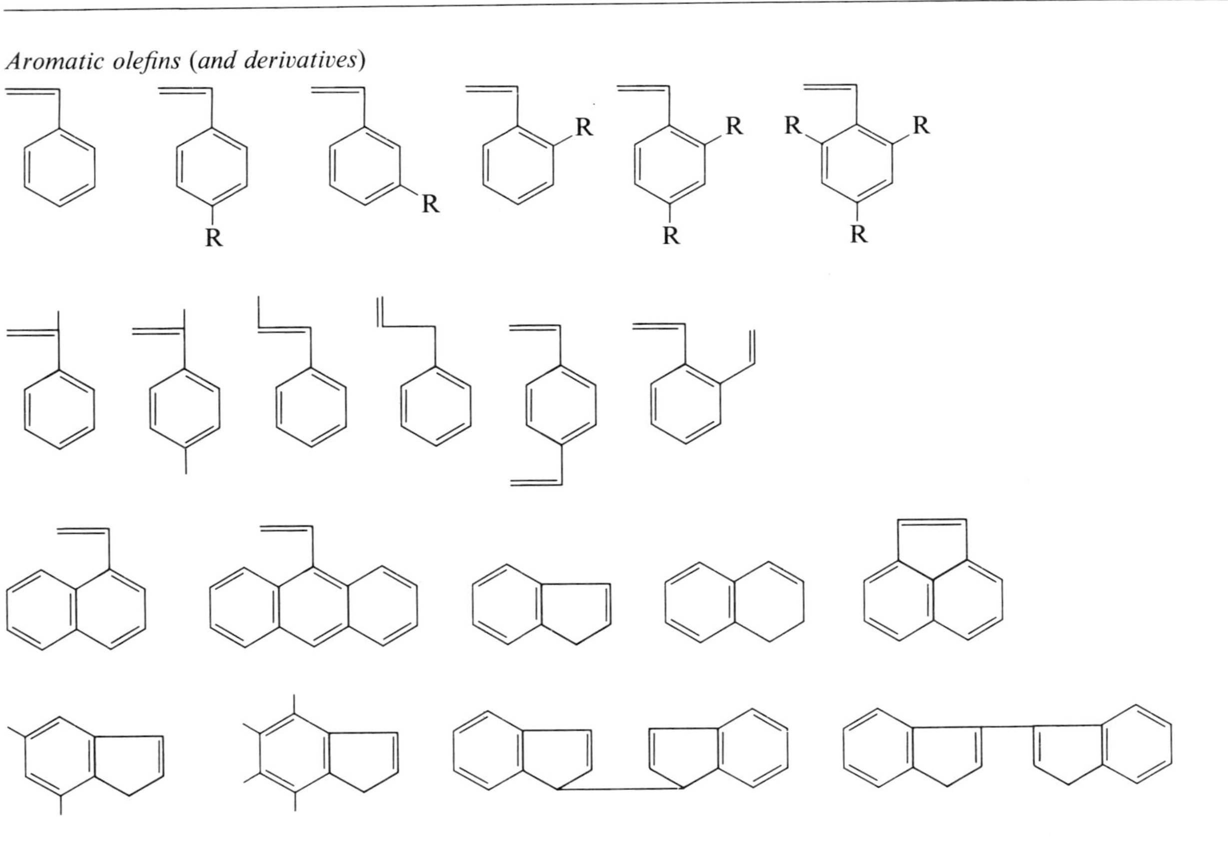

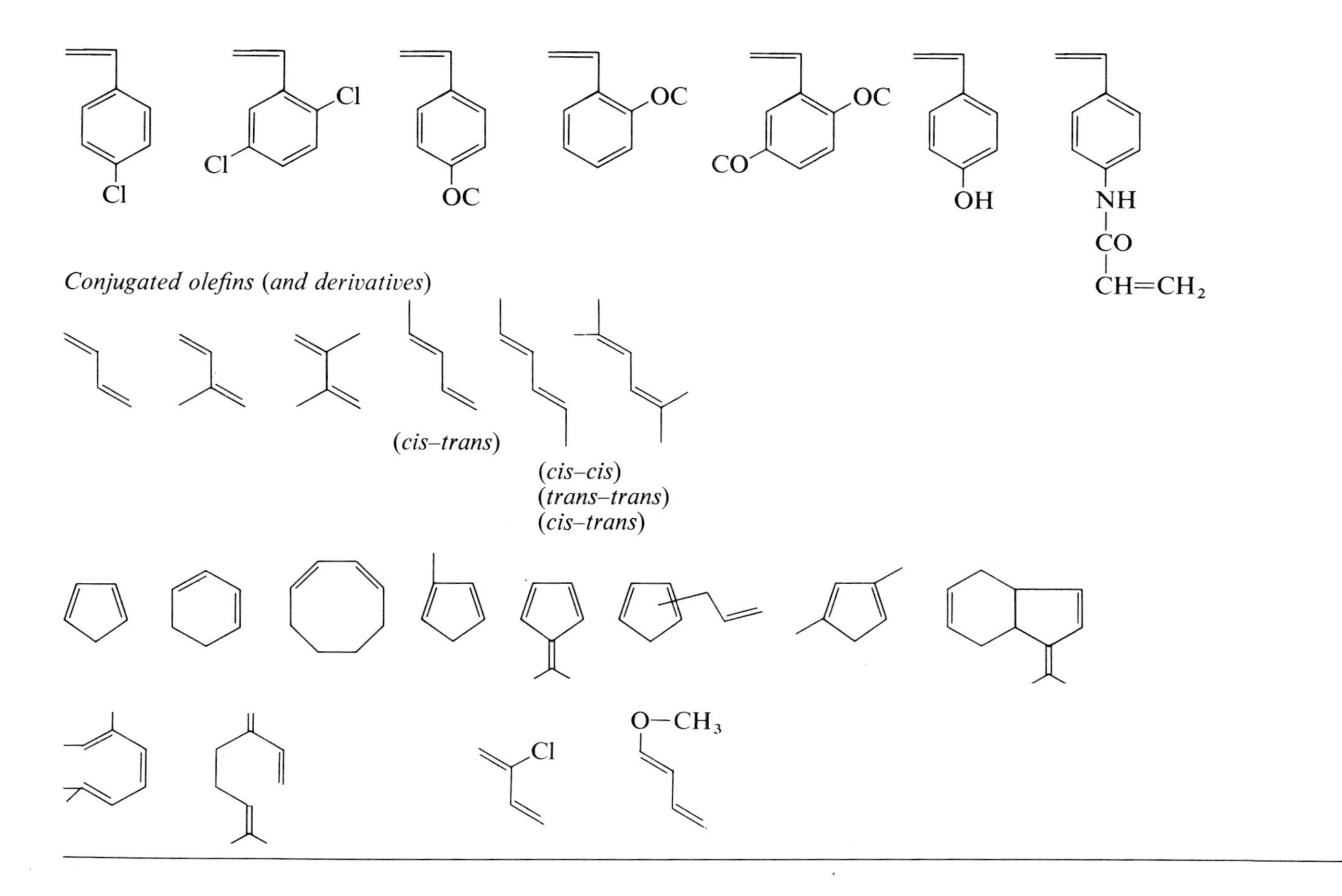
Cl
Cl
Cl
OC
OC
CO
OC
OH
NH
CO
CH=CH2
Conjugated olefins (and derivatives)
(cis–trans)
(cis–cis)
(trans–trans)
(cis–trans)
Cl
O–CH3

Table 2.1 (continued)

Heteroatom-containing compounds

Vinyl ethers (and derivatives)

$CH_2{=}CH{-}O{-}R$ R = methyl, ethyl,
n-, isopropyl,
n-, sec-, iso-, t-butyl,
n-hexyl, 2-ethylhexyl,
decyl, cetyl, octadecyl, phenyl, etc.

$CH_2{=}C(R){-}O{-}R$

R·CH=CH·OR (*cis,trans*)

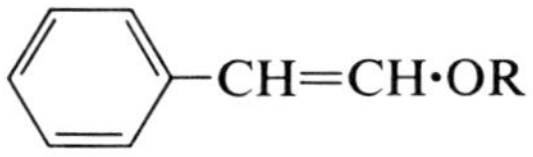

C_6H_5–CH=CH·OR (*cis,trans*)

RO·CH=CH·OR (*cis,trans*)

$CH_3{-}CH{=}CH{-}O{-}CH_2{-}CH_2Cl$

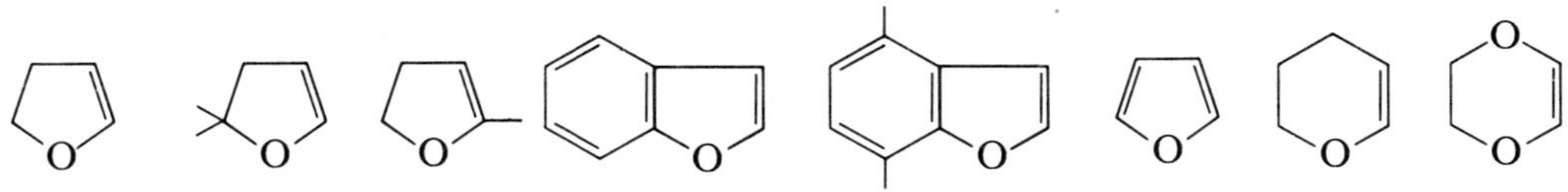

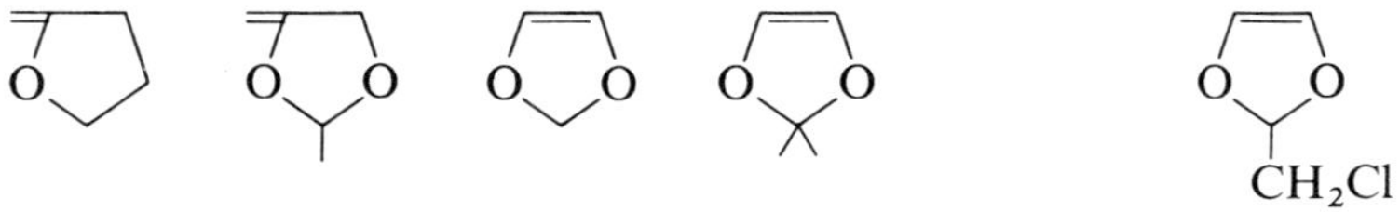

CH_2Cl

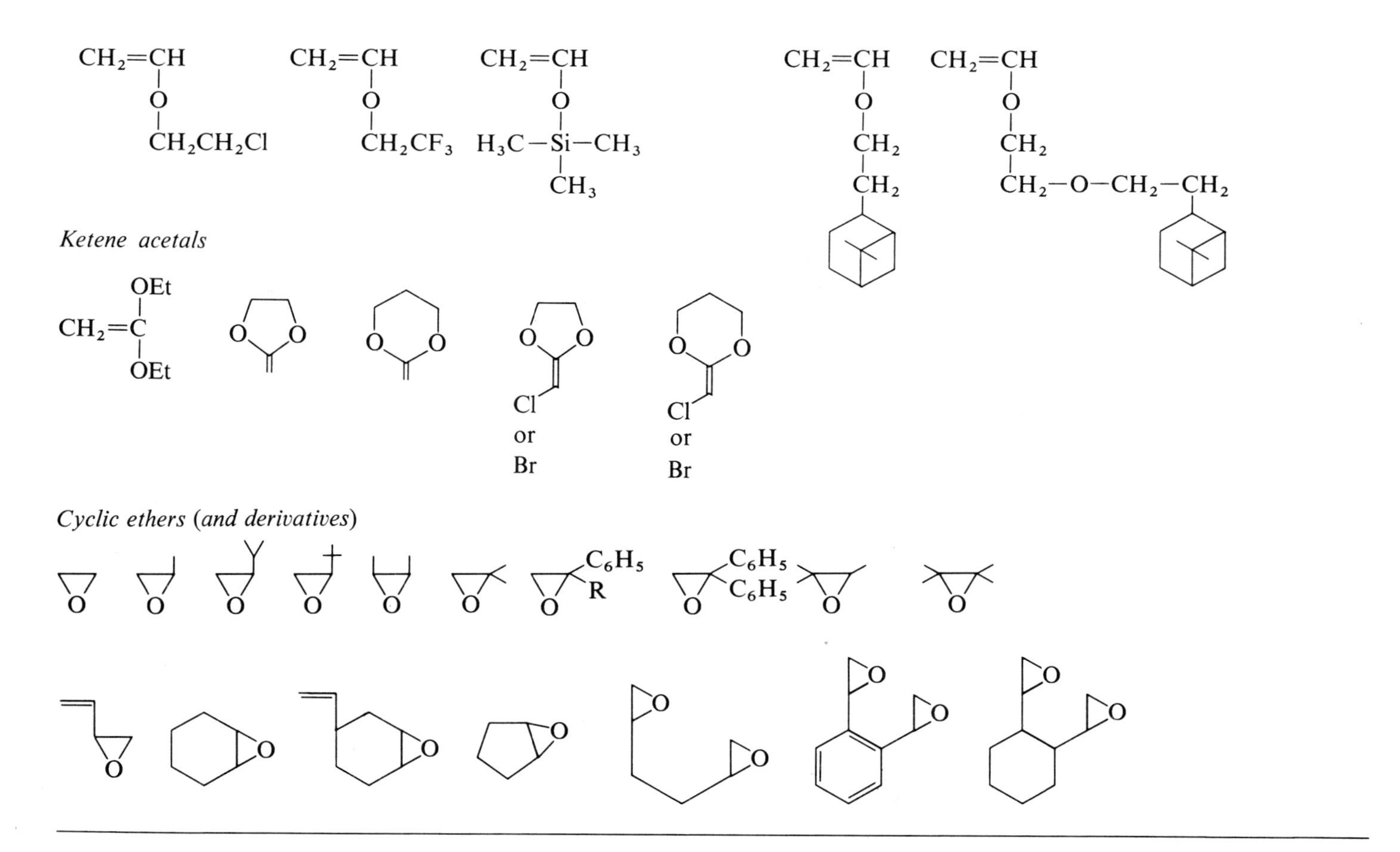

CH₂=CH
O
CH₂CH₂Cl
CH₂=CH
O
CH₂CF₃
CH₂=CH
O
H₃C−Si−CH₃
CH₃
CH₂=CH
O
CH₂
CH₂
CH₂=CH
O
CH₂
CH₂−O−CH₂−CH₂
Ketene acetals
OEt
CH₂=C
OEt
O O
O O
O O
Cl
or
Br
O O
Cl
or
Br
Cyclic ethers (and derivatives)
O
C₆H₅
R
C₆H₅
C₆H₅

Table 2.1 (continued)

Cyclic ethers (and derivatives) (continued)

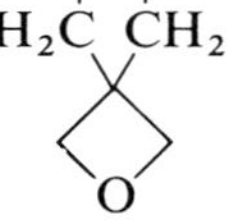
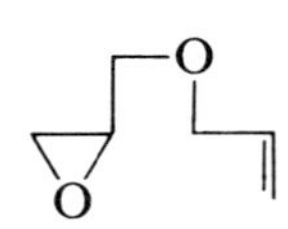
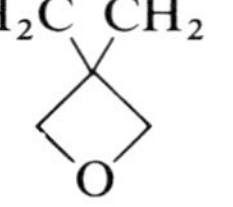

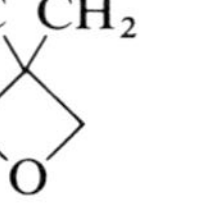
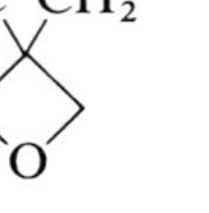
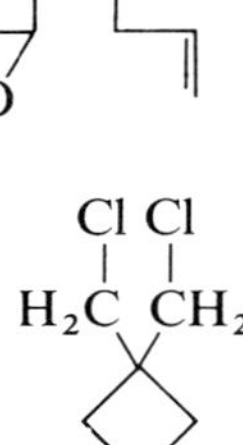

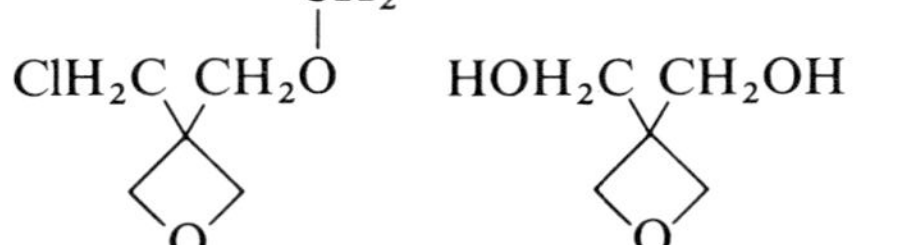
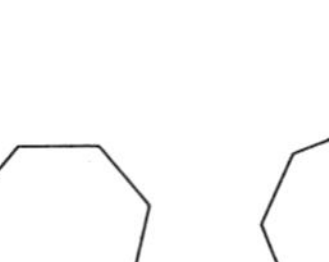
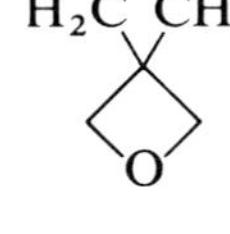
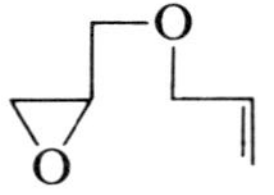

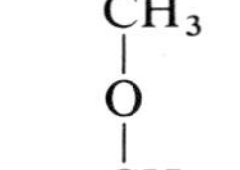
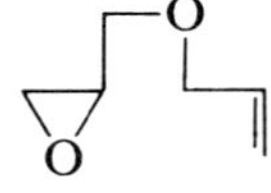
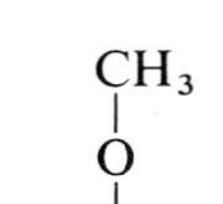
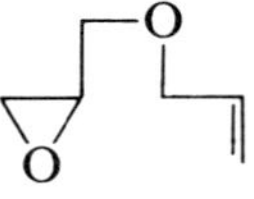

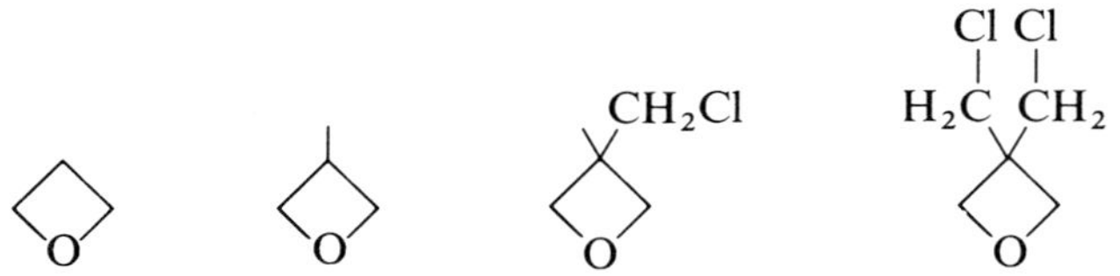

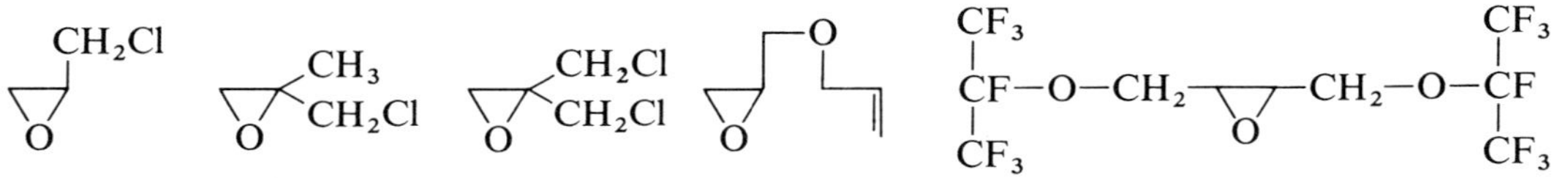

(also with F, Br and I)

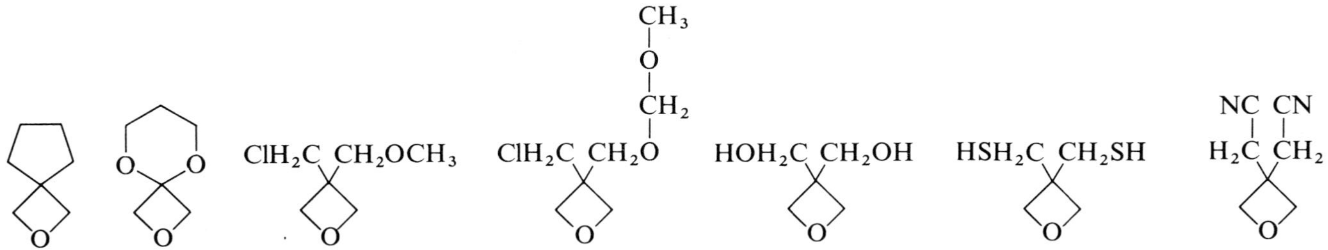

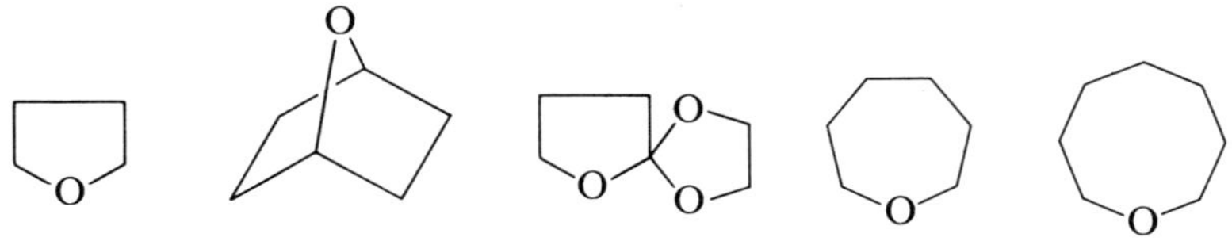

Cyclic formals and acetals

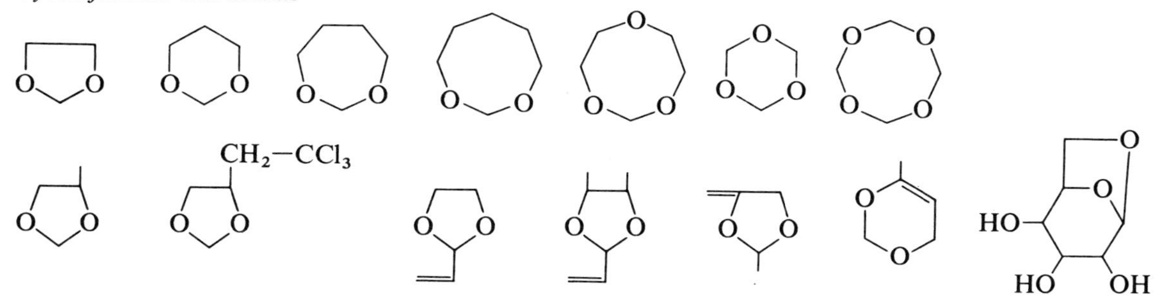

CH_2-CCl_3

HO HO OH

Carbonyl compounds: aldehydes and ketenes

CH_2O CH_3CHO RCHO CHO CHO

$HC(=O)-(CH_2)_{2\ or\ 3}-CH=O$

$OHC-CH_2-CH(R)-CH_2-CHO$

$OHC-CH=CH-CHO$

CHO CHO CHO CHO CHO

$CH_2=CH-CHO$ $NC-CH_2CH_2-CHO$ CHO

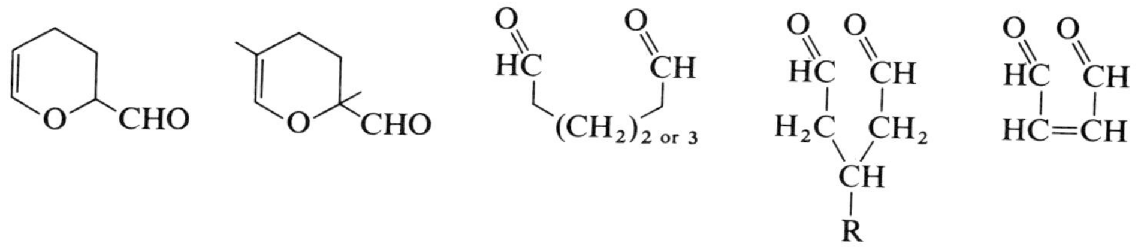

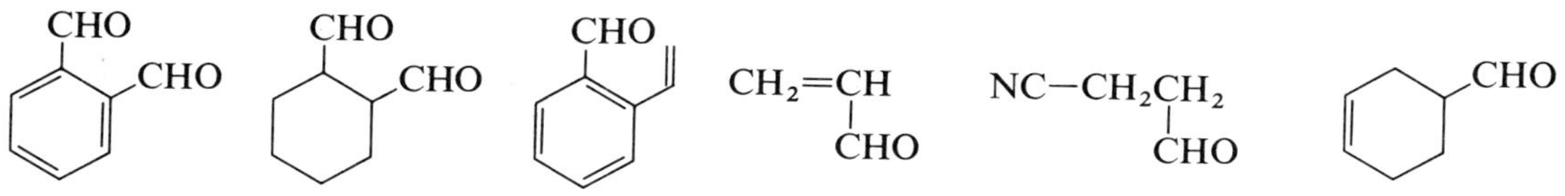

CCl_3-CHO $CHCl_2-CHO$ $CH_2Cl-CHO$ C_3F_7-CHO

Table 2.1 (continued)

Carbonyl compounds: aldehydes and ketenes (continued)

$CH_2{=}C{=}O$ $(CH_3)_2C{=}C{=}O$ $CH_2{=}C{-}CH_2{-}C{=}O$ (ring through O)

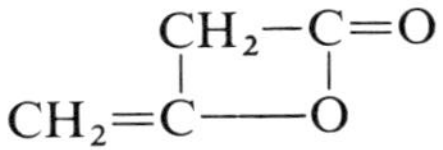

Cyclic esters: lactones, lactides, carbonates and oxalates

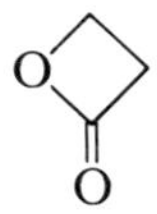

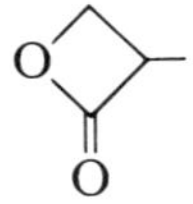

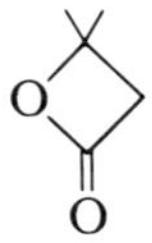

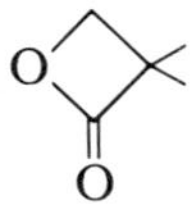

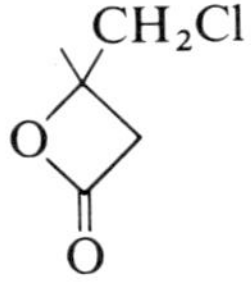

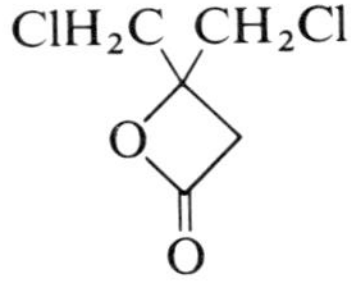

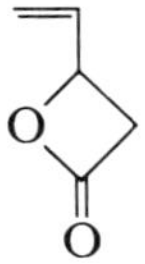

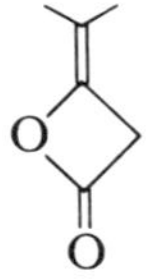

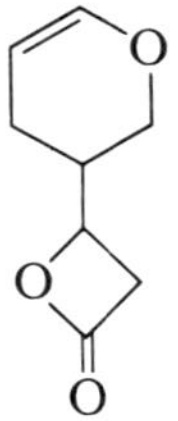

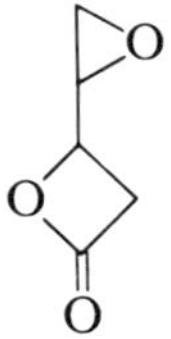

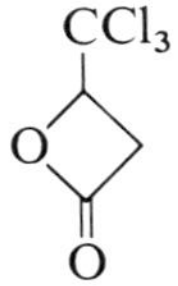

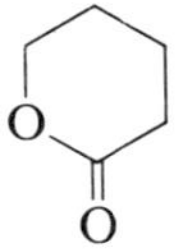

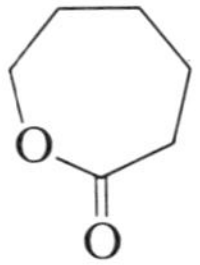

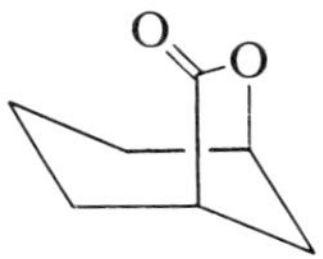

CH_2Cl ClH_2C CH_2Cl

CCl_3

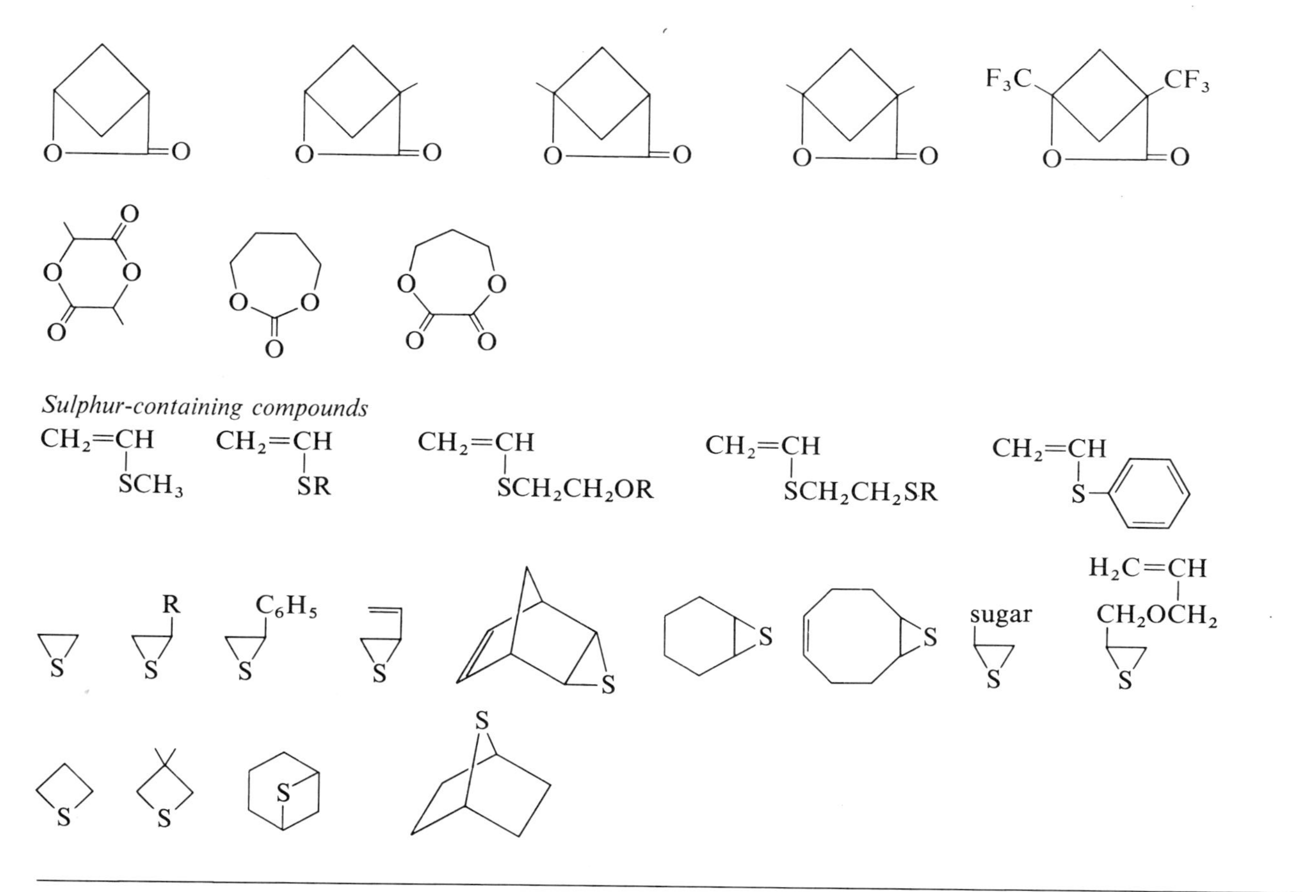
F3C
CF3
O
Sulphur-containing compounds
CH2=CH
SCH3
SR
SCH2CH2OR
SCH2CH2SR
S
R
C6H5
sugar
H2C=CH
CH2OCH2

Table 2.1 (continued)

Sulphur-containing compounds (continued)

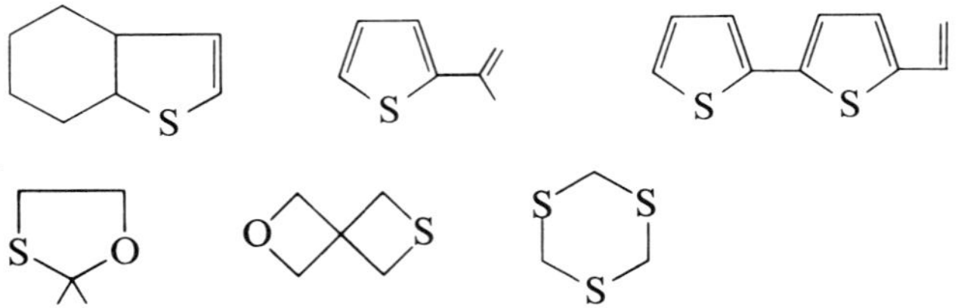

Nitrogen-containing compounds

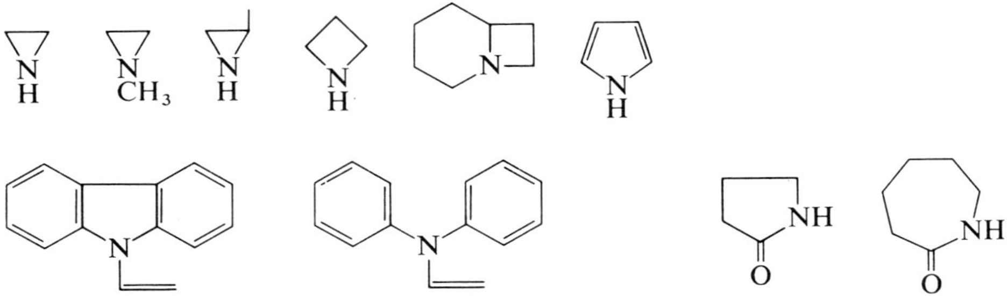

Miscellaneous compounds

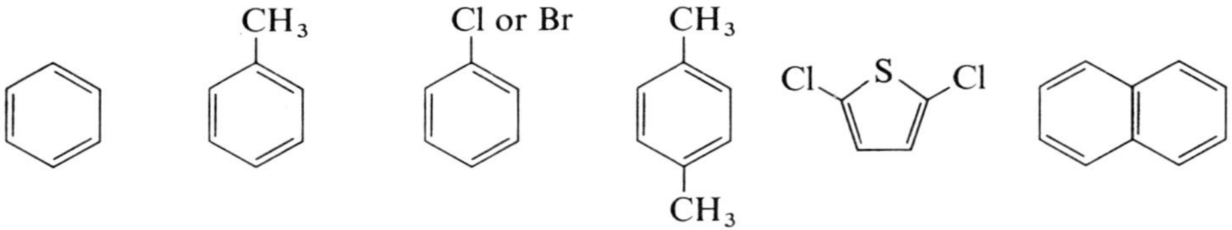

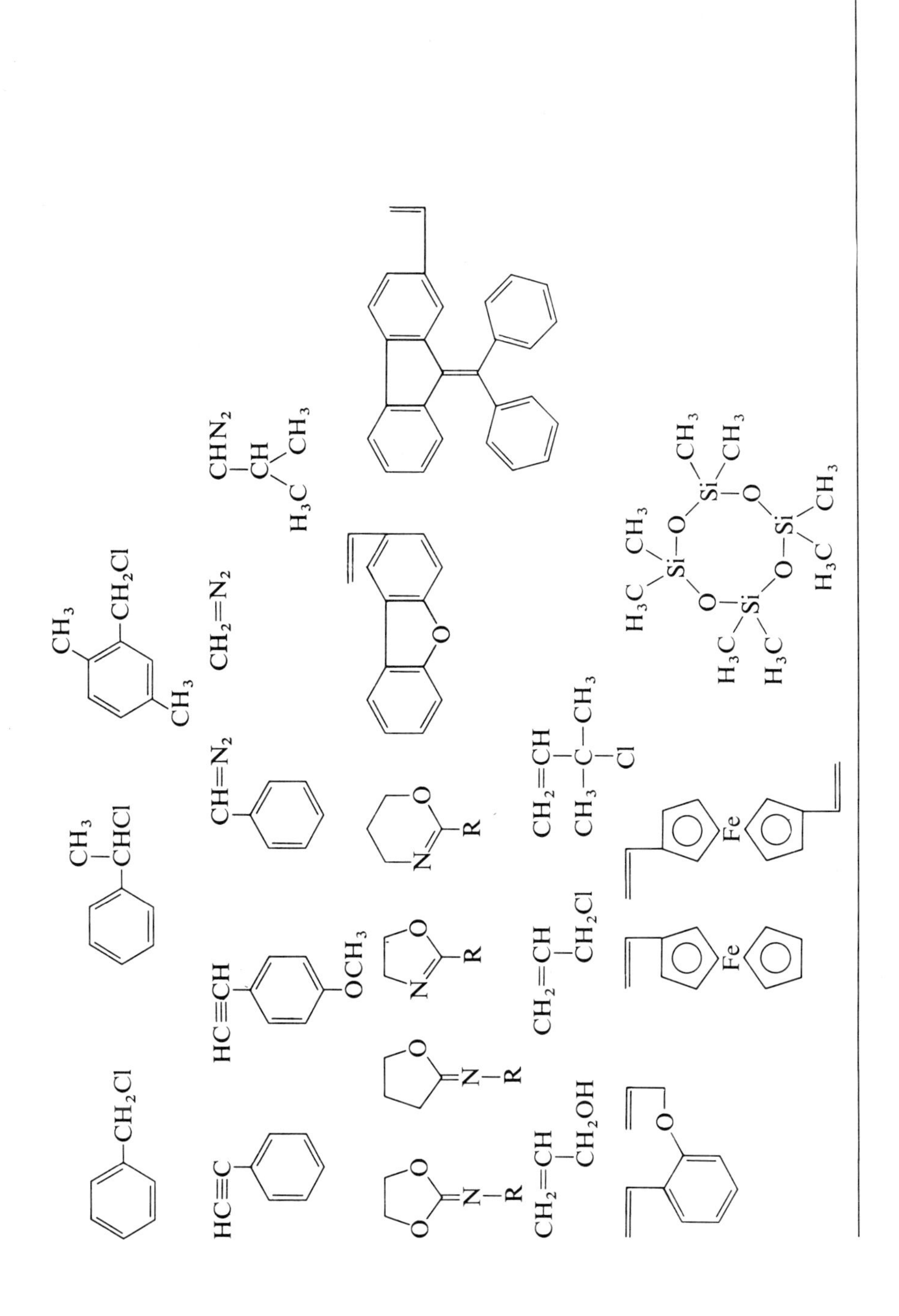

One answer to these questions is that cationic polymerisations provide an inexpensive way for the synthesis of a large number of useful high polymers, (Vistanex, Oppanol B, butyl rubber, Celcon, Penton). The number of monomers amenable to cationic initiation is literally in the hundreds, and certainly much larger than those polymerisable by anionic or anionic coordination mechanism. Among the various polymerisation mechanisms, radicals, of course, are able to initiate the polymerisation of the largest number of monomers, but cationic polymerisations are second in this respect. Table 2.1 is a compilation of essentially all the structures that have been described in the scientific literature to give high polymers (or at least oligomers) by cationic techniques.

Another answer to the above question is that many specific high polymers can be obtained *only* by cationic mechanism. And, in the same vein, a large number of specific structures can *only* be obtained by cationic processes. Thus, high molecular weight polyisobutylene, polytetrahydrofurane, polyisobutylvinyl ether, poly-2-chloroethylvinyl ether, poly-*p*-methoxystyrene, polycyclopentadiene, poly-2,5-dimethylhexa-2,4-diene, polyketene diethylacetal, polytrimethylsilyvinyl ether, etc. can only be synthesised by cationic means and among specific polymer structures the following ones can only be prepared by this technique.

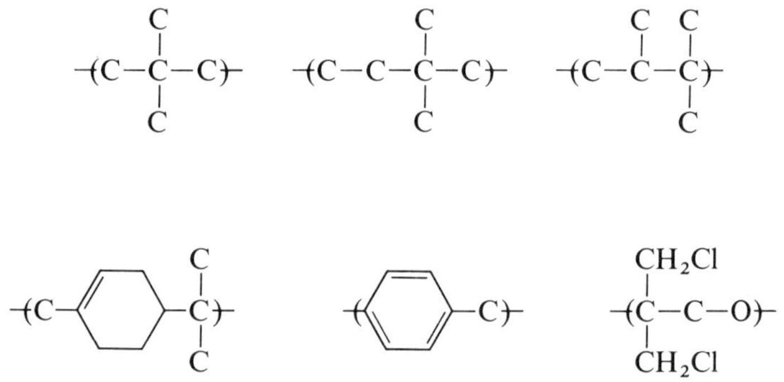

```
       C                 C               C  C
       |                 |               |  |
  -(C--C--C)-   -(C--C--C--C)-    -(C--C--C)-
       |                 |                  |
       C                 C                  C


                 C                                  CH2Cl
                 |                                  |
  -(C--<cyclohexane>--C)-   -(<benzene>--C)-    -(C--C--O)-
                 |                                  |
                 C                                  CH2Cl
```

Still another answer is that many useful oligomers (lubricating-oil additives, etc.) are only accessible by cationic processes. Cationic reactions are of highest practical importance not only for the synthesis of high polymers and/or oligomers but are also the backbone of many of the basic petrochemical processes, i.e. cracking, alkylation, isomerisation, re-forming, etc.

Last, but not least, there is the intellectual challenge which keeps this field constantly advancing.

Cationic polymerisation is the subject of a multi-authored book[1] that summarises information available till 1961–1962 and of an authoritative chapter in a book published in 1964[2]. Since that time several review papers concerning various aspects of this field have appeared. Thus, Higashimura has discussed rate constants[3], Zlamal has written on mechanisms[4], Kennedy, and Ketley and Fisher have summarised isomerisation polymerisations[5,6], Saegusa has surveyed metal alkyl catalysts[7] and Plesch has dealt with pseudocationic polymerisations and other aspects of this field[8].

This Author will critically review significant work in carbenium ion polymerisations of hydrocarbons during the past 5 years, up to and including 1970. In addition, some of the fundamentals of this field will be discussed in the light of recently acquired information. Cationic coordination polymerisations and polymerisations by oxonium, sulphonium, ammonium ions and cation radicals and/or charge-transfer complexes fall outside the scope of this survey.

Suggestions as to the terminology of the positive carbon cations (carbenium ions) and initiator–co-initiator complexes will be made and put to use.

For the purposes of this Review, it is useful to organise the field of carbenium ion polymerisations by elementary events, i.e. initiation, propagation, termination, transfer, etc.

2.2 A NOTE ON TERMINOLOGY

2.2.1 Carbenium ions

Very recently Olah[9] proposed the use of the term carbenium ions to describe trivalent carbo-cations, the parent of which is CH_3^+; the term carbonium ion is reserved for pentavalent positive carbo-cations derived from CH_5^+. This terminology is in line with accepted chemical nomenclature according to which the suffix *-onium* denotes a system in which the coordination number of the central atom is raised by one unit, e.g. sulphonium, oxonium, ammonium, phosphonium. (German organic-chemical literature still occasionally employs the term carbenium ions (see, for example reference 10). This Author accepts this organic-chemical terminology and recommends its use in polymer chemistry.)

2.2.2 Initiator–co-initiator systems

It was discovered in 1950, and often demonstrated since then (for a review see Reference 4), that pure Friedel–Crafts halides are unable to initiate cationic polymerisations and that they require the presence of suitable Brønsted acids (proton sources or protogens) or other sources of cations (cationogens, usually alkyl halides) for initiation. Indeed, certain Brønsted acid/Friedel–Crafts halide combinations, for example HF/BF_3, $HBr/AlBr_3$, $H_2O/TiCl_4$, RX/AlR_2X, etc. are among the strongest acids known and are most effective cationic initiators. It has been shown that the true initiating species are the proton or carbenium ion source, HX or RX, respectively, and, consequently, these materials and not the Friedel–Crafts halides should be considered as *initiators*. Indeed, the function of the Friedel–Crafts halide is, in the first approximation, to remove and stabilise by coordination the halide ion; for example:

$$HF + BF_3 \longrightarrow H^+BF_4^-$$
$$H^+ + C{=}C \longrightarrow H{-}C{-}C^+ \xrightarrow{+M} \text{etc.}$$

Accordingly, the correct terminology is to designate the Friedel–Crafts halide as the *co-initiator* (or less properly as the co-catalyst) and the source of the true initiating species, the protogen or cationogen, HX or RX, respectively, as the *initiator*. This Reviewer has adopted this terminology.

The term 'promoter' is sometimes used, particularly in the American patent literature, to designate protogenic or cationogenic materials in conjunction with Friedel–Crafts metal halides. This terminology is misleading and should not be used[8].

Various other terminologies (e.g. dual acids[11], *syn*-catalysts[8]) that have occasionally been used to designate initiator/co-initiator pairs can be replaced by the more descriptive initiator/co-initiator nomenclature. Natta *et al.*[12] coined the term 'modified Friedel–Crafts catalysts' to denote halides of multivalent metals in their highest valence state in which the halogens are partly substituted by organic groups. This terminology confuses further the rather confused terminology of Friedel–Crafts catalysts. As Friedel–Crafts catalysts have not been defined satisfactorily (see, for example, the Preface to *Friedel–Crafts and Related Reactions*[2]), the modification of such a vague definition becomes meaningless.

2.3 CARBENIUM ION MECHANISMS

2.3.1 Initiation

The elucidation of the details of cationic polymerisation initiation remains, after about 25 years of intensive research, one of the most important problems in this field. The bulk of research, past, present and probably future, has been and will be carried out with systems involving metal halides of the Friedel–Crafts type. While a number of other cationic initiator systems are known (i.e. strong mineral acids, acidic solids, I_2, high-energy radiation) research with these is of less significance. Systematic research on initiation started in 1946–1947 when Evans and Polanyi discovered the concept of co-initiation[11] and proposed their now classical initiation mechanism:

$$BF_3 + YH \longrightarrow F_3BYH$$

$$F_3BYH + CH_2{=}\underset{\displaystyle CH_3}{\overset{\displaystyle CH_3}{\underset{|}{\overset{|}{C}}}} \longrightarrow F_3BY^- + CH_3{-}\underset{\displaystyle CH_3}{\overset{\displaystyle CH_3}{\underset{|}{\overset{|}{C^+}}}}$$

$$CH_3{-}\underset{\displaystyle CH_3}{\overset{\displaystyle CH_3}{\underset{|}{\overset{|}{C^+}}}} + CH_2{=}\underset{\displaystyle CH_3}{\overset{\displaystyle CH_3}{\underset{|}{\overset{|}{C}}}} \longrightarrow CH_3{-}\underset{\displaystyle CH_3}{\overset{\displaystyle CH_3}{\underset{|}{\overset{|}{C}}}}{-}CH_2{-}\underset{\displaystyle CH_3}{\overset{\displaystyle CH_3}{\underset{|}{\overset{|}{C^+}}}} \text{ etc.}$$

Their proposal has been proved to be substantially correct by a generation of subsequent researchers. In retrospect it appears that this work signalled the end of the 'pre-historic' period and the dawn of the 'classical' period in cationic polymerisation research. In many respects we are still in this classical period, as many workers all over the world are still diligently searching for well-defined initiator/co-initiator systems to explore further the boundaries of this concept. This research often involves 'stopping' experiments, i.e. experiments carried out under the highest attainable state of purity (high vacuum, baked glassware, etc.) and is regarded as an indication of the absence of cationogenic materials (usually water), the true *initiating* species. Such experiments are very difficult to perform. Also the danger always remains that in spite of lengthy baking in vacuum, chemisorbed moisture, hydroxyl groups on the surface of the glass ($\text{—}\overset{|}{\underset{|}{\text{Si}}}\text{—OH}$), or other protogens remain on the surface of the glass and are slowly released, or water that had been trapped in the bulk of the glass during its manufacture is slowly diffusing out into the reaction. As 'negative' experiments are involved, i.e. the *absence* of water or other protogens is to be proved, only the complete absence of initiation is of diagnostic value, and a mere slowing down of the rate is of questionable significance. In spite of these experimental difficulties, thanks to courageous workers, the list of well-characterised initiator–co-initiator systems is steadily increasing.

Previous reviewers have assembled exhaustive lists of carefully investigated systems in which the necessity of initiating protogens or cationogens has been established (see References 2, 4, 8 and references therein).

An excellent discussion on aspects of cationic initiation is included in Plesch's recent review[8]. Although this review is only 3 years old, it is already obsolete in many respects and needs correction in others.

Plesch reviewed the work of Eastham and co-workers[13], and on the basis of Eastham's findings up to 1966 concluded that the original simple co-initiation mechanism of Evans *et al.* might not be correct and that more than one complex may be involved in initiation. In a recent paper, Eastham has re-examined his earlier proposals[13] and on the basis of new evidence has concluded that his earlier mechanism cannot apply, and has reverted to the original classical carbenium ion mechanism.

In his review, Plesch[8] considered three classes of initiations: (i) systems in which the necessity for cationogenic agents has been established by conclusive stopping experiments, (ii) systems in which stopping experiments have been inconclusive, i.e. exhaustive drying resulted only in a slowing down of the polymerisation and (iii) systems in which stopping experiments have indicated that protogenic impurities are *not* required for initiation, for example styrene/$SnCl_4$ (CH_2Cl_2), styrene/$TiCl_4$ (CH_2Cl_2), isobutylene/$AlCl_3$ (CH_2Cl_2) and isobutylene/$AlBr_3$ (n-heptane). To explain the observations associated with the third class of system, Plesch examined four possibilities:

(a) The Hunter–Yohé mechanism, for example,

$$AlCl_3 + CH_2{=}C(CH_3)_2 \longrightarrow {}^{-}AlCl_3 \longleftarrow CH_2-\overset{CH_3}{\underset{CH_3}{\overset{|}{\underset{|}{C^{+}}}}}$$

Table 2.2 Polymerisation of olefins by self-initiation in two-component systems

Monomer	*Catalyst*	*Solvent*	*Temp. range/°C*	*Remarks*	*References*
Isobutylene	$AlBr_3$	n-heptane	+20 to −60	H_2O is retarder; $-d[M]/dt = k[I]^2[M]$	15
Isobutylene	$AlEtCl_2$	n-heptane	+21 to −55	H_2O is inhibitor; $-d[M]/dt = k[I]^2[M]$	16
Isobutylene	$AlBr_3 \cdot TiCl_4$ etc.	n-heptane	−13	rate increases on $TiCl_4$ addition non-steady state kinetics proposed	17
Isobutylene	$AlBr_3 \cdot TiCl_4$ etc.	n-heptane	−13	'active' and 'inactive' (or 'passive')	32
Isobutylene	$AlBr_3 \cdot TiCl_4$ etc.	n-heptane	−14	Friedel–Crafts halides investigated	17 33
Isobutylene	AlI_3	n-heptane	−14		17
Isobutylene	$GaCl_3$	n-heptane	−14		17
Isobutylene	$GaBr_3$	n-heptane	−14		17
Isobutylene	$TiCl_4$	CH_2Cl_2 or bulk	−72 to −78	'polymerisation by condensation'	28, 29, 30
Isobutylene	$SnCl_4$	C_2H_5Cl	−78.5	measurable polymerisation rate even after exhaustive drying	65
Isoprene	$AlEtCl_2$	n-heptane and benzene	+20 to −18 +20	$-d[M]/dt = k[IM][M]$	18
Isoprene	$AlBr_3$	benzene, toluene and n-heptane	+21	electrical conductivity and u.v. spectra	20
Isoprene	$AlEtCl_2$	n-heptane	+21		34
Isoprene	$AlBr_3$	n-heptane or toluene or benzene	+21		34

Isoprene	$AlCl_3$	n-heptane	+21	initiator in suspension	34
Isoprene	$AlCl_3$	n-heptane or benzene	+21	re-sublimed $AlCl_3$	21
Indene	$TiCl_4$	CH_2Cl_2	−70	$TiCl_4$ is active alone, rate increases in presence of H_2O or HCl	25
Indene	$SnCl_4$	CH_2Cl_2	−30	$SnCl_4$ is active alone, rate increases in presence of H_2O	26, 27
α-Methylstyrene	$TiCl_4$	CH_2Cl_2	+10 to −72	$TiCl_4$ is active alone, rate increases in presence of H_2O, HCl or added $TiCl_4$	31
α-Methylstyrene	$SnCl_4$	C_2H_5Cl	+55	questionable dryness	66
α-Methylstyrene	$AlEt_2Cl$	CH_3Cl	−50	controls: isobutylene and styrene need initiator under the same conditions	53
Cyclopentadiene	$TiCl_3OBu$	CH_2Cl_2	−43 to −70	$TiCl_3OBu$ is active alone, rate increases in the presence of H_2O, and strongly increases with HCl, second $TiCl_3OBu$ addition also positive	23, 24
Cyclopentadiene	$AlEt_2Cl$	CH_3Cl	−50	controls: isobutylene, styrene need initiator under the same conditions	53
5-Methyoxyindene	BF_3	CH_2Cl_2	> −30		37
Propylene	BF_3	CH_2Cl_2	−35	$d[M]/dt = k[BF_3][BF_3 \cdot CH_3OH]^{\frac{1}{2}}[M]$	42
β-Methylstyrene	BF_3	$(CH_2Cl)_2$	+25	dimerisation is first order in both olefin and BF_3	45
But-2-ene	BF_3	$(CH_2Cl)_2$	+25	isomerisation rate = $k[BF_3][BF_3 \cdot H_2O]$	38, 40

is unlikely because of unfavourable thermochemical estimates and the absence of independent evidence.

(b) Self-dissociation, for example,

$$2TiCl_4 \rightleftharpoons TiCl_3^+ + TiCl_4^-$$

$$TiCl_3^+ + CH_2{=}\underset{\displaystyle C_6H_5}{\underset{|}{CH}} \longrightarrow TiCl_3{-}CH_2{-}\underset{\displaystyle C_6H_5}{\underset{|}{\overset{+}{C}H}}$$

is unlikely because of unfavourable thermochemical estimates with regard to the second step of the above two reactions.

(c) Solvent co-catalysis, for example,

$$TiCl_4 + CH_2{=}\underset{\displaystyle C_6H_5}{\underset{|}{CH}} + CH_2Cl_2 \longrightarrow ClCH_2{-}CH_2{-}\underset{\displaystyle C_6H_5}{\underset{|}{\overset{+}{C}H}} \cdot TiCl_5^-$$

may be possible on the basis of thermochemical considerations and because of the availability of direct proof in similar systems[14].

(d) Electron-transfer initiation, for example,

$$2TiCl_4 + CH_2{=}\underset{\displaystyle C_6H_5}{\underset{|}{CH}} \longrightarrow TiCl_3 + TiCl_5^- + CH_2{-}\underset{\displaystyle C_6H_5}{\underset{|}{\overset{+}{C}H}}$$

is possible at least for reducible metal halides. While this mechanism is pure speculation at the moment, it does not appear to have a prohibitively unfavourable ΔH value.

The above classification is a useful, but insufficient, development in the understanding of initiation mechanisms. First of all, there is little justification for keeping classes (i) and (ii) separate and they should be combined. An unassailable objection is that systems in which drying resulted only in a reduction of rate may well have come to a stand-still under perfectly cation-free conditions. The amounts of cationogenic impurities necessary for initiation may have been so small that efforts at purification were doomed from the beginning. This reasoning leads to a classification of initiation by impurity requirements, i.e., systems which require for initiation the presence of impurities (Plesch's classes (i) and (ii)) and those which do not (class (iii)): an obviously unsatisfactory proposal.

Secondly, the very interesting recent findings that initiation occurs in the isobutylene/$AlBr_3$/n-heptane[15], isobutylene/$AlEtCl_2$/n-heptane[16], isobutylene/AlI_3/n-heptane[17], isoprene/$AlEtCl_2$/n-heptane[18] systems etc., (cf. Table 2.2) cannot be fitted in any of the above classes. Indeed, a large number of carefully purified systems exist in which initiation occurs even in the absence of a separately added cationogen. As summarised in Table 2.2 these systems include non-reducible as well as reducible metal halides. These results will now be discussed, and a simple classification of carbenium ion initiation will be proposed. Also, a new hypothesis to explain initiation in the systems compiled in Table 2.2 will be developed.

Chmelir *et al.*[15] have found that $AlBr_3$ can polymerise isobutylene in n-heptane solution in the absence of any added chemicals. The overall kinetics were found to obey the relationship $-d[M]/dt = k[AlBr_3]^2[M]$. The purity of the system was indicated by preliminary successful stopping experiments with BF_3 and $TiCl_4$. In a later publication, it was described that certain mixed Friedel–Crafts halides polymerise isobutylene more efficiently than $AlBr_3$ alone, and $AlBr_3 \cdot TiCl_4$ was found to be particularly effective[17]. The authors examined a large number of Friedel–Crafts halides (BX_3, GeX_3, TiX_4, $GeCl_4$, SnX_4, VCl_4, $AsCl_3$, $SbCl_3$ and AlX_3, where X = Cl, Br or I) and found that only $AlBr_3$, AlI_3, $GaCl_3$ and $GeBr_3$ are true initiators (active components, according to Reference 17) and that the other halides examined are unable to induce polymerisation alone (passive components). Other experiments showed that when an aliquot of a *per se* inactive $TiCl_4$ is added to a relatively slowly progressing polymerisation of isobutylene initiated with $AlBr_3$, heat evolution occurs and the rate suddenly increases[19]. This is illustrated in Figure 2.1.

Very recently the same group examined the polymerisation of isobutylene with $AlEtCl_2$ in a non-polar solvent[16] and found, again, that polymerisation

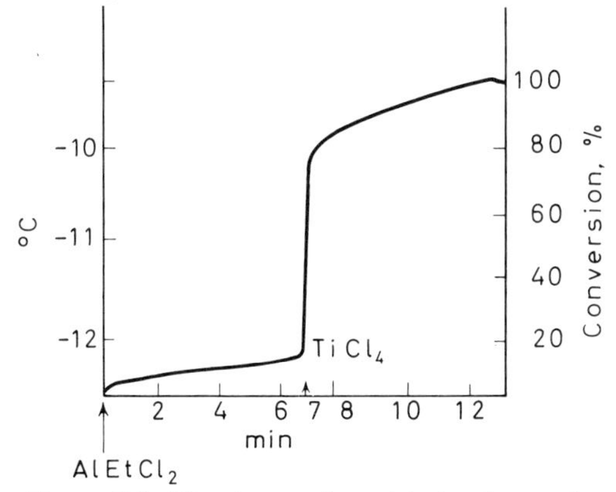

Figure 2.1 Time–conversion of isobutylene polymerisation; [isobutylene] = 0.1 mol l^{-1}, $[AlBr_3]$ = 8.8×10^{-5} mol l^{-1}, $[TiCl_4] = 8.1 \times 10^{-3}$ mol l^{-1} (From Marek and Chmelir[19], by permission of J. Wiley)

was induced by $AlEtCl_2$ alone and there was no necessity for the addition of a third chemical. Significantly, it was noted that water had an inhibiting effect: as soon as water was added to a polymerisation (0.15 mol l^{-1} isobutylene and 4×10^{-4} mol l^{-1} $AlEtCl_2$ plus 5×10^{-5} mol l^{-1} water in heptane (saturated)) the reaction stopped. As with the $AlBr_3$ system investigated earlier, the initial overall rate was found to be first order in monomer and second order in catalyst.

Similarly, Matyska *et al.*[20] found in experiments conducted under high-vacuum conditions that pure $AlBr_3$ induces the polymerisation of isoprene in n-heptane, benzene or toluene solvents. Electrical conductivity and ultraviolet spectroscopy measurements indicated the formation of ions as soon as the monomer was mixed with the aluminum halide. Further, Kössler *et al.*[18] have stated that pure $AlEtCl_2$ is able to polymerise isoprene in n-heptane or in benzene in experiments conducted under vacuum. Matyska *et al.*[21] have found that re-sublimed $AlCl_3$ polymerises isoprene to white solids in n-heptane or benzene (isoprene 0.87 mol l^{-1}, $AlCl_3$ 0.3 g, 21 °C, yield: 2.2 g after 10 min in benzene and 0.5 g after 180 min in n-heptane, Table 2.1, reference 21) while under essentially identical vacuum (anhydrous) conditions $TiCl_4$ was found to be totally inactive.

The Czechoslovak authors have attempted to explain their findings[18–20]; this Reviewer, however, cannot accept their proposals. For example, Marek and Chmelir[19] state that titanium has been detected in purified polyisobuylene prepared with $AlBr_3+TiCl_4$, and consequently assume that the initiating species was $TiCl_3^+ AlBr_4^-$. Apart from the objectionable halide stoichiometry suggested for the alleged initiating ionic species, it is difficult to accept the evidence that Ti was found in the polymer as being anything else but an indication that titanium catalyst residues occur in the product. It is virtually impossible to come to a different conclusion because otherwise one has to accept the implication that Ti—C bonds are established during polymerisation, and that they remain unharmed and survive the aggressive purification procedure which consisted of repeated precipitation of the polyisobutylene from a heptane solution into water and sulphuric acid. The initiation mechanism proposed[15] and developed[16, 18, 19, 22] by the Czechoslovak authors is based on the idea of self-dissociation of Friedel–Crafts halides. For example, with the $AlBr_3$–isobutylene system, initiation is visualised as follows[15]:

$$2AlBr_3 \rightleftharpoons AlBr_2^+ AlBr_4^-$$

$$AlBr_2^+AlBr_4^- + CH_2{=}C(CH_3)_2 \rightarrow AlBr_2{-}CH_2{-}C^+(CH_3)_2\,AlBr_4^-$$

or, with the mixed metal halide $AlBr_4 \cdot TiCl_4$ systems mixed self-dissociation is postulated to give $TiCl_3^+AlBr_3Cl^-$ followed by initiation proper[22]. It is difficult to see appreciable self-dissociation occurring in non-polar media followed by initiation. This mechanism has been criticised by Plesch on the basis of thermochemical estimates[8]. Indeed, in their last publication Lopour and Marek[17], on the basis of new data obtained with mixed Lewis acids, conclude that '. . . the actual catalytically active materials in the two-component system of the AlX_3/MeY_n type are exclusively the mixed halides of aluminum, and not at all some complexes of the two components (for example $TiCl_3^+AlBr_3Cl^-$ [19]) as has been assumed in earlier papers' and conclude, 'In contrast to earlier views . . . we came to the conclusion that the initiation with these two component systems is very complicated, and that a whole series of mixed aluminum halides and catalytically active complexes of different concentration and activity may be involved in this process' (English translation of reference 17).

The salient findings of the French school[23–31] are now summarised. Sigwalt and his co-workers examined with greatest care the details of initiation of isobutylene–$TiCl_4$ and cyclopentadiene–$TiCl_3OBu$ systems, all in methylene chloride solvent. The painstaking care of these workers to exclude the last possible traces of moisture in their experiments is truly remarkable. It is this Reviewer's impression that this group of workers carried out their cationic polymerisation research under the highest attainable anhydrous conditions. Most interesting and puzzling results were obtained with the isobutylene/$TiCl_4(CH_2Cl_2)$ system[28–30]. Monomer and solvent were condensed at -70 °C in one arm of an H-shaped reactor and a phial containing $TiCl_4$ was broken magnetically under the surface of the liquid phase. A not insignificant amount (6–38%) of polymerisation ('pre-polymerisation') occurred which was attributed to the presence of unscavenged impurities (moisture). This pre-polymerisation step then consumed all unscavenged co-initiators and/or promoters.

Absolute dryness in such systems was indicated by the fact that when a second phial containing $TiCl_4$ was broken in a pre-polymerised system, no further polymerisation occurred. After pre-polymerisation, the liquid phase could be tipped over into the second arm of the H-shaped vessel without further polymerisation taking place. Significantly, however, polymerisation occurred when the reactants in the first arm of the H-shaped vessel were vaporised (by removing the cooling jacket) and re-condensed in the second arm. Polymerisation occurred on vaporisation–condensation even when the walls of the second compartment were coated with sodium metal. Similarly, polymerisation also ensued on re-condensation in a bulk system, i.e. in the absence of CH_2Cl_2. The final polymerisation in the second arm always resulted in high yields and high molecular weight products. Polymerisation also occurred in experiments in which $TiCl_4$ was distilled either rapidly or slowly into condensed and 'pre-polymerised' isobutylene–CH_2Cl_2 mixtures. The absence or presence of light did not seem to affect the experiments. As polymerisation is apparently somehow connected to vaporisation and condensation, the French workers have called this phenomenon 'polymerisation by condensation[28]'. The authors examined several possible explanations for their observations, but no firm proposals were made.

Sigwalt and co-workers also examined the polymerisation of indene with $TiCl_4$ and $SnCl_4$ in CH_2Cl_2 at low temperatures, again, under most carefully dry conditions[25–27]. A common feature of these experiments was that slow polymerisation occurred even in the absence of purposely-added initiators (H_2O or HCl), although the rates always strongly increased upon the addition of these protogens. Fundamentally similar results were also obtained with α-methylstyrene/$TiCl_4(CH_2Cl_2)$ as well: polymerisations always proceeded even in the absence of H_2O or HCl, particularly rapidly at lowest temperatures (-72 °C). At higher temperatures ($+10$ or -30 °C) the addition of H_2O or HCl, or a second addition of purest $TiCl_4$, increased the rate.

Sigwalt and co-workers[23, 24] also examined the low-temperature polymerisation of cyclopentadiene induced by $TiCl_3OBu$ in CH_2Cl_2. The salient feature of this research was that the polymerisation of cyclopentadiene could not be stopped even after exhaustive drying. Among other measures, the technique of 'pre-polymerisation' was always used in this work to eliminate the last traces of moisture or other protogens. In spite of the most careful drying, slow but definite polymerisation occurred when a second phial of $TiCl_3OBu$ was broken in the liquid phase of a 'pre-polymerised' system. The addition of H_2O or HCl in pre-polymerised systems resulted in rapid increases in the rate. Extended kinetic investigations were carried out which concentrated only on the effect of H_2O on the polymerisation[24].

The trivial argument that further purification, i.e. drying, would have eliminated residual impurities and stopped the polymerisations summarised in Table 2.2 cannot be invoked: the rate data are reproducible and self-consistent. For the same reason, the results cannot be lightly dismissed[4] by assuming that, for example, $AlBr_3$ may require different kinds of impurities (e.g. HBr) for initiation than BF_3 or $TiCl_4$. Rather, in view of the apparent reproducibility of the observations and the consistency of published phenomena (cf. Table 2.2) another concept, a new mechanism, is needed to account for all the available data.

Clearly, the time has come to abandon the notion that the presence or addition of third component (commonly called initiator) is necessary for the initiation of cationic polymerisations induced by Friedel–Crafts halides. A new classification of initiators is now proposed, and an attempt will be made to develop a theory that explains the initiation of certain olefins (cf. Table 2.2) which do not require the presence or assistance of a separately added cationogen.

2.3.2 Self- and co-initiation

Cationic initiation involving Lewis acids of the Friedel–Crafts halide type may be sub-divided as follows:

(a) Self-initiation with two-component (monomer–initiator) systems. Monomer–metal halide systems which are capable of initiation in the absence of extraneous or added chemicals belong to this class.

It is proposed that self-initiation in two-component (monomer–initiator) systems occurs by the removal of an allylic hydrogen atom, in the form of a hydride ion, from the nucleophilic monomer by strong Lewis acids:

$$—C{=}C—C—H + MX_n \rightarrow —C{=}C—C^+MX_nH^-$$

The allylic hydrogen is provided by the monomer itself, e.g., isobutylene, isoprene, α-methylstyrene or indene; MX_n denotes $AlBr_3$, AlI_3, $GaCl_3$, $TiCl_4$, etc.

(b) Co-initiation with three-component (monomer–initiator–co-initiator) systems. Monomer–metal halide systems that require the participation of a third component for initiation belong to this class. This class can readily be sub-divided according to the nature of the required third component into (i) initiation with protogens, e.g. Brønsted acids such as H_2O, HCl, CH_3COOH, etc., and (ii) initiation by cationogens (carbenium ion sources), e.g. alkyl halides.

This classification facilitates the orderly presentation of important observations pertaining to initiation. It will become apparent, however, that this classification is not based on fundamentals, and that the factor which determines the ultimate success or failure of initiation appears to be the acidity of the Friedel–Crafts halide relative to that of the monomer.

First, we will examine the self-initiation proposition in light of published information, and then subsequently, we will discuss the two co-initiation systems.

2.3.3 Self-initiation

Chmelir *et al.*[15] found that isobutylene can be polymerised by extremely pure $AlBr_3$ in n-heptane solution (cf. Table 2.2). Their kinetic results indicate a second-order dependence on initiator and a first-order dependence on monomer. These basic facts are readily explained by the self-initiation concept:

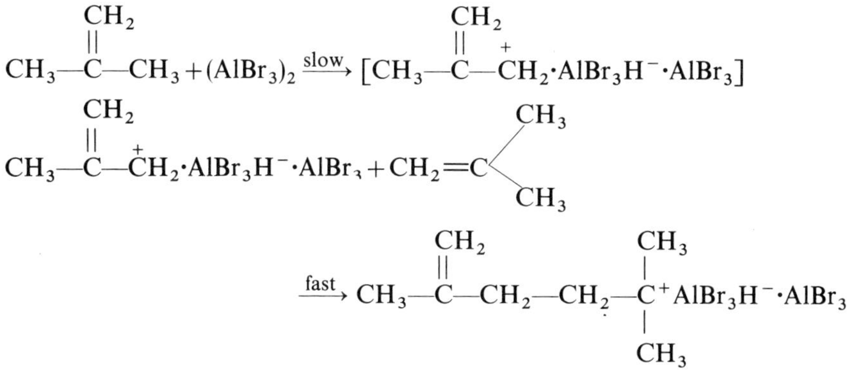

$$CH_3-\underset{\displaystyle CH_2}{\overset{\|}{C}}-CH_3 + (AlBr_3)_2 \xrightarrow{\text{slow}} [CH_3-\underset{\displaystyle CH_2}{\overset{\|}{C}}-\overset{+}{C}H_2\cdot AlBr_3H^-\cdot AlBr_3]$$

(In the printed scheme the $=CH_2$ groups are drawn above the central C atoms.)

$$CH_3-\overset{CH_2}{\overset{\|}{C}}-\overset{+}{C}H_2\cdot AlBr_3H^-\cdot AlBr_3 + CH_2{=}C(CH_3)_2$$

$$\xrightarrow{\text{fast}} CH_3-\overset{CH_2}{\overset{\|}{C}}-CH_2-CH_2-\overset{CH_3}{\underset{CH_3}{C^+}}AlBr_3H^-\cdot AlBr_3$$

Self-initiation occurs by the generation of an allylic carbenium ion via hydride transfer from the isobutylene to the dimeric Lewis acid. This process can be viewed as an acid–base reaction in which the monomer is the base (nucleophile) and the Friedel–Crafts halide the acid (electrophile), and the newly formed carbenium ion and the counter-ion are the conjugate acid and base, respectively. In this self-initiation mechanism the conventional difference between monomer and initiator becomes somewhat diffuse; the monomer provides its own initiator, the first carbenium ion.

The same basic mechanism also fits all the other findings of various Czechoslovak groups[16–20, 32–34]. Self-initiation can readily be envisaged in the isoprene–$AlBr_3$ systems where allyl hydride transfer would yield a favourable resonance-stabilised structure

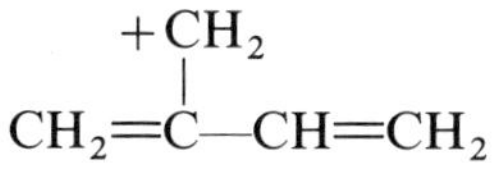

$$CH_2{=}\overset{+CH_2}{\overset{|}{C}}-CH{=}CH_2$$

Hydride transfer is indicated by the appearance of conductivity and characteristic spectra upon mixing isoprene with $AlBr_3$ [20].

In line with these thoughts, the important phenomenon which needs to be explained in the experiments by the Czechoslovak group is not so much the self-initiation of isobutylene by purest $AlBr_3$, but the observation that the rate of polymerisation of a slowly propagating isobutylene–$AlBr_3$ system increases suddenly upon the addition of $TiCl_4$. An explanation could be that the $TiCl_4$ component helps to break up the aluminium bromide dimer Al_2Br_6 into $AlBr_3$, thus creating a stronger Lewis acid, i.e. better hydride acceptor, than the dimer.

All the findings of the French researchers can also be explained by the self-initiation concept. The key phenomenon which needs explanation in this respect is the fact that $TiCl_4$ appears to be active only when condensed from the vapour phase. Cheradame and Sigwalt[28] emphasise this, and state that '... the catalyst is associated with itself in solution and it dissociates when in the vapour phase. From this it can be assumed that the $TiCl_4$ monomer molecule is able to induce direct initiation'. They further state that unpublished evidence is available to prove self-association in solution[28]. On the basis of these facts, self-initiation can readily be invoked, and the only assumption one has to make is that the unassociated $TiCl_4$ is a sufficiently strong Lewis acid for hydride abstraction, whereas the associated (less acidic) titanium tetrachloride which exists in the condensed phase is not. The results of Sigwalt's other experiments with indene, α-methylstyrene and cyclopentadiene monomers in the liquid phase at low temperatures are also in agreement with the self-initiation concept. The problem here is to explain why isobutylene/$TiCl_4$ does not initiate the polymerisation in the liquid phase, but indene and α-methylstyrene do. We recall that both indene and α-methylstyrene are polymerised by $TiCl_4$ or $SnCl_4$ alone[25,31] though H_2O and/or HCl augment the rates. The explanation probably involves the relative nucleophilicity of these monomers: it is much easier to remove an allylic hydrogen by a Lewis acid from the stronger bases indene and α-methylstyrene than from the weaker base isobutylene. The stability of the allylic carbenium ions

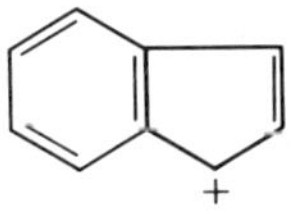

[indenyl cation structure] or $\overset{+}{C}H_2{-}C({=}CH_2){-}C_6H_5$

are most likely greater (thermodynamic acidity) and their rate of formation faster (kinetic acidity) than that of the

$$\overset{+}{C}H_2{-}C({=}CH_2){-}CH_3$$

ion. Thus, even the less-acidic associated $TiCl_4$ or $SnCl_4$ are able to remove hydride ions from the former olefins, whereas the more-acidic 'free' $TiCl_4$ is needed to transfer hydride ions from isobutylene.

Sigwalt's results obtained with the cyclopentadiene–$TiCl_3OBu$ system can also be explained with the self-initiation mechanism[23,24]. According to the French investigators, cyclopentadiene polymerises under the driest conditions, and the introduction of a second aliquot of $TiCl_3OBu$ into a 'pre-polymerised' system results in further, albeit slow, polymerisation. According to the self-initiation hypothesis, initiation occurs via:

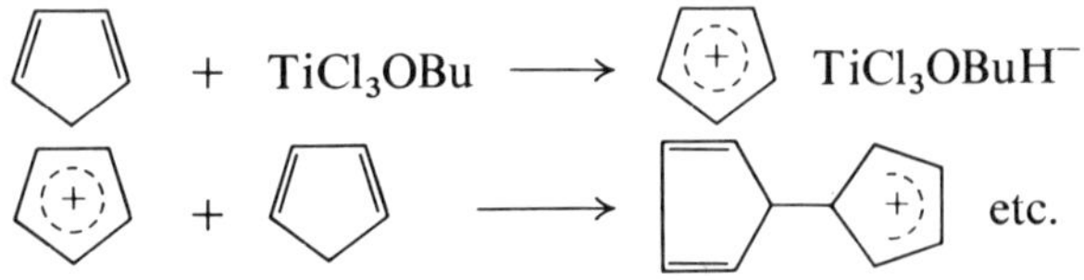

[cyclopentadiene] + $TiCl_3OBu$ ⟶ [cyclopentadienyl cation (+)] $TiCl_3OBuH^-$

[cyclopentadienyl cation (+)] + [cyclopentadiene] ⟶ [cyclopentenyl–cyclopentadienyl cation (+)] etc.

Cyclopentadiene is quite basic and could readily give up a hydride ion to even a less-acidic Lewis acid like $TiCl_3OBu$.

Sauvet, Vairon and Sigwalt[35] examined the dimerisation of 1,1-diphenylethylene with $TiCl_4$ in CH_2Cl_2 in the 0 to -78 °C temperature range. When the $[DPhE]:[TiCl_4]$ ratio was higher than ~ 100, no dimerisation occurred except in the presence of a purposely-added protogen (HCl or H_2O). These findings are in agreement with the self-initiation concept. However, when the $[DPhE]:[TiCl_4]$ ratio was less than ~ 10, dimerisation occurred even in the absence of a separately-added proton source. This last observation cannot be explained by any meaningful initiation theory. A possibility may be perhaps that with this particular dimer a mechanism of the Hunter–Yohe type occurs: $\overset{\delta-}{TiCl_4} \leftarrow CH_2 - \overset{\delta+}{C}(C_6H_5)_2$ (the formation of a tertiary carbenium ion with two aromatic substituents is energetically very favourable) whereas initiation is less favourable with higher complexes such as $TiCl_4 \cdot n$DPhE where $n = 2,3$, etc.

While these experiments with the DPhE–Lewis acid systems are only one year old, it should be mentioned that Evans and his school investigated the dimerisation of DPhE in the presence of a variety of acids and published their results in at least 11 papers over the period 1955–1962[36]. Somewhat contrary to Sigwalt's findings, Evans stated that DPhE cannot be dimerised with $TiCl_4$ in the absence of a protogen[36]. Indeed, the dimerisation of DPhE with $SnCl_4$, $SbCl_3$ and $TiCl_4$ required the separate addition of a protogen (HCl or H_2O). It should, however, be pointed out that Evans *et al.* did not carry out their dimerisation studies over the low $[DPhE]:[TiCl_4]$ concentration range.

Marechal[37] reported that below ~ -30 °C BF_3 does not polymerise 5-methoxyindene, and that a complex is formed between BF_3 and 5-methoxyindene. Upon heating the system above this threshold temperature, rapid polymerisation occurs without the addition of a protogen. These observations could be explained by assuming that first the 5-methoxyindene $\cdot BF_3$ complex dissociates upon heating above -30 °C, and subsequently the free BF_3 removes an allylic hydrogen from the indene thus generating the initiating species:

BF_3 ; CH_3–O ; Δ ; CH_3–O ; $+ BF_3$; CH_3–O ; + ; $\cdot BF_3H^+$

Marechal also stated that while the indene derivative does not need a separately added proton source for initiation, 3,4-dimethoxystyrene, a monomer which does not have an allylic H, required a protogen, such as HCl, HF or CCl_3COOH for initiation.

The work of Eastham and his co-workers, published in a series of 10 publications, should be mentioned in this context[38]. Amongst many other things, this worker studied under most carefully anhydrous conditions the isomerisation of but-2-ene and the polymerisation of propylene with BF_3 and its complexes. It was found[39] that at the practical limit of drying the erratic rate of *cis*-$\rightleftharpoons$*trans*-but-2-ene isomerisation became slow and reproducible, and that slow isomerisation proceeded with BF_3 alone even in the absence of water. The rate of isomerisation was much accelerated in the presence of suitable amounts of moisture, and could be described[40,41] by the expression:

$$d[\text{isomer}_B]/dt = k[BF_3][BF_3 \cdot H_2O][\text{isomer}_A]$$

This law could be interpreted as a composite expression for a self-initiated and co-initiated process:

Self-initiated isomerisation

$$\underset{\substack{| \\ CH_3}}{HC}{=}\underset{\substack{| \\ CH_3}}{CH} + BF_3 \rightleftharpoons \left[\overset{\delta+}{C}H_2 \text{=====} CH \text{=====} \underset{\substack{| \\ CH_3}}{\overset{\delta+}{C}H} \; BF_3H^- \right] \rightleftharpoons \overset{\substack{CH_3 \\ |}}{CH}{=}\underset{\substack{| \\ CH_3}}{CH} + BF_3$$

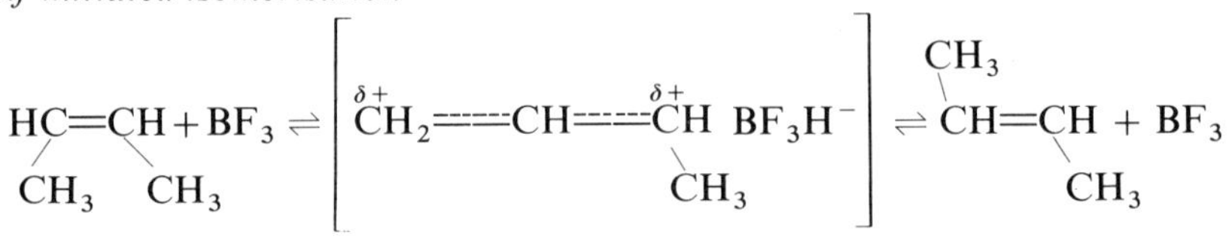

Co-initiated isomerisation:

$$\underset{\substack{| \\ CH_3}}{CH}{=}\underset{\substack{| \\ CH_3}}{CH} + BF_3 \cdot H_2O \rightleftharpoons \left[\overset{\delta+}{C}H_3 \cdots CH \cdots \underset{\substack{| \\ CH_3}}{\overset{\delta+}{C}H} + BF_3OH^- \right] \rightleftharpoons \overset{\substack{CH_2 \\ |}}{CH}{=}\underset{\substack{| \\ CH_3}}{CH}$$

$$+ BF_3 \cdot H_2O$$

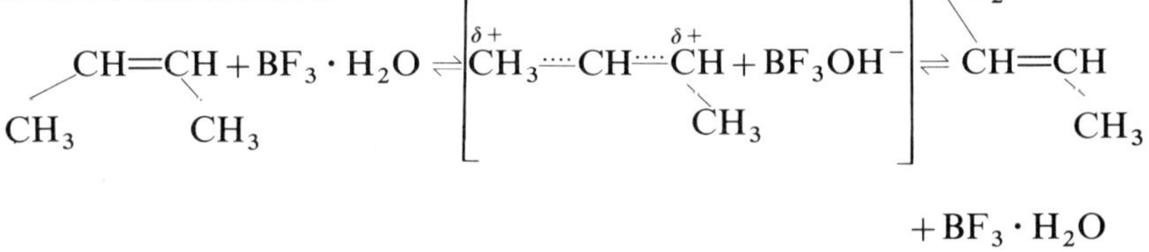

of which the first process is much slower than the second. An important clue may also be gleaned from close examination of Eastham's paper on the polymerisation of propylene[42]. A careful series of experiments indicate that a slow polymerisation proceeds in the presence of BF_3 alone, and that the rate accelerates with $BF_3 \cdot CH_3OH$ complexes (Figure 1 in Reference 42).

In the course of these investigations it was also observed that u.v.[43] and visible[44] chromophores were slowly formed when butenes in ethylene dichloride were mixed with BF_3 and $BF_3 \cdot CH_3COOH$. On the basis of careful work, the conclusion was advanced that the coloured ionic species were allyl carbenium ions which formed by hydride transfer from the butenes to adventitious carbenium ion impurities in the system:

$$CH_3{-}CH{=}CH{-}CH_3 + R_3C^+ \longrightarrow CH_3{-}CH \cdots \overset{+}{C}H \cdots CH_2 + R_3CH$$

In line with the self-initiation proposal, it may also be postulated that the BF_3 was the allylic hydride abstractor.

Recently Eastham and co-workers[45] studied the dimerisation of β-methylstyrene with BF_3. The rate of dimerisation was very sensitive to the presence of impurities. The rates became reproducible when the glass reactor was baked for hours at 500 °C under vacuum. Significantly, dimerisation still occurred under the most carefully dried conditions, even in the absence of an added protogen. The authors advanced the theory that non-scavengable protons (—Si—O—H) could have been involved in initiation. Another possibility is that β-methylstyrene provides its own initiator by:

$$CH_3—CH{=}CH—C_6H_5 + BF_3 \longrightarrow \overset{\delta+}{C}H_2 \cdots CH \cdots \overset{\delta+}{C}H—C_6H_5 \cdot BF_3H^-$$

The fact that the rate of dimerisation was found to be first order in both olefin and BF_3, is in agreement with the self-initiation hypothesis.

Whilst no direct proof exists for the self-initiation mechanism, there is circumstantial evidence to corroborate it. First of all, hydride transfer is a well-established and often-postulated process in carbenium ion chemistry. This process is particularly favourable and may occur extremely rapidly[46] where a resonance-stabilised (e.g. allylic) carbenium ion can be formed. Carbenium ions are capable of removing suitable hydride ions, and in turn generate more stable carbenium ions. As carbenium ions are strong acids in the Lewis sense, it is conceivable that certain metal halides that are strong Lewis acids could also assist in the removal of negative hydrogens. The hydride ion of NaH may coordinate with BF_3; thus, $NaBF_3H$ was prepared[47] by reacting BF_3 with NaH in ether at −70 °C. The B—H and Al—H bond strengths are 79 and 68 kcal mol^{-1}, respectively, not insignificant values.

Secondly, the formation of allylic carbenium ions from olefins (e.g. but-1-ene, but-2-ene) has recently been demonstrated spectroscopically on the surface of Lewis acidic solids[48]. For example, the vibrational spectrum of the species formed from but-1-ene on silica–alumina was indistinguishable from the spectra of the butenyl ion ($CH_2{=}CH—\overset{+}{C}H—CH_3$) obtained from 3-chlorobut-1-ene in fused $SbCl_3$ [48]. In this case the surface, a strong Lewis acid, removes a hydride ion from a suitable olefin, the base.

Thirdly, available reliable quantitative information, for example, rate = $k[AlBr_3]^2$[isobutylene[15,16] and rate = $[BF_3]$ [β-methylstyrene][45], etc. is in agreement with the mechanism of self-initiation.

Thermochemical calculations are of little value as corroborating evidence as condensed phases are involved and the effect of solvation (with the monomer) and/or aggregation of charged species of various structures is difficult to assess. Also, the ultimate fate of the hydride ion is a matter of conjecture. It is conceivable that $AlBr_3H^- \cdot AlBr_3$, could, for example, be involved in the deprotonation of carbenium ions (R^+) to give H_2 and olefins ($R^=$) by

$$R^+AlBr_3H^- \cdot AlBr_3 \longrightarrow R^= + H_2 + (AlBr_3)_2$$

This possibility could be examined by careful H_2 analysis.

Self-initiation via H^- transfer to Friedel–Crafts halides is most probably a slow process, particularly in non-polar media, and is easily masked (out-competed) by more facile initiation mechanisms, for example, initiation with

H_2O, HX or suitable RX. Therefore, observations of self-initiation experiments should be carried out for extended periods (for hours, perhaps for days) particularly with propylene, which not only gives the least-stabilised unsubstituted allylic carbenium ion ($C{=}C{-}C^+$) amongst all potential olefins but also gives a relatively unstable secondary cation on propagation. Isobutylene is much better in this respect: while the self-initiated centre-substituted allylic carbenium ion arising from isobutylene is not significantly more stable than the unsubstituted one obtained from propylene, the propagating species is a very stable tertiary cation. But-1-ene is also a more reactive monomer than propylene because the self-initiated cation is a substituted allylic carbenium ion $CH_2{=}CH{-}\overset{+}{C}H{-}CH_3$ of relatively high stability. Obviously isoprene, α-methylstyrene, indene and cyclopentadiene, the other monomers which show self-initiation, are superior hydride sources. This is in agreement with observations that α-methylstyrene and indene self-initiate with $TiCl_4$ in the condensed phase, whereas isobutylene does not, and that cyclopentadiene self-initiates with the presumably low-acidity $TiCl_3OBu$[23, 24]†.

Self-initiation via H^- transfer is most likely a slow process relative to other types of initiations, and it results in unsaturated end-groups that are structurally very similar, indeed, almost indistinguishable from end-groups arising in common chain-transfer to monomer (proton transfer). For example, in the case of self-initiation with isobutylene, the head-group would be

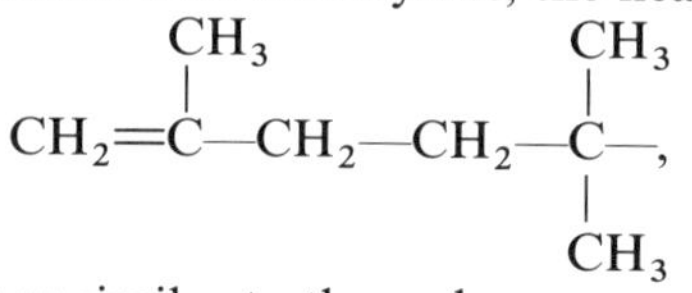

$$CH_2{=}\underset{}{\overset{CH_3}{\overset{|}{C}}}{-}CH_2{-}CH_2{-}\underset{\underset{CH_3}{|}}{\overset{CH_3}{\overset{|}{C}}}{-},$$

a structure which is very similar to the end-group

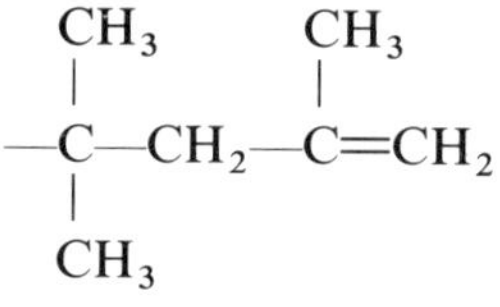

$$-\underset{\underset{CH_3}{|}}{\overset{CH_3}{\overset{|}{C}}}{-}CH_2{-}\overset{CH_3}{\overset{|}{C}}{=}CH_2$$

that is formed on chain transfer.

Some 20 years ago, Fontana and Kidder[49] polymerised propylene and but-1-ene with $AlBr_3$, and reported that reactions only occurred in the presence of suitable initiator ('promoter') such as HBr. This observation, at first sight, contradicts the self-initiation theory; however, on closer reading of Fontana's publication[49] it is noted that the HBr or RBr in these studies were the true promoters in the sense that they merely increased the rates of the polymerisations which proceeded, albeit much more slowly, in their absence (Table 1 in reference 49). In contrast, the authors state in their introduction[49] '. . . it was also demonstrated in one experiment that no reaction

†There is an ambiguous report[34] which could be interpreted that $SnCl_4$ is able to initiate the polymerisation of isoprene. On careful reading of this publication this Author concluded that the polymerisation of isoprene with $SnCl_4$ described in Reference 34, was carried out in moist n-heptane (Table 1 in Reference 24). Indeed Lopour and Marek[17] found $SnCl_4$ a 'passive component' for the polymerisation of isobutylene.

occurred at low temperatures under anhydrous conditions and in the absence of added promoter'. This contradiction cannot now be resolved (no experimental details were given) and confirmatory evidence is needed to assess the validity of this statement.

In a series of papers in 1967, the present Author proposed a theory of termination for cationic polymerisation[50]. According to that theory, termination occurs by the removal of H^- by the propagating carbenium ion from suitable olefins:

$$\sim\sim C^+ + \;\rangle C{=}C{-}\overset{|}{\underset{|}{C}}{-} \longrightarrow \sim\sim CH + \;\rangle C{=}C{-}\overset{+}{\underset{|}{C}}{-}$$

terminating agent — too stable to propagate

This theory has been very useful in controlling the molecular weights of polymers and could also be of value in the functionalisation of polymers. Extensive data have shown that allylic self-termination, while undoubtedly present, is not important with isobutylene (in n-pentane with $AlCl_3(CH_3Cl)$ at $-78\,°C$):

$$\sim\sim\overset{CH_3}{\underset{CH_3}{C^+}} + H{-}\underset{|}{\overset{|}{C}}{-}\overset{|}{C}{=}\overset{|}{C}{-} \rightleftharpoons \sim\sim\overset{CH_3}{\underset{CH_3}{CH}} + {-}\overset{+}{\underset{|}{C}}{-}\overset{|}{C}{=}\underset{|}{\overset{|}{C}}$$

but could very well operate with other monomers including propylene. The present theory of self-initiation via H^- transfer is closely related to the earlier self-consistent theory of termination via H^- transfer: in this sense monomers may provide their own initiating and terminating agents.

As with any other theory, only correct predictions will establish the value of the self-initiation theory. For this theory to be correct only monomers which contain at least one available allylic hydrogen atom could provide their own initiator. Thus styrene (its ring-substituted derivatives) or 3,3-dimethylbut-1-ene should not be capable of initiation by this route. The geometric availability of the allylic hydrogen atom could also be of importance, and it is conceivable that in monomers such as 3,3-diphenylprop-1-ene the allylic H is sterically unavailable and initiation is therefore prohibited.

In this context it is of interest that neither the French nor the Czechoslovak authors, who were most active in this area, ever mentioned the polymerisation of styrene, although the phenomenon of self-initiation with isobutylene was discovered as early as ~1962.

Workers who studied the polymerisation behaviour of styrene under carefully dried conditions (for example: styrene—BF_3—CCl_4 (reference 51), styrene–$AlCl_3$ (reference 52)) state that this monomer cannot be polymerised with Friedel–Crafts halides except in the presence of a protogen or cationogen.

If the theory of self-initiation is correct, small amounts of good hydride donors in conjunction with Friedel–Crafts halides could initiate the polymerisation of less-favourable hydride donors. For example, cyclopentadiene or indene added to a quiescent isobutylene/$TiCl_4$ system in the liquid phase could initiate the polymerisation of the latter. In these cases cyclopentadiene or indene function as initiators.

With these thoughts in mind, this Author has carried out some experiments with α-methylstyrene and cyclopentadiene with $AlEt_2Cl$ as Lewis acid in methyl chloride solvent at −50 °C [53]. Rapid polymerisation occurred when either of these monomers were mixed with $AlEt_2Cl$; however, no polymerisation took place with isobutylene and/or styrene. Apparently with α-methylstyrene and cyclopentadiene, self-initiation via H^- transfer is favourable because of the formation of relatively stable allylic carbenium ions

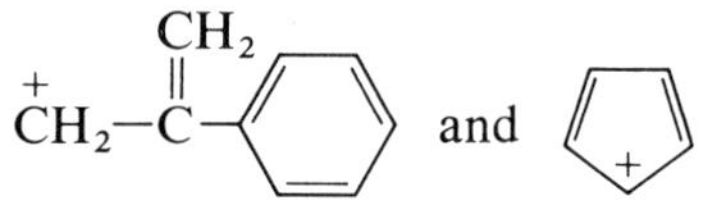

$$\overset{+}{C}H_2{-}\underset{}{\overset{CH_2}{\overset{\|}{C}}}{-}C_6H_5 \quad \text{and} \quad C_5H_5^+$$

but initiating species do not form in the less-favourable isobutylene case.

For further details on self-initiation consult Reference 53a.

Finally, still in the area of self-initiations, Plesch's proposal of electron-transfer initiation with reducible transition metal compounds[8] is a valuable idea and should merit further consideration, for example:

$$2TiCl_4 + CH_2{=}\underset{\substack{|\\R}}{CH} \longrightarrow TiCl_3 + TiCl_5^- + \dot{C}H_2{-}\underset{\substack{|\\R}}{\overset{+}{C}H}$$

At the moment, this purely hypothetical initiation mechanism would fall *pro-forma* into the class of self-initiation with two-component systems. Electron spin resonance (e.s.r.) spectroscopy could be employed to investigate this proposal.

2.3.4 Co-initiation with protogens

As proposed above, co-initiation can be sub-divided according to the nature of the initiator employed (proton sources or cation sources) in conjunction with the Lewis acid co-initiator:

$$MX_n + HX + C{=}C \rightleftharpoons [\text{complex}] \rightleftharpoons HCC^+MX_{n+1}^-$$

$$MX_n + RX + C{=}C \rightleftharpoons [\text{complex}] \rightleftharpoons RCC^+MX_{n+1}^-$$

From the *conceptual* point of view, not much progress has been made in this field since the late forties when the by now classical investigations of British workers[11] showed conclusively that Friedel–Crafts halides require the presence of a third component, usually moisture, for initiation (*vide supra*). This is particularly true for systems that require the addition of protogens (e.g. Brønsted acids). Most research in this area concerned the definition of the rules which govern the interaction between the metal halide and protogen (for example, optimum MX_n : HX ratios, effect of solvent, temperature) and the search for new MX_n : HX combinations. It seems that barring a theoretical breakthrough this is the most explored and consequently least challenging field for further work in the field of cationic initiation.

2.3.5 Co-initiation with cationogens

More fundamental progress has been made in co-initiation with catiogenic initiators, for example, monomer–alkyl aluminium halide–alkyl halide systems. The proposal that cationic polymerisations may be initiated by cationogens (e.g. RX) in conjunction with strong Lewis acids goes back to Pepper[54], and the first conclusive evidence for this concept was provided by Colclough and Dainton[14] who showed that the polymerisation of styrene in nitrobenzene or in 1,2-dichloroethane (but not in CCl_4) solutions may be initiated by alkyl halides such as t-butyl chloride, isopropyl chloride and 1,2-dichloroethane assisted by $SnCl_4$ as a co-initiator. The relative initiator efficiencies of these three alkyl halides, measured by the relative rates of polymerisation, were found to be 138:5.5:1, respectively. The initiation mechanism proposed[14] was:

$$RCl + SnCl_4 \xrightarrow{\text{slow}} R^+SnCl_5^-$$

$$R^+SnCl_5^- + M \xrightarrow{\text{fast}} RM^+ + SnCl_5^-$$

where R = t-butyl, etc.

The study of initiation with alkyl halides and common Lewis acids such as $SnCl_4$, $TiCl_4$, etc. has to be carried out under high-vacuum anhydrous conditions because of the extreme moisture sensitivity of these systems. Consequently, progress in this area was also slow until the realisation that the dialkylaluminium halide–alkyl halide systems, $AlR_2X/R'X$, are much less sensitive to protogenic impurities and consequently better suited to experimentation in an inert-gas atmosphere (e.g. nitrogen-filled dry box and 'syringe-technique'). A large amount of research has been carried out recently with these compounds[55]. The field of cationic polymerisations induced by metal alkyls has been surveyed by Saegusa[56] and that by aluminium alkyls by Kennedy[57] in books that appeared within the last 2–3 years. New information that became available since the appearance of these publications will now be discussed.

The polymerisation of isobutylene, styrene, etc. and the co-polymerisation of isobutylene–isoprene mixtures induced by various alkylaluminium–alkyl halide combinations has been investigated in detail[57]. For example, the following initiation mechanism has been proposed for the $AlEt_2Cl$/t-BuCl system:

$$\text{t-BuCl} + AlEt_2Cl \rightarrow \text{t-Bu}^+AlEt_2Cl_2^-$$
$$\text{t-Bu}^+ + \text{monomer} \rightarrow \text{t-Bu—M}^+ \text{ etc.}$$

This initiator system gives high molecular weight products at relatively high temperatures with a variety of olefins[57]. It is remarkable that alkylaluminium–alkyl halide systems are efficient cationic initiators at all, as independent experiments have shown that in the absence of a monomer the alkylation of t-Bu by $AlEt_3Cl^-$ or $AlMe_3Cl^-$ is extremely rapid.

$$\text{t-BuCl} + AlEt_3 \rightleftharpoons [\text{t-Bu}^+AlEt_3Cl^-] \rightarrow \text{t-BuEt} + AlEt_2Cl$$

In other words, propagation must be much faster than alkylation, a process that would necessarily lead to termination.

In general, it appears that under suitable conditions cationically polymerisable olefins may be initiated by a variety of carbenium ions obtained from alkyl halides and alkylaluminium compounds:

$$AlR_2'Y + RX \rightarrow R^+ + AlR_2'YX^-$$

where R and R′ are alkyl groups, X = halogen and Y = alkyl, halogen or hydride[57]. The heterolytic bond strength of the R—X bond in the organic halide is important in determining initiation. Alkyl halides having low heterolytic R—X bond strengths will be efficient initiators (e.g. tertiary, allyl or benzyl halides). However, if the bond strength is very low the carbenium ion that is obtained is very stable (trityl ion) and initiation of olefins could be slowed down significantly. Thus, for every olefin monomer there will be a preferred medium carbenium ion stability, and the initiating efficiency will be greatest when the stability of the initiating ion is, on balance, the closest to that of the propagating species. For example, t-butyl halides will be among the best initiating alkyl halides for the polymerisation of isobutylene[58] or the benzyl halides for styrene[59, 60]. Details of this chemistry have been discussed in some depth[57].

Conceptually very similar investigations have been conducted by Kagiya *et al.*[61] who polymerised styrene in bulk with silver salt–organic halides at 0 °C. In these experiments a silver salt ($AgClO_4$, $AgBF_4$, etc.), was dissolved in styrene and mixed with an alkyl halide RX in styrene solution. Initiation was pictured in two steps:

$$RX + AgD \rightarrow \overset{\delta+}{R}—\overset{\delta-}{D} + AgX$$

$$\overset{\delta+}{R}—\overset{\delta-}{D} + C{=}\underset{\substack{|\\ C_6H_5}}{C} \rightarrow R\quad C\quad \underset{\substack{|\\ C_6H_5}}{\overset{\delta+}{C}}—\overset{\delta-}{D}$$

In fundamental agreement with the AlR_3 systems above, the most effective alkyl halides in conjunction with $AgClO_4$ were Ph_3C—, Ph_2CH—, $\underset{|}{PhCH}—CH_3$, i.e., halides with the lowest ionisation potentials.

Cesca *et al.*[62] polymerised isopropylidene-3*a*,4,7,7,7*a*-tetrahydroindene

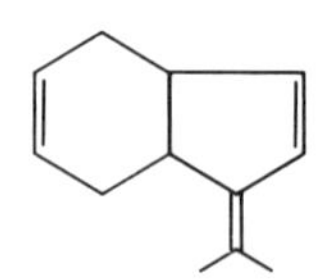

with the $AlEt_2Cl$/t-BuCl system; however, these workers employed a 1:3 co-initiator:initiator ratio and pre-mixed the chemicals, an obviously unsatisfactory technique.

Dialkylaluminium halides in conjunction with a protogen (H_2O) or cationogen (t-BuCl) have been found effective initiating systems for the polymerisation of 2-methylpenta-1,3-diene to give a largely (>90 %) *trans*-1,4 polymer of respectable molecular weight[63].

2.3.6 Co-initiation with polymeric cationogens

Interesting new products have been obtained by extending the principle of cationic co-initiation beyond small molecules to larger polymer molecules[64]. The principle of this chemistry was the recognition that in conjunction with certain alkylaluminium compounds not only small alkyl halides but suitable polymeric halides may also be used as initiating species:

$$\mathrm{PCl + AlEt_2Cl \rightarrow [complex] \rightarrow P^+ + AlEt_2Cl_2^-}$$

$$\mathrm{P^+ + monomer \rightarrow P{-}M^+ \xrightarrow{M} etc.}$$

where P is a polymer containing allylic, benzylic or tertiary chlorides.

Experiments along these lines yielded interesting new graft systems. For example, poly(isobutylene-g-styrene) was prepared by initiating the polymerisation of styrene with a carbenium ion generated by the interaction of diethylaluminium chloride and a chlorinated isobutylene–isoprene copolymer (chlorobutyl rubber, a commercial product), the structure of which is, in the main:

$$\sim\sim CH_2-\underset{\displaystyle CH_3}{\overset{\displaystyle CH_3}{\overset{|}{\underset{|}{C}}}}-CH=\overset{\displaystyle CH_3}{\overset{|}{C}}-\underset{\displaystyle Cl}{\underset{|}{CH}}-CH_2-CH_2-\underset{\displaystyle CH_3}{\overset{\displaystyle CH_3}{\overset{|}{\underset{|}{C}}}}\sim\sim$$

This polymer contains an allylic chlorine which can be readily removed by complexation with $AlEt_2Cl$ to give a macromolecular carbenium ion:

$$\sim\sim CH_2-\underset{\displaystyle CH_3}{\overset{\displaystyle CH_3}{\overset{|}{\underset{|}{C}}}}-CH=\overset{\displaystyle CH_3}{\overset{|}{C}}-\underset{+}{CH}-CH_2-CH_2-\underset{\displaystyle CH_3}{\overset{\displaystyle CH_3}{\overset{|}{\underset{|}{C}}}}\sim\sim$$

$$AlEt_2Cl^-$$

Under suitable conditions this species can initiate the polymerisation of olefins such as styrene to give the following branched structure:

$$\sim\sim CH_2-\underset{\displaystyle CH_3}{\overset{\displaystyle CH_3}{\overset{|}{\underset{|}{C}}}}-CH=\overset{\displaystyle CH_3}{\overset{|}{C}}-\underset{\begin{array}{c}|\\ CH_2\\ |\\ C_6H_5-CH\\ |\\ CH_2\\ |\\ C_6H_5-CH\\ |\end{array}}{CH}-CH_2-CH_2-\underset{\displaystyle CH_3}{\overset{\displaystyle CH_3}{\overset{|}{\underset{|}{C}}}}\sim\sim$$

Since alkylaluminium compounds of the type AlR_2X and AlR_3 induce polymerisation *only* in the presence of suitable cationogenic initiators, polymerisation *must* start on the pre-formed backbone, and homo-polymer formation by direct initiation is eliminated. Chain transfer may yield homopolymer but experimentally it has been shown that under suitable conditions grafting efficiencies of >90% can be obtained.

By the same technique, *p*-chlorostyrene, *p*-methylstyrene and α-methylstyrene have also been grafted onto chlorinated butyl rubber[64]. Besides the chlorinated butyl rubber backbone, a series of other chlorine-containing backbones have also been used to synthesise interesting grafts, for example, isobutylene was grafted on to polyvinylchloride. The key to the synthesis of poly(vinylchloride-g-isobutylene) is the realisation that polyvinylchloride contains a small but important quantity (~3%) of tertiary or allylic chlorines. These active chlorines arise by chain transfer during the free-radical synthesis of polyvinyl chloride. Subsequently, these chlorine atoms can be removed by complexation with AlR_3 and the PVC-carbenium ions are used to initiate the polymerisation of suitable olefins:

$$\mathrm{PVC} + \mathrm{AlR_3} \longrightarrow (\mathrm{PVC})^+\mathrm{AlR_3Cl^-}$$

$$(\mathrm{PVC})^+ + \mathrm{CH_2{=}C(CH_3)_2} \longrightarrow \mathrm{PVC{-}CH_2{-}\overset{+}{C}(CH_3)_2} \longrightarrow \text{etc.}$$

The active sites on the PVC chain are not secondary cations (i.e. carbenium ions formed by removing secondary chlorines) because isopropyl chloride is a relatively sluggish initiator in conjunction with AlR_3.

In addition to chlorinated butyl rubber and polyvinyl chloride, other backbones such as neoprene, chlorinated ethylene–propylene copolymer, etc. have also been used successfully to prepare grafts: poly(chloroprene-g-isobutylene), poly(chlorinated ethylene–co-propylene-g-styrene), etc. The physical properties of these graft systems are at the moment under intensive investigation and very few details are available.

2.3.7 Propagation

2.3.7.1 Addition polymerisations

Up-to-date reviews[3–5, 8] have exhaustively discussed the current status and problems in cationic propagation. Thus, Higashimura[3] reviewed the field from the point of view of rate constants. He reached some important conclusions and pointed out some glaring deficiencies in our understanding of propagation. For example, by summarising and comparing literature data, he established quantitatively the manner in which k_p is affected by the nature of the initiator and the electron-donating ability of the substituent, and noted that k_p values for many cationic polymerisations are smaller than those for free-radical propagations (Table 2.3). As anticipated by Pepper[2], Higashimura concluded that cationic polymerisations are fast not because the rate constants k_p are large but because of the much higher concentration of growing chain-ends in cationic systems than in free radical polymerisations (10^{-4} to 10^{-3} mol l^{-1} versus ~10^{-8} mol l^{-1} respectively). Also, the frequency factor A_p in ion pair propagating cationic polymerisation is much smaller than that for radical polymerisation because the counter-ion reduces the freedom of the monomer in the transition state. However, this picture does not explain the high A_p values found for $HClO_4$-initiated styrene polymerisation.

Plesch[8] commented in detail on the nature of the propagating species and discussed possible consequences of multiple chain carriers (free ions, ion pairs, etc.) having different concentrations and rate constants on the polymerisation.

We are at the threshold of important advances in the understanding of cationic propagation, for interest in this field is growing rapidly and experi-

Table 2.3 Values of k_p in cationic polymerisations
(Ethylene chloride solvent, temperature = 25–30 °C)
(From Higashimura[3], by courtesy of Marcel Dekker)

Monomer and initiator	k_p
Styrene	
$HClO_4$ ($\varepsilon = 9.72$)	1020
H_2SO_4 ($\varepsilon = 9.72$)	456
$SnCl_4$	25.2
I_2	0.22
radicals	2380
I_2	
p-methoxystyrene	350
p-methylstyrene	5.7
p-chlorostyrene	0.071
vinyl isobutyl ether	390
2-chloroethyl vinyl ether	290
$HClO_4$	
p-chlorostyrene	290

mental work has started to elucidate the nature of the growing cationic species. Progress is most likely to come from comparisons with the field of anionic polymerisation, where the definition of ion-pairs, solvated ions, ionic aggregates etc. has already reached a highly sophisticated level[67].

From such a comparison, it is worth while inquiring into the validity of cationic rate data. In certain anionic systems it has been found that free and associated ions have greatly different propagation rate constants, and that the rates are much higher for free ions than for associated ion-pairs[67]. In addition, free ions and various ion-pairs co-exist in cationic systems and available rate data give at best only average rates for the entire spectrum of ionic species, the relative concentrations being strongly dependent on the experimental conditions.

Recent literature indicates that the whole spectrum of cationic species from free ions to 'almost covalent' growing sites may exist under particular conditions. Thus the polymerisation of isobutylene induced by high energy γ-rays probably proceeds via free carbenium ions[68]. A comparison of polyisobutylene molecular weights obtained by irradiation and by chemical initiation with BF_3, $AlCl_3$ or $AlEtCl_2$ revealed that the molecular weights of the former were significantly higher than those of the latter. As irradiation-induced polymerisation proceeds through the activity of free carbenium ions (the electron (counter-ion) is thought to be in a 'hole' during the lifetime of the cation) and chemical initiation is propagated by ion-pairs, differences in molecular weight indicate different relative rates for the elementary steps. Since in this system molecular weights are mainly dependent on the magnitude of the ratio $k_p/k_{tr, M}$, the rate of propagation (transfer to monomer) is probably larger (smaller) in the free carbenium ion system than in the ion pair system. The fact that various chemical initiators give different co-polymerisation reactivity ratios with the same monomer pair is also strongly indicative of the presence of dissimilar ionic species in these systems[3].

At the opposite end of the spectrum which starts with free ions, are pseudo-cationic polymerisations[69]. In pseudo-cationic polymerisations the propagating site is thought to be largely covalent and propagation is visualised as occurring through repetitive monomer insertion into the ester bond; for example, in the styrene/$HClO_4$ system:

$$\sim CH_2-\underset{C_6H_5}{CH}-ClO_4 + CH_2=CH\cdot C_6H_5 \longrightarrow \sim CH_2-\underset{C_6H_5}{CH}-CH_2-\underset{C_6H_5}{CH}-ClO_4$$

There is strong evidence that ions do not exist *during* propagation (no colour, no conductivity, no effect of water on the rate).

Between propagation by free ions and pseudo-cationic polymerisations is the whole spectrum of growing species with various ionicities. Indeed what is badly needed in cationic polymerisation is a suitable method for the characterisation of growing cationic species. Again, the solution to this problem is much farther advanced in anionic systems. While spectacular progress has been made in the last decade in the characterisation of stable carbenium ions, particularly by n.m.r. spectroscopy, the characterisation of the more important unstable (reactive) cations in these polymerisations still awaits a suitable technique. This problem is extremely difficult because of the low concentration of active cations in a polymerising system.

A large amount of fundamental research work relative to problems of carbenium ion propagation, has been carried out by Japanese scientists in Kyoto during the last few years. This most prolific group was originally assembled by Professor Okamura and lately is under the direction of Professor Higashimura. Amongst other things, this group is active in the elucidation of the polymerisation behaviour of α,β-disubstituted olefins, both hydrocarbons and ethers. In a publication[70] which summarises their work up to 1968, amongst other conclusions the following points are emphasised:

(a) A β-methyl substituent in styrene reduces its activity by a factor of 10–20.

(b) A β-methyl substituent in vinyl ethers, in contrast, increases its reactivity considerably. This is attributed to electronic effects.

(c) β-Alkoxystyrenes Ph—CH=CH—O—R, behave as vinyl ethers and not as styrenes.

(d) *trans*-β-Methylstyrene is ~1.4-times more reactive than the *cis* isomer toward a styrene carbenium ion.

(e) Various propenyl alkyl ethers, $CH_3CH{=}CH{-}OR$ where R = methyl, ethyl, isopropyl, n-butyl and t-butyl, have been polymerised by $BF_3 \cdot OEt_2$ as a homogeneous initiator and $Al_2(SO_4)_3/H_2SO_4$ as a heterogeneous initiator. Invariably crystalline products were obtained with the *trans* ether and $BF_3 \cdot OEt_2$, whereas amorphous materials were formed with the $Al_2(SO_4)_3/H_2SO_4$ system. The situation was just the opposite with the *cis* ether: crystalline polymers were obtained with the heterogeneous system and amorphous ones with the homogeneous initiator.

In more recent publications[71] (the series now consists of 13 papers[71]) the authors extended their investigations to other α,β-disubstituted olefins, for example, $CH_3{-}CH{=}CH{-}O{-}CH_2CH_2Cl$ (Reference 72) and R—O—CH =CH—OR (reference 71). One of the most interesting fundamental results in these publications, in the opinion of this Reviewer, is the difference of products obtained from styrene derivatives by radical and carbenium ion mechanisms. In radical polymerisations, β-methylstyrene does not give polymer because of steric compression in the transition state, whereas in cationic polymerisations, β-methylstyrene or β-methylalkylethers give high polymers. These observations indicate fundamental and characterisable differences in the transition states between these two propagation mechanisms.

The strong interest in the fundamentals, particularly kinetics, of cationic polymerisations by the Kyoto group is attested by the large number of related papers recently published (for example, a series of four papers on isobutylene[73], nine on cyclic dienes[74] and two on aspects of copolymerisation[75]).

The effect of electric fields on the polymerisation of various monomers by various initiators has been investigated in depth by Ise and his group. Since 1965, this author has published a series of 17 papers plus some reviews on the subject[76,77]. In cationic systems (e.g. styrene–$BF_3 \cdot OEt_2$, α-methylstyrene–I_2) in the presence of an electric field, the reaction rates are usually strongly enhanced whilst molecular weights are only slightly affected or not at all. According to Ise's careful analysis of the data, increased rates in the presence of an electric field are not due to new reaction mechanisms but to increased propagation rate constants. In other words, the field affects the entropy and not the activation energy of polymerisation. No electro-initiation is involved and the initiation rate constant remains unchanged; the main effect of the field is to increase the degree of dissociation of the active propagating species.

A very interesting rate phenomenon was found in the styrene–$BF_3 \cdot OEt_2$ system at 25 °C, investigated in toluene, toluene–dichloroethane, dichloroethane–nitrobenzene and nitrobenzene media[78]. The results are shown in Figure 2.2. At very low and at very high d.c. values the presence of an external electric field has virtually no effect on the rate; however, at medium d.c. values the field strongly affects the rate of polymerisation. An explanation is that at low d.c. values in toluene, the carbenium ion–counter-ion pairs are held together strongly by electrostatic forces and the external field has no effect on the rate in the presence of tight ion-pairs. A similar situation occurs at high d.c. values in nitrobenzene, where solvated free ions now exist and the external field again does not affect the rate observed. At medium d.c. values in 1,2-dichloroethane, however, the external field is able to increase the concentration of free ions and the rate of polymerisation is accelerated. In a similar fashion, higher rates of polymerisation and higher molecular weight

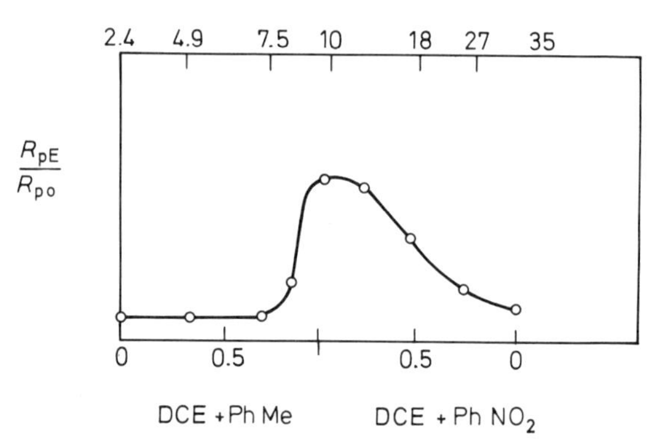

Figure 2.2 Field effect and solvent composition; styrene–$BF_3{\cdot}OEt_2$, 25°C; $E = 0.25$ kV cm^{-1}; R_{pE} = rate of polymerisation in the presence of an electric field: R_{po} = rate of polymerisation in the absence of an electric field; DCE = 1,2-dichloroethane
(From Sakurada *et al.*[78], by courtesy of Marcel Dekker)

polyanethols (poly-β-methyl-p-methoxystyrene) have also been obtained in $BF_3{\cdot}OEt_2$-induced polymerisations in ethylene dichloride solution in the presence of a 0.5 kV cm^{-1} electric field[79].

Electro-polymerisation has also proved useful in controlling the rate of the polymerisation of isobutyl vinyl ether[80]. This monomer exhibits a strong tendency to 'flash' polymerisation with common Friedel–Crafts halide initiators.

Isomerisation polymerisation is another important area where cationic propagation occurs. Such polymerisation may be depicted by the equation

$$n\mathrm{A} \rightarrow \text{—}(\mathrm{B})_{\overline{n}},$$

the well characterised 3-methylbut-1-ene system providing a typical example. The propagation process at low temperatures involves an isomerisation step with the following mechanism:

$$\mathrm{C{=}C{-}\underset{C}{\overset{C}{CH}} + R^{+} \rightarrow \sim R{-}C{-}\overset{+}{C}{-}\underset{C}{\overset{C}{CH}} \rightarrow \sim R{-}C{-}\overset{H}{C}{-}\underset{C}{\overset{C}{C^{+}}} \longrightarrow \sim R{-}C{-}C{-}\underset{C}{\overset{C}{C}}\sim}$$

After quite a large number of publications during the period 1962–1968 dealing with a variety of interesting isomerisation polymerisations, there has been a noticable slowing-down of research of late. This field has been recently extensively summarised[5, 6].

Among the latest publications are those of Kennedy, Plesch and Magagnini[81, 82] who have criticised conclusions of other workers who claimed that isomerisation polymerisation occurs with *p*-methylstyrene, e.g.

$$-CH_2-\overset{+}{C}H(C_6H_4)CH_3 \longrightarrow -CH_2-CH_2(C_6H_4)\overset{+}{C}H_2$$

and *o*-isopropylstyrene, e.g.,

$$-CH_2-\overset{+}{C}H(C_6H_4)C(H)(CH_3)-CH_3 \longrightarrow -CH_2-CH_2(C_6H_4)\overset{+}{C}(CH_3)-CH_3 .$$

Dutch workers[85] have examined the low-temperature polymerisation of 1,1-dialkylethylenes and found that both 2,3-dimethylbut-1-ene and 3,3-dimethylbut-1-ene give the same repeat unit, the first by an intramolecular hydride migration, the second by a methide shift. The methide shift had been found earlier by Kennedy and co-workers[86]:

$$CH_2{=}C(CH_3)-C(H)(CH_3)-CH_3 \xrightarrow{\sim H^-} {-}\!\left(CH_2-C(CH_3)(H)-C(CH_3)(CH_3)\right)_n\!{-} \xleftarrow{\sim CH_3^-} CH_2{=}CH-C(CH_3)(CH_3)-CH_3$$

Russian researchers[87] recently examined the isomerisation polymerisation of 3,3-dimethylbut-1-ene, and established that the nature of the initiator affects the polymerisation: $AlCl_3$ and BF_3 gave backbones of essentially completely rearranged units, whereas less rearrangement occurred with $TiCl_4$.

Very recently, Sartori and co-workers have shown[88] that hydride and methide migration polymerisation occurs consecutively in the 4,4-dimethylpent-1-ene system. These authors found that this monomer which, theoretically can give rise to at least three repeat structures, gives predominantly structure (3).

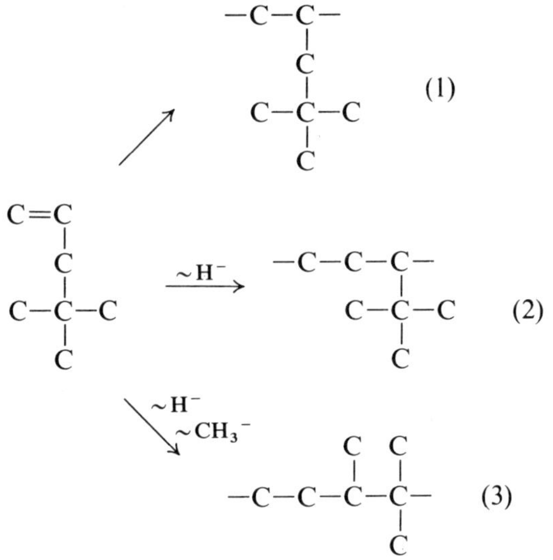

```
                                —C—C—
                                   |
                                   C
                                   |            (1)
                     ↗          C—C—C
                                   |
                                   C
C=C
  |
  C           ~H⁻         —C—C—C—
  |          ——→                |
C—C—C                         C—C—C      (2)
  |                             |
  C                             C
            ↘ ~H⁻
              ~CH₃⁻            C  C
                               |  |
                         —C—C—C—C—      (3)
                                  |
                                  C
```

It was concluded that polymerisation proceeds by:

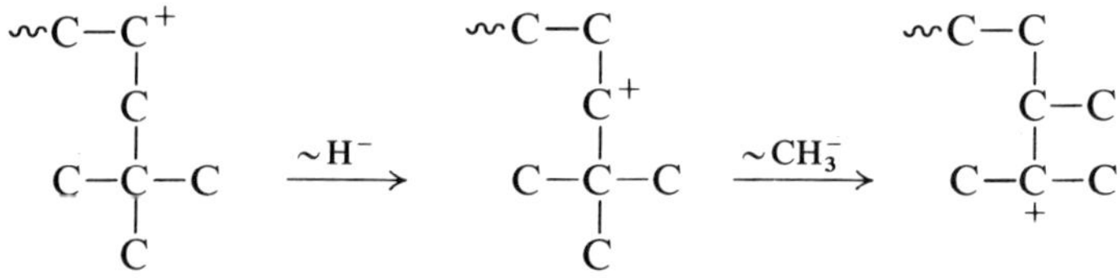

```
~C—C⁺                ~C—C                 ~C—C
   |                    |                    |
   C                    C⁺                   C—C
   |        ~H⁻         |       ~CH₃⁻        |
C—C—C      ——→       C—C—C      ——→       C—C—C
   |                    |                    +
   C                    C
```

The driving force of the hydride shift is provided by the t-butyl group and that of the methide migration by the higher stability of the tertiary cation.

Some interesting isomerisations have recently been published for bicyclic systems[89, 90]. For example, the following mechanism has been proposed for the polymerisation of *endo*-dicyclopentadiene[89]:

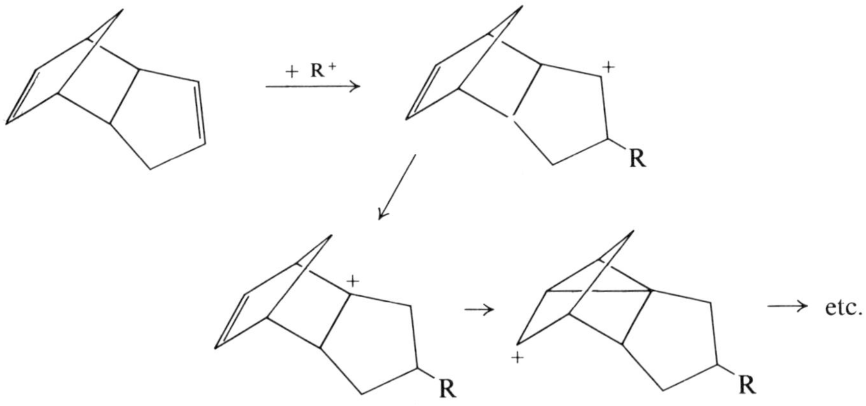

+ R⁺ →

\+

R

\+

R

→

\+

R

→ etc.

Attempts to prepare polymers with the repeat structure

$$-CH_2-CH_2-\underset{\displaystyle C_6H_5}{\underset{|}{CH}}-$$

have not been successful[5]. Preliminary results from the Author's laboratory indicate, however, that a similar repeat unit can be prepared from *p*-methoxy-allylbenzene:

$$CH_2{=}CH-CH_2-C_6H_4-OCH_3 \xrightarrow{\sim R^+} \sim CH_2-\overset{+}{C}H-CH_2-C_6H_4-OCH_3 \xrightarrow{\sim H^-} \sim CH_2-CH_2-\overset{+}{C}H-C_6H_4-OCH_3 \xrightarrow{+M} \text{polymer}$$

The isomerisation polymerisation of 1-vinyl-1-methyl-2,2-dichloropropane with WCl_6, $TiCl_4$ and $SnCl_4$ has also recently been proposed. According to Pinazzi *et al.*[91], the preponderant mechanism might be as follows:

$$CH_2{=}CH-C_3H_2(CH_3)Cl_2 \xrightarrow{R^+} R-CH_2-\overset{+}{C}H-C_3H_2(CH_3)Cl_2$$

or

$$R-CH_2-CH(\text{cyclobutyl cation, } +\text{–}CH_3, \text{ ring } CCl_2) \rightarrow R-CH_2-CH(\text{cyclobutyl, } CH_3, CCl_2) \xrightarrow{-HCl} R-CH_2-CH(\text{cyclobutenyl, } CH_3, Cl)$$

$$R-CH_2-CH(\text{cyclobutyl cation with } Cl_2C, +\text{–}CH_3) \rightarrow R-CH_2-CH(\text{cyclobutyl with } Cl_2C, CH_3) \xrightarrow{-HCl} R-CH_2-CH(Cl-C{=}\text{cyclobutenyl}, CH_3)$$

In a companion paper, these authors examined the cationic polymerisation of spiro-pentane[92] and concluded that this reaction gives rise to cyclised structures similar to that of cyclised poly-1,4-isoprene.

In addition to being a most interesting field of polymerisation chemistry, isomerisation polymerisation has fundamental diagnostic significance: the occurrence of isomerisation during polymerisation may be viewed as virtually direct proof for the existence of carbenium ions during propagation.

2.3.7.2 *Condensation polymerisations*

Aromatic polycondensations by carbenium ions are a relatively small but potentially commercially significant field of polymer science. Continued research interest is insured because of potential uses of low cost monomers, e.g. benzyl chloride. For the purposes of this survey it is convenient to sub-divide these systems in terms of the number of carbon atoms n between the aromatic rings: polyphenyls ($n = 0$), and polybenzyls ($n = 1$).

Polyphenyls ($-C_6H_4-$) were discovered in 1962[93] and research has been active in this area since that time. A comprehensive review of the synthesis, properties and applications of polyphenyls and related materials has appeared very recently[94]. Much of the development research is based on the observation that benzene can be polymerised to poly(p-phenylene) under mild conditions (0.5 h at 35 °C) with Lewis acid–oxidising agent combinations (e.g., $AlCl_3 \cdot CuCl_2$). The polymerisation mechanism is visualised as follows[94]:

$$AlCl_3 + H_2O(\text{traces}) \rightleftharpoons \text{`}H^{\delta+}AlCl_3OH^{\delta-}\text{'}$$

$$C_6H_6 + H^+ \longrightarrow \text{(benzenium ion)} \xrightarrow{+\ C_6H_6} \text{(dimer cation)}$$

$$\xrightarrow[-\ 2HCl]{CuCl_2} \text{(biphenyl cation)} \xrightarrow{C_6H_5} \text{(trimer cation)}$$

$$\xrightarrow[-\ 2HCl]{CuCl_2} \text{(terphenyl cation)}$$

etc., termin.

$$H\!-\!\left[C_6H_4\right]_n\!-\!H$$

Polyphenylene is available commercially and work is currently in progress to improve its processing and fabricating technologies.

Polybenzyls $(\text{—}C_6H_4\text{—}CH_2\text{—})_n$ have been known since the classical work of Friedel and Crafts, and the field has been briefly summarised by Mortillaro[95] and by Kennedy and Isaacson in 1966[96]. In the latter paper, the synthesis of the first linear crystalline polybenzyl and its derivatives was also described. This general synthesis employed aluminium chloride in ethyl chloride as a

$$C_6H_5\text{—}CH_2Cl + MX_n \rightarrow \left[C_6H_5\text{—}\overset{+}{C}H_2MeX_nCl^-\right]$$

$$\xrightarrow[\text{low temp.}]{C_6H_5CH_2Cl} C_6H_5\text{—}CH_2\text{—}C_6H_4\text{—}CH_2Cl + HCl \text{ etc.}$$

diluent at the lowest temperatures and exploited the small, but important, activation energy differences among the various ring positions to direct alkylation only to the *para* position.

Some of the products and their characteristics are shown in Table 2.4. These findings have largely been confirmed by Finocchiaro *et al.*[97, 98] together with

Table 2.4 Characteristics of crystalline polybenzyls
(From Kennedy and Isaacson[96], by courtesy of Marcel Dekker)

Formula monomer	*Formula* polymer repeat unit	*Co-initiator*	*Temp/* °C	*Conv./* %	$\bar{M}_n$	*M.pt/*°C	*Solub. in toluene*
$C_6H_5\text{—}CH_2Cl$	$\text{—}C_6H_4\text{—}CH_2\text{—}$	$AlCl_3$	−134	13.4	~3000	142	yes
$(CH_3)_2C_6H_3\text{—}CH_2Cl$	$\text{—}(CH_3)_2C_6H_2\text{—}CH_2\text{—}$	$AlCl_3$	−78	61.0		148–168	no
$C_6H_5\text{—}CHCl(CH_3)$	$\text{—}C_6H_4\text{—}CH(CH_3)\text{—}$	$AlCl_3$	−130	10.8	2350	205	yes

the fact that the linear polybenzyl exhibits much higher thermostability (stable up to ~400 °C) than the branched material prepared at higher temperatures. Aspects of the polycondensation of benzyl chloride and its derivatives has recently been investigated by Italian workers in a series of papers[99] and by other investigators[100–103].

One of the latest developments in cationic polycondensations is the disclosure of a synthesis of arylene isopropylidene polymers. Thus, Fritz[104] reported that bisarylene halides can be polycondensed with special catalysts to give useful high-molecular-weight plastics. For example:

$$\mathrm{Cl{-}C(C)_2{-}C_6H_4{-}C(C)_2{-}Cl} + \mathrm{C_6H_5{-}O{-}C_6H_5} \xrightarrow{\text{cat.}} \left[\mathrm{{-}C(C)_2{-}C_6H_4{-}C(C)_2{-}C_6H_4{-}O{-}C_6H_4{-}}\right] + \mathrm{HCl}$$

The catalyst is prepared from a trityl salt (Ph_3CMX_n), a typical Friedel–Crafts halide and a nitro compound, e.g. $Ph_3C{\cdot}AsF_6{\cdot}AlBr_3{\cdot}PhNO_2$. The solvating ability of the nitro compound is claimed to be essential for obtaining high-molecular-weight products.

2.3.8 Transfer

Since its introduction into cationic polymerisations[105], the concept of transfer has played a central role in shaping this field of chemistry. As the main single factor determining product molecular weights in carbenium ion polymerisation, the elucidation of the details of transfer was, is and will be, one of the most important goals of serious students of this discipline. Good up-to-date presentations of some of the problems of transfer are given in References 2 and 3.

The essence of the problem presented by transfer may be stated briefly as follows: It is difficult to prepare high-molecular-weight materials by cationic mechanisms because the very reactive propagating carbenium ion easily transfers (chain-breaking) its charge to a series of other nucleophiles in addition to monomer in the system (solvent, counter-ion, pre-formed polymer, impurities). The very characteristics, the great reactivity of the growing species, which makes cationic reactions in general, and cationic polymerisations in particular, such useful and versatile techniques, sets its own limitations. High reactivity means low selectivity, which in turn is equivalent to broad product spread in small molecule chemistry and low molecular weights in polymerisation technology. A variety of transfer mechanisms (see above) compete with propagation but, fortunately, the former usually have higher activation energies than the latter so that enhanced molecular weights can be obtained by lowering the polymerisation temperature ('freezing-out' transfer).

Not much progress has been made in this field from the conceptual point of view during the last two to five years.

In a similar manner to the situation discussed in relation to propagation, researchers are turning their attention to exploring the effect of various ionicities, free ions, ion pairs, etc. on the mechanism of chain transfer. The analysis of the irradiation- and chemical-induced polymerisation of isobutylenes by Kennedy *et al.*[68] indicates that the counter-ion might be aiding monomer transfer:

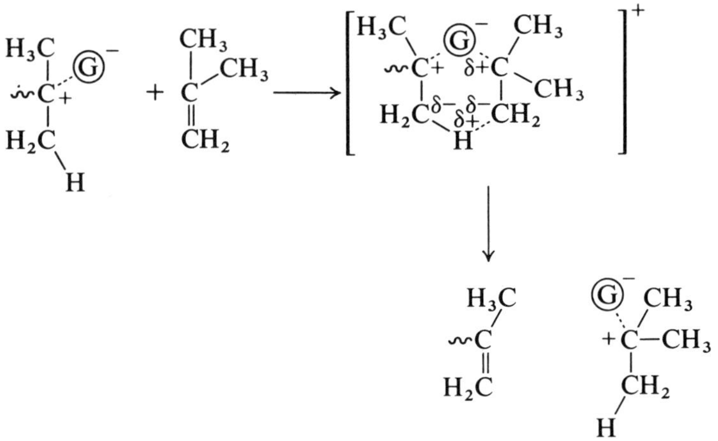

$$
\begin{array}{c}
H_3C \\
| \quad \text{Ⓖ}^- \\
\sim C^+ \\
| \\
H_2C \\
\quad \backslash \\
\quad H
\end{array}
+
\begin{array}{c}
CH_3 \\
| \; / CH_3 \\
C \\
\| \\
CH_2
\end{array}
\longrightarrow
\left[
\begin{array}{ccc}
H_3C & \text{Ⓖ}^- & CH_3 \\
\sim C^+ & & \delta + C \quad CH_3 \\
| & & | \\
H_2C^{\delta-} & \overset{\delta-}{\underset{H}{\delta+}} & CH_2
\end{array}
\right]^+
$$

$$
\downarrow
$$

$$
\begin{array}{c}
H_3C \\
/ \\
\sim C \\
\| \\
H_2C
\end{array}
\qquad
\begin{array}{c}
\text{Ⓖ}^- \; CH_3 \\
/ \\
+C-CH_3 \\
| \\
CH_2 \\
/ \\
H
\end{array}
$$

and that its removal (free ions) would increase molecular weights. Higashimura[3] also agrees that chain transfer may occur with associated carbenium ion–counter-ion pairs, for example, in isobutylene or styrene polymerisations. The mechanism of monomer transfer in pseudo-cationic polymerisations, i.e. under conditions where the growing end is an (almost) covalent ester, has not yet been considered.

A novel method for the characterisation of transfer activity exhibited by added transfer agents has been developed[106]. This technique is of particular interest for industrial applications where the characterisation of the controlling effect of various compounds on the molecular weight is desired. The method has been worked out in detail with isobutylene, and several pure chain-transfer agents (e.g. allyl chloride, t-butyl chloride) have been identified. Pure transfer agents are chemicals which are able to reduce the molecular weight without affecting polymer yield[107]:

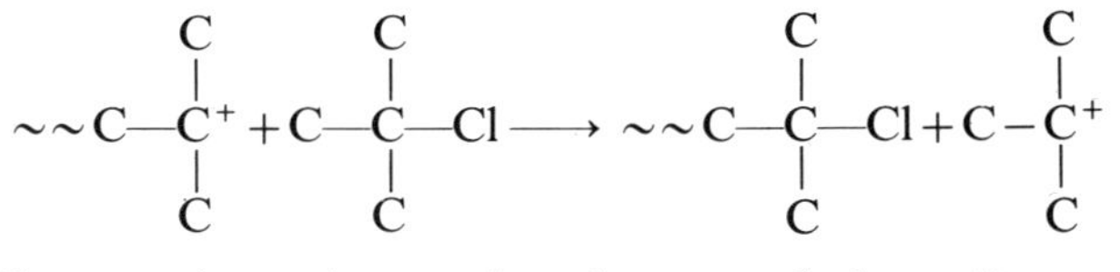

$$
\begin{array}{c} \\ \\ \sim\sim C- \end{array}\!\!\begin{array}{c} C \\ | \\ C^+ \\ | \\ C \end{array} + C - \begin{array}{c} C \\ | \\ C \\ | \\ C \end{array} - Cl \longrightarrow \begin{array}{c} \\ \\ \sim\sim C- \end{array}\!\!\begin{array}{c} C \\ | \\ C \\ | \\ C \end{array} - Cl + C - \begin{array}{c} C \\ | \\ C^+ \\ | \\ C \end{array}
$$

The significance of transfer reactions has recently been discussed by this Author in the framework of a symposium on Unsolved Problems in Polymer Science[108].

2.3.9 Termination

Termination is the complete and permanent destruction of propagating ability and should not be confused with chain breaking. Besides transfer, termination is the most important side-reaction in cationic polymerisation; the main reactions are initiation and propagation. Whilst transfer determines product molecular weights, termination controls ultimate conversions (yields).

Experimentally it is often observed that cationic polymerisations terminate in the presence of unconverted monomer. Why do these polymerisations stop? The answer to this question is far from solved and in spite of its scientific–technological significance not much direct research has been devoted to it during the past few years. Termination is often attributed to impurities, or else to nebulous spontaneous termination processes, the mechanism and/or chemistry of which are completely unknown. Impurity or spontaneous termination are very difficult to separate kinetically and therefore many k_t data are, at best, 'apparent' values.

2.3.9.1 'Immortal' polymerisations

Most cationic polymerisations should indeed have no termination at all. Nonetheless, the experimental demonstration of the absence of termination is very difficult to achieve because of the extreme degree of purity required. In the absence of impurities and in the presence of the monomer as the sole nucleophile in the system, polymerisation should progress until monomer depletion is complete. The carbenium ion cannot 'die', it is 'immortal' provided no nucleophile (besides the monomer) is available for its destruction, so that fresh aliquots of monomer should be polymerised at the same rate as the previous ones. This does not necessarily mean that the molecular weights will increase with conversion, since transfer is still operating. This is a fundamental difference between 'living' species observed with anionic and certain oxonium systems and carbenium ion polymerisations: whereas in 'living' polymerisations conversions *and* molecular weights increase hand-in-hand up to 100% conversion, in 'immortal' cationic polymerisations only conversions increase while molecular weights remain the same, as they are determined by transfer.

2.3.9.2 Termination by certain counter-anions

The counter-anion in the system may or may not be a terminating nucleophile. It should not be terminating with BF_4^- $TiCl_4^-$, etc., because the equilibrium

$$\overset{|}{\underset{|}{C}}{}^{+}\,BF_4^- \rightleftharpoons \sim\overset{|}{\underset{|}{C}}F + BF_3$$

lies strongly toward the left. It may be terminating with $TiCl_4OCOCCl_3^-$ (polymerisation of isobutylene initiated with $CCl_3COOH—TiCl_4$[109]) because the ester formed is stable:

$$\sim C^+TiCl_4OCOCCl_3^- \rightarrow \sim C—OCOCCl_3 \qquad TiCl_4$$

Evidence for this termination is provided by the detection of Cl_3CCOO groups in the polymer by infrared spectroscopy.

2.3.9.3 *Termination by hydride loss*

Another true termination has been proposed by Meier[110], and further development by Fontana and Kidder[111] who visualise the formation of non-propagating resonance-stabilised allylic carbenium ions by hydride loss:

$$\sim\underset{\substack{|\\ R}}{CH}-CH_2-\underset{\substack{|\\ R}}{\overset{+}{C}H}\xrightarrow{\sim R^+}\sim RH+\sim\underset{\substack{|\\ R}}{\overset{+}{C}}-CH_2-\underset{\substack{|\\ R}}{\overset{+}{C}H}\xrightarrow{\substack{CH_2=CH\\ |\\ R}}\sim\underset{\substack{|\\ R}}{\overset{\delta+}{C}}\cdots CH\cdots\underset{\substack{|\\ R}}{\overset{\delta+}{C}H}+\begin{matrix}H_3C\\ |\\ HC^+\\ |\\ R\end{matrix}$$

Evidence for this kind of termination has been provided by kinetic considerations[112].

2.3.9.4 *Termination by hydride capture*

Another type of allylic termination has recently been proposed by Kennedy and Squires[113] on the basis of kinetic investigations:

$$\sim\sim\underset{|}{C^+}+\overset{|}{\underset{|}{C}}=\underset{|}{C}-\overset{|}{\underset{|}{C}H}\longrightarrow\sim\sim\overset{|}{\underset{|}{C}H}+\overset{\delta+}{C}\cdots\underset{|}{C}\cdots\overset{\delta+}{C}$$

These authors examined the effect of a large number of compounds on the yield (and molecular weight) of polyisobutylene ($AlCl_3 \cdot CH_3Cl$ in pentane diluent, -78 °C) through use of the following equation:

$$\frac{W_0}{W_p} = 1+\frac{k_7}{k_5}X$$

where W_0 and W_p are the weights of polymer obtained in the absence and presence of a poison; k_7 and k_5 are the rate constants for allylic termination with poison and spontaneous termination, respectively, and X is the poison concentration. They defined the 'poison coefficient' as the slope of the plot W_0/W_p versus X and examined this quantity for some 40 hydrocarbons[104]. In contrast to transfer agents, which only depress molecular weights but do not affect yields (see above), poisons are chemicals that decrease yields but do not affect molecular weights. Pure poisons for the above system are, for example, propylene, but-1-ene, pent-1-ene[114].

2.3.9.5 *Termination by alkylation*

Model experiments aimed at elucidating the mechanism of transfer and termination are currently being carried out in the Author's laboratory.

Non-polymerisable olefins, alkyl halides and alkylaluminium compounds are being used. These investigations strongly indicate a new termination by an alkylation mechanism:

$$\sim\overset{|}{\underset{|}{C}}{}^{+} + \left[-\overset{|}{\underset{|}{Al}}-R\right]^{-} \longrightarrow \sim\overset{|}{\underset{|}{C}}-R + -\overset{|}{\underset{|}{Al}}$$

Strong circumstantial evidence for this type of termination has been provided by the reaction between t-butyl chloride + $AlMe_3$ which gives rise to neopentane $C(CH_3)_4$ and $AlMe_2Cl$, or in general[115]:

$$AlR_3 + R'Cl \longrightarrow AlR_2Cl + RR'$$

The results of chemical and structure analysis of the products obtained in the reaction between t-butyl chloride and trimethylaluminium-2,4,4-trimethylpent-1-ene were, in part, interpreted as follows[116]:

$$\text{t-BuCl} + AlMe_3 \xrightarrow[-78\,^{\circ}C]{CH_3Cl} \text{t-Bu}^{+}AlMe_3Cl^{-} \longrightarrow \text{t-BuMe}$$

t-Bu^{+}AlMe$_3$Cl^{-} + C=C(C)–C–C(C)(C)–C ⟶

$$\text{t-Bu}-C-\overset{\overset{C}{/}}{C}{+}-C-\overset{|}{\underset{|}{C}}(C)-C \xrightarrow{AlMe_3Cl^{-}} \text{t-Bu}-C-\overset{C}{\underset{|}{\overset{|}{C}}}-Me,\ \ -C-\overset{|}{\underset{C}{C}}(C)-C \quad (C_{13})$$

$\xrightarrow{-H^{+}\ \ AlMe_3Cl^{-}}$ t-Bu–C=C(=C)–C–C(C)(C)–C (3 isomers)

⟶ $CH_4 + AlMe_2Cl$

The detection of the C_{13} product, i.e.

$$C-\overset{C}{\underset{C}{\overset{|}{\underset{|}{C}}}}-C-\overset{C}{\underset{C}{\overset{|}{\underset{|}{C}}}}-C-\overset{C}{\underset{C}{\overset{|}{\underset{|}{C}}}}-C)$$

is again strong indication of the veracity of the termination by alkylation mechanism. In view of the above competitive reactions it is indeed surprising that many olefins can be readily polymerised to high molecular weight products with alkylaluminium compounds[117].

A general scheme for the mechanism, with particular reference to termination operating in olefin polymerisation initiated by AlR_3/RCl systems, may be proposed as follows:

$$AlR_3 + R'Cl \xrightarrow[\text{Slow}]{k_i} [R'^+AlR_3Cl^-]$$

$$[R'^+AlR_3Cl^-] \xrightarrow{k_t} R'R + AlR_2Cl$$

$$[R'^+AlR_3Cl^-] \xrightarrow[k_p]{+M} R'M^+AlR_3Cl^-$$

$$R'M^+AlR_3Cl^- \xrightarrow{k_t} R'MR + AlR_2Cl$$

$$R'M^+AlR_3Cl^- \xrightarrow[k_p]{+M} R'MM^+ AlR_3Cl^-$$

$$R'MM^+ AlR_3Cl^- \dashrightarrow R'Polymer\ R + AlR_2Cl$$

$$R'MM^+ AlR_3Cl^- \xrightarrow[k_p]{+M_n} Polymer^+AlR_3Cl^-$$

(Dotted lines lead from each AlR_2Cl back to $AlR_3 + R'Cl$.)

Competition between alkylation and propagation starts as soon as the first ion pair is formed. Propagation prolongs the life of the carbenium ion but ultimately alkylation will terminate it. Termination gives rise to an unsaturated hydrocarbon ($R\sim\sim R'$) and AlR_2Cl. The driving force of this reaction is the formation of an Al—Cl bond. The newly formed AlR_2Cl can react with unconsumed alkyl halide R'Cl in the system (indicated by the dotted lines in the scheme). The discussions of reactions arising beyond this point is outside the scope of this Review.

2.4 CONCLUSIONS

Substantial progress has been made in the field of cationic polymerisations in general, and in carbenium ion initiated polymerisations in particular, over the past 2–5 years. In the area of initiation, the well-established notion that all Friedel–Crafts halides require a 'co-initiator' must be revised and new concepts of initiation will be needed which fit all the new data. A proposal has been made by this Author to explain self-initiation in the absence of obvious proton or carbenium ion sources, e.g. H_2O or RX. Progress in the field of co-initiation with cationogens was made by the discovery and development of certain new alkylaluminium systems, for example, $R_2AlX/R'X$. The advantages offered by these alkylaluminium co-initiators are convenience of experimentation, high molecular weight products at relatively high temperatures, controlled initiation leading to novel graft copolymers and partially-controllable termination mechanisms.

In the area of propagation, progress is being made by emphasising the role and the effect of various ionic species on the rate, molecular weight and its distribution. Pseudo-cationic polymerisation was discovered during this period. In pseudo-cationic systems the propagating species is viewed as a covalent ester (mostly $—ClO_4$) and the concept of ions is completely abandoned. Between propagating free ions (irradiation-induced polymerisation) and covalent esters (pseudo-cationic systems) is the entire spectrum of ionicities, the exploration of which is under way.

Significantly, research on the cationic polymerisation of α,β-disubstituted olefins demonstrated that high-molecular-weight polymers can be obtained from these monomers. The rate of certain cationic polymerisations has been increased by applying an external electric field and the effect has been explained by an increased free ion concentration. Interesting publications have dealt with aspects of isomerisation polymerisation: 2,3-dimethylbut-1-ene and 3,3-dimethylbut-1-ene give the same polymer, the former by a H^- shift and the latter by a CH_3^--shift mechanism.

In cationic condensation polymerisations, progress has been made in the continued exploration of polyphenyls. Interesting linear, crystalline polybenzyls ($—C_6H_4—CH_2—$) and derivatives thereof have been prepared by low-temperature $AlCl_3$-induced polymerisation, and a communication has given a synthesis of $—C(CH_3)_2—C_6H_4—C(CH_3)_2—C_6H_4—O—C_6H_4—$, a plastic with interesting physical properties.

Not much conceptual progress has been made in the understanding of the important transfer processes. The molecular weights of polyisobutylenes synthesised by γ-irradiation were found to be much higher than those obtained by chemical initiation which could be an indication that transfer is reduced in the absence of counter-ion. A similar phenomenon has also been observed with polyisobutylvinyl ether.

Most conventional carbenium ion polymerisations should be 'immortal' in the absence of nucleophilic species (besides the monomer); however, impurities usually mask this phenomenon. The following available true termination mechanisms are most characteristic of carbenium ion polymerisations: termination by (i) certain counter-anions, (ii) hydride loss, (iii) hydride capture and (iv) alkylation.

References

1. Plesch, P. H. (Ed.) (1963). *The Chemistry of Cationic Polymerization.* (New York: The MacMillan Co.)
2. Pepper, D. (1964). *Friedel–Crafts and Related Reactions* Ed., Olah, G. A., Vol. II. 1293, (New York: Interscience Publishers)
3. Higashimura, T. (1969). *Structure and Mechanism in Vinyl Polymerization* Ed., Tsuruta, T. and O'Driscoll, K. F., 313, (New York: Marcel Dekker, Inc.)
4. Zlamal, Z. (1969). *Kinetics and Mechanisms of Polymerization* Ed., Ham, G. E., Vol. I, 231, (New York: Marcel Dekker, Inc.)
5. Kennedy, J. P. (1967). *Encycl. Polymer Sci. Techn.,* **7,** 754
6. Ketley, A. D. and Fisher, L. P. (1969). *Structure and Mechanism in Vinyl Polymerization,* Ed., Tsuruta, T. and O'Driscoll, K. F., 449, (New York: Marcel Dekker)
7. Saegusa, T. (1969). *Structure and Mechanism in Vinyl Polymerization,* Ed., Tsuruta, T. and O'Driscoll, K. F., 283, (New York: Marcel Dekker, Inc.)
8. Plesch, P. H. (1968). *Progress in High Polymers,* **2,** 137
9. Olah, G. A. (1971). *J. Amer. Chem. Soc. Div. Petrol. Chem. Prepr.,* **16,** C11
10. Franzen, V. (1958). *Reaktionsmechanisimen,* 115 (Heidelberg: Alfred Hüttig Verlag)
11. Evans, A. G. and Polanyi, M. (1947). *J. Chem. Soc.,* 252

12. Natta, G., Dall 'Asta, G., Mazzanti, G., Giannini, U. and Cesca, S. (1959). *Angew Chem.*, **71**, 205
13. Roberts, J. M., Katovic, Z. and Eastham, A. M. (1970). *J. Polymer Sci., A-1,* **8,** 3503
14. Colclough, R. O. and Dainton, F. S. (1958). *Trans. Faraday Soc.*, **54,** 886
15. Chmelir, M., Marek, M., and Wichterle, O. (1967). *J. Polymer Sci., Part C,* **16,** 833
16. Solich, J. M., Chmelir, M. and Marek, M. (1969). *Collect. Czech. Chem. Commun.*, **34,** 2611
17. Lopour, P. and Marek, M. (1970). *Makromol. Chem.*, **134,** 23
18. Kössler, I., Stolka, M. and Mach, K. (1963). *J. Polymer Sci., Part C,* **4,** 977
19. Marek, M. and Chmelir, M. (1968). *J. Polymer Sci., Part C,* **23,** 223
20. Matyska, M., Svestka, M. and Mach, K. (1966). *Collect. Czech. Chem. Commun.*, **31,** 659
21. Matyska, M., Mach, K., Vodehnal, J. and Kössler, I. (1965). *Collect. Czech. Chem. Commun.*, **30,** 2569
22. Chmelir, M. and Marek, M., (1968). *J. Polymer Sci., Part C,* **22,** 177
23. Vairon, J. and Sigwalt, P. (1971) *Bull. Soc. Chim. Fr.*, 559
24. Vairon, J. and Sigwalt, P. (1971). *Bull. Soc. Chim. Fr.*, 569
25. Cheradame, H., Hung, N. A. and Sigwalt, P. (1969). *Compt. Rend.*, **268,** 476
26. Polton, A. and Sigwalt, P. (1970). *Bull. Soc. Chim. Fr.*, 131
27. Polton, A. and Sigwalt, P. (1969). *Compt. Rend.*, **268,** 1214
28. Cheradame, H. and Sigwalt, P. (1970). *Bull. Soc. Chim. Fr.*, 843
29. Cheradame, H. and Sigwalt, P. (1965). *Compt. Rend.*, **260,** 159
30. Cheradame, H. and Sigwalt, P. (1964). *Compt. Rend.*, **259,** 4273
31. Branchu, R. B., Cheradame, H. and Sigwalt, P. (1969). *Compt. Rend.*, **268,** 1292
32. Chmelir, M. and Marek, M. (1968). *J. Polymer Sci., Part C,* **22,** 177
33. Chmelir, M. and Marek, M. (1967). *Collect. Czech. Chem. Commun.*, **32,** 3047
34. Gaylord, N. G., Matyska, B., Mach, K. and Vodehnal, J. (1966). *J. Polymer Sci., A-1,* **4,** 2493
35. Sauvet, G., Vario, J. P. and Sigwalt, P. (1970). *Bull. Soc. Chim. Fr.*, 4031
36. Elliott, B., Evans, A. G. and Owens, E. D. (1962). *J. Chem. Soc.*, 689, and ten previous papers
37. Totrai, J. P., Mayen, M. and Marechal, E. (1971). *Macromol. Preprints,* I.U.P.A.C., Boston, July, 1971, p. 172, and personal communication
38. Woolhouse, R. A. and Eastham, A. M. (1966). *J. Chem. Soc. B,* **33,** and nine previous papers
39. Eastham, A. G. (1956). *J. Amer. Chem. Soc.*, **78,** 6040
40. Clayton, J. M. and Eastham, A. M. (1957). *J. Amer. Chem. Soc.*, **79,** 5368
41. Clayton, J. M. and Eastham, A. M. (1961). *Can. J. Chem.*, **39,** 138
42. Szell, T. and Eastham, A. M. (1966). *J. Chem. Soc. B,* 30
43. Roberts, J. M. (1965). *J. Chem. Soc.*, 310
44. Roberts. J. M. (1965). *J. Chem. Soc.*, 1761
45. Armstrong, V. C., Katovic, Z. and Eastham, A. M. (1971). *Can. J. Chem.*, **49,** 2119
46. Deno, N., Peterson, N. and Saines, H. W. (1960). *Chem. Rev.*, **60,** 7
47. Goubean, J. and Bergman, R. (1950). *Z. Anorg. Allg. Chem.*, **263,** 69
48. Leftin, H. P. (1968). *Carbonium Ions,* Vol. I, 393, (New York: Interscience)
49. Fontana, M. C. and Kidder, G. A. (1948). *J. Amer. Chem. Soc.*, **70,** 3745
50. Kennedy, J. P. and Squires, R. G. (1967). *J. Macromol. Sci. Chem.*, **A1,** 995, and six previous papers
51. Clark, D. (1953). *Cationic Polymerisation and Related Complexes,* Ed., Plesch, P. H., 99, (Cambridge: Heffer)
52. Jordan, D. O. and Treolar, F. E. (1961). *J. Chem. Soc.*, 737
53. Kennedy, J. P. (1970). Unpublished results
53a. Kennedy, J. P. (1972). *J. Macromol, Sci.*, **A-6,** 329
54. Pepper, D. C. (1954). *Quart. Rev. Chem. Soc.*, **8,** 88
55. Kennedy, J. P. and Gillham, J. K. Unpublished results
56. Saegusa, T. (1969). *Structure and Mechanism in Vinyl Polymerization,* Ed., Tsuruta, T. and O'Driscoll, K., 283, (New York: Marcel Dekker)
57. Kennedy, J. P. (1968). *Polymer Chemistry of Synthetic Elastomers,* Ed., Kennedy, J. P. and Tornquist, E. G., Vol. I., Chap. 5A (New York: Interscience Publishers)
58. Kennedy, J. P. (1966). *Int. Symp. Macromol. Chem. Tokyo-Kyoto,* Abstract 2, 104
59. Kennedy, J. P. (1969). *J. Macromol. Sci.*, **A-3,** 861

60. Kennedy, J. P. (1969). *J. Macromol. Sci.*, **A-3,** 885
61. Kagiya, T., Izu, M., Maruyama, H. and Fukui, K. (1969). *J. Polymer Sci., Part A,* **7,** 917
62. Cesca, S., Roggero, A., Palladino, N. and DeChirico, A. (1970). *Makromol. Chem.*, **136,** 23
63. Cuzin, D., Chauvin, Y. and Lefebre, G. (1967). *Europ. Polymer J.*, **3,** 367
64. Kennedy, J. P. (1971). *Macromol. Preprints,* I.U.P.A.C., Boston, July, 1971, p. 105
65. Norrish, R. G. W. and Russell, K. E. (1952). *Trans. Faraday Soc.*, **48,** 91
66. Dainton, F. S. and Tomlinson, R. H. (1951). *J. Chem. Soc.*, 151
67. Szwarc, M. (1968). *Carbanions, Living Polymers and Electron Transfer Processes,* (New York: Interscience Publishers)
68. Kennedy, J. P., Shinkawa, A. and Williams, F. (1971). *J. Polymer Sci., A-1,* **9,** 1551
69. Gandini, A. and Plesch, P. H. (1965). *J. Chem. Soc.*, 4765
70. Mizote, A., Higashimura, H. and Okamura, S. (1968). *Pure Appl. Chem.*, **16,** 457
71. Higashimura, T., Masamoto, J. and Okamura, S. (1971). *Polymer J.*, **2,** 153
72. Higashimura, T., Ohsumi, Y., Okamura, S., Chujo, R. and Kuroda, T. (1969). *Makromol Chem.*, **126,** 99
73. Okamura, S., Higashimura, T., Imanishi, Y., Yamamoto, R. and Kimura, K. (1967). *J. Polymer Sci., C,* **16,** 2365 and three previous publications
74. Imanishi, Y., Matsuzaki, K., Yamane, T., Kohiya, S. and Okamura, S. (1969). *J. Macromol. Sci. Chem.*, **A3,** 249
75. Masuda, T. and Higashimura, T. (1971). *Polymer J.*, **2,** 29 and one previous paper
76. Takaya, H., Hirohara, H., Nakayama, M. and Ise, N. (1971). *Trans. Faraday Soc.*, **67,** 119 plus 16 previous papers
77. Ise, N. (1969). *Advan. Polymer Sci.*, **6,** 347
78. Sakurada, I., Ise, N. and Hayashi, Y. (1967). *J. Macromol. Sci.-Chem.*, **A-1,** 1039
79. Cerrai, P., Andruzzi, F. and Giusti, P. (1969). *Chim. Ind. (Milano),* **51,** 687
80. Funt, B. L. and Blain, T. J. (1971). *J. Polymer Sci.*, **A-1, 9,** 115
81. Kennedy, J. P., Magagnini, P. and Plesch, P. H. (1971). *J. Polymer Sci.*, **A1, 9,** 1635
82. Kennedy, J. P., Magagnini, P. and Plesch, P. H. (1971). *J. Polymer Sci.*, **A1, 9,** 1647
83. Prischchepa, N. D., Goldfarb, Yu., Ya., Krentsel, B. A. and Shishkina, M. V. (1967). *Vysomolek Soedin.*, **9,** 2426
84. Aso, C., Kunitake, T. and Shinkai, S. (1969). *Kobunshi Kagaku,* **26,** 280
85. Von Lohuizen, O. E. and DeVries, K. S. (1968). *J. Polymer Sci.*, **C-16,** 3943
86. Kennedy, J. P., Elliott, J. J. and Hudson, B. E. (1964). *Makromol. Chem.*, **79,** 109
87. Maltsev, V. V., Plate, N. A., Azimov, T. and Kargin, V. A. (1969). *Polymer Sci., USSR,* **11,** 248
88. Sartori, G., Lammens, H., Siffert, J. and Bernard, A. (1971). *Polymer Letters,* **9,** 599
89. Cesca, S., Priola, A. and Santi, G. (1970). *Polymer Letters,* **8,** 573
90. Corner, T., Foster, R. G. and Hepworth, P. (1969). *Polymer,* **10,** 393
91. Pinazzi, C. P., Pleurdeau, A. and Brosse, J. C. (1971). *Makromol. Chem.*, **142,** 259
92. Pinazzi, C. P., Brosse, J. C. and Pleurdeau, A. (1971). *Makromol. Chem.*, **142,** 273
93. Kovacic, P. and Kyriakis, A. (1962). *Tetrahedron Letters,* 467
94. Speight, J. G., Kovacic, P. and Koch, F. W. (1971). *J. Macromol. Sci.-Revs. Macromol. Chem.*, **C5(2),** 295
95. Mortillaro, L. (1968). *Mater. Plast. Elastomeri,* **34,** 26
96. Kennedy, J. P. and Isaacson, R. B. (1966). *J. Macromol. Chem.*, **1,** 541
97. Finocchiaro, P. and Passerini, R. (1968). *Ann. Chim. (Roma),* **58,** 418
98. Finocchiaro, P. (1968). *Boll. Sci. Fac. Chim. Ind. Bologna,* **26,** 255
99. Montaudo, G., Finocchiaro, P. and Passerini, R. (1967). *Ann. Chin. (Roma),* **56,** 1006 and four previous papers
100. Spanier, E. J. and Caropreso, F. E. (1969). *J. Polymer Sci. A*-1, **7,** 2679
101. Parker, D. B. V., Davies, W. G. and South, K. D. (1967). *J. Chem. Soc.*, 471
102. Parker, D. B. V. (1969). *European Polymer J.*, **5,** 93
103. Grassie, N. and Meldrum, I. G. (1969). *European Polymer J.*, **5,** 159 and one previous paper
104. Fritz, A. (1971). *Polymer Preprints,* **12,** 232
105. Eley, D. D. and Richards, A. W. (1949). *Research,* **2,** 147
106. Kennedy, J. P. and Squires, R. G. (1967). *J. Macromol Sci.-Chem.*, **A1,** 995 and six previous papers
107. Kennedy, J. P., Bank, S. and Squires, R. G. (1967). *J. Macromol. Sci.-Chem.*, **A1,** 977
108. Kennedy, J. P. (1971). *Polymer Preprints,* **12,** 4

109. Plesch, P. H. (1950). *J. Chem. Soc.,* 543
110. Meier, R. L. (1950). *J. Chem. Soc.,* 3656
111. Fontana, C. M. and Kidder, G. A. (1948). *J. Amer. Chem. Soc.,* **70,** 3745
112. Fontana, C. M. (1959). *J. Phys. Chem.,* **63,** 1167
113. Kennedy, J. P. and Squires, R. G. (1967). *J. Macromol. Sci.-Chem.,* **A1,** 805
114. Kennedy, J. P. and Squires, R. G. (1967). *J. Macromol. Sci.-Chem.,* **A1,** 831
115. Kennedy, J. P. (1970). *J. Org. Chem.,* **35,** 532
116. Kennedy, J. P. and Gillham, J. (1971). *Polymer Preprints,* **12,** 463
117. Kennedy, J. P. (1969). *J. Macromol. Sci.-Chem.,* **A3,** 885 and four preceding papers

3
Morphology of Lamellar Polymer Crystals

A. KELLER
University of Bristol

3.1 INTRODUCTION

It is about $3\frac{1}{2}$ years since I wrote my extensive review in *Reports on Progress in Physics*[1]. Its scope and purpose as laid out in its Preface applies unaltered for the present Survey: to provide a critical review of certain circumscribed aspects of polymer crystals which is readable for the uninitiated, and at the same time to provide 'stop press' reporting for the specialist. As the present version has to be both significantly shorter and has 3–4 years more to cover, I have had to be even more concise and selective. I have tried not to save space unduly at the expense of the introductory Sections, as I consider these essential for newcomers and outsiders if they are to appreciate what the subject is all about. Neither did I want to sacrifice the readibility of the Survey and provide merely a kind of source work or catalogue. It follows that space-saving had to be at the expense of the breadth of subject matter, which here is confined to the simplest lamellar crystals. Further, it had to be also at the expense of certain analytical details for which the reader is referred to earlier reviews[1–3, 171], to textbooks[4–6] and eventually to the original sources.

Within the restricted subject matter, the framework of the previous review[1] has been maintained, not for my convenience, but because I consider this still the most appropriate. Even if there appears to be some repetition as a consequence it is hoped that the attentive reader will notice that this is for a purpose: namely, wherever such a repetition occurs it serves to prepare the ground for the presentation of the more recent developments which follow. It will be noticed further that the reporting of the latest material does not feature as an addendum but is closely integrated with the recapitulation of the preceding developments. Thus it is hoped that the Review succeeds in bringing out trends with common and continuous underlying threads. In fact, only subject matters and individual works have been selected where common, underlying motives are discernible.

It is recognised that such a method of selection may involve some arbitrariness, and unavoidably significant topics and individual works may have been omitted. Neither is the length devoted to individual works or topics a measure of their intrinsic significance. In an effort to reconcile extreme condensation with readability the pure linguistics, i.e. the number of words needed to express a given conception, was often the factor which determined the length of a given item. Neither is the number and frequency of the references meant to be a measure of the significance of the contributions made by individual authors and schools. In order to keep the Bibliography within reasonable bounds sometimes groups of papers on a given subject from the same school are represented by only one example, in the hope that this will provide the key to the rest for those who consult it.

With a few exceptions, the illustrations chosen may well appear 'standard', as they have featured repeatedly in earlier reviews. Their purpose is to convey the basic concepts to readers for whom the subject is still unfamiliar, and also to provide fixed points of reference for the text, rather than represent new material in themselves. I can only hope that the first impression of '*déjà vu*' which these familiar pictures are likely to create will be at least partially dispelled by the actual reading of the text.

3.2 CRYSTALLINITY—GENERAL CONCEPTS

The material of this Review consists of molecules which are long chains. It is a characteristic of such chains that in the amorphous state they will take up their most probable configuration, which is the random coil. Such coils can be extended many times. If chain mobility permits, thermal motion will re-establish the random configuration when the stress is released, which is the basis of the long-range elasticity of elastomers (rubbers). Dependent on the molecular mobility, and hence temperature, such materials will exhibit the full spectrum of behaviour from a hard glass to a viscous fluid. In addition to these essentially amorphous characteristics, such long-chain molecules can crystallise, which will drastically influence the properties of the material. This crystallisation and the resulting structures will be the subject of the Review. (For general background see standard textbooks, e.g. Reference 7.)

Crystallinity implies a crystalline lattice. This will be first defined. If the chains are of regular chemical (including stereochemical) composition, they possess a periodic structure. If straightened, this gives rise to a one-dimensional periodicity. If several chains are lying parallel with their internal periodicities in register, the atoms will define a three-dimensional lattice, the smallest regularly repeating three-dimensional unit being the unit cell. The size and geometry of this unit cell and the arrangement of the atoms within it can be determined by the traditional methods of x-ray crystallography, in the case of polycrystalline polymers from powder and fibre patterns. Such structure determination consists of two facets, the first related to the configuration of the chain and the second to the packing of the chains. The first is the more intrinsic property as it is determined by valence forces. Different polymers of course have different crystal structures. The crystal structure of polyethylene, the polymer to be referred to most frequently, is illustrated in

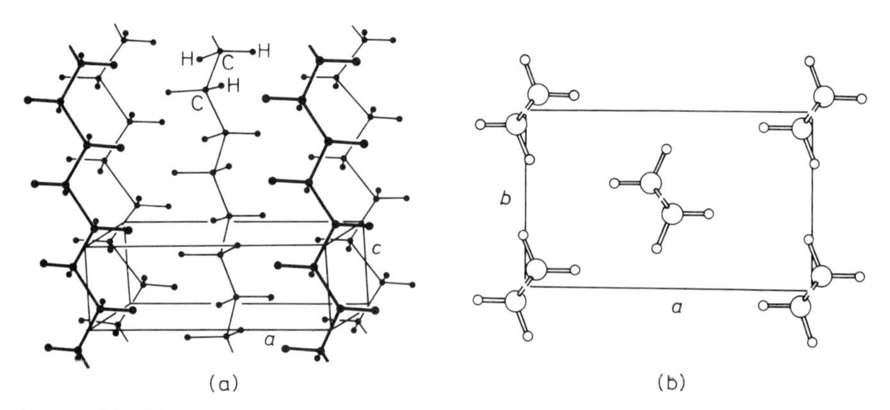

Figure 3.1 The crystal structure of polyethylene, (a) general view (b) projection along *c*-axis (chain direction)
(From Reference 8, by courtesy of Bunn, C. W. and Elsevier)

Figure 3.1. In other polymers, the chains may have other conformations, for example they may form helices of various kinds, and it is these helices which are parallel giving rise to the crystal lattice.

With the atomic positions determined we still do not know what is the size and shape of the crystals, and we only know that they must be so small that they evade detection by say direct inspection of a solid polymer with the optical microscope. Neither do we know how the chain length is related to the crystals. This is the subject of crystal texture studies as opposed to the traditional determination of atomic structures. Texture studies acquire special importance for high polymers and will be a subject of this Review.

From the earliest, the crystalline elements have been envisaged as elongated along the chain direction (fibrous units). Further, from the fact that a crystalline polymer does not behave as a usual crystalline object but retains some of the characteristics of the corresponding amorphous material (long-range elasticity, pliability), it was inferred that it must also contain an amorphous component, which was substantiated by the fact that several of its properties, such as density and heat of fusion, did not reach the value corresponding to that of a fully crystalline material. The model embodying all these features, termed fringe micelle, envisaged a molecularly-interconnected array of small crystals. This has been illustrated so frequently that it will not be repeated here, particularly as it has been substantially modified by subsequent developments. The most essential concept originating from this era is the degree of crystallinity, which can be determined by a suitable crystallinity-sensitive property. Understandably, great importance is attached to this quantity, as through it we can define a property which lies between that of the purely amorphous and purely crystalline phases. Nevertheless, important as this measure may be, it cannot define crystallinity uniquely as it depends to some extent on the technique which is used as an indicator of crystallinity. Further, even when confining ourselves to one technique, a given degree of crystallinity can be attained by a variety of different morphologies, and hence does not define a sample uniquely.

Crystallinity in polymers can be induced essentially in three ways:

(a) thermally, i.e. by supercooling of a melt or solution,

(b) by molecular orientation, and

(c) by polymerising the monomer under conditions which leads to *in situ* crystal formation.

It is to be noted that only the first has an equivalent in simple substances, and in consequence it is this mode of crystallisation which has been studied most extensively. This will be the subject of the present Review. Methods (b) and (c) are unique to polymers. While there are important historical precedents, they have been taken up from the modern viewpoint only recently. Mention of method (b) will be made only while information on (c) is so sporadic that in spite of its obvious significance it will not be included here.

3.3 ON MORPHOLOGY IN GENERAL

3.3.1 Concepts and methods

Morphology is the study of external form. It is the first stage in most sciences, usually superseded by subsequent developments on the molecular and atomic structures. The study of morphology in polymers, however, has some special added significance. This is due to the circumstance that here morphological and molecular features merge. It happens that observational information on sizes and shapes enlarges our knowledge on the molecular level; in fact, as will be seen below, it has served to bring about decisive progress in this direction. The reason for this lies in the fact that the morphological features in polymeric substances are of small dimension while the molecules are huge in comparison with those in simple substances, with the result that there is not such a large gap between the one and the other as is common in more traditional subjects. Neither is the study of the morphology confined to pure viewing. Admittedly all available microscopic techniques are applied, such as all aspects of optical microscopy (polarising optics, phase contrast, interference microscopy etc.) and of electron microscopy (transmission, replication, scanning etc.). But in addition, diffraction, specially x-ray, techniques are also involved such as can provide information on shapes and size of the morphological entity. This becomes feasible in view of the small sizes, which brings the corresponding diffraction effects within the range of small-angle x-ray scattering techniques. The combination of direct viewing and diffraction, as for example transmission electron microscopy and low-angle x-ray scattering, proved to be a powerful method in exploring the morphology. Viewing can also be combined with the study of the atomic structure, as in the case of combined electron microscopy and electron diffraction, which brings the study of morphology in close combination with the whole body of structural crystallography.

Details of techniques will not be included here. It will only be stated that technical developments are essential for progress in this field. This involves not only better instruments but improved techniques for making the samples accessible to morphological observation. Also the interpretation of the observations needs constantly to be improved upon. For example, in the case of electron microscopy, the sample interior needs to be dispersed or opened up to make it accessible for imaging by the electrons. But at the same time the beam, while forming an image, significantly alters the material under examination, an alteration which needs constant evaluation. This should suffice to show that even the primarily observational part of morphological work is not merely a direct description of what is seen when the object in question is placed under the viewing instrument.

3.3.2 Morphological entities

To facilitate the survey, the principal morphological entities will be enumerated at the onset. These are the entities that can be directly observed by the appropriate viewing instruments. There are various ways in which these can be grouped. Here we shall define two basic classes namely lamellar and fibrillar units from which all other more complex morphological entities can be derived. It is the lamellae which form the main subject of the present Review. Nevertheless at this point the entire spectrum of units will be introduced.

3.3.2.1 Lamellar crystals

As recognised around 1957[9–11] these are the products of crystallisation, forming under quiescent or unstressed conditions, either from solution or from the melt on appropriate undercooling. The simplest unit is the single lamella most readily obtained from dilute solution (Figure 3.2). These develop into thicker aggregates through spiral terrace growths centred on screw dislocations (Figure 3.3). The consecutive terraces or layers are generically part of the same crystal although usually not in perfect register with each other. They may splay out in more-or-less irregular fashion giving rise to sheaves (Figure 3.4). It is such sheaves which under normal conditions are the first recognisable morphologies in crystallisation from the melt. By a continued fanning growth mechanism such sheaves develop into spherulites, which are radially-symmetrical growth features (except for the central sheaf region) (Figure 3.5). The concentric banding observed in many cases

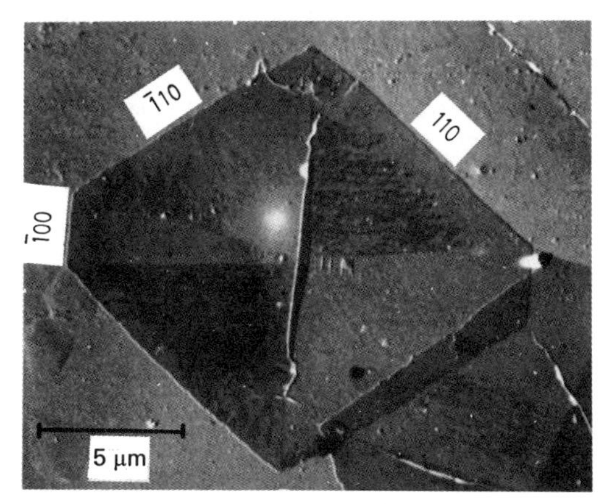

Figure 3.2 Typical monolayer polyethylene crystal. The central pleat is due to a hollow pyramid collapsing and the underlying sectorisation is revealed by diffraction contrast. The prism face indices are indicated (Electron micrograph from Bassett, D. C., Frank, F. C. and Keller, A.[12], by courtesy of Taylor and Francis)

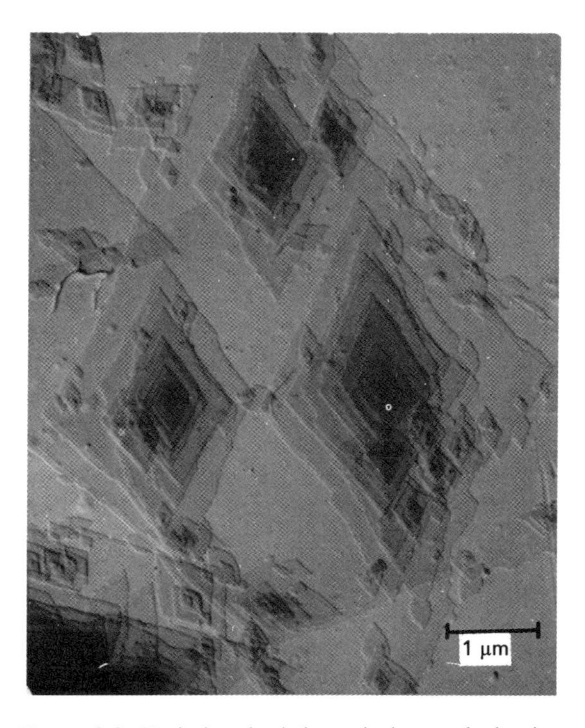

Figure 3.3 Typical polyethylene single-crystal showing spiral multilayer growth centred on screw dislocations (Electron micrograph from Keller, A. and O'Connor, A.[13], by courtesy of The Faraday Society)

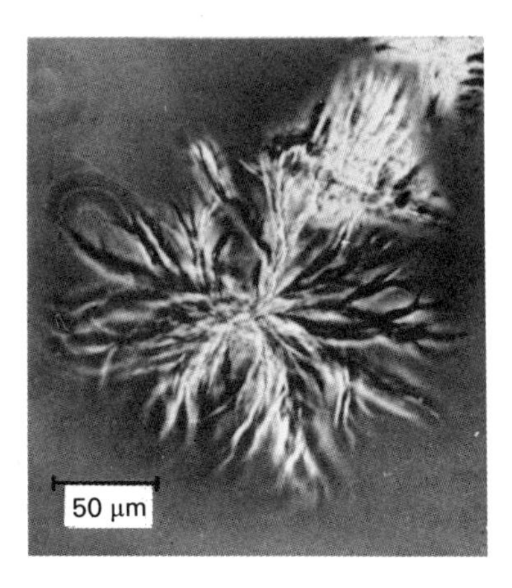

Figure 3.4 Sheaving layer structures in multilayer crystals of polyethylene. The layers are viewed edge-on and are seen as fibres while the sheaf floats in the mother liquid
(Photomicrograph from Bassett, D. C., Keller, A. and Mitsuhashi, S.[14], by courtesy of Interscience)

Figure 3.5 Banded spherulites of poly(trimethylene glutarate) grown between glass surfaces and viewed between crossed polaroids
(From Keller, A.[172], by courtesy of Interscience)

Figure 3.6 Surface of a sample of spherulitic polyethylene containing a region of banded spherulite. Electron micrograph, replica
(Based on Fischer, E. W.[11]. By courtesy of Professor E. W. Fischer)

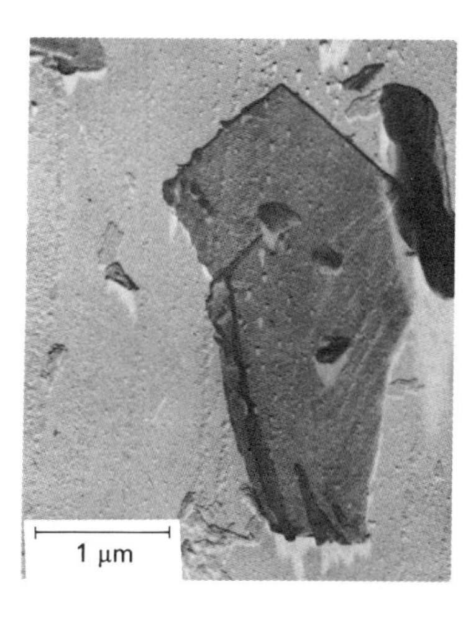

Figure 3.7 Layer fragment from bulk polyethylene dispersed with fuming nitric acid; *b* axis (see Figure 3.1) vertical
(Electron micrograph from Keller, A. and Sawada, S.[16], by courtesy of Alfred Huethig Verlag)

reflects a periodic twisting of the birefringent units. In morphological terms, this periodicity is to be envisaged as an air-screw-like twisting of ribbon-like lamellae which themselves grow outwards in radial directions, with the helicoidal windings in phase all around a given radius. The detailed morphology of such spherulites can only be revealed at the sample surfaces (Figure 3.6). Methods revealing the sample interior drastically interfere with the structure (Figures 3.7 and 3.8 provide some illustrations) and present interpretive problems.

The spherulites represent the largest units in the morphological hierarchy, and it is these which constitute the usual melt-crystallised specimens. They

Figure 3.8 Twisted lamellar aggregate of polyethylene extracted from a bulk sample containing banded spherulites with the aid of nitric acid digestion. The twist period corresponds to the original banding seen under the polarising microscope
(Electron micrograph from Keller, A. and Sawada, S.[16], by courtesy of Alfred Huethig Verlag)

meet along straight or hyperbolic boundaries and represent a kind of grain structure. It must be remembered that these grains are not single crystals as in the case of metals, but complex lamellar aggregates.

Our knowledge of the formation of multilayers and of layer aggregation into more complex sheaf and spherulitic structures is largely descriptive. Most of the exhaustive work is explicitly or implicitly centred on the structure and formation of an individual lamella. At this stage the scope of the Review will be delineated further: it will essentially be concerned with the individual basic crystal layer.

3.3.2.2 Fibrous crystals

Even if not the subject of this Review, these must at least be placed on the map. Traditionally, fibrous crystals have been envisaged as the basic units of crystallisation, as this appears to be compatible with the alignment of long-chain molecules. This is underlined by the fibre-forming tendency of both synthetic and natural long-chain materials, and the importance such fibres possess in technology and in the living world. It is a comparatively recent

Figure 3.9 Fibrillar crystals with lamellar overgrowth (shish-kebabs) of polyethylene grown in an agitated solution.
(Electron micrograph from Pennings, A. J.[17], by courtesy of Dr. A. J. Pennings and Pergamon Press)

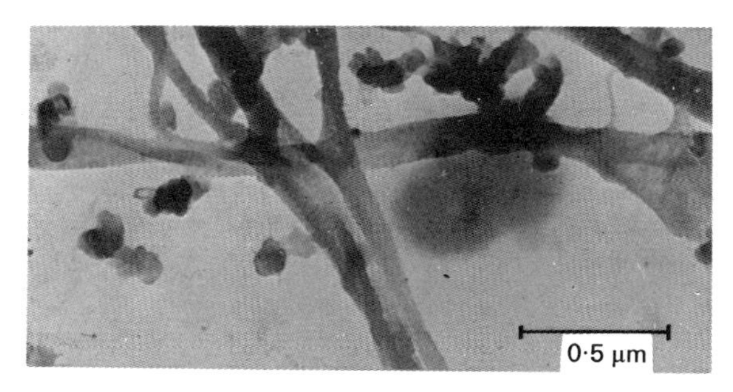

Figure 3.10 The fibrous backbones of polyethylene 'shish-kebabs' obtained by crystallisation from agitated solution at sufficiently high temperature to prevent subsequent platelet deposition
(From Keller, A. and Machin, M. J.[18] with contribution by Willmouth, by courtesy of Interscience)

recognition that an external orienting influence is required for their formation; otherwise lamellar units result.

Lamellar and fibrillar crystal forms are not completely separate categories as they can exist in various combinations. A particularly striking one is the so-called 'shish-kebab' structure which consists of a central fibre and a transverse lamellar overgrowth (Figures 3.9 and 3.10). The shish-kebabs are all important constituents of a polymer crystallised under stress[18], or during flow in a longitudinal (as opposed to the usual transverse) velocity gradient[19], both of which elongate the molecules.

3.3.3 Alternative classifications

Other classifications are also adopted. The one most frequently used is based on the molecular configuration, and accordingly groups the different morphologies into folded- or extended-chain types. In general, lamellae contain folded and fibrous extended-type chains. Nevertheless the assignment of morphologies to molecular conformations is not unique. Thus, fibrous

Figure 3.11 Extended chain-type crystal morphology in polyethylene crystallised under 4.8 kbar hydrostatic pressure. Electron micrograph; replica of a fracture surface. The broad banding corresponds to the cross-sectional view of lamellae with the chain direction along the fine striations
(From Prime, R. B., Wunderlich, B. and Melillo, L.[20], by courtesy of Wunderlich, B. and Interscience)

crystals obtained from pre-oriented molecules usually contain some folded chain constituents (responsible for the structures in Figure 3.10), whilst lamellae grown from unoriented melt under high hydrostatic pressure consist of essentially extended-chain molecules (Figure 3.11).

Another frequently adopted criterion is whether the system has crystallised from solution or from the melt. This latter classification is usually imposed by practicability. The morphological entities can be obtained in isolation only from solution, and it is under these conditions that they can be most definitively studied. As the entities arising from the melt are not available in dispersion, information on the morphologies of melt-crystallised systems is necessarily always less definitive. The usual course of research therefore is to establish certain basic features from solution crystallisation first and then apply this newly-gained knowledge to crystallisation phenomena from the melt. This traditional division will be abandoned in the present Review. Although features arising from crystallisation from solution and melt will be invoked without distinction as required by the subject matter under discussion, it must be remembered nevertheless that evidence gained on solution-grown crystals is usually on a stronger footing.

3.4 BASIC ASPECTS OF LAMELLAR CRYSTALS

3.4.1 Single crystals and chain folding–general principles

Lamellar crystals were first obtained from dilute solution, in which form they are best observable. The usual method of preparation is to cool a solution to an appropriately low temperature when the crystals precipitate forming a suspension. These suspended particles can be examined in a variety of ways. The most straightforward is to sediment them on a microscope slide in which form they can be examined by the optical microscope, usually by phase or interference contrast. If sedimented first on an appropriate substrate, they can also be examined by electron microscopy in direct transmission. In this way the layers can be directly observed. They are usually of the order of 100 Å in thickness, and to all appearances uniform, displaying well-defined crystallographic facets (Figure 3.2). Electron diffraction patterns obtainable on individual crystals while under observation in the electron microscope confirmed the single-crystal nature of the lamellae, and also identify the orientation of the unit cell within them. The most significant feature is the chain orientation, which is either perpendicular or at a specified large angle to the basal surface of the crystal.

The observation quoted immediately raises the dilemma of how a chain many times longer than the layer thickness and of non-uniform length, as chains usually are in the polymer, can be confined within the lamellae in the above orientation. The answer is provided by the familiar chain-folded model as shown in Figure 3.12. According to this model the chains must fold back and forth at regular intervals so that the lattice geometry between the straight stems is preserved. Furthermore, the fold length must be sufficiently uniform to give rise to the smooth layer observed by electron microscopy. Figure 3.12 was drawn with sharp, hairpin-like folds with adjacent chain re-entry and of strictly uniform length. This is a plausible but not a necessary consequence of the basic observation. Alternative ways of achieving such a folded structure have been proposed, and have been a subject of much discussion for a number of years. We shall return to this specific topic later.

Lamellar crystal growth and the underlying chain folding is a general behaviour of long-chain molecules. It is important to recognise that not only the simple polyethylene molecule with its planar zigzag chain configuration in the crystal but also chemically more complex molecules with bulky side-groups undergo folding. If the chain configuration in the crystal lattice is helical, then the whole helical chain will fold. It is becoming gradually

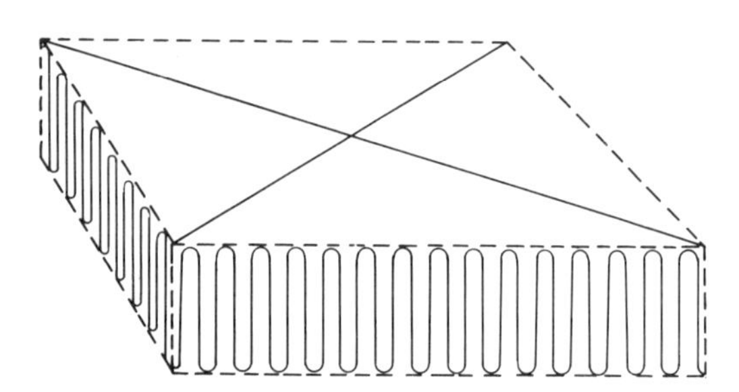

Figure 3.12 Diagrammatic representation of chain-folding in polymer crystals (the folds are drawn sharp and regular). The origin of sectorisation is indicated

apparent that complex biological substances such as polypeptides or even DNA can be obtained in folded forms, and may occur as such in Nature[21, 22].

The most important characteristic of a chain-folded system is the fold length. As apparent from the above argument, the fold length is equivalent to the layer thickness, and hence it can be determined by measuring this crystal dimension. Thus we have the unique situation where a crystal dimension is associated with a molecular quantity, the fold length. Consequently this is an example where the study of crystal structure, in the traditional sense, and that of crystal texture become closely related.

The layer thickness, and hence the fold length, can be measured directly from shadow length in the electron micrographs, but more easily and representatively it can also be determined by low-angle x-ray scattering. In this method, crystals are sedimented in the form of a macroscopic mat. Such mats give rise to discrete x-ray reflections at small angles in several orders. The large spacings corresponding to such small diffraction angles are in quantitative agreement with the measured layer thickness determined by electron microscopy, and in cases where the layers all sediment horizontally they are in an orientation which corresponds to the basal surfaces of the lamellae[1–4, 13]. From this it follows that they result from the one-dimensional periodicity produced by the stacking of layers, and hence provide a measure of the fold length.

Diffraction effects due to large periodicities had been observed in bulk polymers at a much earlier stage, but it is due to the recognition of lamellar crystals, and the ensuring chain-folded model in crystals grown from dilute solution, that they can now be assigned to chain folding under these more general circumstances. The complementary application of electron microscopy and diffraction methods, especially that of low-angle x-ray diffraction, should be noted: this combination forms the principle armoury of morphological studies in high polymers.

It should however be pointed out that neither of these methods measures a molecular property directly; they only disclose some morphological features associated with chain folding. It is the object of most recent developments to obtain direct molecular information concerning chain folding.

3.4.2 Variations in the fold length

3.4.2.1 The effect of supercooling

It has been shown that the fold length is not an invariable quantity, but is determined by the crystallisation temperature. Higher crystallisation temperatures produce higher fold lengths[13] (Figure 3.13). Subsequent studies using different solvents have revealed that the principal variable is not the crystallisation temperature itself but the supercooling[23, 24] (Figure 3.13). This supercooling-dependence is surprisingly reproducible and practically independent of occasional irregularities in the molecule, such as occur in materials which are nominally identical but of different manufacture. However, if such irregularities become numerous so as to affect the supercooling, as for example side-groups do in copolymers of polyethylene with propylene and butene, the long spacing will be affected due to the alteration in the supercooling[25].

The long period seems to be unaffected by concentration and by the usual variation in molecular weight[26]. For short molecules, however, the molecular weight will have an effect on the dissolution temperature and on the supercooling, which will therefore influence the long period[27]. It is an important characteristic of the fold length that it is not affected by the thickness of the crystal at previous stages of growth. Thus if pre-grown crystals are used as

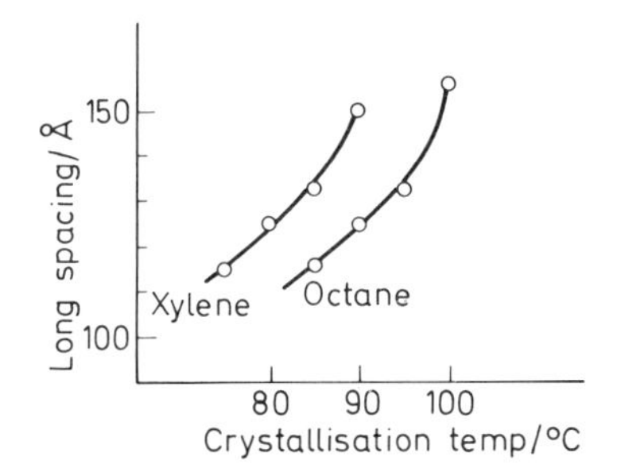

Figure 3.13 Long spacings in polyethylene as a function of crystallisation temperature for crystallisation from two solvents, xylene and octane
(From Kawai, T. and Keller, A.[23], by courtesy of Taylor and Francis)

seeds, or if the supercooling is changed during growth, the fold length will correspond to the supercooling prevailing at the appropriate stage of crystal growth. Morphologically this is demonstrated by the occurrence of upward or downward steps in the crystal according to whether the supercooling is respectively lower or higher than in the preceding stages of growth.

The supercooling-dependence of the fold length is one of the most firmly established and quantitatively reproducible behaviours of long chains. It is the foundation on which much of our present appreciation of the subject is based.

3.4.2.2 *Annealing behaviour*

Once formed, crystals can subsequently increase their fold length in an irreversible manner by raising the temperature beyond that of the original

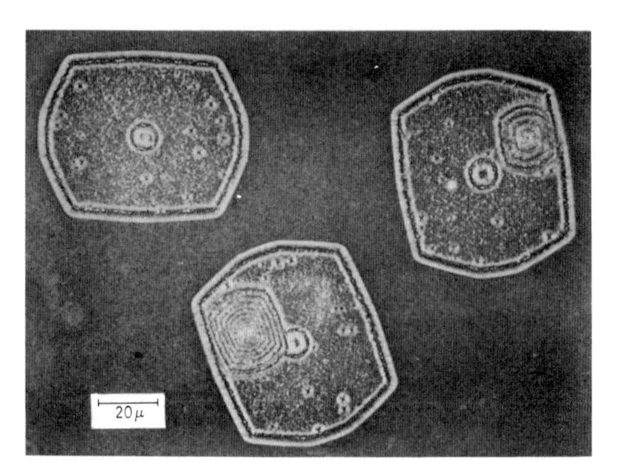

Figure 3.14 Crystals of a low molecular-weight fraction of poly(ethylene oxide) (POE 6000) obtained from the melt between glass surfaces by use of the self-seeding method. Edges and some specific sites along the surface are made visible by the initiation of spherulite growth at these discontinuities, the material being drawn from the surrounding melt (self-decoration). The thickening in the central region starts at a later stage of growth and is interpreted as isothermal re-folding to longer fold length and in the present case corresponds to the extended chain length
(From Gonthier, A. and Kovacs, A. J.[34], by courtesy of Dr. A. J. Kovacs)

crystallisation[28]. The effect is most clear-cut when the heat treatment is performed on dried crystals or aggregated crystal mats. The evidence for fold-length increase is as follows. Isolated layers are seen to become thicker by shadow-length measurements in the electron microscope[13, 28], while electron diffraction reveals that the chains retain their unaltered orientation perpendicular to the lamellar surface. Clearly, the lamellae can only become thicker through an increase in the fold length. This is substantiated by the increase in the long period as revealed by low-angle x-ray reflection on sedimented mats, the technique usually employed for the study of annealing. The long-period increase is more pronounced at higher annealing temperatures. The same end result is obtained when crystals are heated while in suspension. In this case, however, the situation is more complicated, as it is coupled with complete dissolution and re-deposition of the chains[29,30].

Long-spacing increase can occur also at the crystallisation temperature, in which case it follows or is concurrent with the primary chain-folded crystallisation[31]. There is no observation of such an effect in solution-grown

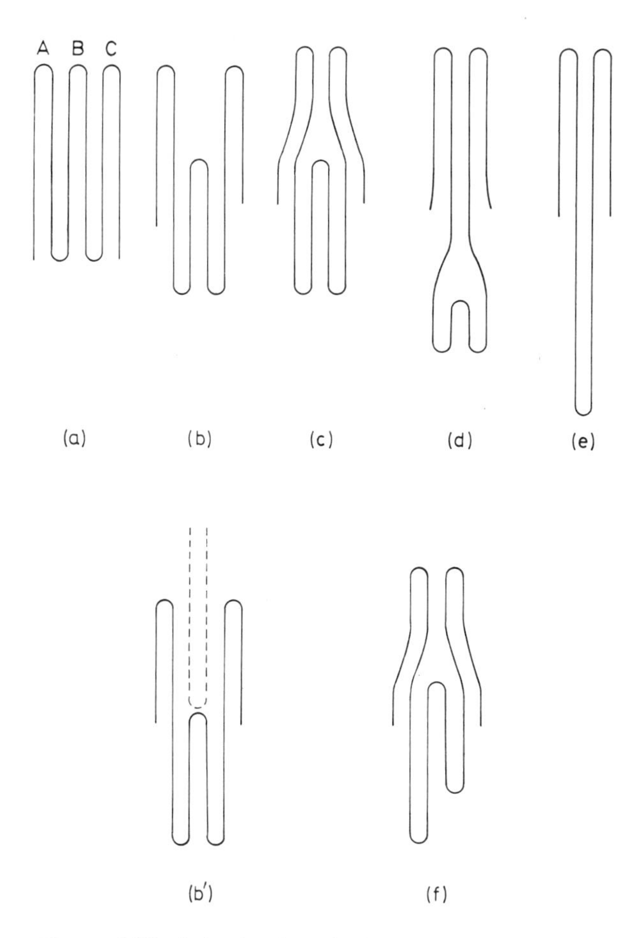

Figure 3.15 A simple scheme for fold-length increase; (a)–(e) show the origin of preference for a doubled fold length and (b′) a possible influence of superposed layers
(From Dreyfuss, P. and Keller, A.[42], by courtesy of Interscience)

crystals, but it can play a significant part in crystallisation from the melt, where it is often associated with secondary crystallisation[32,33]. It is envisaged that if the temperature is high enough the chains start to re-fold immediately after they have formed the usual chain-folded crystals. Striking visual evidence of this effect on individual melt-grown crystals has been obtained quite recently[34] (Figure 3.14). This chain re-folding following primary crystallisation takes up a particularly central position in crystallisation studies under hydrostatic pressure[35,36], a subject we shall not go into in this Review.

There is no unanimity as regards the mechanism of fold-lengths increase. A very high chain mobility is clearly required. According to some views, complete melting and recrystallisation is involved[37], but, according to the majority of evidence, melting if it occurs can only be partial[1,38]. Molecular orientation is certainly preserved, in fact it is part of the evidence for chain re-folding[1,28]; even the crystallographic orientation along directions which are lateral with respect to the chains are preserved (see Reference 1).

The time- and temperature-dependence of the fold-length increase has been extensively studied[38]. Only the conclusions will be recapitulated here. In the best-explored polyethylene system at a given temperature there is first a large sudden fold-length increase followed by a slower process which displays a logarithmic dependence with time. This whole process is accelerated by increasing the temperature.

The actual molecular motions involved have usually been envisaged as a threading of the full chain through the crystal[40] pulling its ends with it[39], a motion which would result from cooperative jumps within the highly-mobile chain. While this accounts for the logarithmic time-dependence, and for some effects on the mechanical behaviour, the very strong molecular-weight dependence it predicts[40] is certainly not observed. Further, the pulling in of chain-ends is contradicted by experiments on the chemical accessibility of end-groups, according to which most of the ends are outside the crystal to begin with, Section 3.8.3.3, and stay there during the course of chain re-folding[41].

In addition to the effects listed, a preference for fold-length increase by a factor of two has been observed in a number of circumstances. This was first noticed in polyamides[61], and then on revision of polyethylene data during the rapid initial jump. The simple scheme in Figure 3.15 seems capable of accounting for these effects[42]. The principle point is that the chains are only envisaged as rearranging locally by pulling out two or a limited number of folds at a time. This obviates the necessity for the full chain threading in the same direction simultaneously. Amongst others, this mechanism leads naturally to an energetic preference for the double fold length, while it implies defects due to folds buried within the crystal interior during the intermediate stages. As will be seen later, such buried folds have recently emerged as general features even in unannealed crystals, Section 3.8.3.2(c).

3.4.3 Some remarks on theory

This Section provides only the briefest sketch and is included to enable an appreciation of the general subject matter which follows.

The unexpected phenomenon of chain folding requires explanation, of course. The principle facts to be accounted for are: why do chains fold, why is folding uniform within the limits observed, why does the fold length vary with crystallisation temperature in the manner observed and why does the fold length increase on annealing? Historically, the theories which have been advanced are of two kinds, equilibrium and kinetic, of which the latter is now much more in prominence.

3.4.3.1 Equilibrium theories

This class of theory, based on lattice dynamical considerations, envisages a state of minimum free energy corresponding to a particular extension of the lattice along the chain direction[43]. Folding of the chains is one of the ways by which the crystal size can be limited to the required value. The observed trend with crystallisation temperature can thereby be accounted for, but not the exact relation between fold length and supercooling. Reversibility with temperature cycling is also expected as opposed to the irreversible increase observed on annealing. (Recently, long spacings which vary reversibly with temperature have in fact been observed[44–49]. The interpretation, however, tentative as it is[45, 46], relies more on surface melting (Section 3.8.2.4(b)) than on a reversible variation of the fold length.)

3.4.3.2 Kinetic theories

According to kinetic theories, the observed chain-folded crystals correspond to the structure with maximum crystallisation rate rather than maximum thermodynamic stability; the latter corresponds to chains which are fully extended. Thus, folding should occur since in this way crystallisation can proceed faster, and at a particular supercooling there should be a fold length at which the rate is maximum. This will be the fold length which determines the overall crystal thickness. It follows that the chain-folded crystal will not be in its lowest state of free energy, but it can tend towards this state by increasing the fold length, an effect which takes place subsequent to the primary chain deposition particularly on heat treatment (Section 3.4.2.2).

The various kinetic theories consider the simple model of a new chain which folds up along an existing prism face[50–53]. The thickness of the substrate face should be immaterial as according to experiment the fold length is determined by the prevailing supercooling and not by the crystal thickness along which the deposition occurs (Section 3.4.2.1). Indeed all the theories successfully incorporate this particular behaviour. As the chain folds there will be two competing effects: the increase in free energy as the first segment deposits, and the gain in stability as the chain folds back on itself and covers up the exposed side of the neighbouring stem, thus forming more crystal. The free energy increases with increasing length of the segment (more surface is created), while formation of longer fold stems leads to greater stability by virtue of the formation of more crystal. It will be apparent that these two opposing trends will lead to a fold length at which the deposition rate is maximum. This will be the most probable fold length which will determine the thickness of the layer.

The individual theories work out these consequences in a quantitative form. Although different in their mode of formulation they all arrive essentially at a similar expression for the most probable fold length L^*

$$L^* = \frac{2\sigma_e T_m^0}{\Delta H \Delta T} + \delta L \tag{3.1}$$

Here T_m^0 is the melting (or dissolution) temperature of the fully extended chain crystal, ΔH is the heat of fusion, ΔT the supercooling and σ_e the surface free energy along the basal plane of the crystal, which includes the work required to create a fold. At all except extreme supercoolings, δL is a small quantity (about $\sim 15\%$ of the first term); the different theories give it a different explicit form. It follows that under the supercoolings which normally prevail it is essentially the first term on the right-hand side of equation (3.1) which determines the fold length. Its inverse dependence on supercooling describes the form of the observed curves shown in Figure 3.13, and by a reasonable choice of the parameter σ_e (around 50–150 erg cm^{-2} for polyethylene) the experimental curve can be numerically fitted. In spite of the latitude in σ_e, this can be considered as a remarkable success in view of the complexity of the system and the comparative simplicity of the model.

It is to be noted that the first term on the right-hand side of equation (3.1) has the same form as the one which describes the size of a critical nucleus. In fact, it has the same significance as it describes the lowest fold length which has any stability at all, a length which in the usual chain-folding process is only slightly exceeded (i.e. by δL). It is apparent that the actual lamellar thickness is to a close approximation controlled by the two-dimensional nucleation which is involved in the deposition of the chain-folded ribbon. The remarkable additional feature, however, is that here the crystals retain this thickness, hence the critical nucleus size determines the final texture of the material, in contrast to traditional substances where the crystals continue to grow in all directions, and all relation between the critical nucleus size and the final crystal is lost. It follows that the correspondence of equation (3.1) with the measured lamellar thickness is not such a trivial matter as it may appear from some recent controversies[54, 55]. In fact, the model of adjacently re-entrant deposition of chains as embodied in the theories, is essential to account for the whole phenomenon. Whether such a model is unique, and whether others could also lead to the same result, is a different question; the fact is that such alternative theories do not exist at present. The significance of these rather polemical remarks will be apparent in Section 3.8.

The molecular model in the above kinetic theories involves the uninterrupted deposition of a single chain of infinite length. The simplifications in this model are self-evident. In practice, amongst others, the deposition of any given chain will terminate and a new one will start. Some of the consequences will be apparent from Figure 3.16. Even in the special case, when the depositing chain ends exactly at the lamella surface, it is not likely that the new chain will continue depositing at one of its ends (Figure 3.16a), but will start its deposition somewhere along its length (Figure 3.16b). It is apparent that regular folding can only proceed at one side of the deposited segment; the other side will dangle free in solution (Figure 3.16c). If the first chain does not end exactly at the surface or its deposition is prematurely interrupted (Figure 3.16d) there will be two such free dangling ends. Such hairs or cilia have important implications as regards the regularity of the fold surface, and will be discussed in Section 3.8. It is mentioned here because such loose ends can also fold into the crystal and contribute to its growth. In this way they will compete with new molecules ariving at the crystal face for continuing the chain-folded deposition, and thus will affect the kinetics

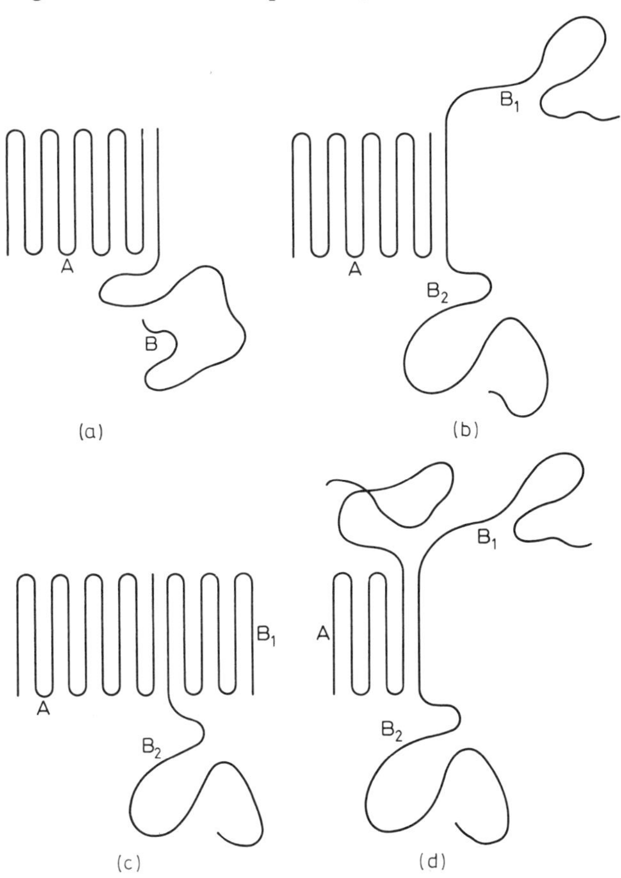

Figure 3.16 Schematic representation of the origin of cilia formation and of the resulting surface looseness
(From Keller, A.[2], by courtesy of Dietrich Steinkopt Verlag)

of crystal growth. A high cilia density, which depends primarily on molecular weight, can largely shield the growing crystal surface from contact with molecules which are fully within the solution. Amongst others, these are conditions which are treated in a quantitative manner by the latest elaboration of the kinetic theories due to Sanchez and DiMarzio[56].

3.4.4 Comparison of chain folding in different polymers: invariant long spacings

As mentioned above, the long period–supercooling dependence such as that shown in Figure 3.13 is the most solidly established experimental observation on chain folding. Not only polyethylene obeys these relations but also a variety of other polymers which have been examined from this point of view. Amongst others these were poly(oxymethylene)[57], poly-4-methylpentene[57], several polyalkenamers[58] (polyethylene-type compounds with regularly spaced double bonds) and guttapercha[59], to mention only studies on the morphologically well-established products of crystallisation from solution.

Polymers with large identity periods, such as polyesters, are particularly interesting as here the fold length increases in a stepwise, discontinuous manner[60]. This, in fact, is to be expected as only in this case can the folded stem, after having folded over, come into crystallographic register with its neighbour, i.e. only in this case can the crystal lattice be preserved.

However, a whole class of compounds, namely polyamides, does not display this behaviour: they always precipitate with the same fold length irrespective of the crystallisation temperature[61]. It appears that each compound has a characteristic fold period from which it cannot be induced to depart by alteration of supercooling. It is not clear at present whether a new principle is involved or, what is more likely, whether it is merely a variant of the established behaviour where the fold length varies only by large steps due to the large chemical repeat unit, and that these steps need to be induced by large changes in supercooling such as are not readily achieved in practice. Irrespective of which is the case, the fact that there exists a fold length which is insensitive, if not invariant, to conditions of crystallisation enables different polyamides to be compared. In the nylon 66, 68, 6.10, 6.12 series, which has been studied to the greatest extent, the characteristic fold length corresponds to four monomer units. This means that the characteristic fold length increases in proportion to the monomer length: hence we have an example where the fold length is a function of the chemical structure.

As a constant number of repeat units means a constant number of hydrogen bonds, the most obvious inference is that it is the hydrogen bonds which determine the fold length and that a certain fixed number of them is required for folding. In the above system, if all of them are formed, their number is 16. It was found that this figure applied to a large number of other nylons, including such short ones as nylon 3[62].

When comparing the above nylons with say polyethylene, it is found that the characteristic fold length is lower than any realisable fold length in the latter (in nylon 66 the fold length is 53 Å – when allowing for chain obliquity this corresponds to a maximum straight stem length of 68 Å as compared to 100–180 Å in polyethylene). This suggests that hydrogen bonding is an inducement to folding by leading to a larger decrease in the volume free energy which has to pay for the fold, and hence leading to the shorter fold length. Indeed, the fold length seems to be unusually low in all hydrogen-bonded systems as, for example, synthetic polypeptides[63] and silk fibres spun by certain insects[64] when in a chain-formed arrangement. (Here the number of hydrogen bonds lies in the range of values quoted above for polyamides.)

It is clear that the constancy of monomer units, i.e. hydrogen bonds per fold length, cannot be maintained without limit as the monomer length is increased, otherwise the fold length of polyethylene would be eventually exceeded. In fact the characteristic fold length for nylon 10.10 and 12.12 does drop to three monomer units (12 hydrogen bonds) in accordance with this expectation[62, 65].

The above examples illustrate a new trend with further possibilities: the effect of interchain forces on chain folding. In other words, we are departing from the inert, chemically featureless molecule of polythene which serves as a model substance for the behaviour of a long chain *per se*. By going over to chains of increasing complexity and chemical specificity we are approaching biologically significant macromolecules, where this specificity itself is of prime interest.

In addition to hydrogen bonding or dipole interaction, ionic forces can influence or facilitate the folding behaviour of long chains which leads to the possible chain-folding behaviour of polyelectrolytes. An important study in this direction has been carried out by Keith and collaborators on salts of polyglutamic acids where hydrogen bonding and ionic interaction play a simultaneous part[63]. Here it was found that bivalent cations promoted chain-folding crystallisation by providing bridges between the acid residues of adjacent chain segments.

3.5 CRYSTAL GROWTH

3.5.1 Growth rates

3.5.1.1 General

In the theory discussed in the previous Section, only the most probable fold length and its distribution has been considered and not the deposition rate, and hence the growth rate, of the crystal itself. Of course the rate of segment deposition is an integral part of the theory, as it is the resultant of the deposition and detachment which determines the fold length. If now the nucleation of a new strip along an otherwise smooth crystal prism face is the growth rate-determining factor, the same expressions can be used to describe the rate of crystal growth as are used to describe the nucleation rate.

The measurement of growth rates was one of the earliest aims of crystallisation studies, which was first achieved by dilatometric analysis and later by direct observation of spherulites, once the significant part played by the spherulites had been recognised. The principal finding from these studies was that the growth rate (R) depends on the temperature as

$$R \sim e^{-Q/\Delta T} \tag{3.2}$$

which is the characteristic dependence of secondary nucleation on supercooling (ΔT = supercooling, Q = constant)[66]. The discovery of chain-folded single-crystal growth and the success of the kinetic theory gives an obvious and self-evident structural basis to this deduction of spherulitic growth controlled by secondary nucleation. At the same time it establishes a link between studies on the development of spherulites and growth of chain-folded single crystals with far-reaching implications.

The general expression for the rate of nucleation, and hence for nucleation-controlled growth, is

$$R = R_0 e^{-\Delta F/kT} \times e^{-\Delta G/kT} \quad (3.3)$$

Here R_0 contains constants pertaining amongst others to the geometry of the situation, and the second exponential term contains the work required to form a nucleus of critical size (ΔG). This term involving the work needed to form a nucleus of critical size is a decreasing function of the supercooling, and hence will in itself increase the rate with decreasing temperature. The first exponential term contains the interfacial transport term, i.e. the barrier which a segment has to overcome in order to be attached. The influence of temperature on ΔF (the corresponding activation free energy) is very small at low supercoolings where the nucleation term alone will govern the growth rate. At high supercoolings, however, such as in a viscous medium pertaining for crystallisation from the melt, it may become very large and reverse the temperature dependence of the growth rate. This is the reason for the observed reduction in the growth rate as the glass transition temperature is approached in bulk systems.

Numerical evaluation of the ΔF term exists, but except when the absolute magnitude of the rate is of interest for its own sake, it is only of consequence for the crystallisation behaviour from the melt. A comprehensive discussion of this matter has been given by Price[67], and a salient experimental study including the evaluation of the factor ΔF in the case of spherulite growth in isotactic polystyrene has been presented recently by Suzuki and Kovacs[68].

Confining our attention to chain-folded single-crystal growth, ΔG in equation (3.3) can be written as

$$\Delta G = \frac{4b\sigma . \sigma_e T_m^o}{\Delta H \Delta T} \quad (3.4)$$

by inserting the work required to form a secondary chain-folded nucleus as expressed by the theory[69]. As noted, this contains the product $\sigma \times \sigma_e$, (where σ and σ_e are the respective surface free energies of side and basal faces of the nucleus and b is a cross-sectional dimension of the chain). Even without recourse to absolute values of the growth rate, the measurement of the temperature-dependence provides a test for equation (3.3), and provides a method for assessing $\sigma \times \sigma_e$. This is the principal preoccupation of much of the work on growth rates.

3.5.1.2 Supercooling dependence

Here we shall only be concerned with studies where the growth of crystals is observed directly. Material on this subject is comparatively scarce owing to the difficulty of following the growth of crystals continuously. This could be done in the special case of poly(ethylene oxide), which can crystallise at room temperature, permitting direct observations under the microscope under room-temperature conditions[70]. In this way the temperature-dependence expressed in equation (3.3) may be verified directly. In addition, from plots of log R *v.* ΔT a value for $\sigma \times \sigma_e$ may be obtained from equations (3.3) and

(3.4). It was found that $\sigma \times \sigma_e$ was closely similar for crystals grown from dilute solution and from the bulk (the latter deduced from analogous observations on spherulites). The similarity of the form of the temperature-dependence of the growth rate in spherulites and single crystals has already been indicated (equation (3.2)). We now see further that even the numerical values of the parameters are similar, suggesting that basically identical molecular processes might pertain.

The same conclusion as above has been reached previously[71] from a study of polyethylene by sampling the crystallising solutions at certain intervals. For some accidental reason, the origin of which is now clear (see below), the crystals were all simultaneously nucleated enabling the determination of the growth in successively sampled specimens. Again the temperature-dependence conforms to equation (3.3) which also permitted the determination of $\sigma \times \sigma_e$. This was found to be 600 ergs cm^{-2}, in close agreement with values obtained from subsequent analogous studies on spherulites by the same authors, with the important implications already stated.

The study of growth rates of solution-grown crystals, however, can only be placed on a systematic basis if the simultaneous formation of crystals, and hence a uniform crystal population, can be reproducibly ensured. This was made possible by the self-nucleated crystal-growth procedure[72, 73] to be enlarged upon in Section 3.5.2 enabling systematic growth-rate studies on the individually observed crystals as a function of different variables.

Recent studies on the growth of crystals of isotactic polystyrene suggests the involvement of variables not taken into account by simple considerations. It is known that polystyrene is difficult to crystallise, mainly because the crystallisation is very slow, even at supercoolings of about 150 °C (rather exceptional for crystallisation from solution). Keith, Vadimsky and Padden[74] succeeded in speeding up this growth by using poor solvents. The difference in growth rates amounted to more than three orders of magnitude which cannot be accounted for by uncertainties in the actual values of the parameters nor by differences of any of the thermodynamic parameters featured in equation (3.3) in the different solvents. Conformational differences of the molecules in the solution phase are suspected as being responsible for these large variations in growth rate.

3.5.1.3 Concentration dependence

Another line of enquiry which utilises sampling of uniform size crystals at different stages of growth is the investigation of the dependence of growth rate on concentration. This dependence can be described by

$$R \propto c^{\alpha} \tag{3.5}$$

where c is the concentration and α a constant. According to the first studies on unfractionated linear polyethylene, $\alpha = \frac{1}{3}$, thus the growth rate is a remarkably insensitive function of concentration[75]. Similar results were reported soon afterwards from different sources[76, 77]. Diffusion as a rate-controlling factor could be ruled out by simple consideration of the numerical values. Under these conditions, an exponent α less than 1 must mean that the deposition of molecules does not increase in proportion to their rate of arrival, and hence to their number present in the solution. Such a situation would arise if there is a barrier to the deposition of the chains at the crystal interface so that the number of chains in the solution ceases to be the rate-determining factor. In particular, a crowding of molecules at the crystal face is envisaged so that the concentration there does not depend on the number of molecules in the interior of the solution. Such crowding could arise in our particular system due to the fact that the full deposition of a chain already attached may take some appreciable time, during which the portion still in solution shields the surface from the arrival of new molecules. Alternatively, a significant length of chain portion may stay in solution corresponding to a state of equilibrium for which attachment and detachment rates are equal. It is also possible that long chain segments could be excluded as cilia due to other chains depositing nearby. Such a cilia may fold into the crystal at a later stage but in the meantime would shield the face from newly arriving molecules. The origin of these cilia has been outlined above in connection with Figure 3.16 and is dealt with in the latest theories of chain-folding by Sanchez and DiMarzio[56]. This theory deals in a quantitative manner with the growth rate as influenced by cilia.

All these effects, which give rise to crowding of dissolved chains (or portions thereof), are expected to be more enhanced with increasing molecular weight. In agreement with this picture, the concentration-dependence of the growth rate decreases further when passing on to extremely high molecular weights and under these circumstances α becomes $\frac{1}{4}$[78]. Conversely, the concentration dependence ought to increase for shorter chains. This again is observed. The change is quite dramatic at a molecular weight of 10 000, provided that the fraction is sharp. Under these conditions crystallisation from octane yields $\alpha = 1$ at 94.5 °C and $\alpha = 2$ at 96 °C [78]. A value of $\alpha = 1$ implies that the growth rate is proportional to the number of molecules arriving at the face, thus there is no hindrance, while $\alpha = 2$ implies that the simultaneous arrival of two molecules is the rate-controlling step. The latter conclusion could also be interpreted as meaning that a given chain is not long enough to form a stable secondary nucleus and that for this reason the contribution of a second chain is needed. The fact that this stage is reached only with short chains, and at high temperature, i.e. at low supercoolings, is consistent with this picture.

3.5.2 Self-seeding

Establishment of this phenomenon relies on the combination of the following two factors:

(a) Existence of a crystallisation memory. This means that crystallisation rate and texture depends on the state of crystallinity prevailing before the system was fused or dissolved, in spite of the fact that the molten or dissolved polymer has had all the characteristics of a true melt or solution. Thus recrystallisation isotherms, for example as measured dilatometrically, are dependent on the temperature to which the melt or solution has been heated. Again, when crystallisation of a spherulitic sample is observed microscopically, the spherulites appear at the same positions where they were observed prior to melting[6]. All this means of course that nuclei survive the melting or dissolution process.

(b) Pre-determined nuclei can be destroyed by heating the melt or solution to a sufficiently high temperature (e.g. recrystallisation isotherms become independent of the melt temperature), and can be regenerated by repeating the whole experimental cycle. This of course implies that the nuclei in question are crystalline constituents of the polymer and not extraneous heterogeneities.

The subject studied most extensively originates from some dilatometric experiments on the crystallisation rates of polyethylene from solution[72]. Here a crystallisation memory has been observed up to 13 °C beyond the dilatometrically-recorded dissolution temperature (97 °C), i.e. nuclei must persist within the temperature range of 97–110 °C in diminishing number with increasing temperature. This was supported by observations of the resulting crystals: the crystals were all of highly uniform size, the latter increasing with increasing crystallisation temperature consistent with a reduction in the number of nuclei. This effect proved to be of consequence both for controlled crystal growing[79–81] in service of other studies, and for the study of this particular facet of nucleation itself. In view of the fact that the longest molecules are involved in this nucleation phenomenon[82], self-seeding promises to provide a highly specific tool for molecular-weight characterisation, particularly at the highest end of the usual distribution which is most difficult to study by conventional methods[83–85].

3.6 MOLECULAR WEIGHT FRACTIONATION ON CRYSTALLISATION

Interrelation between chain-folded crystallisation and molecular weight has been clearly apparent for a long time. The principal question to be answered is: does molecular fractionation occur during crystallisation, and if so, in what form? The problem arises equally for crystallisation from the melt and from solution. It can be approached most directly in the case of solutions, as here crystallised and non-crystallised fractions can be physically separated and examined. For crystallisation from the melt, the problem is only amenable along a less-direct route, and in the special case of extended chain crystals formed under high hydrostatic pressure[35, 36].

It is evident that in extreme cases a certain amount of fractionation must occur. If we take a mixture of paraffins and long-chain polymers the latter will precipitate in xylene say below 70 °C, while the former stays in solution. The issue becomes more problematic when a molecular-weight distribution such as that existing in the usual high polymer is considered. The question is: do chains of different length precipitate simultaneously or do the long ones come down first and the rest follow successively? The issue is not self-evident, as crystallisation is a kinetic phenomenon and the higher mobility of the short chains, and hence their greater ability to reach the crystal, has to be taken into account simultaneously with the greater thermodynamic driving force for the longer chains. In spite of such problems a certain amount of fractionation has become apparent during the course of experimentation, and attempts to describe the situation analytically have been made[86, 87]. Kawai[87], in particular, tried to account for the phase separation purely on thermodynamic grounds, but in this case comparison with experiment had to rely on the scanty information available at that time.

Definitive approach has become possible only recently through the work of Sadler[88] who in the case of polyethylene crystallising from xylene separated the components which crystallised from those which remained in solution for a range of crystallisation temperatures, and determined the full molecular-weight distribution of both by means of gel permeation chromatography (g.p.c.). Thus well-founded experimental material became available for interpretation. For the evaluation of the result the thermodynamic method along the lines suggested by Kawai was adopted, but with substantial modifications. The justification for using the thermodynamic approach rested on the observation that the same separation was achieved whether the final state was approached from a less-crystalline state (by crystallisation) or from a more crystalline state (by dissolution) at a given temperature.

As a basic criterion for separation it was considered, following the lines of Kawai, that in equilibrium the thermodynamic potential (μ) of a component of a particular degree of polymerisation P is the same in the liquid (L) as in the solid phase (S) as referred to a common reference state (0) namely

$$\mu_P^{\mathrm{L}} - \mu_P^0 = \mu_P^{\mathrm{S}} - \mu_P^0 \tag{3.6}$$

The left-hand side may be readily expressed in terms of standard solution theory, while the right-hand side is given by

$$\mu_P^{\mathrm{S}} - \mu_P^0 = (\mu_P^{\mathrm{S}} - \mu_P'^{\mathrm{S}}) - (\mu_P'^{\mathrm{S}} - \mu_P^0) \tag{3.7}$$

with $\mu_P'^{\mathrm{S}}$ referring to the monodisperse crystalline phase.

The first term on the right side involves the entropy of mixing in the solid. For this, the reasonable assumption was made that each molecule is a chain-folded ribbon which is to be mixed with others randomly throughout the crystal. The second term in equation (3.7) is the free energy difference between the monodispersed liquid and crystal.

By developing these expressions a partition is obtained for any degree of polymerisation between liquid and solid phase which may be readily expressed in terms of the fraction of material, e.g. in the liquid phase, (f_{L}), a quantity directly comparable with experiment. The expression obtained for f_{L} was found to be of the correct form and could be fitted to the experimental f_{L} *v.* P curve virtually exactly using molecular parameters which were as reasonable as could be expected at the present state of knowledge. The surprising feature of this agreement is that it implies an equilibrium between crystal and solution. This may appear physically unrealistic as it is reasonable to expect that the crystal interior would be sealed-off by newly-depositing layers. Nevertheless, any adjustment to allow for this situation upset the observed agreement.

Thus, we have a situation where the subject of fractionation is at last extensively documented and the results can be accounted for quantitatively and predicted by analytical expressions. Nevertheless, the basis on which the interpretation rests is contrary to *a priori* physical expectations, a point requiring further evaluation.

3.7 THE FOLD PLANE

3.7.1 Sectorisation

Growth of the lamellae occurs predominantly along their lateral faces. From the start, growth was visualised as a folding-up of molecules along these faces forming folded ribbons lying parallel to the prism faces (Figure 3.12). It was immediately obvious that this should lead to a sub-division of the crystal into distinct sectors, there being as many sectors as there are prism faces distinguished by the plane of folding (Figure 3.12). When first proposed this situation had no precedent in crystal growth, and consequently when it was verified it was considered convincing evidence for the above model.

The evidence has been reviewed elsewhere[3]. Nevertheless, in view of the uniqueness of this feature, the essentials will be recapitulated. In many cases the distinctness of the sectors becomes visible by virtue of a hollow pyramidal morphology. In fact the crystals are like tents, the different sectors forming the panels. Under exceptional circumstances these hollow pyramids can be made directly visible. In most instances, however, these pyramids flatten during the normal course of sample preparation. In the simplest case, the crystal will have to wrinkle as a natural consequence of this flattening. Amongst others, characteristic central pleats appear (Figure 3.2), a result which can serve as an aid to the re-constitution of the original pyramidal morphology. In fact it was through these pleats that the pyramidal morphology was first inferred[3].

For a further appreciation of the actual images seen, some knowledge of the molecular features is needed. It has been established in a variety of ways that the chain is parallel to the original pyramid axis and identically so throughout the whole crystal. This means that facets of the pyramids are oblique with respect to the chains, i.e. that the different folds must be at different heights, in other words that they are staggered. The sectorisation can then be understood as visualising different staggers in different sectors, the stagger being the same but reversed in sign in sectors which are crystallographically identical but diametrically opposed in a layer (e.g. 110 and $\bar{1}\bar{1}0$ in Figure 3.2). This in turn has been interpreted as resulting from a systematic stagger due to special space requirements of the fold. More precisely, the folds protrude beyond their projected stem area so that they occupy some of the cross-sectional area of the neighbouring stems. As a result such stems, together with their folds, need to be shifted up or down, which if it occurs systematically gives rise to the oblique basal face observed. The slopes of the pyramids arising in this way are in the range of 25–35 degrees.

Now if a pyramidal crystal collapses and the structure remains unaltered, then the basal plane will lie flat on the substrate with the consequence that the chains will be inclined and in different directions within the different sectors. As sectors with different chain inclinations will satisfy different diffraction conditions, the conditions for sector distinction by diffraction contrast have been established. An example of this is apparent in Figure 3.2.

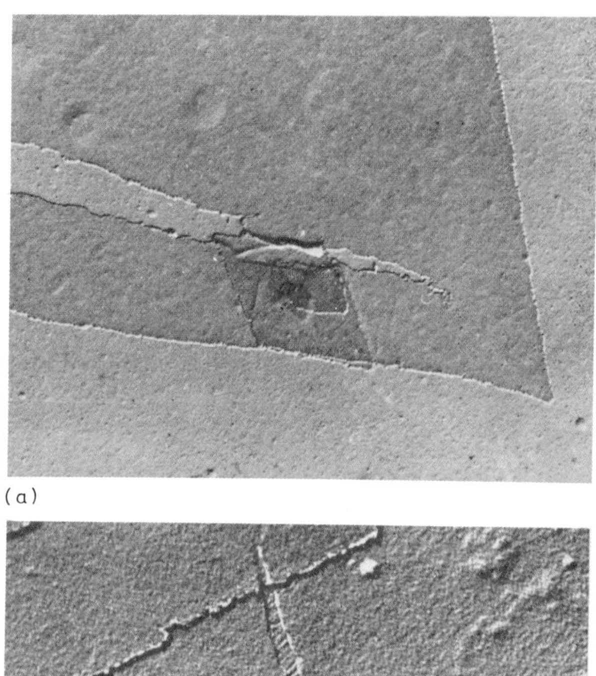

(a)

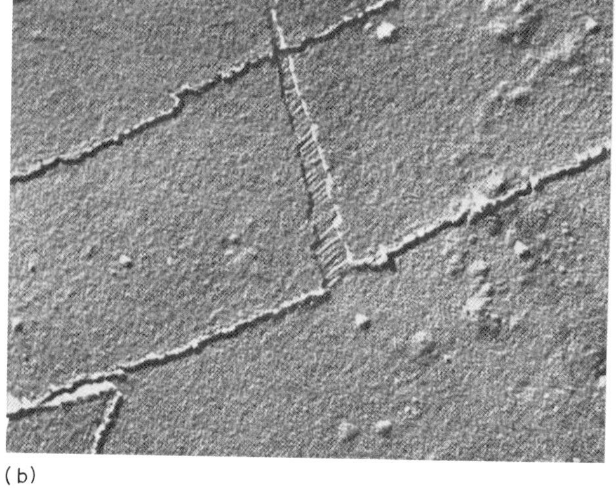

(b)

Figure 3.17 Sectorisation as revealed by fracture in a single-crystal of polyethylene. (a) Fracture parallel and (b) perpendicular to the {110} plane bounding the corresponding sector. The clean fracture in (a) and the threads pulled in (b) should be noted.
(Lindenmeyer[90] micrograph by Holland. By courtesy of Dr. P. H. Lindenmeyer and Interscience)

The hollow-pyramidal morphology is only the simplest, and hence the most illustrative, consequence of basal plane obliquity and of the underlying fold stagger. Other, more complex, examples have been reviewed previously[3]. Neither is the mode of flattening described above the only way a pyramid or other non-planar crystal types can collapse. An alternative way of flattening involves the removal of the stagger, i.e. by plastic deformation within the layer itself. Clearly, if the chains can slide past each other within the crystal until the folds are all level, the pyramid will become flat within an unaltered basal area. In this case neither morphological nor diffraction-contrast evidence will indicate that we have had a pyramid to begin with.

Some crystals do grow apparently flat to begin with, but nevertheless display the sectorisation principle in other more subtle ways. Thus very shallow pyramids (slope 1–2 degrees) can arise through shear of the sub-cell, which is of different sign for the different sectors with the folds remaining level, as opposed to much steeper pyramids (slope 20–35 degrees) arising from fold staggering[89]. The detection of these relies on more sophisticated observations not to be reviewed here[1,89].

Finally, crystals can grow genuinely flat but nevertheless still reveal sectorisation by other effects. There is no clear-cut criterion for the occurrence or absence of a non-planar morphology and the underlying fold stagger. Nevertheless, accumulation of the evidence suggests that at least in the much-studied polyethylene system a very low molecular weight material, where a molecule only contains a few folds, will grow in the flat morphology. It is possible that the alternation of folds with chain ends at the fold surface ensures that there is enough space to accommodate the additional area required by the fold.

Amongst numerous other manifestations of sectorisation, one of the most striking ones is the distinction of sectors by cleavage properties. The crystals cleave readily along planes which are parallel to the prism faces in each sector. Thus the cleavage planes form a concentric system[90]. This is one of the clearest indications for a preferred fold direction. Figure 3.17 contrasts cracking along a cleavage plane with that along a direction perpendicular to it, the pulling of threads in the latter case being clearly visible.

3.7.2 Influence on crystal habits

The habit problem in a monolayer polymer crystal is automatically divided into two aspects by the different intrinsic characteristics of the basal plane and the prism faces. The former is directly related to folding, and as such has been the subject of the discussion so far. The prism faces are determined by the packing of the straight stems. As in the case of polyethylene, this is identical with that in paraffins: to a first approximation one would expect the prism face development in polymer crystals to follow the same pattern as in paraffinoid substances. For example, in the case of polyethylene the basic pattern emerges that the {100} face increases continuously at the expense of the {110} face as either the temperature or concentration or both are raised[1].

Descriptions of the external prism face features are numerous (e.g. Heijde[91], and Patel[92]) and cannot be fully reviewed here. It will only be recalled that a situation involving a new principle arises when the relation of the fold to the sub-lattice is considered. As discussed above, the prism face developments define the fold plane. While in a paraffin there is no structural limitation to the way in which prism faces develop, with folding the continuity of a folded ribbon when passing from one sector to the next creates a special issue even in a planar crystal[93]. In addition, the matching requirements of the sectors arising from the pyramidal and related morphologies have also to be met (the different type of panels of the tents have to match)[1,23]. Thus we have a case where the fold surface and prism face developments interact.

3.7.3 The problem of elongated crystals

All the foregoing remarks were concerned with crystal growth with constant shape. If the prism face habit changes during the course of growth, say due to change in concentration, the sector shapes will be modified in accordance with the altered prism face development. Thus we can have kinked or curved sector boundaries. In addition new sectors can develop or the development of existing ones may cease, resulting in buried sectors within the crystal. All these are variants on the theme of sectorisation which will not be pursued further here.

A completely new issue arises, however, in case of lath- or streamer-type crystal growth where the crystals do not grow at constant shape but only at the tip. This is a particularly important habit form, as it is the characteristic mode of crystallisation from the bulk and is the form which leads to the development of spherulites. This raises the question of the plane of folding in such crystals. The obvious suggestion is that folding occurs along the leading facets at the tip ($\{110\}$ in case of Figure 3.7) and spherulite growth is usually represented in these terms[69]. Nevertheless there are effects which suggest that folding may occur along the longest side-faces of the laths. Amongst these is the clean splitting of lath-shaped polypropylene crystals, which occurs only along the lath direction[94] (and not concentrically as implied by Figure 3.12, (which depicts crystals grown with constant shape, or parallel to the tip faces if these were the fold planes)[69]. This involves a new issue, namely the possibility of chains folding into the solution (as opposed to folding along the prism face). Clearly this would only be possible if several chains folded up simultaneously. The fact that in polyethylene lath-shaped crystals arise in concentrated solutions and from the melt is at least consistent with this possibility.

3.8 THE FOLD SURFACE

3.8.1 The general issue

The subject of the fold surface is the most problematic and controversial in the whole field, and is central to the entire subject of polymer crystallisation. It has been extensively reviewed by the present author on previous occasions[1,2]. New evidence however has since become available. Subjects under this general heading include such interconnected issues as the structure of the fold, degree of crystallinity of the crystal, nature of the crystal defects and of the amorphous material within the crystal, the role of chain-ends and the effect of finite molecular lengths in the case of short chains, structural origin of the melting range, some features of long-spacing changes on heating, degradation pattern of single-crystals and several others. The lack of overall agreement which will be apparent from this survey is largely due to the fact that there is no model for the fold surface which is equally satisfactory from the point of view of all the above topics. The diverging views which feature

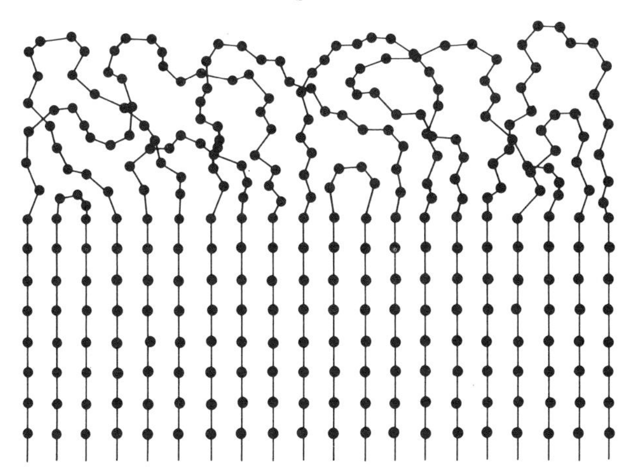

Figure 3.18 Model of an intrinsically disordered fold surface in the representation of Fischer[100] after Flory[170]
(By courtesy of Professor E. W. Fischer and Dietrich Steinkopt Verlag)

in the literature arise more often from the way different investigators weigh the various pieces of evidence than from actual faults in experimentation.

To simplify the discussion the essential issue can be stated in its most extreme form as follows: is the fold surface crystallographically defined down to the molecular level (as in Figure 3.12) and if so what is its structure? Alternatively, does the fold surface represent a completely disordered and hence amorphous layer (e.g. Figure 3.18) as suggested by some evidence to be listed below? If so, how is this to be envisaged in molecular terms? It must be stated at the onset that the truth is not likely to be quite so clear-cut and the alternatives not necessarily so mutually exclusive.

3.8.2 On disordered fold surfaces

3.8.2.1 Evidence for disorder

The amount of disorder is usually assessed by traditional determinations of crystallinity. The methods themselves are based on measurements of specific volume, heat of fusion, x-ray diffraction, infrared absorption and n.m.r. or other properties which have different values for the amorphous and crystalline states and hence can furnish a value for the degree of crystallinity (e.g. see Miller[95]). The results will not be surveyed here in any detail. A short account has been given previously by the present author[1] and there is a recent comprehensive review by Mandelkern[54] which should be consulted for further particulars.

The essential point to summarise here is that all polymeric materials (with the exception of some extended-chain crystals obtained under pressure[35, 36]) display an appreciable amount of deficiency in crystallinity. This has always been a familiar situation for melt-crystallised material. In the case of solution-grown single crystals, however, it has led to the apparently paradoxical concept of a partially-crystalline single crystal. Thus, the usual polyethylene single crystals are only about 75–90% crystalline and the remaining 25–10% is usually referred to as the 'amorphous' content.

The next question is where and how is this amorphous content accommodated by the crystal? In view of the fact that we have a morphologically defined crystal entity we can ask this question in more concrete terms than was possible earlier in the case of polycrystalline bulk material.

3.8.2.2 On the possible site of disorder

It is a well-established fact that a polymer crystal cannot be perfect even within its interior. Although the crystal layer may appear morphologically perfect over regions of many micrometers, the x-ray reflections due to lateral spacings $\{hk0\}$ are comparatively broad, indicating that the crystal does not diffract coherently throughout. From the crudest estimate, using the broadening of the interference function due to limited crystal extension (Scherrer formula), a lateral dimension of about 300 Å is obtained for the coherently-scattering domain in a polyethylene single crystal. More rigorous treatments due to Hosemann and collaborators[96], which separate the line-broadening into components due to lattice distortions (paracrystallinity) and to limited crystal size, arrive at a block-structure of essentially the same magnitude. These authors are of the view that the crystal contains blocks delineated by twist boundaries.

It is common knowledge that for longer fold lengths, as achieved by heat annealing, the $hk0$ reflections sharpen, indicating an increase in the coherently-scattering domain size not only along the chain but also in the lateral direction. According to rigorous line-shape analysis[97], the domain shape remains unaffected but its lateral size increases in proportion to the crystal thickness. It is suggested that the fundamental block conforms to an equilibrium shape corresponding to the Wulff criterion.

The nature of the block walls is an open question. It is likely to be the site of a variety of crystal defects (jogs, kinks[98, 99], which in turn give rise to dislocations). Important as this problem of crystal defects may be, it has become apparent for a variety of reasons that it is not likely to be a major contributory factor to the crystallinity defects under discussion. It is implied by correlations observed between the fold length and the density[38] that the deficiency in crystallinity lies principally along the fold surface. Through a similar procedure, Mandelkern[54] arrives at the explicit conclusion that defects in the interior do not contribute significantly to the measured enthalpy deficit. Consequently it is the fold surface to which we shall turn to next in our quest for the origin of the amorphous material.

3.8.2.3 Amorphous crystalline layer structure from low-angle x-ray diffraction

As stated earlier, the angular position of the low-angle x-ray maxima define the thickness of the lamellae. This includes both the crystal lattice and the interface which produces the electron-density fluctuation responsible for the scattering. In itself it does not say how the total thickness is divided into these two components. Nevertheless such information can be extracted by detailed analysis of the intensity distribution and of the absolute scattered intensity. The former enables the determination of the correlation function which gives a measure of how the total periodicity is divided into its components, a two-component layer structure being assumed. The latter, in addition, allows the determination of the electron-density differences involved. The electron density in the crystal lattice being known, the knowledge of the corresponding quantity at the interfaces follows.

It is not possible to give details either of procedure or results of the extensive work on this subject, only the broadest conclusions will be quoted. The results performed on ethylene in the bulk form[102, 103] and in the form of single crystals[100, 101] are fully consistent with a two-phase structure and in fact require such a structure. The electron density obtained for the interface structure corresponds to that calculated for the amorphous phase. The amorphous characteristics of the interface are further brought out by the variation of the scattering power with temperature. By these means a glass transition may be diagnosed at the surface of polyethylene single crystals[104].

3.8.2.4 Melting behaviour in relation to the fold surface structure

(a) *The melting range*—It is general knowledge that crystalline polymers possess a melting range instead of a sharply defined melting point. The following factors may be responsible for this effect either in isolation or in combination.

(i) Fractionation according to chemical regularity or according to molecular length, or both together, where the different species segregate in different crystals or crystal portions. These crystals or crystal portions will then have different melting points giving rise to a melting range, as observed.

(ii) Crystals with different thicknesses, hence fold lengths, are present where those with shorter folds melt at a lower temperature according to the relation

$$T_m = T_m^0\left(1-\frac{2\sigma_e}{L\Delta H}\right) \tag{3.8}$$

(for symbols see Section 3.4.3.2).

This polydispersity in fold length could arise by non-isothermal crystallisation. Even when all crystals have the same fold length to begin with, the fold length could increase non-uniformly when the crystals are heated during the course of the usual melting point determination.

(iii) Lamellar crystals melt down gradually from the surface downwards. The possibility of this particular effect is of special concern to us here as it is intimately linked to the problem of what the fold surface is like.

(b) *Interfacial melting* – Arguments relating to a melting process proceeding from layer surface downwards are of two kinds. Some are based on the interpretation of experimental data, while others involve *a priori* theoretical predictions.

Experimental indications – It has been observed that the intensity of the low-angle x-ray reflections increases reversibly on heating and cooling of the crystals[46, 100, 105]. (The crystals had been appropriately stabilised against irreversible fold-length increase by prior heat-treatments at temperatures which were higher than those reached by the above temperature cycling.) This reversible intensity increase must be due to an increase in the electron density difference between crystal core and interface responsible for the reflection. Effects due to differences in thermal expansion having been taken into account, a notable change in electron difference is left, and the suggestion is that this is due to an increase in the amorphous component along the crystal surface. Accordingly the crystal should melt from the surface downwards.

The uniqueness of this inference, however requires renewed scrutiny in view of the observation that the $00l$ reflections in paraffins (which are analogous to the low-angle reflections in polymers, except that in paraffins they are related to the extended chain-length and correspond to the true lattice periodicity) display a similar reversible intensity variation on temperature cycling[106]. In this case the effect in question cannot be due to the formation of amorphous material, and creation of interfacial gaps due to motion of whole chains along their lengths[106] and formation of kinks along the chains within the lattice are invoked[107]. Clearly these possibilities also apply to polyethylene. The distinction between these two effects which are potentially capable of affecting the low-angle x-ray intensities, namely surface melting – a possibility specific to chain-folded polymers – and increased chain mobility within the lattice which does not require the existence of a fold surface, is a subject of current topicality.

A priori arguments – *A priori* reasoning in support of the existence of interfacial pre-melting is based on the following principle. Consider a large loop at the fold surface. The configurational entropy of such a loop will be reduced because its ends are fixed at the two entry points in the crystal which is equivalent to saying that the chain is in tension. If we make the loop longer while keeping the end-separation fixed, the tension is decreased hence its entropy increased. As the end points are fixed, the loop will be enlarged only if the crystal is thinned. Thus we lose crystal but at the same time we gain entropy due to the increase in the size of the loop. This additional entropy ΔS_D adds to the usual melting entropy ΔS associated with detachment of segments of an otherwise unrestricted chain, and the combined entropy change $\Delta S + \Delta S_D$ may more than compensate for the increase in enthalpy due to the reduction in the amount of crystal.

Conditions for this interfacial amorphous–crystalline equilibrium have been worked out in considerable quantitative detail[108–110]. (It should be remembered that the equilibrium is confined to within a particular lamellar structure which itself is not in a state of true equilibrium, the latter corresponding to crystals with full chain extension.) The entropy change ΔS_D for a given loop length decreases for decreasing separation between points of entry in the crystal, hence the above pre-melting phenomenon should be less pronounced for loops with closer re-entry distances. For the closest re-entry distances, and certainly for adjacent re-entry, the effect should not only be absent but, on the contrary, ΔS_D should be negative. Thus the crystal surface consisting of such loops should not only not pre-melt but should be superheatable with respect to the body of crystal. (This is already implicit in Zachman's first paper[108], a qualitative explanation of it is contained in Reference 1.) In conclusion, it might appear that interfacial pre-melting, as far as it occurs, requires not only loose loops but loops of widely-separated re-entry points, typically 30 Å apart. This would invalidate a model of the kind shown in Figure 3.12, which is in contradiction with a variety of facts, in addition to making the model of simple chain deposition on which the highly successful kinetic theories are based inapplicable.

A model which reconciles these viewpoints considers an assembly of loops with adjacent re-entry but of differing contour lengths (as opposed to treatments in References 108–110, which consider lengths of only one kind at a time). This assumption gives rise to an additional entropy term which takes into account the fact that loops of different lengths may be arranged in different ways[111]. Due to this term, a broad melting range associated with interfacial melting should also be possible when the re-entry points are close, including the case when they are adjacent. The fold surface here implied would arise if the stems of different folds terminated at different heights which is predicted by theory[53], and is supported by experiment (Section 3.8.3.2(c)).

By focusing attention on adjacent re-entry it is not implied that non-adjacent re-entry does not occur. In fact by the considerations in Figure 3.16 such loops must arise in certain numbers even on the most extreme view of a regular, adjacently re-entrant fold structure. Increasing re-entry distance increases the configurational restrictions. Consequently, such loops with widely-separated entry points will be expected to have a large contribution to interfacial pre-melting, and increasingly so with increasing re-entry distance. Thus a few loops with widely-separated ends could have a disproportionately large effect in producing the symptoms of pre-melting along the crystal interface[111].

3.8.2.5 Surface swelling

Four entirely independent methods have revealed that the surfaces of solution-grown polyethylene crystals can swell when brought into contact with appropriate liquids.

N.M.R. spectroscopy – It was observed that the proton resonance absorption line, which in polymer single crystals is broad as expected for a crystalline solid, develops a narrow component when exposed to swelling agents[112–114]. This implies increased mobility of chains consistent with swelling of a portion of the crystal. As true crystals themselves will not swell, the existence of amorphous material is indicated.

Liquid uptake – Swelling could be directly ascertained by determining the weight of sedimented mats before and after exposure to swelling agents[115]. Appreciable liquid uptake was observed.

Long period changes – In mats of solution-grown crystals the long period, as determined by the low-angle x-ray scattering, increased reversibly on contact with certain liquids[116]. As a well-defined limiting value was reached, this is not due to one stage in a disintegrating sediment but to a well-defined amount of liquid uptake between the crystal interfaces. The effect increased with the known swelling power of the liquid. The largest change was 35 Å observed with decalin. This is far larger than could be associated with absorption, and implies swelling of material along the crystal surface. The amount of swelling also depended on crystallisation conditions and on molecular weight. The latter effect was particularly pronounced: it varied from 5 to 35 Å on going from molecular weights of 3×10^3 to 10^6.

Electron microscopy – Variations along the crystal surface could be directly observed by means of electron microscopy using a special decoration technique[117]. This consisted of vacuum-depositing gold on the crystal sufficiently thinly so as not to cover the surface uniformly. At this stage the crystal surface was seen to be covered with black dots which corresponded to gold particles. These particles could decorate edges and steps with great sensitivity which in itself provided an important method for the study of surface topography. Our present concern however is the grain density along the step-free surface. On the whole, this is uniform although small differences between crystal periphery and interior can sometimes be seen when the decoration pattern is always coarser in the central portion. On subsequent flooding of the decorated crystals with a swelling agent (xylene) the decoration pattern can coarsen drastically (Figure 3.19) which may be associated with mobilisation of the chains along the crystal surface. In the usual crystal, this effect is not uniform throughout: the coarsening is pronounced in the interior and may be absent along the periphery (Figure 3.19). Detailed quantitative work involving fractions (Section 3.6) characterised by g.p.c. revealed that the mobilisation behaviour is a function of the molecular weight[118]. Thus fractions below 8000 do not show the effect, and above it they do so to an increasing extent with molecular weight. Accordingly the differences in Figure 3.19 are associated with fractionation (Section 3.6), the finer texture corresponding to the low molecular-weight material which deposits along the periphery.

From a comparison of the results of the four methods the following pattern emerges. N.M.R. spectroscopic and liquid-uptake determinations reveal swelling but do not define specifically how the portion of the material capable of swelling is related to the crystal. This is defined by low-angle x-ray scattering and electron microscopy which identifies the crystal surface as the location of this material. In addition, the latter methods reveal that the ability to swell is not an invariant property but depends on a number of factors, in particular on molecular weight. This has two main implications. Firstly, it is consistent with a picture such as that depicted in Figure 3.16 which accounts for the origin of surface elements with extra looseness. The formation of these is expected to be promoted by longer chains. Secondly, it also implies that there is no unique fold-surface structure. This is most conspicuously demonstrated by electron microscopy, as here different kinds of regions can be seen even within the same crystal layer (Figure 3.19).

Quantitative interpretation of all these effects is more problematic.

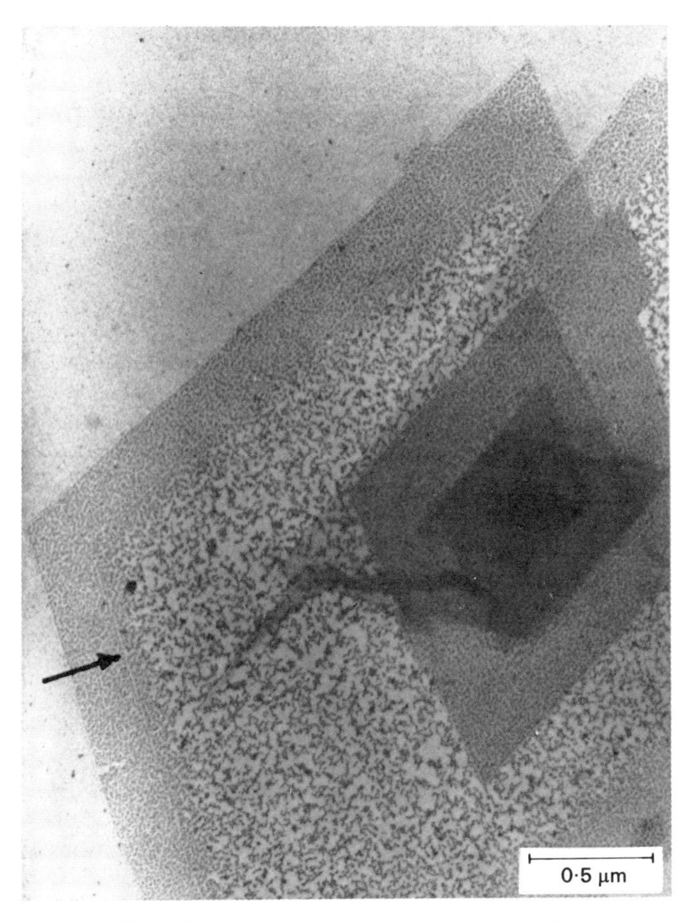

Figure 3.19 Polyethylene crystal, decorated with gold and subsequently flooded with xylene revealing mobilisation of the fold surface in the layer interior (coarse decoration pattern). The finer decoration pattern at the periphery and in the overgrowth layers characterises the unmobilisable surface; this was the pattern over the entire crystal before flooding. A step due to deliberate change in growth temperature is also brought out by the decoration (arrowed)
(From Blundell, D. J., Sadler, D. M. and Keller, A.[118])

Only n.m.r. spectroscopy can in principle give direct information on the amount of polymer involved, but there are serious problems concerning line separation. The weighing and low-angle x-ray techniques can measure the amount of liquid involved; the x-ray technique in particular is very convenient for comparing specimens and grading them with meaningful figures.

Electron microscopy is qualitative but a most sensitive and rapid diagnostic tool.

3.8.2.6 *Selective surface degradation*

It has been observed that strong oxidising agents such as fuming nitric acid[114, 119, 120] and, more recently, ozone[121] can selectively digest the crystals. The technique was first used for morphological studies in the bulk phase[119] where it served to disperse the massive sample into its characteristic lamellar elements (see Figures 3.7 and 3.8). In fact this was the first really convincing demonstration of lamellae within the interior of the bulk material.

The technique has been applied to solution-grown single crystals as an aid to the study of their structures[114, 120]. It was observed that the crystal was attacked along the basal planes and was thinned down to a certain extent, after which the attack stopped. This is usually attributed to a highly vulnerable surface layer and a resistant crystal core, and is considered as chemical evidence for the existence of an amorphous–crystalline sandwich structure. It has become apparent recently that this classification cannot be maintained on the basis of the above evidence alone, as the reaction can also be stopped by the accumulation of reaction products without necessarily involving a structure of intrinsically greater chemical resistance[122]. It follows that quantitative evaluation of the true surface structure by this method relies on a more sophisticated approach to be outlined in Section 3.8.3.2(c).

3.8.3 Structural definition of the fold surface

3.8.3.1 *On the structure of the fold*

(a) A priori *considerations*—As already stated in Section 3.8.1, in view of the regular appearance of the crystals there was at first no reason to assume anything but crystallographic regularity along the fold surface. The pyramidal nature implying a systematic stagger of the folds[3, 4] and the crystallographic interaction between layers, as revealed by the occasionally observed dislocation networks at the layer interfaces[123] (Figure 3.20) (see later), were all consistent with crystallographic regularity, and hence a well-defined fold structure. Attempts therefore have been made to define the energetically most-favourable fold configuration in terms of atomic positions in a chain, other developments notwithstanding. These attempts range from simple tracing of hairpin-like carbon–carbon paths in a diamond lattice so as to make the stems equivalent to the planar zigzag of polyethylene within a crystal lattice (Frank, quoted in Reference 124) to the application of the more sophisticated methods of conformational analysis[125, 126]. The latest analysis, which includes not only the rotation around single bonds but also variations in bond angles[126], leads to particularly simple fold structures where most of the departures from the planar zigzag (hence most of the fold) are within three to four C—C bonds, with only much smaller and rapidly diminishing departures spreading into the body of the crystal.

(b) *Experimental attempts to define the structure of the fold* – The possibility of determining the structure of the fold, if such a unique structure exists, in terms of atomic positions is presently remote. Nevertheless other, rather less direct, experimental investigations have been made with this objective in mind.

One such method relies on infrared spectroscopy and has been applied by Koenig and his collaborators[127–129]. We must distinguish here between two distinct stages. Firstly, the establishment of a correlation between infrared

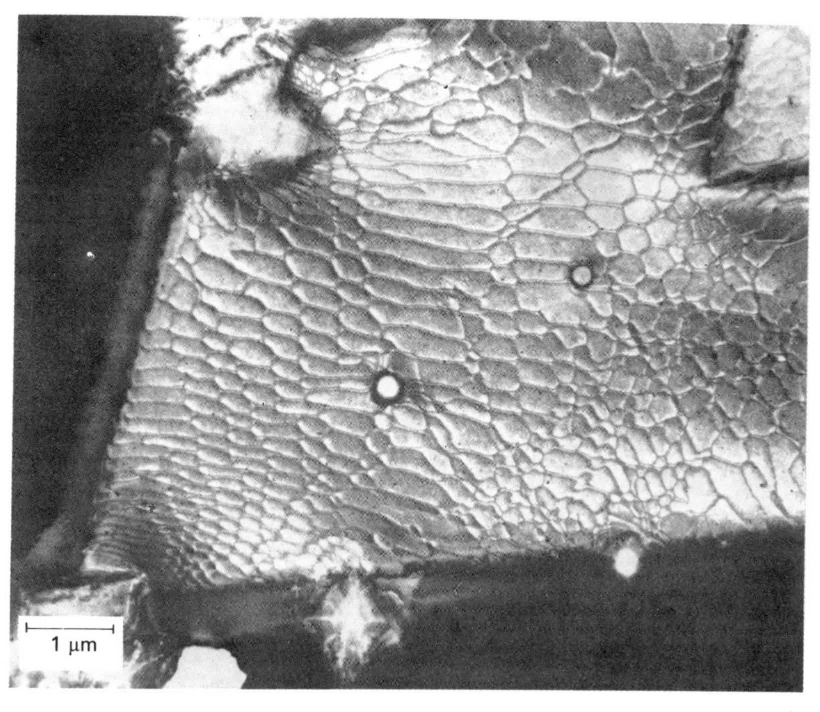

Figure 3.20 Dislocation network between two chain-folded polyethylene crystal layers Bright-field electron micrograph; the contrast is reversed on printing.
(Holland, V. F. and Lindenmeyer, P. H.[173], by courtesy of Dr. P. H. Lindenmeyer and the American Association for the Advancement of Science)

spectra and chain folding, and secondly definitive deductions regarding the structure of the fold itself. The mere fact that correlations between the strength of certain spectral lines and the number of chain folds do seem to exist already implies that the structure of the fold is specific enough to influence the spectrum. This is an important inference which in itself is not affected by possible uncertainties in the assignment of the spectral features in question to specific bond configurations.

The correlations referred to are based on the following general procedures. Crystals of different fold periods, as assessed by x-ray low-angle scattering, are prepared and their i.r. spectra examined for any correlation between spectral features and number of folds, bearing in mind that thicker layers mean fewer folds. Two kinds of correlation have been found in different polymers: correlation of intensity ratios of assigned bands and the correlation of actual intensities of hitherto unassigned bands with the inferred number of folds. The former is the only one which can be found in polyethylene[127] and corresponds to the intensity ratio of two bands, one arising from two successive bonds in *gauche* configuration and the other from random *gauche* bonds. It is argued that major differences in the intensities of these two bands are expected between a regularly folded structure and a random chain configuration.

The second kind of correlation is more specific as it associates the previously unassigned band, which is related to the inferred fold content, to a feature of the fold itself. Such assignments could be made in polyamides[128] and polyethylene terephthalate[129]. The establishment of the correlations is not always quite straightforward as two variables are involved: the number of the folds and the perfection of the folds. Both can vary when changing the fold length, e.g. by heat treatment, and the corresponding effects have to be identified and separated in the spectra.

Another attempt to define the fold structure relies, to a certain extent at least, on specific chemical reactivity. The most prominent example is the work on poly(vinylidene chloride). Here single crystals are reacted with pyridine[130]. This reaction proceeds in two stages at successively reduced rates, corresponding to the successive removal of one and then a second hydrogen per fold, with the consequent formation of pairs of conjugated double bonds. This immediately indicates that two closely situated hydrogens, and only such hydrogens, are involved, which allows some specific statements of the fold structure. Other experiments along these lines include the bromination of polyethylene single-crystals[131] and the epoxidation of poly(*trans*-1,4-butadiene) crystals[132].

3.8.3.2 *Information on the fold stem*

Evidence under this heading relates more directly to structural arrangements concerning the straight stems. In most cases this is also of consequence for the structure of the folds, as information on the stem length and stem separation defines the folds, at least within certain limits.

(a) *Spectroscopic evidence for adjacent re-entry and on fold plane type*— The subject under this heading has been mainly studied by Krimm and coworkers[133–135] and represents perhaps the most definitive experimentation on the whole topic. The effects featuring in these studies are particular infrared absorption bands in polyethylene which show splitting due to interaction between vibrations of the two chains in the unit cell, such as CH_2 bending with components at 1463 and 1473 cm^{-1}, and CH_2 rocking with corresponding bands at 720 and 731 cm^{-1}. The studies in question also employ deuterated polyethylene where the above bands will, of course, have appropriately altered frequencies but otherwise will be similarly split. As the interaction producing the split requires interaction between equivalent species in a mixture of polyethylene and perdeutero-polyethylene, the split for each species will depend on their nearest-neighbour environment. With unconnected chain traverses, as in a mixture of paraffins, this nearest-neighbour environment will only depend on the proportion of the two constituents. However, when the straight stems are molecularly connected, as in a chain-folded system, it will also depend on the type of folding; conversely it should be possible to use the splitting behaviour of the bands as indicator of how the chains fold. The expected effects for random mixing and for various kinds of folding have been worked out theoretically and then compared with experiments.

Predictions for random mixing were tested on paraffins, and found to be in complete agreement with expectation but at variance with the observations on polyethylene. The results on polyethylene were in complete quantitative agreement with folding which is adjacently re-entrant. Thus even when the concentration of one of the polyethylene species was very low, band splittings such as are expected from isolated chain-folded ribbons of this species were observed, where each stem has two neighbouring stems of its own kind.

This method can also distinguish between folding along {110} and {100} planes. In fact, when a single-crystal sample, possessing initial {110} folds, was heat-annealed, the fold plane changed into {100} by this test. {100}-folding was also found to be characteristic of material crystallised from the melt, adjacent re-entry being necessarily implied. This is a significant finding, as adjacent re-entry in the melt is most debatable by *a priori* reasoning.

The above conclusions rely on the absence of phase separation of the deuterium- and proton-containing phases. The possibility of such a phase separation has been pointed out recently in view of observed melting-point differences[55]. Krimm *et al.*, however, maintain that such a phase separation can be identified when it occurs and does not affect the above evidence[136].

(b) *Spectroscopic measure of stem length*—This Section refers to a recent but very promising development. It has been known for some time that the fundamental symmetrical, accordion-type stretching vibration of a straight paraffin chain is Raman active, and gives rise to a Raman frequency which is inversely proportional to the length of the coherently vibrating chain[137]. Thus the appropriate region of the Raman spectrum can give information on the straight-chain length in paraffins. It is apparent that if applied to sufficiently long chain lengths this method should also tell us how long the straight stems are in a chain-folded crystal of polyethylene. However, in view of the inverse proportionality with chain length, this brings us into the very low frequency range of the Raman spectrum ($\sim$10–30 cm^{-1}). With recent advances in laser technology this kind of work has just become possible. The first work reported[138] has already obtained a direct correspondence between the straight-chain length as assessed spectroscopically and the crystal thickness as measured by low-angle x-ray scattering for a series of crystals of different fold lengths. The spectroscopic values are slightly lower than the ones obtained by x-rays (by about 20 Å per repeat period, i.e. 10 Å per fold surface). Whether this small difference is due to an 'amorphous' portion of the fold, or to more detailed considerations in the decoupling of the vibration is a matter of current investigation[139].

(c) *Stem-length distribution by selective degradation*—The use of selective degradation as an aid to structural studies of polymer crystals has been referred to in Section 3.8.2.6. It was stated there that under the influence of concentrated HNO_3, and more recently ozone, the crystals are gradually thinned[114, 119, 120] (see also Ref. 1). In this way the fold surface is shaved off until extended dicarboxylic acids of uniform length which traverse the full thickness of the thinned layer are obtained and are correspondingly shortened as the layer thins further[122]. Our principal interest here will be the molecular-weight distribution at the stage when folds are being cut.

It has been observed from the g.p.c. chromatograms that molecules do not shorten uniformly but that a peaked molecular weight distribution develops[140–142] (Figure 3.21). This immediately suggests that the folds are cut,

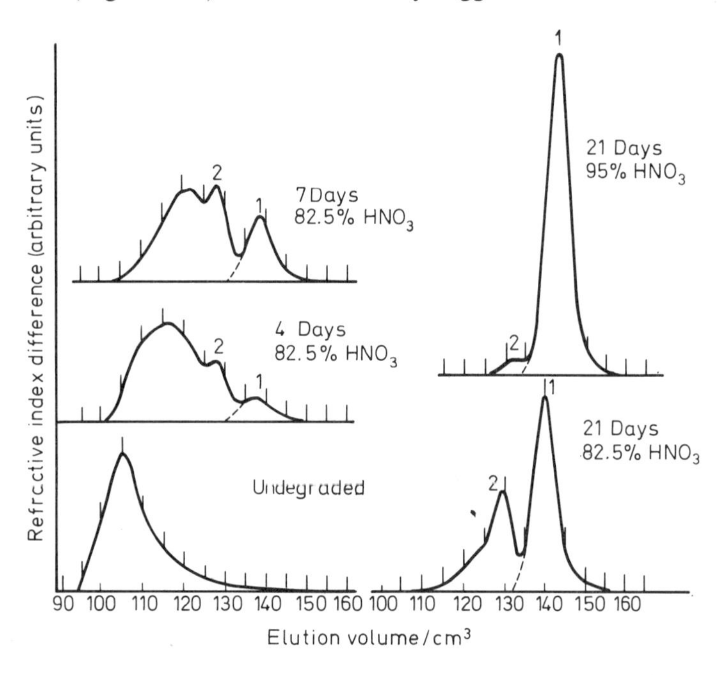

Figure 3.21 A series of gel permeation chromatograms obtained from polyethylene crystals in increasing stages of degradation with nitric acid. The detected refractive index difference, a measure of the amount of material, is plotted against elution volume (larger elution volume corresponds to lower molecular weight)
(From Keller, A. and Udagawa, Y.[122], by courtesy of Interscience)

and single, double and higher multiple-chain traverse lengths arise defining a quantised molecular weight distribution. The peaks change in relative height, the lowest molecular weight peak becomes gradually more dominant until it is the only peak remaining. This peak corresponds to the unfolded dicarboxylic acid referred to above. The quantised distribution, and what it entails, is an important observation as it provides chemical proof of chain folding. Thus, the appearance of a peaked chromatogram in degraded polyethylene can be used as a diagnostic tool for the chain-folded structure even in its broadest aspect. Such tests have in fact been performed on a range of sample types including annealed crystals and melt-crystallised material, both in its unoriented form and as drawn fibres, when chain-folded structures could thus be defined[142–144]. Only a schematic representation of the principles will be attempted here with regard to this still developing subject.

Once in possession of a calibration[145], the molecular weight, and hence the extended length, corresponding to the g.p.c. peaks may be defined at the appropriate stage of chain cutting. The detailed analysis has been confined to the most clearly defined single and double traverse peaks (peaks 1 and 2 in Figure 3.21).

One of the most general features (not recognised in the first works[142,143]) is that both peaks shift continuously towards the shorter chain-length region in the chromatogram during the course of degradation[144,147,148] while their areas vary in a manner already indicated, in fact in accordance with random fold-cutting statistics[146,147]. Both peaks 1 and 2 (Figure 3.21) shift, but broadly retain their ratio in terms of molecular length (for minor changes see below). This peak-shift is interpreted in the following way: the folds are at uneven heights in the layer and, as the layer thins, those closest to the surface are cut first producing single and double traverses. As these traverses are shortened further – which involves converting the double traverse into a single traverse – folds situated deeper within the layer are exposed to attack which gives rise to double and single traverses of shortened chain length. As long as peak 2 persists there are double traverses, hence folds, present. The stage at which peak 2 disappears, and where only single traverses are present, corresponds to the limiting depth to which folds can be found within the crystal.

The most important consequence of these findings is the establishment of a fold-length fluctuation – predicted first by the kinetic theories[53]. Further, such experiments provide a means of defining the surface roughness quantitatively. In the first place, it defines the depth down to which buried folds can be found, and secondly it gives the distribution of folds across the surface layer which contains the folds. According to the latest work reported, the maximum depth amounts to 20–30 Å per surface[147,149]. However, the distribution is not uniform. For example[149], for a layer 150 Å thick, a very few (up to 5 %) of the folds extend practically across the full layer thickness (as defined by the low-angle x-ray reflections), about 15–20 % spread over a depth of up to 8 Å, 65 %, i.e. the majority, are within the comparatively narrow range of depths at 8–14 Å and only the remaining small fraction (~ 10–15 %), extending over a wide range of depths into the crystal, spreads down to the maximum depth of around 30 Å.

An additional informative quantity is the ratio of single to double traverses. If this figure is close to 1:2 and the peaks are sharp, there can be little material in the fold which will thus approximate to a hairpin. If however, for the same ratio the peaks are broad, then a similar conclusion can only be assumed with certain qualifications[147] (for the implications of peak broadening see Ref. 150). Ratios smaller than 1:2 imply appreciable fold looseness under all circumstances. When taken in conjunction with some earlier work[142], the latest results indicate the following[149]:

The ratio is broadly in the range of 1:2 and remains so throughout, which has already been implied above when stating that peaks 1 and 2 shift together during layer thinning. Some changes, however do occur in this ratio during the shift of the peaks. At the early stages of thinning the ratio is noticeably smaller by about 15–20%, and gradually converges to 1:2 as the thinning proceeds. Over the intermediate range, when around 70% of the folds are cut, it is smaller by about 5–6% than 1:2. For the folds which are cut last, the ratio is close to 1:2. (Earlier less-detailed works gave an average of 5% departure allowing for error limits[142, 147].)

Combining these results with those quoted above suggests that folds which are close to the layer surface contain appreciable looseness, the majority at intermediate level are essentially sharp with a small amount of looseness, while those deep down must be very sharp indeed.

From this it follows that the centre of the crystal is virtually fold-free. As we go outwards, we meet folds in increasing number. When they are few they could count as crystal defects. When they are not surrounded by a sufficient number of neighbouring stems they would collapse and become equivalent to amorphous material even when their re-entry is adjacent. Folds terminating within the crystal must be adjacently re-entrant for reasons of simple geometry. Large, loose, non-adjacently re-entrant loops and hairs may however contribute to the outermost layer, in fact such loops would have to be located at the periphery. These are the chain elements which could swell to the largest extent (Section 3.8.2.5) although any chain element which has a component parallel to the lamellar surface could contribute to this swelling effect.

3.8.3.3 Chain ends and cilia

As the molecule is not of infinite length, there will be chain ends which will necessarily represent discontinuities in the polymer crystal. Whichever way it is accommodated it will cause crystallinity deficiency. If the deposition of a chain is interrupted in the way illustrated in Figure 3.16 a loose cilia may result, in which case the end will be outside the crystal in any event. But even if the chain deposition can occur unhindered the situation is unlikely to be as in Figure 3.16 a–c, i.e. the chain will only exceptionally be able to complete a full stem. The question arises whether the fractional stem will fold into the lattice giving rise to a defect or stay outside as a loose hair; alternatively it may form a shorter hairpin in which case it contributes further to the unevenness of the fold length.

The problem could be tackled experimentally in the case of linear polyethylene (Marlex) possessing a double bond at one of its ends. This double bond can be identified by its characteristic infra-red spectrum. It was observed that when polyethylene single crystals with fully accessible surfaces were exposed to ozone then the number of terminal double bonds diminished very rapidly to about 10% of its initial value, remaining constant thereafter[151]. The effect was complete before any molecular degradation became apparent. From a variety of conclusive checks it was shown that the accessibility of terminal double bonds signifies that the chain ends are outside the lattice. It follows that in a single crystal around 90% (this figure can vary slightly with sample type) will be on the crystal surface. It could be shown further that this is a statistical property of the chain deposition. Corroborative evidence has also been obtained by reactions in suspension[152].

Loose hairs or cilia have been referred to repeatedly in the foregoing discussion. They possess the characteristics of loose amorphous material. The fact that chain-ends are found outside the crystal implies in itself that they could be associated with loose chains dangling in the solution, although this does not give direct information on the loose chain portion to which the ends are attached. This is provided by an experiment along an entirely different route[153].

This experiment is essentially a continuation of those discussed in Section 3.8.3.2(a), involving a mixture of ordinary and deuterated polyethylene. However, in this case the two are not mixed molecularly but crystallised separately as two distinct populations of single-crystals, one built of the usual and the other of deuterated polyethylene. Infrared spectra exhibited band splits characteristic of each (Section 3.8.3.2(a)). If however, these two crystal suspensions are mixed and sedimented together then according to the proportion of crystal constituents, and the filtering and storage conditions, a reduction in the band splitting was observed. Such a reduction in band splitting implies a reduction in nearest-neighbour contact between each molecular species, and, as described previously, occurs when chain portions containing hydrogen and deuterium are mixed within the lattice. Based on this observation the authors infer that cilia from one crystal can penetrate into layers with which they are in contact. This inferrence is supported by a plausible quantitative estimate regarding the number of cilia involved. It is apparent that the uneven fold surface inferred above should be particularly inducive to cilia penetration as it provides the gaps between the stems where the cilia could be accommodated. Further, such interpenetration could provide mechanical strength. In fact macroscopic crystal aggregates can show remarkable coherence and mechanical strength[154–156] which could be due to this reason.

3.8.3.4 Diffraction evidence in polyamides and polypeptides

The study of materials other than the much-investigated polyethylene system provides new possibilities for the central problem of how chains fold, in addition to giving information specific to the new materials themselves. Polyethylene is the closest approximation to the abstraction of a long chain because it is so unspecific along its full length. Other, chemically more-specific molecules represent more specialised cases, yet because of their greater specificity offer new opportunities for the study of their chain-folding properties.

The polypeptide glutamic acid can be made to fold in the form of lamellar single crystals by the aid of bivalent cations which form bridges between carboxyl groups situated along adjacent stems[63]. The structure of these crystals can be closely determined by x-ray diffraction, which fixes the position of the atoms along the chain segments and the cations between them. Although information on the atoms within the fold is not directly available, the fixing of the stem structure leaves little variability for the fold. This arises from the specificity of the chain where both ionic and hydrogen bonds have to be satisfied. Accordingly, the fold must be sharp and the position of the individual atoms can be determined within certain limits.

Polyamides clearly fall in between polypeptides and polyethylene, as in fact they do between most synthetic and biological materials. They are hydrogen-bonded but also contain varying lengths of methylenic sequences. The same result will hold for their folding behaviour of which some has already been referred to in Section 3.4.4. Thus, it has been stated that, say, in the series nylon 66–6.12 the lamellar thickness corresponds to the length of four monomer units[61]. For example, in nylon 66 the lamellar thickness is 53 Å while the chemical repeat period along the lamellar normal – allowing for obliquity – is 13 Å. We see that in contrast to polyethylene the chemical repeat period forms a substantial fraction of the lamellar thickness. This has important consequences for the diffraction behaviour, a feature which it should share with other compounds of large chemical identity period.

It is obvious that such a chain cannot fold at any random position as can polyethylene, because the chain has to come into crystallographic register with its neighbouring stems. This itself will restrict the usual considerations concerning the fold.

The fact that we have only a very few repeat units within a layer can give rise to novel diffraction effects[65, 157]. Consider a lamella in isolation. This in the case of nylon 66 will have an internal periodicity of four repeats. The corresponding diffraction pattern will be equivalent to that given by a grating with four slits (to quote an example from elementary optics), which in addition to the strong reflections associated with the grating period will give weak subsidiary maxima (in this case two between each pair of primary maxima). Such a situation can be achieved if the layers while all parallel are so irregularly spaced that higher orders of the fundamental stacking period of the layers die out, in which case at sufficiently high diffraction angles the pattern given by an individual layer alone will be seen. The subsidiary maxima referred to above have in fact been observed under such conditions[65, 157]. This in itself implies that crystallographic regularity extends across nearly the full lamella (which approximates to four monomer units across), hence there is no room for substantial fold looseness[157]. Detailed analysis of such subsidiary maxima can provide much further information. Thus it tells us that in nylon 66 the chains fold in the diacid half of the chemical repeat[157], by which the fold structure, if not fully determined, is at least closely circumscribed. This method of analysis, based on subsidiary diffraction maxima, holds much promise for future developments.

If we introduce the concept of fold-length fluctuations, then the combined requirement of a well-defined fold portion (acid fold) and the long chemical repeat will mean that even the smallest fluctuation will have to be very large, as the chain cannot fold just anywhere and maintain the lattice. In fact it will have to comprise a full monomer length. This has certain assessable consequences in the diffraction pattern which in fact have been observed[157].

3.9 SHORT POLYMERIC CHAINS

The subject of short chains is of fundamental interest as it should provide a link between the behaviour of traditional organic chain molecules, such as paraffins, and typical high polymers. This topic also provides further information on the nature of the fold surface in general, hence it links up with the preceding Section. Only the briefest survey can be given here.

The principal issue under this heading is the commencement of chain folding. Paraffins, or oligomers in general, crystallise in an extended chain configuration. As we increase the length, at some stage the chains will begin to fold. As will be apparent later there is no universal answer as to the onset of this folding behaviour; it depends to some extent on whether crystallisation occurs from the melt or from solution. In general for the same polymer, crystallisation from the melt favours chain extension up to much higher chain lengths than in the case of crystallisation from solution. Also the type of end-group has a significant influence as clearly the end region here becomes a non-negligible portion of the chain. Studies on this topic are greatly hampered by the lack of availability of materials of well-characterised chain length.

Different patterns have emerged from different investigations. Early work on oligomeric amides[158–161] of strictly uniform length, and with rather bulky end-groups, has led to the conclusion that the chains stay straight up to a certain length beyond which they start to fold back. This implies fractional stem lengths within the lattice. This applies to crystallisation from the melt[158–160] and from solution[161].

A large body of work on the crystallisation of sharp fractions of poly(oxyethylene) from the melt has revealed a different pattern[162–165]. Here the fold length can only be changed in a quantised manner by varying the crystallisation temperature or by annealing. For a fraction with M_n = 12 650 the realisable x-ray long periods correspond to fractions 1.0, $\frac{1}{2}$, $\frac{1}{3}$ and $\frac{1}{4}$ of the extended chain length for the peak value of the molecular weight distribution[164]. This means that the chain is either completely extended or folds once, twice and three times, with one, two, three and four full traverses respectively. This behaviour implies that the chains are of equal lengths. According to these authors the small, but finite, polydispersity of the material would be taken care of by cilia formation; the material in excess to the nearest integral fold length would be lying on the fold surface as a loose hair.

Recent investigations on the crystallisation behaviour of short-chain polyethylene fractions[27, 167] have led to the following results. When crystallised from solution, the fold length was found to vary in a continuous fashion as in the case of the high molecular weight material (Figure 3.13). At the same time, selective degradation tests (Section 3.8.3.2(c)) revealed pronounced fold-length fluctuations, and the ozone test for end groups (Section 3.8.3.3) indicated that the ends are in an accessible position at the surface. The inference is that each short chain is straight or completes one, two etc. folds according to its length and the prevailing crystallisation temperature. The pronounced fold-length fluctuations would then be the consequence of the polydispersity of the sample. Indeed in a sample with practically no polydispersity and for one particular fold length, which corresponded to one half of the chain length and also proved to be a preferred layer thickness on precipitation from solution, the fold surface was found to be even by the degradation test[168].

When the end groups were polar (carboxyl) they also tended to keep the fold length invariant, possibly due to interaction between them at the crystal surface. This may perhaps link up with the behaviour of poly(oxyethylene) outlined in the penultimate paragraph which possessed hydroxyl end groups.

Finally this work on short-chain polyethylenes unexpectedly accounted for the interfacial dislocations observed sporadically on previous occasions (Figure 3.20). As emphasised repeatedly previously[1,2,123] the dislocation phenomenon is due to the fact that consecutive crystal layers are slightly rotated, but interaction between these layer surfaces attempts to ensure crystal continuity between the layers (for details see Ref. 1). Previously this interaction has been quoted as support for crystallographic regularity along the fold surface. Recent work[166,167] has revealed that the prerequisites for this dislocation phenomenon are short molecules which only fold once or twice and polydispersity in chain length. At the same time a pronounced fluctuation of the fold length was observed by the test quoted in Section 3.8.3.2(c). The conclusion is that the chain ends from one layer mesh with the gaps which arise from the fluctuation of the fold stem-length in the others. This interlocking would then ensure the crystallographic continuity required. This interpretation accounts for all the known facts concerning these crystals. We see further that the picture is also consistent with the general conditions relating to the concepts of fold-length fluctuation, ciliary penetration etc. developed in the preceding Sections.

In conclusion, a few statements about crystallisation from the melt will be made. It has been mentioned that chain extension can persist to higher chain lengths in the melt than during crystallisation in solution. With fractions of molecular weight up to 10 000, chain extension has been observed and a considerable amount of fractionation was noted[169]. The method of observation here includes the examination of fracture surfaces where the lamellar structures can be seen as striation resembling strata in a rock face. Under normal circumstances, at molecular weights greater than 10 000, the chains will fold. If however crystallisation is conducted under high hydrostatic pressure, then the chain extension can be preserved to much higher chain lengths; thus for 5000 atmospheres, or higher, chain extension persists up to several microns (Figure 3.11)[20,35,36]. These are the so-called extended-chain type crystals which hardly display any polymeric behaviour at all. The materials are brittle and behave like polycrystalline substances. They melt close to the extrapolated melting points for infinite chain length. Further, density and heat of fusion measurements indicate the highest crystallinities recorded so far. The reason why hydrostatic pressure should induce such chain extension is problematic and is subject to much current work not to be reviewed further here.

3.10 CONCLUDING REMARKS

At this point external boundary conditions force this Review to a close. It will have become apparent from casual remarks that vast tracts in the field of crystal morphology have been left untouched. Even within the narrow subject matter of the Review much has been left untold. Adhering to the intentions outlined in the Preface, I have tried, in what has been included, to discern or establish unifying trends. I hope that here and there such trends have in fact emerged. Without them the whole subject would be merely an undigestible store of individual pieces of knowledge continuously swelled by the increasing tide of scientific literature, a threat hanging over this as over so many subjects. I also hope that such common threads, no matter how tenuous, will help to hold together the knowledge which exists, and thus secure our grasp over what has been achieved. At the same time it is to be hoped that these threads possess sufficient persistence to direct future inquiries in the field of polymer crystals along profitable channels.

Note added in proof

I wish to draw attention to some further material concerning Section 3.8.3.2(a) relating to arguments about the spectroscopic proof of adjacent fold re-entry, which has a very central position in the field. In a paper to appear in 'Macromolecules', Stehling, Ergos and Mandelkern present detailed arguments on the phase separation in hydrocarbon–deuterocarbon systems which if applicable to the polyethylene experiments would weaken or even invalidate the evidence for adjacent fold re-entry in references 133–136. This argument has been countered by Krimm and Ching (to appear in Macromolecules) who maintain the validity of their original point of view in favour of adjacently re-entrant folds. In the course of new work they show that '*not only are their*' (i.e. Stehling, Ergos and Mandelkern) '*claims unsubstantiated but a detailed analysis of their results provides additional support for the conclusion of Bank and Krimm that random re-entry is a negligible contribution to chain folding in dilute solution grown crystals (110) folding predominating*' (quotation from Krimm and Ching). Without adding comments of my own I conclude this addendum on this note. I wish to remark that all my advanced information as presented here originates from Krimm and Ching to whom I wish to express my thanks.

References

1. Keller, A. (1968). *Reports on Progress in Physics,* **32,** Part 2, 623
2. Keller, A. (1969). *Kolloid Z.u.Z. Polymere,* **231,** 386
3. Keller, A. (1964). *Kolloid Z.u.Z. Polymere,* **197,** 98
4. Geil, P. H. (1963). *Polymer Single Crystals,* (New York: Interscience)
5. Mandelkern, L. (1964). *Crystallisation of Polymers,* (New York: McGraw Hill)
6. Sharples, A. (1966). *Introduction to Polymer Crystallisation,* (London: Edward Arnold)
7. Meares, P. (1965). *Polymers: Structure and Bulk Properties,* (London: Van Nostrand)
8. Hill, R. (1953). *Fibres from Synthetic Polymers,* (Amsterdam: Elsevier)
9. Till, P. H. (1957). *J. Polymer Sci. A,* **24,** 301
10. Keller, A. (1957). *Phil. Mag.,* **2,** 1171
11. Fischer, E. W. (1957). *Z. Naturforsch.,* **12a,** 753
12. Bassett, D. C., Frank, F. C. and Keller, A. (1963). *Phil. Mag.,* **8,** 1755
13. Keller, A. and O'Connor, A. (1957). *Discuss. Faraday Soc.,* **25,** 114
14. Bassett, D. C., Keller, A. and Mitsuhashi, S. (1963). *J. Polymer Sci.A,* **1,** 763
15. Keller, A. (1959). *J. Polymer Sci.,* **39,** 151
16. Keller, A. and Sawada, S. (1964). *Makromol. Chem.,* **74,** 190
17. Pennings, A. J. (1966). *Proc. Int. Conf. Crystal Growth, Boston,* 389 (Oxford: Pergamon P.)

18. Keller, A. and Machin, M. J. (1967). *J. Macromol. Sci. B*, **1**, 41
19. Pennings, A. J., Van der Mark, J. M. A. A. and Booij, H. C. (1970). *Kolloid Z.u.Z. Polymere*, **236**, 96
20. Wunderlich, B., Prime, R. B. and Melillo, L. (1969). *J. Polymer Sci. A-2*, **7**, 2091
21. Padden, F. J. and Keith, H. D. (1965). *J. Appl. Phys.*, **36**, 2987
22. Giannoni, G., Padden, F. J. and Keith, H. D. (1969). *Proc. Nat. Acad. Sci.*, **62**, 964
23. Kawai, T. and Keller, A. (1965). *Phil. Mag.*, **8**, 1203
24. Nakajima, A., Hayashi, S., Korenaga, T. and Sumida, T. (1968). *Kolloid. Z.u.Z. Polymere*, **222**, 128
25. Holdsworth, P. J. and Keller, A. (1967). *J. Polymer Sci. B*, **5**, 605
26. Kawai, T., Hama, T. and Ehara, K. (1968). *Makromol. Chem.*, **113**, 282
27. Keller, A. and Udagawa, Y. *J. Polymer Sci. A-2*, in the press
28. Statton, W. O. and Geil, P. H. (1960). *J. Appl. Polymer Sci.*, **3**, 357
29. Holland, V. F. (1964). *J. Appl. Phys.*, **35**, 59
30. Blackadder, D. A. and Schleinitz, H. M. (1966). *Polymer*, **7**, 603
31. Hoffman, J. D. and Weeks, J. J. (1965). *J. Chem. Phys.*, **42**, 4301
32. Peterlin, A. and Roeckl, E. (1963). *J. Appl. Phys.*, **34**, 102
33. Peterlin, A. (1964). *J. Appl. Phys.*, **35**, 75
34. Gonthier, A. and Kovacs, A. J. (1971). *Symposium on Morphology of Polymers, Prague*, Preprint A1, and private communication
35. Prime, R. B. and Wunderlich, B. (1969). *J. Polymer Sci. A-2*, **7**, 2061
36. Rees, D. V. and Bassett, D. C. (1971). *J. Polymer Sci. A-2*, **9**, 385
37. Kawai, T. (1965). *Makromol. Chem.*, **84**, 290
38. Fischer, E. W. and Schmidt, G. F. (1962). *Angew. Chem. Int. Ed. Engl.*, **1**, 448
39. Hoffman, J. D., Williams, G. and Passaglia, E. (1966). *J. Polymer Sci. C*, **14**, 173
40. Peterlin, A. (1965). *Polymer*, **6**, 25
41. Keller, A. and Priest, D. J. (1970). *J. Polymer Sci. B*, **8**, 13
42. Dreyfuss, P. and Keller, A. (1970). *J. Polymer Sci. B*, **8**, 253
43. Peterlin, A. and Fischer, E. W. (1960). *Z. Phys.*, **159**, 272
44. O'Leary, K. and Geil, P. H. (1967). *J. Macromol. Sci., B*, **1**, 147
45. Burmester, A. F. (1970). *Report T.R. No. 163*, Division of Macromolecular Science, Case Western Reserve University, Cleveland
46. Kavesh, S. and Schultz, J. M. (1971). *J. Polymer Sci. A-2*, **9**, 85
47. Point, J. J., Gilliot, M., Dosière, M. and Goffin, A. (1971). *IUPAC Symposium on Morphology of Polymers, Prague*, Preprint G.1
48. Keller, A. and Pope, D. P. (1971). *J. Materials Sci.*, **6**, 453
49. Dawkins, J. V., Holdsworth, P. J. and Keller, A. (1968). *Makromol. Chem.*, **118**, 361
50. Lauritzen, J. I. and Hoffman, J. D. (1960). *J. Res. Nat. Bur. Stand. A*, **64**, 73
51. Price, F. P. (1961). *J. Chem. Phys.*, **35**, 1884
52. Frank, F. C. and Tosi, M. (1961). *Proc. Roy. Soc. A*, **263**, 323
53. Lauritzen, J. I. and Passaglia, E. (1967). *J. Res. Nat. Bur. Stand. A*, **71**, 261
54. Mandelkern, L. (1970). *Progress in Polymer Science*, Ed. Jenkins, A. D. **2**, 201, (Pergamon: Oxford)
55. Mandelkern, L. (1971). *XXIII IUPAC Symposium, Boston, Preprint*, **2**, 794 and lecture delivered
56. Sanchez, I. C. and DiMarzio, E. A. *Macromolecules*, in the press, and private communication
57. Bassett, D. C., Dammont, F. R. and Salovey, R. (1964). *Polymer*, **6**, 579
58. Keller, A. and Martuscelli, E. (1971). *Makromol. Chem.*, **141**, 189
59. Keller, A. and Martuscelli, E. (1972). *Makromol. Chem.*, **151**, 189
60. Stejny, J., Atkins, E. D. T. and Keller, A. Unpublished results
61. Dreyfuss, P. and Keller, A. (1970). *J. Macromol. Sci. B*, **4**, 811
62. Dreyfuss, P. and Keller, A. Unpublished results
63. Keith, H. D., Giannoni, G. and Padden, F. J. (1969). *Biopolymers*, **7**, 775
64. Geddes, A. J., Parker, K. D., Atkins, E. D. T. Beighton, E. (1968). *J. Mol. Biol.*, **32**, 343
65. Dreyfuss, P., Keller, A. and Willmouth, F. M. *J. Polymer Sci. A-2*. In the press
66. Mandelkern, L. (1958). *Growth and Perfection of Crystals: Proceedings of an International Conference on Crystal Growth, Cooperstown, New York*, Ed. Doremus, R. H., Roberts B. W. and Turnbull, D., 467, (New York: John Wiley & Sons)
67. Price, F. P. (1969). *Nucleation*. Ed. Zettelmoyer, A. C., 405. (New York: Marcel Dekker)

68. Suzuki, T. and Kovacs, A. J. (1970). *Polymer J.* (Japan), **1,** 82
69. Hoffman, J. D. (1964). *S.P.E. Trans.,* **4,** 315
70. Nardini, M. J. and Price, F. P. (1966). *Crystal Growth: Proc. Int. Conf. on Crystal Growth, Boston,* Ed. Peiser, H. S., 395 (Oxford: Pergamon)
71. Lindenmeyer, P. H. and Holland, V. F. (1964). *J. Appl. Phys.,* **35,** 55
72. Blundell, D. J., Keller, A. and Kovacs, A. J. (1966). *J. Polymer Sci. B,* **4,** 481
73. Kovacs, A. J. and Manson, J. A. (1966). *Kolloid Z.u.Z. Polymere,* **214,** 1
74. Keith, H. D., Vadimsky, R. G. and Padden, F. J. (1970). *J. Polymer Sci. A-2,* **8,** 1687
75. Blundell, D. J. and Keller, A. (1968). *J. Polymer Sci. B,* **6,** 433
76. Johnsen, V. and Lehmann, J. (1969). *Kolloid Z.u.Z. Polymere,* **230,** 317
77. Seto, T. and Mori, N. (1969). *Reports on Progress in Polymer Physics in Japan,* **12,** 157
78. Keller, A. and Pedemonte, E. To be published
79. Blundell, D. J. and Keller, A. (1968). *J. Macromol. Sci. B,* **2,** 337
80. Kovacs, A. J., Lotz, B. and Keller, A. (1969). *J. Macromol. Sci. B.,* **3,** 385
81. Wittmann, J. C. and Kovacs, A. J. (1970). *Ber. Bunsenges. Phys. Chem.,* **74,** 901
82. Blundell, D. J. and Keller, A. (1968). *J. Macromol. Sci. B.,* **2,** 301
83. Keller, A. and Willmouth, F. M. (1970). *J. Polymer Sci. A-2,* **8,** 1443
84. Keller, A. and Sadler, D. M. (1970). *J. Polymer Sci. A-2,* **8,** 1457
85. Carr, S. H., Keller, A. and Baer, E. (1970). *J. Polymer Sci. A-2,* **8,** 1467
86. Booth, C. and Price, C. (1966). *Polymer,* **7,** 85
87. Kawai, T. (1967). *Makromol. Chem.,* **102,** 125
88. Sadler, D. M. (1971). *J. Polymer Sci. A-2,* **9,** 779
89. Bassett, D. C. (1964). *Phil. Mag.,* **10,** 595
90. Lindenmeyer, P. H. (1963). *J. Polymer Sci. C,* **1,** 5
91. Heijde, van der H. B. (1967). *J. Polymer Sci. A-2,* **5,** 225
92. Patel, G. N. and Patel, R. D. (1970). *J. Polymer Sci. A-2,* **8,** 47
93. Burbank, R. D. (1960). *Bell Syst. Tech. J.,* **39,** 1627
94. Morrow, D. R., Sauer, J. A. and Woodward, A. E. (1968). *J. Polymer Sci. C,* **16,** 3401
95. Miller, R. L. (1966). *Encyclopedia of Polymer Science and Technology,* **4,** 449 (New York: John Wiley & Sons)
96. Hosemann, R. (1970). *Ber. Bunsenges. Phys. Chem.,* **74,** 755
97. Čačkovič, H., Hosemann, R. and Wilke, W. (1969). *Kolloid. Z.u.Z. Polymere,* **234,** 1000
98. Pechhold, W. and Blasenbrey, S. (1970). *Kolloid. Z.u.Z. Polymere,* **241,** 955
99. Blasenbrey, S. and Pechhold, W. (1970). *Ber. Bunsenges. Phys. Chem.,* **74,** 784
100. Fischer, E. W. (1969). *Kolloid Z.u.Z. Polymere,* **231,** 458
101. Fischer, E. W., Goddar, H. and Schmidt, G. F. (1967). *J. Polymer Sci. B,* **5,** 619
102. Vonk, C. G. and Kortleve, G. (1967). *Kolloid Z.u.Z. Polymere,* **220,** 19
103. Kortleve, G. and Vonk, C. G. (1968). *Kolloid Z.u.Z. Polymere,* **225,** 124
104. Fischer, E. W., Kloos, F. and Lieser, G. (1969). *J. Polymer Sci. B,* **7,** 845
105. Nukushina, Y., Ito, Y. and Fischer, E. W. (1965). *J. Polymer Sci. B,* **3,** 383
106. Sullivan, P. K. and Weeks, J. J. (1970). *J. Res. Nat. Bureau Stand.,* **74A,** 203
107. Fischer, E. W. (1971). *Pure and Applied Chem.,* **26,** 285
108. Zachmann, H. G. (1964). *Z. Naturforsch.,* **19a,** 1397
109. Zachmann, H. G. (1969). *Kolloid. Z.u.Z. Polymere,* **231,** 504
110. Fischer, E. W. (1967). *Kolloid Z.u.Z. Polymere,* **218,** 97
111. Zachmann, H. G. and Peterlin, A. (1969). *J. Macromol. Sci. B.,* **3,** 495
112. MacCall, D. W. and Anderson, E. W. (1963). *J. Polymer Sci. A,* **1,** 1175
113. Fischer, E. W. and Peterlin, A. (1964). *Makromol. Chem.,* **74,** 1
114. Blundell, D. J., Keller, A. and O'Connor, T. (1967). *J. Polymer Sci. A-2,* **5,** 991
115. Blackadder, D. A. and Lewell, P. A. (1970). *Polymer,* **11,** 147
116. Udagawa, Y. and Keller, A. (1971). *J. Polymer Sci. A-2,* **9,** 437
117. Bassett, G. A., Blundell, D. J. and Keller, A. (1967). *J. Makromol. Sci. B,* **1,** 161
118. Blundell, D. J., Keller, A. and Sadler, D. M. To be published
119. Palmer, R. P. and Cobbold, A. (1964). *Makromol. Chem.,* **74,** 174
120. Peterlin, A. and Meinel, G. (1965). *J. Polymer Sci. B,* **3,** 1059
121. Priest, D. J. (1971). *J. Polymer Sci. A-2,* **9,** 1777
122. Keller, A. and Udagawa, Y. (1971). *J. Polymer Sci. A-2,* **9,** 1793
123. Holland, V. F. and Lindenmeyer, P. H. (1965). *J. Appl. Phys.,* **35,** 3235
124. Keller, A. (1962). *Polymer,* **3,** 393
125. McMahon, P. E., McCullough, R. L. and Schlegel, A. A. (1967). *J. Appl. Phys.,* **38,** 4123

126. Allegra, C., Corradini, P. and Petraccone, P. *Macromolecules.* In the press
127. Koenig, J. L. and Witenhafer, D. E. (1965). *Makromol. Chem.*, **99,** 193
128. Koenig, J. L. and Agboatwalla, M. C. (1968). *J. Macromol. Sci. B*, **2,** 391
129. Koenig, J. L. and Hannon, M. J. (1967). *J. Macromol. Sci. B*, **1,** 119
130. Harrison, I. R. and Baer, E. (1969). *J. Colloid Interface Sci.*, **31,** 176
131. Harrison, I. R. and Baer, E. (1971). *J. Polymer Sci. A-2*, **9,** 1305
132. Stellman, J. M. and Woodward, A. E. (1971). *J. Polymer Sci. A-2*, **9,** 59
133. Tasumi, M. and Krimm, S. (1968). *J. Polymer Sci. A-2*, **6,** 995
134. Bank, M. I. and Krimm, S. (1969). *J. Polymer Sci. A-2*, **7,** 1785
135. Kikuchi, Y. and Krimm, S. (1970). *J. Macromol. Sci. B*, **4,** 461
136. Bank, M. I. and Krimm, S. (1970). *J. Polymer Sci. B*, **8,** 143; (1971). *Discussion Remarks, IUPAC Symposium, Boston*
137. Mitzushima, S. I. and Shimanouchi, T. (1949). *J. Amer. Chem. Soc.*, **71,** 1329
138. Peticolas, W. L., Hibler, G. W., Lippert, J. L., Peterlin, A. and Olf, H. G. (1971). *Appl. Phys. Lett.*, **18,** 87
139. Peterlin, A., Olf, H. G., Peticolas, W. L., Hibler, G. W. and Lippert, J. L. (1971). *J. Polymer Sci. B*, **9,** 583
140. Blundell, D. J., Keller, A., Ward, I. M. and Grant, I. J. (1966). *J. Polymer Sci. B*, **4,** 781
141. Winslow, F., Hellman, M. Y., Matreyek, W. and Salovey, R. (1967). *J. Polymer Sci. B*, **5,** 89
142. Williams, T., Blundell, D. J., Keller, A. and Ward, I. M. (1968). *J. Polymer Sci. A-2*, **6,** 1613
143. Williams, T., Keller, A. and Ward, I. M. (1968). *J. Polymer Sci. A-2*, **6,** 1621
144. Sadler, D. M., Williams, T., Keller, A. and Ward, I. M. (1969). *J. Polymer Sci. A-2*, **7,** 1819
145. Frank, F. C., Ward, I. M. and Williams, T. (1968). *J. Polymer Sci. A-2*, **6,** 1613
146. Williams, T. and Ward, I. M. (1969). *J. Polymer Sci. A-2*, **7,** 1585
147. Keller, A., Martuscelli, E., Priest, D. J., Udagawa, Y. (1971). *J. Polymer Sci. A-2*, **9,** 1807
148. Ward, I. M. and Williams, T. (1971). *J. Macromol. Sci.* B5, 693
149. Keller, A., Patel, G. N. and Stejny, J. To be published
150. Williams, T., Udagawa, Y., Keller, A. and Ward, I. M. (1970). *J. Polymer Sci. A-2*, **8,** 35
151. Keller, A. and Priest, D. J. (1968). *J. Makromol. Sci. B*, **2,** 479
152. Witenhafer, D. E. and Koenig, J. L. (1969). *J. Polymer Sci. A-2*, **7,** 1279
153. Bank, M. I. and Krimm, S. (1969). *J. Appl. Phys.*, **10,** 4248
154. Holdsworth, P. J. and Keller, A. (1968). *J. Polymer Sci. A-2*, **6,** 707
155. Ishikawa, K., Miyasaka, K. and Maeda, M. (1969). *J. Polymer Sci. A-2*, **7,** 2029
156. Blackadder, D. A. and Lewell, P. A. (1970). *Polymer*, **11,** 125
157. Atkins, E. D. T., Keller, A. and Sadler, D. M. *J. Polymer Sci. A-2.* In the press
158. Zahn, H. and Pieper, W. (1961). *Angew. Chem.*, **73,** 246
159. Zahn, H. and Pieper, W. (1962). *Kolloid Z.u.Z. Polymere*, **180,** 97
160. Kern, W., Davidovits, J., Rautekus, K. J. and Schmidt, G. F. (1961). *Makromol. Chem.*, **43,** 106
161. Baltá-Calleja, F. J. and Keller, A. (1963). *J. Polymer Sci. A*, **2,** 2151
162. Arlie, J. P., Spegt, P. and Skoulios, A. (1965). *C.R. Acad. Sci. Paris*, **260,** 5774
163. Gilg, B., Spegt, P. and Skoulios, A. (1965). *C.R. Acad. Sci. Paris*, **261,** 5482
164. Arlie, J. P., Spegt., P. and Skoulios, A. (1967). *Makromol. Chem.*, **104,** 212
165. Spegt., P. (1970). *Makromol. Chem.*, **139,** 139
166. Sadler, D. M. and Keller, A. (1970). *Kolloid Z.u.Z. Polymere*, **239,** 641
167. Sadler, D. M. and Keller, A. (1970). *Kolloid. Z.u.Z. Polymere*, **242,** 1081
168. D'Ilario, L., Martuscelli, E. and Keller, A. *J. Polymer Sci. A-2.* In the press
169. Anderson, F. R. (1963). *J. Polymer Sci. C*, **3,** 123
170. Flory, P. J. (1962). *J. Amer. Chem. Soc.*, **84,** 2857
171. Ingram, P. and Peterlin, A. (1968). *Encyclopedia of Polymer Science and Technology*, **9,** 204 (New York: John Wiley & Sons)
172. Keller, A. (1959). *J. Polymer Sci.*, **39,** 151
173. Holland, V. F. and Lindenmeyer, P. H. (1965). *Science*, **147,** 1296

4
Stereoregular Polymerisation

I. PASQUON
Istituto di Chimica Industriale, Politecnico di Milano
and
L. PORRI
Istituto di Chimica Organica Industriale, Università di Pisa

4.1 INTRODUCTION

This review essentially describes the most significant results and discussions on stereospecific polymerization published from 1966 to 1970.

A short general review covering this topic up to 1966 is published in the *Encyclopedia of Polymer Science and Technology* under 'Stereoregular Linear Polymers – Preparation'[1], and more detailed ones elsewhere[2,3]. Further and more specific reviews will be referred to in this article.

4.2 STEREOSPECIFIC POLYMERISATION OF NON-ASYMMETRIC α-OLEFINS

The stereospecific polymerisation of α-olefins, may be both isospecific and syndiospecific; syndiotactic polymers of α-olefins have been obtained so far only from propylene. Reviews on these topics have been published in the last few years particularly by Berger *et al.*[4], Boor[5,6], Cossee[7], Coover[8], Henrici-Olive and Olive[9], Hoeg[10], Jordan[11], Pasquon *et al.*[12] and Smith[13].

Internal olefins have been polymerised through isomerisation to isotactic polymers of the corresponding olefins[14]. Such a monomer isomerisation polymerisation can be carried out at a temperature of *c.* 80 °C, by use of the conventional AlR_3–$TiCl_3$ catalysts or catalysts of this type modified by suitable additives[14]

4.2.1 Isospecific polymerisation

The isotactic polymers which have interesting applications are polypropylene and, to a lesser extent, polybutene and poly(4-methylpent-1-ene).

4.2.1.1 Catalytic systems

The best catalysts for the isospecific polymerisation of α-olefins – in particular propylene – are still those discovered in 1954–1955, i.e. the heterogeneous Ziegler–Natta catalysts based on alkyl aluminium compounds and violet

$TiCl_3$. The system $Al(C_2H_5)_2Cl$–δ-$TiCl_3$ – where δ-$TiCl_3$ contains $AlCl_3$ in solid solution – is now extensively used; in fact, this system exhibits high stereospecificity and catalytic activity[15].

The polymerisation is generally accomplished in the presence of a hydrocarbon diluent, but processes without diluent are also known; one of these, proposed by BASF[16], in which violet $TiCl_3$ and AlR_3 are used, gives higher polypropylene yields than the traditional ones, but lower stereospecificity. No practical interest for the stereospecific polymerisation of α-olefins has developed using the catalyst systems based on chromium oxides, which were developed for the polymerisation of ethylene[17–21], and for those promoting high-yield ethylene polymerisation, prepared from $TiCl_4$ and an aluminium alkyl compound supported on a solid substrate (e.g. based either on magnesium compounds or on another bivalent metal)[22–25]. The stereospecificity of such catalyst systems is always low and sometimes very low.

Recent patent and scientific literature[6, 10, 26–28], has reported highly stereospecific catalyst systems obtained by addition of different compounds to the traditional Ziegler–Natta systems, either prepared with organometallic compounds other than aluminium alkyls, or with Ti compounds other than $TiCl_3$. Few of these systems are of practical interest.

With regard to the 3-component catalyst systems consisting of violet-$TiCl_3$–$AlRCl_2$ (where R = alkyl) – Lewis base[8, 10], Zambelli *et al.* and Watt[29–32] clearly demonstrated by i.r., n.m.r. and thermodynamic analysis, that the third component causes dismutation of $AlRCl_2$ with consequent formation of AlR_2Cl, which is the true alkylating agent responsible for catalyst activation. The above investigations were carried out with different Lewis bases. Formation of AlR_2Cl was not observed in experiments carried out by using as third component, e.g. aldehydes[33] or amino-silane compounds[34]. However, the molecules that can react with organometallic compounds such as aldehydes, or those which contain more than one basic-type atom are inconvenient for studies of this type.

The systems prepared from only one transition metal, which either do or do not contain organometallic compounds, although slightly active and often slightly stereospecific, exhibit a particular scientific interest[6, 10]. Among the first systems prepared in the absence of organometallic compounds, were those based on $Ti + C_2H_5Br$, Ti + violet $TiCl_3$, $Ti + I_2$ [35], those consisting of ground $TiCl_2$ [10, 36], and those consisting of mixtures of violet $TiCl_3$ either with or without $AlCl_3$ in solid solution, and trialkylamine either in the presence or in the absence of hydrogen[5]. The last systems are highly stereospecific, but slightly active.

The polymerisation of several olefins has been more recently accomplished by Herwig *et al.* in the presence of catalysts consisting of mixtures of $TiCl_4$ or $VOCl_3$ with allylchromium[37, 38].

Quite interesting results have been obtained by Giannini *et al.*[39] in the field of the polymerisation of propylene and of 4-methylpent-1-ene with some benzyl derivatives of titanium and of zirconium, even free from halogens. These authors found that pure $Ti(CH_2C_6H_5)_3Cl$, $Ti(CH_2C_6H_5)_3Br$, $Ti(CH_2C_6H_5)_3F$ and $Zr(CH_2C_6H_5)_4$, in benzene solution, polymerise propylene – although slowly – to a partially isotactic polymer. $Zr(CH_2C_6H_5)_4$ also polymerises 4-methylpent-1-ene. The interest of these results lies in the complete solubility of these catalysts in aromatic solvents. The above catalysts do not decompose, with formation of a solid phase, before polymerisation as $TiCH_3Cl_3$ does[6] and this is the first example of isospecific polymerisation of α-olefins in the homogeneous phase, in the presence of soluble catalysts.

4.2.1.2 *Stereoregularity of polymers*

The degree of steric purity of isotactic polypropylene was determined by Zambelli, Segre *et al.*[40–42], by comparing the 100 MHz n.m.r. spectra of isotactic poly(*cis*-propylene-1,2,3,3,3-d_5) and of isotactic poly(*trans*-propylene-1,2,3,3,3-d_5). These authors concluded that the residual polypropylene after extraction with boiling n-heptane may reach a degree of steric purity above 98 %. The same conclusion was drawn by Heatley *et al.*[43].

By analysing sterically disordered polymers of these monomers, Heatley and Zambelli[44] also attributed the n.m.r. peaks to tetrads. Flory[45] criticised such results on the basis of conformational determinations; however his experimental support was poor.

Some statistical models have been proposed for the calculation of the degree of stereoregularity of polypropylene. However, these models may be correctly applied only if some parameters, which are at present difficult to evaluate are known[46–49].

4.2.1.3 *Kinetic aspects*

The propagation rate of the polymeric chain of isotactic poly(but-1-ene) obtained in the presence of catalytic systems based on $TiCl_3$ (α, β, γ and δ), $TiCl_2$ or VCl_3 and organometallic compounds, was investigated by Natta, Pasquon, Zambelli *et al.* by checking the variation of the polymer molecular weight *v.* time[50, 51].

The most significant conclusions drawn by these authors may be summarised as follows:

(a) The propagation rate of the isotactic polymeric chains is practically the same for all investigated systems based on $TiCl_2$ or violet $TiCl_3$ (in the α, γ and δ modifications) either with or without $AlCl_3$ in solid solution; the rate is independent of the organometallic compound (Al or Be) used for catalyst preparation and independent of the overall polymerisation rate.

(b) For systems based on β-$TiCl_3$, the propagation rate of the isotactic polymeric chains is lower than that corresponding to the systems based on violet $TiCl_3$, whereas for the VCl_3-based systems, such a rate is much higher.

(c) The activation energy of the chain-propagation process was estimated at *c.* 4.5 kcal mol^{-1} for systems based on violet $TiCl_3$ and at *c.* 2.3 kcal mol^{-1} for those based on VCl_3.

The above results show that the kinetically-determining step of the chain growth, for the isotactic polybutene, is strictly connected with the nature of the transition metal of the catalytic complex, and is independent of the type of organometallic compound; it seems unlikely that the rate-determining step is monomer diffusion.

In a study on propylene polymerisation in the gas phase on monocrystals of α-$TiCl_3$ and $AlMe_3$, Guttman and Guillet[52] conclude instead that the kinetically determining step is monomer diffusion. These authors also determined the rate of the polymer chain growth at the start of the polymerisation; the values obtained are higher by a few orders of magnitude than those previously reported by other authors[12, 53]. Kissin *et al.*[54] found values that are of the same order of magnitude (mmol mol^{-1} ($TiCl_3$)) as those previously reported.

Such discrepancies are partly due to the fact that catalysts and operating conditions cannot be always strictly compared and in particular to the various techniques – mostly indirect – used to determine the active sites.

All data published on this topic, however, suggest that the concentration of active sites for stereospecific polymerisation in the catalyst systems based on violet $TiCl_3$ and aluminium alkyl is only a few mmoles of sites per mole of $TiCl_3$. By considering the overall polymerisation rate of these systems and the average molecular weight of the polymer obtained, the conclusion may be drawn that, under the conditions usually adopted for propylene polymerisation, the time of polymer chain growth does not exceed a few seconds or a few minutes. Hence, in the presence of such catalysts, it seems hardly possible to prepare heteroblock copolymers by sequential polymerisation of different monomers.

4.2.1.4 *Stereospecificity and heterogeneity of the Ziegler–Natta systems*

Natta, Pasquon, Zambelli *et al.* studied the stereospecificity of catalyst systems based on some transition metal halides and on different organometallic compounds in the polymerisation of propylene[50, 55]. By n.m.r. and x-ray analyses of polypropylene fractions, separated from the raw polymer by solvent extraction, they reached the conclusion that each catalyst system contains different types of catalytic complexes leading to the formation of macromolecular segments with an atactic, isotactic and syndiotactic structure, respectively. The relative percentage of these different complexes varies when passing from one system to another and does not only depend on the nature of the transition metal, but also on the nature of the organometallic compound used for catalyst preparation.

Several authors reached similar conclusions with regard to the heterogeneity of the catalytic complexes present in the Ziegler–Natta systems[5, 56–58].

4.2.1.5 *The nature of the isospecific catalytic complexes—Reaction mechanism*

Rodriguez *et al.*[59–61], by studying the reaction between α-$TiCl_3$ and $AlMe_3$ and other aluminium alkyls, and with the aid of electron microscopy, also confirmed the hypothesis previously put forward by several authors[6, 7, 62] i.e. that in Ziegler–Natta violet $TiCl_3$-based systems the active sites in the isospecific polymerisation of α-olefins are arranged on the side surface of the $TiCl_3$ crystals, where Ti atoms are accessible. The same conclusions were also drawn by Kollar *et al.*[63] and by Guttman and Guillet[52]. The latter authors, who observed by electron microscopy, the start of propylene polymerisation on monocrystals of α-$TiCl_3$ in the presence of $AlMe_3$, without solvent, conclude that the {001} planes of α-$TiCl_3$ crystals are completely inactive. According to Rodriguez *et al.*[59,60], such planes are covered by inactive aluminium alkyls derived from a reaction between the aluminium alkyl used for catalyst preparation and the $TiCl_3$ surface. Chirkov *et al.*, who studied the polymerisation in the presence of solvent, reached quite different conclusions[54,64]. They believe that the active sites are essentially arranged on the {001} basal planes of the α-$TiCl_3$ crystals. The experimental methods they used for the evaluation of the active-site concentration may be questioned.

It must be pointed out that, during polymerisation, $TiCl_3$ particles gradually crumble until they become completely dispersed in the polymer formed[6,62,65]. This phenomenon may produce a remarkable change in the $TiCl_3$ surface, but this does not necessarily occur in the runs in which very low amounts of propylene per unit weight of $TiCl_3$ are polymerised, in the absence of a hydrocarbon diluent.

Several authors believe that in the Ziegler–Natta heterogeneous system, the transition metal present in catalytic complexes that are isospecific in the polymerisation of α-olefins has valency 3 [5,66–70].

According to Overberger *et al.*[67,68], for $AlEt_3$–$TiCl_3$ based systems, the macromolecules of amorphous low-molecular-weight polypropylene is formed on sites containing Ti^{IV} adsorbed on the $TiCl_3$ surface.

It was also found that the transition metal present in Ziegler–Natta heterogeneous systems has different valency states (2,3 or 4)[6,71,72].

It must be remembered, however, that some ambiguities may exist in the determination of the valency state of a metal present on the crystal surface. It is to be noted in this respect, that the isospecific catalyst sites, present in the $TiCl_2$- and violet $TiCl_3$-based Ziegler–Natta systems, must be similar, in spite of the different valency states of Ti in the halide, since the rate of polymer chain growth is the same for the two systems[50].

The transition metal of the catalytic complexes present in the Ziegler–Natta homogeneous systems, which are especially active in the polymerisation of ethylene, is supposed to be both in the tetravalent[9,73,74] and trivalent states[75,76].

The existence of catalytic systems that contain only Ti as a metal and are active in the isospecific polymerisation of α-olefins, demonstrates that isospecific catalytic complexes may contain only one type of metal. Some authors extended this hypothesis also to the Ziegler–Natta bimetallic systems, prepared e.g. from violet $TiCl_3$ and an organic aluminium compound[5–7]. According to these authors, the active sites that are present on the surface of solid $TiCl_3$-based catalysts, should exclusively be monometallic and might be represented as follows

```
        R
        |  .Cl
        | .
  Cl—Ti—□      (1)
  Cl/   |
        Cl
```

where □ is an octahedral position on which the olefin is coordinated to the transition metal and R is the alkyl group of the metal alkyl. This hypothesis was supported by Boor[5] and also on the basis of the calculations by Begley and Pennella[77]; according to these authors, the crystal lattice of $TiCl_3$ stabilises the Ti—C bonds and therefore complexation of the site with an alkyl metal is not required in order to make it stable and more active.

However, the hypothesis of the monometallic catalytic complexes does not allow an easy interpretation of some experimental findings. Under given conditions some types of α-$TiCl_3$ in combination with $ZnEt_2$ did not polymerise propylene, but are immediately activated by small amounts of $AlEt_3$, added after $ZnEt_2$ [50, 78]. In these cases, formation of Ti—C bonds occurs even in the absence of $AlEt_3$ since the alkylating power of $ZnEt_2$ is higher than that of $AlEt_3$. This suggests that aluminium alkyl is not simply an alkylating agent.

Other experimental data, which can hardly be interpreted by the hypothesis of exclusively monometallic sites, concern the influence exerted by the nature of the organometallic compound on both stereospecificity and activity of the catalyst systems based on a given transition metal halide[50, 51, 55, 78].

The existence of bimetallic catalytic complexes in Ziegler–Natta homogeneous systems based on Ti and Al compounds which are active in the polymerisation of ethylene[9, 73, 79, 80] and in the oligomerisation of propylene to partially stereoregular products was recently proposed by Henrici–Olive and Olive[81] and by Reichert and Malmann[80].

Bimetallic catalytic complexes of the type:

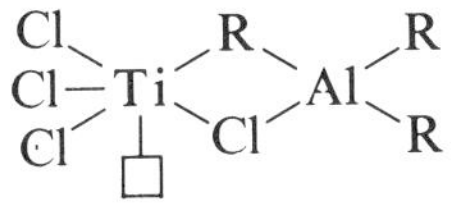

$$Cl_3Ti(\square)(\mu\text{-}R)(\mu\text{-}Cl)AlR_2$$

have been proposed by Rodriguez and van Looy[61] for Ziegler–Natta heterogeneous systems. Chirkow, Kissin and co-workers[54, 64] have recently proposed a model according to which the catalytic bimetallic complex would be situated on the {001} plane of the $TiCl_3$ crystals and would be characterised by two vacancies around the Ti atom. This hypothesis was convincingly contested by Boor[5].

D'yachkovskii, Shilov *et al.*[82, 83] came to the conclusion that, in the case of soluble catalysts from $AlMe_2Cl$ and $(C_5H_5)_2TiCl_2$, which are active in the polymerisation of ethylene, the active species is the $(C_5H_5)_2Ti^+(CH_3)$ cation.

The mechanisms proposed for the interpretation of the isotactic polymerisation of α-olefins have been extensively reviewed and discussed by Boor[5,6].

The mechanism that presently is most widely accepted may be written schematically as follows:

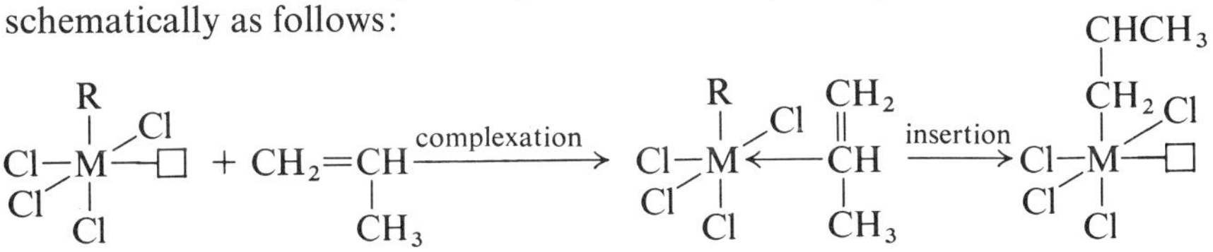

$$Cl_3M(R)\square + CH_2{=}CH(CH_3) \xrightarrow{\text{complexation}} Cl_3M(R)\leftarrow \begin{matrix} CH_2 \\ \| \\ CH(CH_3) \end{matrix} \xrightarrow{\text{insertion}} Cl_3M(CH_2\text{–}CHCH_3\text{–}R)\square$$

where: $\square$ = vacancy; R = alkyl; M = transition metal.

According to this scheme, which may also apply if the catalytic complex is bimetallic, olefin insertion occurs on the transition metal—carbon bond, preceded by complexation on a vacant site of the transition metal. The polymerisation is known to proceed by exclusive *cis*-addition to the double bond[40, 84–86].

According to Cossee[7, 87], Boor[5] and other authors, the kinetically determining step of the overall process, should be olefin insertion on the metal—carbon bond. This hypothesis is not easily adaptable to some experimental results. In particular, it does not easily agree with the linear dependence of the reaction rate on monomer concentration, observed by several authors for wide concentration ranges. More consistent with experimental findings is the hypothesis that the kinetically determining step is olefin complexation on the transition metal[50, 51]. It is also consistent with the fact that, in the case of the isospecific polymerisation of substituted styrenes with Ziegler–Natta catalysts based on Ti halides and aluminium alkyl compounds, the polymerisation rate increases with increase in the electron density of the double bond[88].

However, experimental data supporting the hypothesis of the olefin pre-complexation on the transition metal were obtained by Henrici–Olive and Olive[74], who studied soluble Ziegler–Natta catalysts active in the polymerisation of ethylene.

With regard to the mechanism of formation of isotactic polymers, the hypotheses put forth since 1960 by Cossee and Arlman and later worked out[5–7, 87] are still widely accepted. According to such hypotheses, the isospecific arrangement of the monomer units is due to steric hindrances resulting from the interaction of the monomer unit that polymerises with the asymmetric catalytic site and precisely with substituents (Cl) bound to the transition metal of the catalytic complex.

This hypothesis agrees with the results obtained by Pino and co-workers[102] on the stereoselective polymerisation of racemic α-olefins and with the results obtained by Zambelli[86], through n.m.r. analysis, concerning the nature of steric irregularity of isotactic polypropylene. However, Heatley *et al.*[43], in a study of the same type, reach different conclusions. These discrepancies are probably due to the fact that the polymers observed by the two groups of authors were prepared with different types of catalysts.

According to Rodriguez and van Looy[61], the isotactic arrangement should originate from the interaction between the methyl group of the coordinated propylene and an alkyl group of the aluminium alkyl of the catalytic complex.

In order to explain the formation of isotactic products, the Cossee–Arlman mechanism assumes that after insertion of each monomer unit on the metal–carbon bond, there takes place an inversion of the position around the transition metal atom of the vacancy and polymer chain. Such a migration would be questionable because it probably requires a higher activation energy than that experimentally observed for the overall polymeric chain growth[50, 51].

4.2.2 Syndiospecific polymerisation

The best catalysts for polymerisation of propylene to syndiotactic polymer are those systems obtained from vanadium compounds (e.g. VCl_4 $VOCl_3$,

$V(acac)_3$, $VO(OEt)Cl_2$) and $AlEt_2Cl$. The catalysts are soluble in hydrocarbon solvents and employed at low temperature (e.g. −78 °C), in the presence of a sufficiently weak Lewis base[89–94].

Zambelli, Natta, Pasquon *et al.*[91, 94, 95], by e.s.r. and kinetic analyses and by studying different catalytic systems, came to the conclusion that, in such systems, the catalytically active species consist of VCl_2Et, complexed with aluminium alkyl halides through chlorine-bridge bonds. Catalytically active complexes represent a small fraction (0.1–0.5 %) of the vanadium present in the system.

Also weak Lewis bases, e.g. anisole, can participate in the formation of the catalytic complex[94]. Lehr and Carman also[75, 96] conclude that active species contain VCl_2R. According to these authors, such species would be formed from an inactive trivalent vanadium compound, whereas, according to Zambelli *et al.*, the precursor may be tetravalent. The valency of V in the catalytic system was also investigated by Svab *et al.*[97, 98].

Determinations with ^{14}C-labelled $AlEt_2Cl$ revealed that the reaction occurs by insertion of the monomer units on the V–C bond[91].

The kinetic behaviour of the polymerisation reaction[94] also suggests that monomer-unit insertion on the metal–carbon bond occurs after monomer complexation on the transition metal of the catalytic complex. Also in this case the addition to the double bond is *cis*[84].

The syndiotactic arrangement was attributed to a methyl–methyl interaction between entering monomer and the last polymerised unit[92, 93]. In agreement with this hypothesis, Zambelli, Pasquon *et al.*[99] observed that in ethylene–propylene, ethylene–but-1-ene, and propylene–but-1-ene copolymerisations, carried out in the presence of syndiospecific catalysts, a strong tendency to an alternating sequence distribution exists; on the other hand, in copolymerisations carried out with isospecific systems, a random sequence distribution is observed.

The mechanism proposed by Boor and Youngman requires an insertion of the $V(CH_2\underset{\underset{CH_3}{|}}{CH})_nR$ type. According to Suzuchi and Taakegami[100, 101] insertion would be of the $V(\underset{\underset{CH_3}{|}}{CH}\,CH_2)_nR$ type. Such conclusions have been deduced from the analysis of the reaction products between pent-1-ene and syndiospecific catalytic systems, followed by hydrolysis. It is to be noted however that the active and syndiospecific catalytic species present in the above catalyst systems constitute a minute fraction of the whole amount of vanadium compounds present, several of which contain V—C bonds.

4.2.3 Isospecific polymerisation of asymmetric α-olefins

The polymerisation of asymmetric α-olefins has received much attention in recent years, in view of the advantages that the use of monomers of this type offers in the study of the mechanism of the stereospecific catalysis and of the conformational equilibria of the isotactic macromolecules in solution.

The relevant results obtained from previous work in this field can be summarised as follows[102]: (i) The isotactic polymerisation of racemic olefins has been found to be 'stereoselective', that is, each resulting macromolecule is made up of monomeric units deriving predominantly from one enantiomer of the monomer. (ii) Asymmetric catalysts (e.g. those prepared from optically active organometallic compounds) give optically active polymers, thus indicating that one enantiomer is polymerised preferentially (stereoelective polymerisation). (iii) Isotactic polymers of optically-active α-olefins exhibit a molar rotatory power much higher than that of low molecular weight paraffins having a structure similar to that of the monomeric unit; this has been attributed to the existence in solution of a helical conformation with the spiral predominating in one screw sense.

Work in this field has continued intensively in recent years, mainly by Pino and co-workers. The aim of this work was (i) to examine the quantitative aspects of the stereoselective polymerisation process, which was previously examined on a rather qualitative basis; (ii) to obtain additional data on the stereoelective polymerisation process, and (iii) to get additional proof of the helical conformation of the isotactic optically active macromolecules in solution.

The conformational studies will not be examined here, as they are not strictly related to the polymerisation process. This topic has been reviewed in a recent publication by Pino, Ciardelli and Zandomeneghi[105].

Reviews on stereoselective and stereoelective polymerisation of asymmetric α-olefins, covering the period until 1966 or later, have been published[102–105].

4.2.3.1 *Stereoselective polymerisation*

Previous evidence that the polymerisation of racemic α-olefins was a stereoselective process was mainly based on the fact that the isotactic polymers obtained from (*R*) (*S*) and from (*S*)-4-methylhex-1-ene were identical on x-ray examination[102]. This was considered indicative of the fact that the polymer from the (*R*)(*S*) monomer was not a random copolymer of the two enantiomers, but a mixture of macromolecules, each of which consisted of units derived largely from one enantiomer. It was also shown that the polymer obtained from the racemic monomer could be separated, by column chromatography on an optically-active support into optically active fractions of opposite sign[102]. However, the isolation of optically active fractions could not be considered by itself a proof of the occurrence of a stereoselective process since optically-active macromolecules can, in principle, be obtained also from a statistical copolymer of the (*R*) and the (*S*) enantiomer.

In order to get a better evaluation of the phenomenon of stereoselectivity in the polymerisation process, the chromatographic separation of the polymer obtained from various racemic monomers was systematically investigated[106–108]. It was found that separation depended not only on the prevalence of one enantiomer in the macromolecule, but also on molecular weight and stereoregularity[106]. Only those polymers having the asymmetric carbon atom α or β to the main chain could be separated[106, 108]. Highly crystalline poly-(S)-3-methylpent-1-ene was found to be particularly efficient as the optically active support[106].

A quantitative evaluation of the chromatographic separation was made by comparing the separability of equimolar mixtures of the polymers obtained from the (*R*) and the (*S*) enantiomer with the separability of the polymers obtained from the racemic monomer[108].

The results concerning the separation of poly-(*R*)(*S*)-3,7-dimethyloct-1-ene and poly-(*R*)(*S*)-4-methylhex-1-ene have shown that for these polymers the separability is not significantly lower than that of equimolar mixtures of the polymers from each enantiomer, whereas it is significantly higher than that expected for a statistical copolymer of the two enantiomers[108]. This is indicative of a stereoselective process.

In addition, it has been observed that a polymerisation can be stereoselective even when it is not highly stereospecific, as shown by the fact that the acetone-soluble fraction of poly-(*R*)(*S*)-3,7-dimethyloct-1-ene (amorphous because of low stereoregularity) has a separability higher than that expected for a statistical copolymer. On the basis of the above results Pino and co-workers suggested that the stereoselectivity is to be attributed to the intrinsic dissymmetry of the catalytic centres, rather than to asymmetric induction exerted by the growing chain[107,108]. This hypothesis was corroborated, according to the above authors, by the following findings: (i) The copolymerisation of (*R*)(*S*)-3,7-dimethyloct-1-ene with (*S*)-3-methylpent-1-ene gives isotactic macromolecules consisting prevailingly of homopolymers of (*R*)-3,7-dimethyloct-1-ene and of copolymers of (*S*)-3,7-dimethyloct-1-ene with (*S*)-3-methylpent-1-ene. Analogous results, confirming that the optically active comonomer copolymerises preferentially with the enantiomer of the racemic monomer having the same absolute configuration, were obtained in the copolymerisation of (*R*)(*S*)-3-methylpent-1-ene with (*R*)-3,7-dimethyloct-1-ene[109]. (ii) The copolymerisation of (*R*)(*S*)-4-methylhex-1-ene with C_2H_4 gives copolymers which have been found to result from a stereoselective process. The fact that the stereoselectivity is not affected by the incorporation of ethylene units indicates that it does not depend on asymmetric induction exerted by the growing chain[110].

Recently, studies with model compounds have shown that an optically active olefin coordinates to a metal preferentially in one of the two possible modes, as a result of the asymmetric induction exerted by the asymmetric carbon atom originally present in the molecule. This finding led Pino and co-workers to conclude that if a polymerisation is stereospecific it must be stereoselective[111,112].

4.2.3.2 *Stereoelective polymerisation*

In a series of papers a systematic investigation has been reported of the polymerisation of (*R*)(*S*)-3,7-dimethyloct-1-ene, (*R*)(*S*)-4-methylhex-1-ene and (*R*)(*S*)-3-methyloct-1-ene by asymmetric catalysts obtained from $TiCl_4$ or $TiCl_3$ and an organometallic compound of general formula MeR_n^* [Me = Li, Be, Al, Ga, In, Zn; R* = (*S*)—CH_2—$CH(CH_3)$—C_2H_5, (*S*)—$(CH_2)_2$—$CH(CH_3)$—C_2H_5, (*S*)—CH_2—$CH(CH_3)$—C_6H_5]. The results permit a better evaluation of the phenomenon of stereoelective polymerisation.

The catalysts in which the asymmetric carbon atom of the group R* is in β or γ position with respect to metal have been found to be stereoelective[113–115]. For the system MeR^*_n–$TiCl_4$, the stereoelectivity decreases in the order: $Zn \gg Al > In > Ga$ and is negligible for Li and Be[115].

Only racemic olefins in which the asymmetric carbon atom is α or β to the double bond can undergo stereoelective polymerisation[115]; the enantiomer with the same chirality of the group R* is preferentially polymerised[113–116].

However, the efficiency of the stereoelective polymerisation process with the above systems has been found to be rather modest; the optical purity of the isotactic polymers obtained with the most stereoelective catalysts are not higher than 10.5%. The maximum value observed for the relative polymerisation rate of the enatiomers is *c.* 1.2. On the basis of the above results it was believed[115] that the stereoelective power of the catalyst is attributable to the fact that one or more groups R* are bonded to the Ti atom in the catalytic complex. The presence of the asymmetric groups R* leads to the preferred coordination of the enantiomer having the same chirality as R*.

4.3 POLYMERISATION OF CONJUGATED DIOLEFINS

Most of the work carried out in this field in recent years concerns the polymerisation of butadiene, particularly to *cis*-1,4 polymers. Much attention has been given to the polymerisation by monometallic catalysts, especially with those derived from π-allyl derivatives of transition metals. A new class of polymers, named 'equibinary polydienes', has been obtained from butadiene and isoprene. Several reviews in this field have recently been published[117–120].

4.3.1 Butadiene, isoprene and pentadiene

4.3.1.1 Polymerisation by Ziegler–Natta catalysts

The *cis*-polymerisation of butadiene by catalysts from aluminium alkyls and cobalt compounds has been the subject of several papers. The rate of polymerisation with the $CoCl_2{\cdot}2$ py–$AlEt_2Cl$ system is remarkably increased by electron acceptors ($AlCl_3$, $SnCl_4$, $AlEtCl_2$), which also increase the molecular weight of the polymers obtained. Such electron acceptors, however, do not affect the polymerisation by the $CoCl_2{\cdot}2py$–$AlEtCl_2$ system[121].

Particular features are displayed by cobalt catalysts prepared from alkyl aluminium halides containing Al—O—Al bonds (aloxanes). These compounds are obtained from $AlEt_2Cl$ on reaction with water or distannoxanes $(R_3Sn)_2O$. Catalysts from aloxanes give polymers of higher molecular weight than those obtained by catalysts from $AlEt_2Cl$, $AlEtCl_2$ and $Al_2Et_3Cl_3$. Moreover, they are very active in some olefinic solvents, e.g. *cis*-but-2-ene, in which other catalysts have poor activity[122]. The molecular weight of the *cis*-polybutadiene and the polymerisation rate can be regulated by using catalysts prepared from specific blends of $AlEt_2Cl$ and aloxanes. The above results seem to indicate that aloxanes play an important role in the well-known activation by water of the cobalt-catalysed polymerisation of butadiene[122].

In the polymerisation of pentadiene the cobalt catalysts from aloxanes give *cis*-1,4-syndiotactic polymers in all of the solvents used (aromatic, aliphatic, olefinic). Catalysts from $AlEt_2Cl$ give instead 1,2-syndiotactic polymers in aliphatic solvents and polymers with a *cis*-1,4–1,2 mixed structure in benzene[123].

Recently, a systematic investigation of the polymerisation of pentadiene to *cis*-1,4 isotactic polymers by AlR_3–$Ti(OR)_4$ has been reported[124]; the results agree with those previously published[118].

A new type of catalyst for the vinyl-type polymerisation of conjugated diolefins, consisting of a combination of $Co(SCN)_2{\cdot}2py$ and $(Et_2Al)_2SO_4$, has been reported by Iwamoto and Yuguchi[125]. This system gives 1,2-syndiotactic polymers from butadiene and 3,4-polymers from isoprene.

Kinetic studies on the *cis*-polymerisation of butadiene have been performed[126, 127] with the two homogeneous systems $(Al(Bu^i)_3$–$TiCl_2I_2$ and nickel carboxylate–$BF_3{\cdot}OEt_2$–$AlEt_3$. With both catalysts the polymerisation is first order with respect to monomer and catalyst concentration. Only a small amount of the transition metal is active in the catalysis. No termination has been observed in either case, but transfer with monomer occurs with the boron–nickel system.

The *cis*-polymerisation of isoprene has received relatively little attention. The following systems have recently been proposed for this type of polymerisation: β-$TiCl_3$–$AlR_3(AlR_2Cl)$[28], $Ti(OR)_4$–$AlEtCl_2$[129] and $TiCl_4$-polyiminoalanes[130]. These systems appear to be variants of the classical AlR_3–$TiCl_4$ heterogeneous system to which no alternative has yet been found for the production of high molecular weight *cis*-polyisoprene.

4.3.1.2 *Polymerisation by π-allyl derivatives of transition metals*

A great deal of work has been carried out in recent years on the polymerisation of conjugated diolefins by π-allyl derivatives of transition metals[131–144]. The interest in these catalysts is easily understood if one considers that when a conjugated diolefin is polymerised by a transition metal catalyst the growing chain is bonded to the metal by a π-allylic bond. π-Allyl complexes constitute therefore a good model of the active-site structure. Table 4.1 summarises the results obtained with the most common π-allyl derivatives of transition metals. The stereospecificity, as well as the activity, depend both on the transition metal and on the nature of the ligands bonded to the metal. Thus $(\pi\text{-}C_4H_7)_3Cr$ gives crystalline 1,2-polybutadiene, whereas $(\pi\text{-}C_4H_7)CrCl_2$ gives *cis*-polybutadiene and the combination of $(\pi\text{-}C_4H_7)_3Cr$ with molecular oxygen gives *trans*-polybutadiene[141–143, 152]. $(\pi\text{-}C_4H_7)_4W$ is itself inactive for the polymerisation, but $(\pi\text{-}C_4H_7)_3WCl$ gives 1,2-polybutadiene of high molecular weight[144].

The influence of the anionic ligand has been extensively studied in the case of π-allylnickel derivatives of type π-allyl–Ni–X. These complexes polymerise butadiene to give almost exclusively 1,4-units. Some data concerning the influence of X on the polymer microstructure are summarised in Table 4.2.

Polymers with a high *cis*-content are obtained[145] with anionic ligands having a very high electron-withdrawing power, such as CF_3COO^-, CCl_3COO^-. The nature of the anionic ligand has a remarkable influence not only in the microstructure but also on the rate of polymerisation. Thus (π-C_3H_5) $NiOCOCF_3$ polymerises butadiene much faster than (π-C_3H_5)NiCl. Moreover, in the series of π-allylnickel halogenoacetates a relationship has

Table 4.1 Polymerisation of butadiene by π-allyl derivatives of transition metals

Catalyst	*Polymer microstructure*			*Reference*
	cis %	*trans* %	1,2 %	
(π-C_4H_7)$_3$Cr	—	20	80	141, 152
(π-C_4H_7)$CrCl_2$	90	5	5	141–143
(π-C_4H_7)$_2CrOCOCCl_3$	93	4	3	141–143
(π-C_4H_7)$_3Cr+\frac{1}{2}O_2$	—	90–95	—	141, 152
(π-C_4H_7)$_3$Nb	—	—	100	140
(π-C_4H_7)$_2$NbCl	91	5	4	140
(π-C_4H_7)NiCl	85–90	5–10	—	131, 138
(π-C_4H_7)$_4W+HCl$	—	8	92	144
(π-C_4H_7)$_4Mo+CCl_3COCl$	—	0.5	99.5	183
(π-C_4H_7)$_3$Rh	—	94	6	140
(π-$C_{12}H_{18}$)$RuCl_2+2PPh_3$	54	28	18	139

Table 4.2 Polymerisation of butadiene by π-allyl–Ni–X. Influence of the anionic ligand X on the polymer microstructure

X	*Polymer microstructure*		*Reference*
	cis %	*trans* %	
Cl	85–90	15–10	131, 138
Br	65–75	35–25	131, 138
I	—	100	131, 138
$OCOCF_3$	97	3	168, 169
$O-C_6H_4-NO_2$	97	5	—
$OSO_2\cdot C_6H_4\cdot CH_3$	50	50	184
$O\cdot C_6H_2Br_3$	—	100	184

been found between the electron withdrawing properties of the halogenoacetate anion and the catalytic activity, the latter increasing in the order: $CH_3COO < CH_2ClCOO < CHCl_2COO < CCl_3COO < CF_3COO$ [143, 146, 147].

The catalytic activity of the complex (π-C_3H_5)NiX is greatly influenced by addition of various types of electron acceptors, such as Lewis acids ($AlCl_3$, $AlBr_3$, $TiCl_4$, $SnCl_4$, $ZnCl_2$, etc.)[131, 138], halogen-containing quinones (fluoranyl, chloranyl, etc.)[148, 149, 151], CCl_3COCl, CCl_3CHO, CCl_3COCCl_3 [150]. Ionic complexes or charge-transfer complexes are formed in the reaction. In the case of π-allylnickel iodide or bromide the electron acceptors (A) also change the microstructure of the polymers, favouring the formation of *cis* units[134]. Thus polybutadienes having 90–97 % *cis* content have been obtained from combinations of (π-C_3H_5)NiI with CCl_3CHO (Ni:A = 1:2), CCl_3COCl (Ni:A = 1:2), fluoranyl (Ni:A = 2:1), CCl_3COCCl_3 (Ni:A = 1:1)[150].

Instead of π-allylnickel halides, other compounds such as bis-(π-allyl)nickel complexes, nickel hydrides and bis(cyclopentadienyl)nickel have been used, in combination with Lewis acids, to form active catalysts for the 1,4-polymerisation of butadiene[152,155]. The results are similar to those obtained from π-allylnickel halides.

Addition of electron donors (ethers, alcohols, water, phosphites, etc.) to π-allyl–Ni–X cause a decrease in the *trans* content. π-Allylnickel chloride or bromide, for example, which give prevailingly *cis* polymers in hydrocarbon solvents, give *trans* polymers in ethanol, tetrahydrofuran, or benzene saturated with water[131,153,154,156].

Molecular oxygen or peroxides have been reported to have some influence on the catalytic activity of π-allylnickel chloride[157]. Kinetic studies on the polymerisation of butadiene by π-allylnickel derivatives have been carried out[158].

4.3.1.3 Polymerisation by catalysts not containing metal–carbon bonds

Several new catalysts not containing metal–carbon bonds of σ- or π-allylic type have been reported. Halogen deficient crystals of $CoCl_2$ and $NiCl_2$ were found to be capable of polymerising butadiene to *cis* polymers[159]. The catalytic activity was attributed to the presence of very small amounts of cobalt and nickel monohalides. (Cyclo-octa-1,5-diene)NiX (X = Br, I) polymerises butadiene to *cis* (X = Br) and to *trans* (X = I) polymers[160]. A combination of reduced nickel with organic halides (e.g. benzyl bromide or chloride, benzoyl chloride) or with $MeSiCl_3$ were also found to be active catalysts for butadiene polymerisation[161]. In all the above systems metal–π-allyl bonds are believed to be formed on reaction with monomer.

Several papers deal with the polymerisation of butadiene to *trans* polymers by rhodium salts in aqueous emulsion or in protic solvents[162–165]. The results reported, concerning the influence of additional diolefins, the role of the surfactants and the use of trace compounds, confirm those of previous work. The activity of the rhodium catalysts in protic solvents has been attributed to the presence of a rhodium–allyl bond formed, in a way not yet clarified, on reaction of the rhodium salt with butadiene in the protic solvent.

Several bimetallic catalysts not containing metal–carbon bonds have been described. These include the following systems: $AlCl_3$–$CoCl_2$–Al–arene, $CoCl_2$–$AlCl_3$[166], $Ni(PCl_3)_4$–$TiCl_4$[167], $Ni(CO)_4$–VCl_4, $Ni(CO)_4$–WCl_6 and $Co_2(CO)_8$–$MoCl_5$[155]. All the above systems give *cis*-1,4-polymers from butadiene, with the exception of the $Co_2(CO)_8$–$MoCl_5$ system which gives 1,2-polybutadiene. The nature of the active species in the above systems has not been elucidated, but it is believed that π-allyl derivatives are formed on reaction with the monomer.

4.3.1.4 Polydiolefins having equibinary microstructures

Catalysts have been found that give polymers composed of only two of the various possible isomeric structures, in a practically 50:50 molar ratio from butadiene and isoprene. For these polymers the name 'equibinary polydienes' has been proposed. Up to now the following equibinary polymers have been obtained, with different catalyst systems: poly(*cis*-1,4-*trans*-1,4)butadiene, poly(*cis*-1,4-*trans*-1,4)isoprene, poly(*cis*-1,4–3,4)isoprene, and poly(3,4–1,2)-isoprene.

Poly(*cis*-1,4-*trans*-1,4)butadiene has been obtained by using as catalyst π-allylnickel trifluoroacetate modified with suitable additional ligands, such as CF_3COOH, benzene or styrene[154, 156]. Other additional ligands, such as triphenylphosphite, cause instead the formation of *trans*-1,4-polybutadiene. This shows the great versatility of π-allyl-$NiOCOCF_3$, which, depending on the added ligand, can give three types of polybutadienes, all with a very high 1,4 content

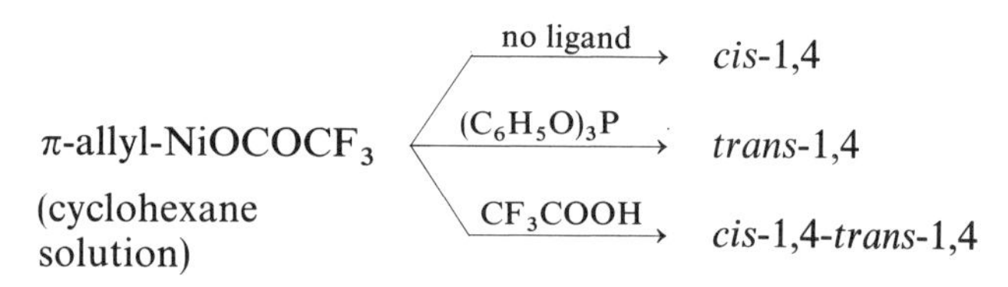

The modification of the catalytic species by the added ligand has a reversible character; the elimination of the compound modifying the stereospecificity (CF_3COOH, benzene) results in a catalyst promoting the *cis* polymerisation.

Polymers of isoprene having a *cis–trans* equibinary structure have been obtained by π-allylnickel trifluoroacetate in aromatic solvents[168]. Cobalt catalysts give two types of equibinary polyisoprenes; the systems CoX_2–$PhMgBr$–CH_3OH (1 :2·2 :2; X = F, Cl, Br, I, acac) and $AlEt_2(OEt)$–CoX_2 (5 :1) give poly(*cis*-1,4–3,4)isoprene, whereas the system CoF_2–$PhMgBr$–hexamethyl phosphoramide promotes the formation of a polymer reported to be 50% 3,4 and 50% 1,2[169–171].

The structure of the equibinary polymers, determined by spectroscopic methods, indicates a binary isomeric composition. Poly(*cis–trans*)butadiene has also been examined by x-rays and the results exclude the presence of long sequences of *trans* or *cis* units.

The formation of the equibinary polymers has been attributed to a modification of the initial symmetry of the catalytic complexes by coordination of the added ligands. The propagation may occur by alternate insertion of the two different structural units in the chain through non-symmetrical coordination positions in the catalytic complexes.

4.3.2 Other diolefins

Some work has been carried out on the polymerisation of conjugated diolefins other than butadiene, isoprene and pentadiene. Stereoregular polymers have been obtained from 2-methylpenta-1,3-diene, 4-methylpenta-1,3-diene, hexa-2,4-diene and 2-cyanobutadiene. The relevant data are reported in Table 4.3.

Table 4.3 Stereospecific polymerisation of some substituted butadienes

Monomer	*Catalyst system*	*Polymer structure*	*Reference*
2-Methylpenta-1,3 -diene* (*trans*-isomer)	AlR_3–$TiCl_4$	100% *cis*-1,4 crystalline‡	172
	$AlEt_3$–$Ti(OR)_4$–VCl_3	100% *trans*-1,4 crystalline‡	172
4-Methylpenta-1,3-diene†	$AlEt_3$–α–$TiCl_3$	100% 1,2 isotactic	173
	$AlEt_3$–$TiCl_4$	100% 1,2 isotactic	173
	$AlEt_3$–VCl_3	100% 1,2 isotactic	173
Hexa-2,4-diene (*trans–trans* isomer)	$AlEt_3$–$TiCl_4$	100% *trans*-1,4 crystalline§	174
	$AlEt_2Cl$–$Co(acac)_2$	100% *trans*-1,4 crystalline	174
	$AlEt_2Cl$–$Ti(acac)_3$	100% *trans*-1,4 crystalline	174
2-Cyanobutadiene	LiBu	100% 1,4 (not reported whether *cis* or *trans*); crystalline	175
	$AlEt_2Cl$	crystalline	175
	$AlEt_3$	crystalline	175

*Amorphous polymers having *trans*-1,4 structure have been obtained by the following catalysts: $AlEt_2Cl$–$CoCl_2$·2py–H_2O; $AlEt_3$–VCl_3; $AlEt_2Cl$–$V(acac)_3$; $AlEt_3$–$VOCl_3$ (Reference 172)

†Amorphous polymers having 1,2 structure have been obtained with the following catalysts: $AlEt_3$–$Ti(OBu)_4$; $AlEt_3$–$VO(OBu)_3$ (Reference 173)

‡It has not been reported whether the stereoregularity is of isotactic or syndiotactic type.

§A *trans*-1,4 erythro di-isotactic structure has been assigned to the polymers. This structure is however inconsistent with the reported identity period of 2.3 Å

4.3.3 Inclusion polymerisation

The polymerisation of diolefins included in a crystalline matrix has recently received renewed attention due to its considerable stereospecificity[176].

The discovery that *trans, anti, trans, anti, trans*-perhydrotriphenylene (PHTP) forms crystalline channel-like inclusion compounds with several substances having a linear structure and in particular with macromolecular substances[177], led Farina and co-workers to study the γ-ray radiation polymerisation of several monomers included in PHTP[178, 179]. Butadiene and 2,3-dimethylbutadiene polymerise in the presence of PHTP to yield crystalline polymers having a *trans*-1,4 structure.

Polymerisation of *cis*- or *trans*-penta-1,3-diene yields crystalline polymers having a *trans*-1,4 isotactic structure[180]. Furthermore, by irradiation of *trans*-penta-1,3-diene included in optically active PHTP, optically-active *trans*-1,4 isotactic polypentadiene was obtained[181]. This is a novel example of an asymmetric synthesis.

Structural studies on polymerisation of 2,3-dichlorobutadiene by the thiourea canal complex have also been reported[182].

4.4 POLYMERISATION OF CYCLO-OLEFINS

Previous work in this field was concerned with the polymerisation of norbornene, cyclobutene, 3-methylcyclobutene and cyclopentene. In recent years the investigation has been extended to the higher homologues of the series. One of the main achievements recently attained is the clarification of the mechanism of polymerisation of this class of monomers.

Since the polymerisation of small-ring cyclo-olefins is in some aspects different from that of the higher homologues, the two classes will be treated separately.

Recent reviews in this field have been published[185, 186].

4.4.1 Small-ring cyclo-olefins

4.4.1.1 Monomers containing a cyclobutene ring

Previous work has shown that cyclobutene can give three types of monomeric units (2–4):

(i) → (2)

(ii) → $\diagup CH_2\diagdown C{=}C \diagup H$ with H and $CH_2\diagup$ (3)

(iii) → H, H, $\diagdown CH_2 \diagup C{=}C \diagdown CH_2 \diagup$ (4)

and that scheme (i) can lead to two different types of crystalline polymers (polycyclobutylenamer-2), which were tentatively assigned an erythro di-isotactic and an erythro disyndiotactic structure[185]. 3-Methylcyclobutene was also polymerised, either via addition polymerisation (scheme (i)) or via ring-opening (schemes (ii) and (iii)). The influence of the transition metal of the catalysts upon polymer structure was investigated[185].

Recently, a systematic investigation of the polymerisation of cyclobutene and 3-methylcyclobutene by catalysts obtained from $VOCl_3$ or VCl_4 and metal alkyls of the 1–4 group has shown that both polymer yield and mode of polymerisation are influenced by the nature of the metal alkyl. Polymer yields decrease in the order: $AlEt_3 > BeEt_2 > MgEt_2 \gg LiBu > ZnEt_2 > SnBu_4$. Catalysts from $AlEt_3$ or $BeEt_2$ give polymers constituted by units of structure (2), whereas catalysts from LiBu polymerise via ring-opening[187].

1-Methylcyclobutene has been found to polymerise only with the aluminium alkyl–WCl_6 system; all the other catalysts employed for the polymerisation of cyclobutene and 3-methylcyclobutene are inactive for the polymerisation of this monomer. The polymers obtained have an irregular structure, similar to that of 1,4-polyisoprenes cyclised by cationic agents[188].

Two bicyclic monomers containing a cyclobutene ring, bicyclo[4.2.0]oct-7-ene and bicyclo[3.2.0]hepta-2,6-diene, have been polymerised by Ziegler–Natta systems, as well as by catalysts from Group VIII metal halides[189]. Only the cyclobutene moiety is active in the polymerisation, which can occur via addition to the double bond or via ring-opening, depending on the catalyst employed, as observed for cyclobutene.

4.4.1.2 *Norbornene and its derivatives*

It was previously shown[186] that norbornene can give three types of monomeric units (5–7):

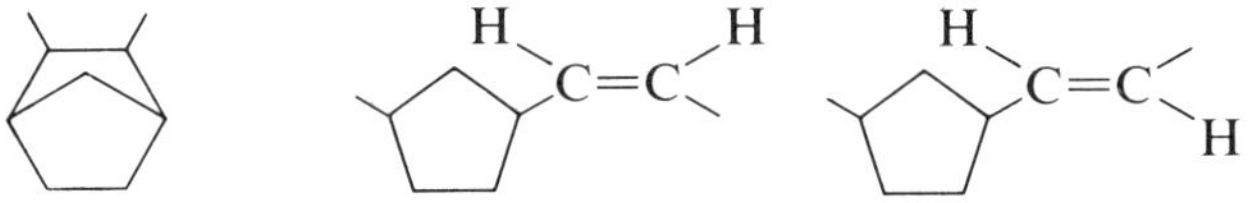

(5) (6) (7)

Catalysts of Ziegler–Natta type, or monometallic catalysts from halides of Group VIII metals were used for the polymerisation. New catalysts have now been investigated as reported in Table 4.4. In addition to norbornene, some 5-substituted norbornenes have also been polymerised. Selectivity was observed in the polymerisation of 5-carboxy- and 5-hydroxy-methyl-norborn-2-ene; only the *exo* isomer was polymerised. This fact, first observed by Michelotti and Carter, indicates that the coordination of the monomer, in this type of polymerisation, is *endo*, rather than *exo*.

Although it is now possible to prepare from norbornene and 5-substituted norbornenes polymers prevailingly constituted by units (5), (6) or (7), it is not clear which are the factors determining the orientation in the polymerisation.

4.4.2 Cyclopentene and higher cyclo-olefins

Previous work had shown that homopolymers of cyclopentene composed of cyclic saturated units could not be obtained, presumably because of steric factors[185].

Homopolymers of cyclopentene, however, were obtained via ring-opening polymerisation, initially by using heterogeneous molybdenum oxides–alumina catalysts[195], and later, in much higher yields, by using combinations of aluminium alkyls and molybdenum or tungsten halides[185]. Catalysts based on Group VIII metals, which are capable of polymerising cyclobutene and norbornene, were found to be inactive for the ring-opening polymerisation of cyclopentene[185].

Recent developments in this field include the following:

(i) the extension of ring-opening polymerisation to cyclo-olefins higher than C_5; (ii) improved catalyst systems and (iii) structural studies on the polyalkenamers and evaluation of their properties.

(i) The feasibility of polymerising cyclo-olefins higher than C_5 was first demonstrated by Natta and co-workers, who polymerised cycloheptene, cyclo-octene and cyclododecene by using catalysts obtained from $AlEt_3$ or $AlEt_2Cl$ and WCl_6, $WOCl_4$, $MoCl_5$ or other Mo compounds[196, 197]. Polyalkenamers of the *trans* type were obtained with all the catalysts used, even

Table 4.4 Catalysts for the polymerisation of norbornene

*Monomer**	*Catalyst*	*Solvent*	*Polymer structure*	*Reference*
R = —H, $—CH_2OCH_3$ $CH_3CO_2—$, $—CO_2CH_3$, $—CH_2OH$	$(NH_4)_2IrCl_6$	Aqueous emulsion	(6)+(7)	190
R = —H	$RuCl_3 \cdot 3H_2O$	Aqueous emulsion	(6)+(7)	190
R = —COOH(*exo*), $—CH_2OH$(*exo*)—$CH_2OCOCH{=}CH_2$	$IrCl_3 \cdot 3H_2O$	Ethanol	(6)+(7)	191
R = —H, $—CH_2OH$, $—CH_2OCH_3$, $—CH_2OAc$	$PdCl_2$	Bulk	(5)	192
R = —COOH	$PdCl_2$	Bulk	(6)+(7)	192
R = —H	$MoCl_5$	Carbon tetrachloride	(7)	192
R = —H	WCl_6	Carbon tetrachloride	(6)+(7)	193
R = —H	$ReCl_5$	Carbon tetrachloride	(6)+(7)	193
R = —H	$[IrCl(en)_2]_2$†	Aqueous emulsion Benzene, Bulk	(6)+(7)	194
R = —H	$[(\pi—C_8H_{11})PdCl]_2$	—	(5)	194

*Where not specified, 5-substituted monomers are a mixture of the *endo* and *exo* isomers
†(en) = cyclo-octene; $(en)_2$ = cyclo-octa-1,5-diene, cyclohexa-1,4-diene

with molybdenum catalysts, which with cyclopentene give the *cis*-polypentenamer. I.R. and x-ray examination showed a regular arrangement of units —CH=CH—$(CH_2)_n$—(n = 5, 6, 10) in the polymer. Cyclohexene was found to be inert to the polymerisation in the presence of all the catalysts used[196].

Calderon and co-workers[198] polymerised medium-ring cyclo-olefins using as catalyst a combination of $AlEtCl_2$ (or $AlEt_2Cl$) and ethanol-modified WCl_6. These authors found that not only simple cyclo-olefins such as cyclo-octene and cyclododecene, but also substituted cyclo-olefins and cyclo-olefins containing more than one double bond in the ring can polymerise via ring-opening. From cyclo-octa-1,5-diene and cyclododeca-1,5,9-triene they obtained linear polymers having a structure equivalent to 1,4-poly-butadiene, and from 3-methyl- and 3-phenylcyclo-oct-1,ene they obtained polymers constituted by units —CH=CH—CHR—$(CH_2)_5$— (R = Me, Ph). These findings indicate that no shift of the double bond occurs during the ring-opening polymerisation. The feasibility of polymerising substituted cyclo-olefins in the absence of secondary reactions suggested the application of the ring-opening polymerisation process for the preparation of perfectly alternating copolymers.

Indeed, Ofstead found that the polymerisation of substituted cyclo-octa-1,5-dienes (1,2-dimethyl-, 1-methyl-, 1-chloro-, 1-ethyl-cyclo-octa-1,5-diene) yielded polymers equivalent to an alternating butadiene–substituted butadiene structure[199]. From 1,2-dimethylcyclo-octa-1,5-diene, e.g. perfectly alternating butadiene–2,3-dimethylbutadiene copolymers were obtained:

$$\text{(1,2-dimethylcyclo-octa-1,5-diene; ring substituents Me, Me)} \longrightarrow -CH_2-CH=CH-CH_2-CH_2-\overset{\overset{\large Me}{|}}{C}=\overset{\overset{\large Me}{|}}{C}-CH_2-$$

(ii) Several types of activators have been proposed for the aluminium alkyl–WCl_6 catalysts[200, 201]. These include peroxides, molecular oxygen, alcohols, phenols, 1,3-dinitro-2,5-dichlorobenzene, t-butyl hypochlorite. The mechanism of this type of activation is difficult to interpret.

Several new types of tungsten catalysts have been proposed by Haas *et al.*[202]. These are (a) combination of a tungsten halide with alkali or alkaline earth metals, with Grignard compounds, with tin alkyls or with HSi compounds, or (b) a combination of tungsten aryl with $TiCl_4$, $SnCl_4$ or BCl_3.

The data of Table 4.5, show that efficiency and stereospecificity of the polymerisation of cyclopentene depend markedly on the catalyst used.

An interesting system for the polymerisation of cyclopentene consists of WF_6 and $Al_2Et_3Cl_3$. Depending on the Al:W molar ratio one can obtain both *cis*-polypentenamer (Al:W $\simeq$ 1) and *trans*-polypentenamer (Al:W $\simeq$ 7), and also all the intermediate *cis*:*trans* ratios[201].

In addition to the molybdenum or tungsten catalysts, systems based on Ta, Nb[203] and Re[204–206] have been proposed in the patent literature for the ring-opening polymerisation of cyclo-olefins.

The presence of small amounts of extraneous olefins[202, 207] and also of

Table 4.5 Influence of the catalyst system on the yield and on the structure of polypentenamer

Catalyst components		*Molar ratio*	*Polypentenamer*		*Reference*
A	B	A/B	*Yield* %	*trans* %	
$AlEt_3$	$TiCl_4$	2.5	1	95	185
$AlEt_3$	$MoCl_5$	2.5	27	1	185
$AlEt_3$	WCl_6	1.4	35	90.4	202
$AlEt_2Cl$	WCl_6	1.5	53	88	185
$RuCl_3$–ethanol			0		
$HSnEt_3$	WCl_6	1.75	35	90.2	185
$Cr(allyl)_3$	WCl_6	0.4	6	27.4	202
$Li_3[W(C_6H_5)_6]$	BCl_3	0.8	6	93.8	202
$Al[W(C_6H_5)_5]$	$TiCl_4$	0.85	7	60	202
$Li_3[W(C_6H_5)_6]$	$TiCl_4$	5	11	87.8	202

diolefins has been found to have a remarkable influence on the molecular weight of the polyalkenamers. The reason for this will be clear from the mechanism of the ring-opening polymerisation (see Section 4.4.3).

(iii) Structural studies and physical investigations on the whole series of polyalkenamers have been the object of several papers[208–210]. The technological evaluation of *trans*-polypentenamer, which exhibits very interesting elastomeric properties, has also been reported[201, 202, 211]. Those topics, however, will not be reviewed here.

4.4.3 Polymerisation mechanism

In the recent years the mechanism of the ring-opening polymerisation of cyclo-olefins has been thoroughly investigated. It was previously suggested that this type of polymerisation proceeded by cleavage of the carbon–carbon single bond adjacent to the double bond. Ring strain energy was believed to be the main contributor to the driving force of the reaction[185].

Scott and co-workers have recently shown[212] that, at least in the case of molybdenum- or tungsten-catalysed polymerisation, this mechanism is not valid. According to these authors the ring-opening polymerisation of medium-ring cyclo-olefins is a special case of olefin metathesis[213] and proceeds through scission of the olefinic bond to produce macrocyclic molecules.

Polymerisation by metathesis involves the following steps: (a) formation of a bis-olefin complex with the tungsten of the catalytic complex, (b) trans-alkylidenation, through formation of an intermediate of uncertain nature (a quasi-cyclobutane intermediate according to some authors[212], a carbene intermediate according to others[185, 215]); (c) olefin exchange, where a monomer replaces one of the double bonds coordinated to the transition metal:

$$\left[(CH_2)_n \begin{matrix} CH \\ \| \\ CH \end{matrix}\right] \rightarrow W' \leftarrow \left[\begin{matrix} CH \\ \| \\ CH \end{matrix} (CH_2)_n\right] \longrightarrow (CH_2)_n \begin{matrix} CH \cdots CH \\ \vdots \;\; W' \;\; \vdots \\ CH \cdots CH \end{matrix} (CH_2)_n$$

$$\downarrow$$

$$\left[(CH_2)_n \begin{matrix} CH{=}CH \\ \\ CH{=}CH \end{matrix} (CH_2)_n\right] \xleftarrow{\text{Cyclo-olefin}} (CH_2)_n \begin{matrix} CH{=}CH \\ \downarrow \\ W' \\ \uparrow \\ CH{=}CH \end{matrix} (CH_2)_n$$

$$\begin{matrix} \downarrow \\ W' \\ \uparrow \\ CH{=}CH \\ \left[(CH_2)_n\right] \end{matrix}$$

(W′ = tungsten of the catalytic complex)

Ring-strain energy is not the main contributor to the driving force of the reaction, since the basic reaction can be carried out with strain-free macrocyclic molecules or even open-chain olefins. The double bonds of the macrocyclic molecules formed in the polymerisation can undergo further metathesis reactions not only with the monomer itself, but also with other double bonds, in another macromolecule (inter-molecular metathesis) or in the same one (intramolecular metathesis). The polymerisation is therefore an equilibrium process.

Such a scheme is supported by the fact that macrocyclic molecules have actually been found in the polymerisation products of cyclo-olefins by the $AlEtCl_2$–C_2H_5OH–WCl_6 system[212, 214].

In the absence of side reactions only macrocyclic molecules are formed. By metathesis of a macrocyclic molecule with an acylic olefin, an open-chain macromolecule is formed. This reaction accounts for the influence of extraneous olefins on the molecular weight[202, 207].

At present, it is not clear whether the metathesis mechanism applies also to the ring-opening polymerisation of small-ring cyclo-olefins by catalysts based on Ru, Os or Ir, which do not polymerise medium-ring cyclo-olefins.

4.5 POLYMERISATION OF ACRYLIC MONOMERS

4.5.1 Acrylic esters

α-Alkylacrylates and some acrylates may be polymerised to isotactic, syndiotactic and stereoblock products, depending on the type of initiator, the nature of solvent (either polar or non-polar) and the temperature[1, 216].

The catalysts used are organometallic compounds, such as lithium, magnesium or cadmium alkyls, lithium aluminium hydrides, Grignard compounds and amidic magnesium, as well as metal sodium, in convenient solvents. These initiators act in the homogeneous phase.

More recently, it was found that Ziegler–Natta systems based on AlR_3–$VOCl_3$ [217] or on AlR_3–$TiCl_4$ [218], as well as $AlEt_3$ [219] polymerise methyl methacrylate to syndiotactic products at sufficiently low polymerisation temperatures. Polymerisation in the presence of $AlEt_3$ is initiated photochemically; in this case the mechanism is of a radical type[219].

An anionic mechanism was proposed for the polymerisation of methyl methacrylate, activated by Et_2AlNPh_2 in toluene at $-50\,°C$, to syndiotactic polymer[220].

Some acrylic esters, such as isopropylacrylate and methyl methacrylate yield syndiotactic polymers even if polymerisations were initiated by the conventional radical means provided that the operating temperature was low ($<0\,°C$) [1, 216, 221].

Highly isotactic poly(methyl methacrylates) were obtained by Blumstein *et al.*[222] by radical polymerisation of a monolayer adsorbed on sodium montmorillonite. Steric regularity of the polymer obtained in this case was attributed to a regular bi-dimensional arrangement of the monomer adsorbed.

Ballard *et al.*[223] have found that the polymerisation of methyl methacrylate is promoted by π-allylchromium compounds through an anionic coordinated mechanism.

Polymer stereoregularity depends considerably on the nature of both ester group and substituents in acrylic ester. Yuki *et al.*[224] observed, for instance, that the polymerisation of trityl methacrylate by Bu^nLi yields a highly isotactic polymer by operating in toluene and in tetrahydrofuran. The same polymer obtained by a radical process at 30 and 60 °C is less stereoregular, although its content of isotactic triads is relatively high. In the case of diphenylmethyl acrylate, the polymer obtained in toluene at high temperature is highly isotactic, whereas that obtained in tetrahydrofuran is highly syndiotactic.

The peculiar behaviour of trityl methacrylate suggests that, in this case, contrary to the hypotheses proposed for methyl methacrylate, stereoregularity may be attributed to the growing chain. For this monomer, the syndiotactic arrangement is not favoured for either anionic or radical polymerisations, owing to the steric bulkiness of the trityl group.

The same authors[225] studied also the influence of substituent on the stereoregularity of poly (α-alkyl acrylates).

Also the presence of heteroatoms (N or O) in the ester group conditions the stereoregularity of poly(methyl methacrylates) obtained with Bu^nLi, Bu^iMgBr, or $LiAlH_4$ in toluene or in tetrahydrofuran at −78 °C [226]. Highly stereoregular polymers, which are isotactic or syndiotactic depending on whether one operates in toluene solution or in the presence of tetrahydrofuran, were obtained by polymerisation of trimethylxylyl methacrylate in the presence of Bu^nLi [227, 228].

Okuzawa, Hirai *et al.*[229, 230] studied the radical polymerisation of 2:1 and 1:1 complexes of methyl methacrylate with $ZnCl_2$, $SnCl_4$ or other Lewis acids, with particular reference to the structure of the polymers obtained. Research in this field was also carried out by Otsu *et al.*[231]. The stereoisomeric composition of the polymers obtained from complexed and uncomplexed monomers, are influenced in a different way by the polymerisation temperature. The stereoregularity of the polymerisation process is conditioned by the structure of the complexes.

Several recent studies of the stereospecific polymerisation of α-alkyl acrylates and acrylates have been concerned with characterisation by i.r. analysis[221, 232], n.m.r. [233–240] or by fractionation techniques of the normal and hydrolysed polymers[241–244] obtained with different initiators.

Several authors[241–244] studied the formation of the so-called 'stereocomplexes' containing isotactic and syndiotactic macromolecules in an iso:syndio ratio of 1:1 and 1:2. These stereocomplexes, which preferably form in weakly-polar solvents, can be hardly distinguished from stereoblocks. In several cases, the stereospecific polymerisation of methyl methacrylate seems to be strictly connected with the formation of stereocomplexes. In particular it was observed that, during the polymerisation carried out at −50 °C with different Grignard compounds, the presence of preformed isotactic chains favours the formation of syndiotactic chains and vice-versa[243].

The polymerisation initiated by lithium alkyls, in particular Bu^nLi has been particularly studied in recent years[245–250]. Korotkov, Krasulina *et al.*[247–249] examined the kinetic aspects of the polymerisation in toluene at temperatures ranging from 0 to −80 °C. They proposed a single reaction mechanism for the formation of isotactic (in hydrocarbon solvents) syndiotactic (in polar solvents) or block poly(methyl acrylates).

Baca, Lim *et al.*[245–246] studied the influence of initiators, based on Li, Na and K tertiary-alkoxides on the stereoregularity of poly(methyl methacrylate).

The effect of amino and of alkoxyalcohols on the structure of polymethacrylate obtained in the presence of Bu^nLi was investigated by Maruhashi and Takida[250].

The systems based on Grignard compounds have also been thoroughly studied[233, 235, 243, 251–257]. In the case of polymerisations of methyl methacrylate initiated by amidic magnesium derivatives, both stereoregularity and activity extensively depend on the nature of the amidic group of the catalyst; in toluene at low temperature (−78 °C) compounds with a piperidine ring yield syndiotactic products, whereas organomagnesium compounds with a pyrazole ring yield highly isotactic products[256, 257]. In these cases, the degree of stereoregularity of the isotactic products does not seem to be influenced either by polymerisation temperature or by the solvent, contrary to what is observed for syndiospecific polymerisation.

Several hypotheses concerning the details of the polymerisation mechanism of methacrylates and of acrylates in the presence of organometallic compounds have been proposed[*, 225, 226, 233, 240, 249, 251, 258, 259].

Particularly interesting are the conclusions concerning the mode of double-bond opening of the monomer, derived from n.m.r. analysis of the polymers. It has been observed by Yoshino *et al.*[239] that in the anionic polymerisation of acrylates the type of opening is not always *cis* as in propylene polymerisation and depends on the monomer and on polymerisation conditions, e.g. the kind of catalyst and polymerisation temperature. For instance, in the anionic polymerisation of methyl acrylate -α-β-d_2 in toluene at −78 °C, with $LiAlH_4$ catalyst *trans* opening occurs and the process is stereoregular with respect both to the $CDCO_2R$ and the CD groups[260]. In the polymerisation of isopropyl acrylate initiated with phenylmagnesium bromide or diphenylmagnesium it has been shown (a) different modes of monomer double-bond opening i.e., the *cis* mode, the *trans* mode, and a 1:1 mixture of them, occur in different temperature ranges separated by narrow boundary regions, (b) the opening mode is determined in the initial very short period, and (c) the initially determined mode persists against temperature change beyond the boundary of its characteristic range[239].

Investigations by Fowells, Bovey *et al.*[240] have shown the differences in α- and β-carbon tacticity between polymers produced under different conditions to be related to the counter-ion of the catalyst, the ratio of Lewis base (diethyl ether or tetrahydrofuran) to initiator, and the temperature during polymerisation.

According to these authors[240] the differences in the configuration of polymers produced under different experimental conditions in the presence of organolithium compounds is due to differences in propagating species which can be present, namely, unsolvated contact ion pairs, peripherally solvated contact ion pairs and solvent-separated ion pairs. These will be present in different proportion depending on the solvent.

It has to be taken into account that the polymerisation mechanism may

---

*A complete literature on this topic, until 1967, is reported by Fujimoto, Kawabata and Furukawa[259].

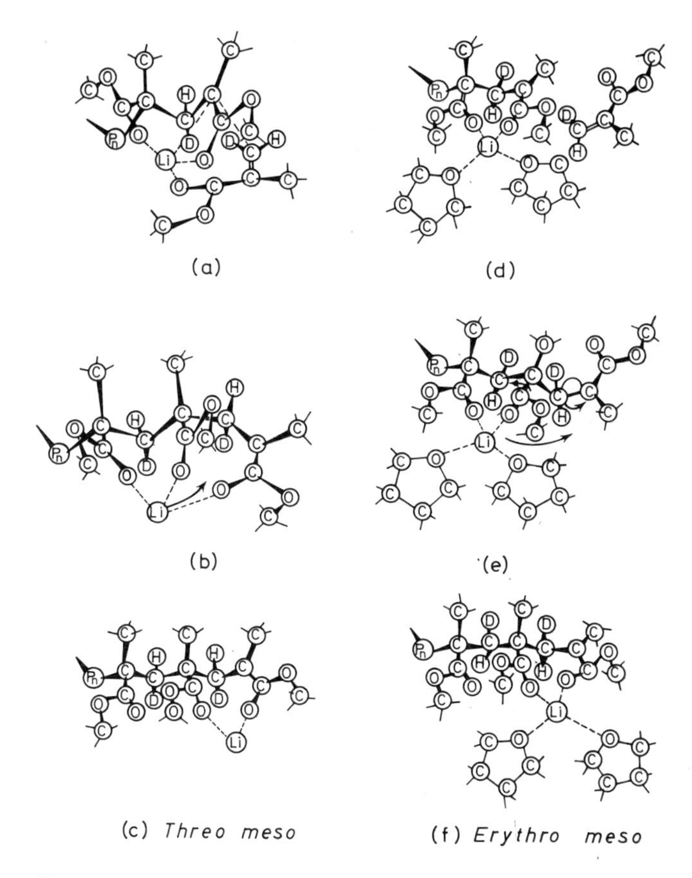

Figure 4.1 (a) An isotactic-like approach of the monomer to the chelated contact ion pair. (b) The new C–C bond has been formed with the methylene D on the same side of the zigzag as the ester function. (c) The Li^+ moves up to the new anion, with concurrent rotation of the new penultimate ester group, forming the same chelated structure as in (a). (d) A syndiotactic-like approach of the monomer to the peripherally solvated contact ion pair. There is no coordination of the monomer carbonyl with the counter-ion, and non-bonded interactions force it into a syndiotactic-like approach. (e) the new C–C bond has been formed. (f) The Li^+ and its peripheral solvent shell moves up to the terminal unit, with concurrent rotation of the ester function. As the new anion resides largely on the carbonyl, there is a simultaneous rotation about the new α,β bond to reduce charge separation. This results in an *erythro–meso* placement, the methylene D being on the opposite side of the zigzag from the ester groups and the α carbon now in an incipient isotactic configuration.
(Reprinted from Fowells, W., *et al.*[240], *J. Amer. Chem. Soc.*, 88 (1967), 1396. Copyright 1967 by the American Chemical Society. Reprinted by permission of the copyright owner.)

be remarkably conditioned by the nature both of the substituent present in the monomer and of those present in the initiator, as in the case of amidic magnesium compounds.

Of particular importance may also be the formation of polymeric stereocomplexes and the presence of preformed macromolecules in the reaction system.

All these different phenomena, combined with the fact that even the nature of the catalytic complexes is often not clearly known, emphasises the difficulties one has to face in proposing reaction mechanisms, which will fit all the experimental data. The schemes of two reaction mechanisms proposed by Fowells, Bovey *et al.*[240] concerning the polymerisation in the presence of Bu^nLi, in hydrocarbon solvent and in THF respectively, are shown in Figure 4.1.

4.5.2 α-Alkylacrylonitriles

The catalyst system for the stereospecific polymerisation of methacrylonitriles and of other α-alkylacrylonitriles to isotactic polymers[261] are based on compounds of alkylmagnesium of alkylberyllium, alkyllithium, alkylzinc, sodium alkylaluminium, lithium alkylaluminium, magnesium alkylaluminium, or on organometallic complexes of the same type as the previous ones, but containing a metal–nitrogen bond, especially of the diphenylamidic type[262, 263].

Joh *et al.*[262, 263] found that the stereospecificities of the latter catalysts are higher than those of the corresponding ones not containing metal–carbon bonds. In such catalysts the diphenylamidic group may play an important role in the steric control in the propagation step.

The same authors[264] found that stereospecificity of $MgEt_2$-based systems may be increased by the addition of ethers. Such polymerisations are generally carried out in the homogeneous phase above room temperature, either in toluene, dioxane or anisole. Their mechanism has not been thoroughly investigated but it seems to be of the anionic coordinated type[265].

Partially stereoregular, isotactic polymers of methacrylonitrile were obtained by radical polymerisation of complexes of the monomer with $ZnCl_2$or $SnCl_4$ [266], at a temperature ranging from −78 to 100 °C.

4.6 POLYMERISATION OF VINYL ETHERS

Vinyl ethers may be polymerised in a stereospecific way generally to isotactic polymers and sometimes even to syndiotactic polymers. The more stereospecific and conventional catalysts used in such polymerisations are of the Friedel–Crafts type, conveniently modified in order to reduce the cationic activity (e.g. $AlBrEt_2$, $AlBr_2Et$, $TiCl_2(acac)_2$. Conventional Friedel–Crafts catalysts, such as $AlCl_3$ or $BF_3{\cdot}OEt_2$ yield either amorphous or partially crystalline products, although they are more active than the modified catalysts.

Several other types of initiators, both homogeneous and heterogeneous, for use in hydrocarbon and in some cases in polar solvents, have been proposed for the stereospecific polymerisation of vinyl ethers[1, 267–269].

Stereoregularity of poly(vinyl ethers) depends greatly on the nature of catalyst, solvent and monomer, on the monomer concentration, and on the polymerisation temperature. The effect of some parameters on stereoregularity of poly(benzyl vinyl ether)[269], and poly(allyl vinyl ether)[270], obtained in the presence of $BF_3 \cdot OEt_2$ has been recently examined by Yuki *et al.*

More recently, investigations have been carried out on catalyst systems consisting of $HCl+AlCl_3$ [271] and modified Ziegler–Natta systems, such as $VCl_3 \cdot LiCl+AlR_3$ [272, 273]. With the latter catalyst, Joh, Yuki *et al.* came to the conclusion that for $Bu^i_3Al–VCl_3 \cdot LiCl$ ratios lower than five, the mechanism was predominantly cationic, whereas for higher values, it might be of the anionic coordinated type. These authors propose this hypothesis on the basis of the observation that $VCl_3 \cdot LiCl$ polymerises isobutyl vinyl ether to isotactic polymer, and styrene to an amorphous polymer, whereas the system $Bu^i_3Al–VCl_3 \cdot LiCl$ at Al:V ratio 6, yields isotactic polystyrene and isotactic poly(vinyl ethers) having high crystallinity.

Furukawa *et al.*[274, 275] studied the effect of β-alkyl and α-alkoxyl substitution on the relative polymerisabilities of α, β substituted unsaturated ethers. They observed that a β-methyl group on vinyl isobutyl ether increases its polymerisability in homogeneous catalysis (with $BF_3 \cdot OEt_2$), while it reduces the polymerisability in heterogeneous catalysis (e.g. with $Al_2(SO_4)_3–H_2SO_4$) probably because of the steric hindrance toward the adsorption of monomers on the catalytic surface.

With the homogeneous catalyst, as the substituents β-alkyl or α-alkoxyl become bulkier, the steric effect of the polymer end apparently overshadowed the electronic effect of substituents on the monomer reactivity.

The same authors[276] studied also the electronic effects of ring substituents on the polymerisability of substituted phenyl vinyl ether. The results obtained agree with Hammett correlation.

Interesting results for the understanding of reaction mechanisms have been obtained by Okamura[277–281] *et al.* from investigations on the polymerisation of β-substituted vinyl ethers. These authors observed[278, 279] that the steric structure of polymers of propenyl alkyl ethers (methyl, ethyl, isopropyl, butyl and t-butyl) depends on the monomer structure and on the polymerisation conditions as shown overleaf.

The different behaviour of the two types of catalysts is essentially due to the different physical state (homogeneous for the boron catalyst and heterogeneous for the aluminium one).

According to the relationship between the monomer and polymer structures proposed by Natta *et al.*[282], one deduces that double-bond opening can be either *trans* (solid lines in scheme) or *cis* (broken lines).

From n.m.r. analysis, of poly(methyl propenyl ether) it has been observed[279] that the tacticities of the α-C were different from those of the β-C in all the polymers obtained. The crystalline polymer obtained from *trans* or *cis* monomer is always threo-di-isotactic. Though the configurations of all α-carbon were isotactic, a small amount of syndiotactic structure was observed in β-C. In the amorphous polymer obtained from *cis* monomer by the homogeneous $BF_3 \cdot OEt_2$ catalyst, the configuration of the α-C was isotactic, but the β-C was atactic. These facts suggest that the type of opening of a monomeric double bond is complicated, or that C—C double bond in an incoming monomer rotates in the transition state.

By ^{13}C n.m.r. spectroscopy, the same authors[280] concluded that the monomer reactivity was not parallel to the π-electron density of the β-carbon in the vinyl ethers, or in the alkenyl ethers. Thus the attack of the carbonium ion on the β-carbon is not the rate-determining step in the cationic polymerisation of vinyl ether derivatives. In this case, a transition state in which

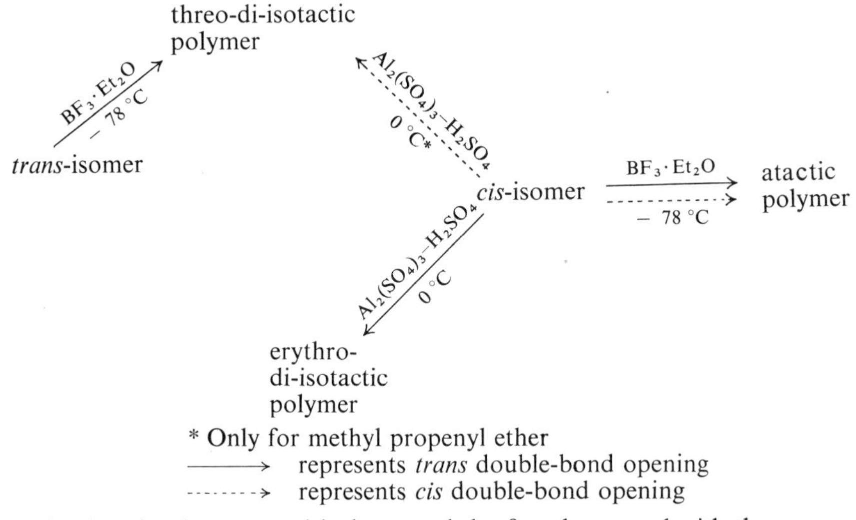

* Only for methyl propenyl ether
——→ represents *trans* double-bond opening
- - - - → represents *cis* double-bond opening

carbonium ion interacts with the α- and the β-carbons, and with the oxygen of the monomer was proposed.

This model can explain the relative reactivity of vinyl ethers and β-substituted vinyl ethers observed in the cationic copolymerisation catalysed by $BF_3 \cdot OEt_2$.

According to Fueno *et al.*[283], the polymeric chain ends should approach the monomer from above the centre of its double bond and interact with both the α- and β-carbon giving a π-complex and gradually forming a bond with the β-carbon of the monomer.

Optically active vinyl ethers with different structures have been obtained by different polymerisation methods, by Pino and co-workers[285–289] and by Liquori and co-workers[284]. Most of the work done in this field concerns conformational studies of the isotactic macromolecules in solution. This topic, which has been reviewed in a recent publication[105], will not be examined here.

Polymerisation of racemic 1-methylpropyl vinyl ether by the stereospecific catalyst system $Al(OPr^i)_3-H_2SO_4$ was shown to be stereoselective[287]. Moreover, copolymerisation of racemic 1-methylpropyl vinyl ether (8) with several optically-active vinyl ethers by the same system, gives products consisting of copolymers of the optically–active comonomer with the enantiomer of (8) having the same configurations and of homopolymers of the other enantiomer of (8)[289]. These facts are analogous to those observed in the polymerisation of racemic α-olefins by Ziegler–Natta catalysts.

4.7 POLYMERISATION OF EPOXIDES

Research in the field of the stereospecific polymerisation of epoxides containing an asymmetric carbon atom, which had been accomplished before 1966, was reviewed by Tsuruta[290], Gurgiolo[291], Furukawa and Saegusa[292] and more recently by Ishii and Saki[293].

Among the earlier catalysts used for the polymerisation – at least partly stereospecific – of epoxides, we mention the following:complex $FeCl_3$–propylene oxide, $Fe(OEt)_3$, organometallic compounds in the presence of water, alcohol or a chelating compound (e.g. AlR_3–H_2O, AlR_3–(acac), ZnR_2–H_2O, ZnR_2–ROH) or alcoholates used in the presence of metal halide (e.g. $Al(OR)_3$–$ZnCl_2$, $Al(OR)_3$–$FeCl_2$, $Ti(OR)_4$–$ZnCl_2$.

The literature also reports systems consisting of an aluminium hydride complex and a chelating compound (e.g. biacetyl monoxime or acetylacetone[294]), decacarbonyldimanganese[295] and others based on $BeEt_2$–H_2O [296]. These catalyst systems which may be homogeneous or heterogeneous are generally employed at room temperature or above, in hydrocarbon or other solvents. Some of them, e.g. the system AlR_3–H_2O (1 :1), are active even at -78 °C.

In the recent years, the systems based on $ZnEt_2$–H_2O and on $AlEt_3$–H_2O, have been studied extensively. The nature of the active species in these polymerisations has not yet been fully elucidated. According to Ishimori, *et al.*[297], the active species in $ZnEt_2$–H_2O systems, are $Et(ZnO)_nZnEt$ $Et(ZnO)_nH$, or $HO(ZnO)_nH$, the zinc–oxygen bonds having disordered würzite structures.

The importance of the catalyst structure, is evident from the observations by the same authors[298] that amorphous $Zn(OMe)_2$ prepared from $ZnEt_2$ and MeOH was very active in the polymerisation whereas crystalline $Zn(OMe)_2$, used in the same conditions was practically inactive. They also demonstrated by using catalytic systems prepared from $ZnCl_2 + MOR$ (M = Li, Ne and K), that the presence of the alkyl–zinc bond in zinc alkyl–alcohol systems is not necessarily an indispensable factor for the formation of high polymers[299].

According to Nakaniwa, *et al.*[300], the catalytic activity of the $ZnEt_2$–H_2O based systems is proportional to the amount of both Zn–O and Zn–Et bonds bound to the catalyst. The active species for the formation of high moleçular weight propylene oxide is thus not a simple Zn—O bond but Zn—O bond co-existing with Et—Zn bond.

According to Ishimori *et al.*[297], the polymerisation proceeds simultaneously through a cationic mechanism involving a process of S_N1-type ring-opening and through an anionic (or coordinated anionic) mechanism involving a process of S_N2-type ring-opening reaction. The latter process can be considered as the main reaction of the propagation step. In this case the $\beta(CH_2O)$ bond of the expoxide is preferentially opened.

The mechanism proposed by Ishimori and Tsuruta implies the formation of a four-centre intermediate. According to Vandenberg[301], who systematically investigated the AlR_3–H_2O system, with or without acetylacetone, both in polar and non-polar solvents, the coordination polymerisation cannot propagate through a four-centre intermediate containing only one metal atom.

For the AlR_3–H_2O system a simple mechanism based on two metal atoms is shown

Bimetallic active species, containing Al and Zn have been also proposed by Kuntz and Kroll[302], who studied catalytic systems based on R_2Al–acac–0,5H_2O–R_2Zn.

Vandenberg[301] also reached some other interesting conclusions which can be summarised as follows:

(i) Epoxides polymerise with inversion of configuration of the ring-opening carbon atom. This conclusion is true for both monosubstituted and symmetrically disubstituted epoxides and applies to all presently known mechanisms of polymerisation, i.e., cationic, anionic and coordination.

(ii) Monosubstituted epoxides with a coordination catalyst polymerise largely by attack on the primary carbon.

(iii) With coordination catalysts, it appears that stereoregularity is controlled by steric hindrance inherent in metal sites.

Studies on the polymerisation of *cis*- and *trans*-2,3-epoxybutanes with cationic catalysts, indicate that methods other than steric hindrance at a catalyst side can control the stereoregularity of epoxide polymerisation. For the above monomer, stereoregularity appears to be due to a difference in steric hindrance between the two enantiomorphs, i.e. it is more favourable sterically to add the same enantiomorph than the opposite one and it does not require a favourable steric site in the catalyst.

In the polymerisation of *meso cis*-oxide with cationic catalyst a disyndiotactic polymer is obtained. Here too the stereoregularity is independent of the catalyst used and it appears that steric factors inherent in the last unit of the growing polymer chain cause the next unit added to have an opposite steric configuration.

Further information on the reaction mechanism has been obtained from the polymerisation of optically active monomers. Inoue *et al.*[303] observed an asymmetric selection of the antipode monomer in the copolymerisation of D- and L-propylene oxide by $ZnEt_2$–H_2O. They interpreted this result by the formation in the initiating reaction of an asymmetric structure in the active species.

The importance of the steric configuration of the monomer unit in the polymerisation of propylene oxide was further evidenced by Livshits *et al.*[304] who observed that the polymerisation rate of optically active isomers in the presence of magnesium oxalate is twice that of the racemic mixture.

Furukawa *et al.*[305–307] by investigations carried out with optically-active catalytic systems (such as (+)bis(2-methylbutyl)zinc–water or $ZnEt_2$–(+) L-amino acids) used in the polymerisation of racemic propylene oxide, confirmed that polymerisation is stereoselective.

Note added in proof

Anderson, Hoover and Vogl have found[308] that various cyclo-olefins (cyclopentene, cyclohexene, 3-methylcyclohexene, cycloheptene, cyclo-octene, and cyclododecene) give polymers by 1,2 addition polymerisation at 300 °C and 65 000 atm (65 kbar). Molecular weights are rather low, as indicated by inherent viscosities of 0.05–0.1 dl/g.

The polymerisation of dicyclopentadiene by various transition metal catalysts has been reported[309]. Different polymers have been obtained depending on the catalyst.

Norbornene, cyclopentene, cyclo-octene and cyclododecene have been polymerised by various catalysts consisting of a combination of a π-allyl derivative of transition metal with a Lewis acid[310, 311]. Cyclopentene has been reported[311] to give addition polymers by catalysts obtained from π-allyl-NiX (X = Cl, I) and $AlBr_3$, $TiCl_4$, $MoCl_5$. This seems to be the first example of addition polymerisation of cyclopentene by transition metal catalysts.

References

1. Pasquon, I. (1970). *'Stereoregular Linear Polymers'* in *Encyclopedia Polymer Science and Technology*, vol. 13. (New York: Wiley)
2. Ketley, A. D., Ed. (1967–1968) *'The Stereochemistry of Macromolecules'*, 3 vols. (New York: Marcel Dekker)
3. Bawn, C. E. H. (1962). *Proc. Chem. Soc. (London)*, 165
4. Berger, M. N., Boocock, G. and Harward, R. N. (1969). *Advan. Catal.*, **19**, 24
5. Boor, J., Jr. (1970). *Ind. Eng. Chem., Prod. Res. Develop.*, 9(4), 437
6. Boor, J., Jr. (1967). *'Nature of the Active Site in the Ziegler-Type Catalyst'*, in *'Macromolecules Reviews'*, vol. 2, Ed. by A. Peterlin, M. Goodman, S. Okamura, B. H. Zimm and H. F. Mark. (New York: Interscience)
7. Cossee, P. (1967). *'Mechanisms of Ziegler-Natta Polymerization II. Quantum—Chemical and Crystal—Chemical Aspects'*, in Stereochemistry vol. 1, Ed. by A. D. Ketley. (New York: Marcel Dekker)
8. Coover, H. W., Jr., McConnell, R. L. and Joyner, F. B. (1967). *Relationship of Catalyst Composition to Catalytic Activity for the Polymerisation of α-Olefins'* in *'Macromolecular Reviews'* vol. 1. (New York: Interscience)
9. Henrici-Olive, G. and Olive, S. (1969). *Advan. Polym. Sci.*, **6**, 421
10. Hoeg, D. F. (1967). *'Mechanisms of Ziegler-Natta Catalysis I. Experimental Foundations'* in *'Stereochemistry of Macromolecules'*, vol. 1, Ed. by A. D. Ketley. (New York: Marcel Dekker)
11. Jordan, D. O. (1967). *'Ziegler–Natta Polymerization'. Catalysis, Monomers and Polymerization Procedures'* in *'Stereochemistry of Macromolecules*, vol. 1, Ed. by A. D. Ketley. (New York: Marcel Dekker)
12. Pasquon, I., Valvasorri, A. and Sartori, G. (1967). *'Copolymerization of Olefins by Ziegler–Natta Catalysts'* in *'Stereochemistry of Macromolecules'*, vol. 1, Ed. by A. D. Ketley. (New York: Marcel Dekker)
13. Smith, W. E. (1969). *'Mechanism of Stereospecific Polymerization of Propylene'* in *'Vinyl Polymers'*, part II, Ed. by G. E. Ham. (New York: Marcel Dekker)
14. Kennedy, J. P. and Otsu, T. (1970). *Adv. Polym. Sci.*, **7**, 369
15. Tornqvist, E. G. M. (1969). *Ann. N.Y. Acad. Sci.*, **155**, 447

16. Badische Anilin und Soda Fabrik A.G., *Belg. Pat.* (Appl. 7 Aug. 1968)
17. Shiba, T., Shih, C. C. and Takashima, K. (1966). *Kogyo Kagaku Zaschi,* **69,** 1003
18. Revillon, A., Daniel, T. C. and Guyot, A. (1968). *J. Chim. Phys. Physicochim. Biol.,* **65,** 857
19. Buniyat-Zade, A. A., Bulatnikova, E. L. and Dalin, M. A. (1968). *Dokl. Akad. Nauk. Azerb. SSR,* **24,** 31
20. Chirkov, V. N., Bulatnikova, E. L. and Buniyat-Zade, A. A. (1970). *Azerb. Khim. Zh.,* **3,** 114
21. Clark, A. (1969). *Advan. Chem. Ser.,* **91,** 387. (Amer. Chem. Soc.: Washington); *ibid.* (1969). *Catal. Rev.,* **3,** 145
22. Stevens, J. (1970). *Hydrocarbon Processing,* **49,** (11), 179
23. Sato, A., Takeda, S., Konotsune, S., Kato, M. and Tanoike, T. *Germ. Pat.* 1 904 815 (Appl. 3 Febr. 1968–9 Jul. 1968)
24. Kashiwa, N., Tokuzumi, T., Otake, H. and Fujimura, H. *Germ. Pat.,* 1 939 074 (Appl. 1 Aug. 1968–30 Dec. 1968)
25. Montecatini Edison S.p.A., *Belg. Pat.* 742 003 (Appl. 20 Oct. 1969); *ibid.* 742 112 (Appl. 24 Oct. 1969); *ibid.* 747 727 (Appl. 20 Mar. 1970); *ibid.* 752 395 (Appl. 23 June 1970)
26. Murray, J., Scharp, M. J. and Hockey, J. A. (1970). *J. Catal.,* **18,** 52
27. Marconi, W., Mazzei, A., Cesca, S. and DeMaldé, M. (1969). *Chim. Ind. (Milan),* **51,** 1084
28. Cesca, S., Santostasi, M. L. and Bertolini, G. (1969). *Chim. Ind. (Milan),* **51,** 1093
29. Zambelli, A., Segre, A. L., Marinangeli, A. and Gatti, G. (1966). *Chim. Ind. (Milan),* **48,** 1
30. Zambelli, A., Gatti, G., Marinangeli, A., Cabassi, F. and Pasquon, I. (1966). *Chim. Ind. (Milan),* **48,** 333
31. Zambelli, A., Giongo, G. M. and Segre, A. L. (1969). *Chim. Ind. (Milan),* **50,** 1185
32. Watt, W. R. (1969). *J. Polym. Sci. A-1,* **7,** 787
33. Watt, W. R. and Fischer, C. D. (1968). *J. Polymer Sci. B,* **6,** 109; *ibid.* (1969). *A-1,***7,** 2815
34. Matsumura, K., Atarashi, Y. and Fukumoto, O. (1969). *J. Polym. Sci. A-1,* **7,** 511
35. Matlak, A. S. and Breslow, D. S. (1965). *J. Polym. Sci. A,* **3,** 2853
36. Benning, C. J., Wszolek, W. R. and Werber, F. X. (1968). *J. Polym. Sci. A-1,* **6,** 775
37. Herwig, H., Gumboldt, A. G. M. and Weissermal, R. V. *U.S. Pat.* 3 501 415 (1970)
38. Idemitsu Kosan Co. Ltd. *Brit. Pat.* 1 184 593 (Appl. Aug. 9, 1966); [Chem. Abstr. (1970), **72,** 122 180]
39. Giannini, U., Zucchini, U. and Albizzati, E. (1970). *J. Poly. Sci. B,* **8,** 405
40. Zambelli, A., Segre, A. L., Farina, M. and Natta, G. (1967). *Makromol. Chem.,* **110,** 1
41. Segre, A. L. (1968). *Macromolecules,* **1,** 93
42. Zambelli, A. and Segre, A. L. (1968). *J. Polym. Sci. B,* **6,** 473
43. Heatley, F., Salovey, R. and Bovey, F. A. (1969). *Macromolecules,* **2,** 619
44. Heatley, F. and Zambelli, A. (1969). *Macromolecules,* **2,** 618
45. Flory, P. J. (1970). *Macromolecules,* **3,** 613
46. Shelden, R. A. (1969). *J. Polym. Sci. A-2,* **7,** 1111
47. Shelden, R. A., Fueno, R. and Furukawa, J. (1969). *J. Polym. Sci. A-2,* **7,** 763
48. Luisi, P.L., Mazo, R. M. (1969). *J. Polym. Sci. A-2,* **7,** 775
49. Elias, H. G. (1970). *Makromol. Chem.,* **137,** 277
50. Pasquon, I. (1967). *J. Pure Appl. Chem.,* **15,** 465
51. Natta, G., Zambelli, A., Pasquon, I. and Giongo, G. M. (1966). *Chim. Ind. (Milan),* **48,** 1298
52. Guttman, J. Y. and Guillet, J. E. (1968). *Macromolecules,* **1,** 461; *ibid.* (1970). **3,** 470
53. Coover, H. W., Jr., Guillet, J. E., Combs, R. L. and Joyner, F. B. (1966). *J. Polym. Sci. A-1,* **4,** 2583
54. Kissin, Yu. V., Mezhikovskii and Chirkov, N. M. (1970). *Europ. Polym. J.,* **6,** 267
55. Natta, G., Zambelli, A., Pasquon, I. and Giongo, G. M. (1966). *Chim. Ind. (Milan),* **48,** 1307
56. Keii, T., Soga, K., Go, S., Takaheschi, A. and Kojima, A. (1966). *J. Polym. Sci. C,* **23,** 453
57. Schnecko, H., Lintz, W. and Kern, W. (1967). *J. Polym. Sci. A-1,* **5,** 205
58. Lovering, E. G. and Wright, W. B. (1968). *J. Polym. Sci. A-1,* **6,** 2221
59. Rodriguez, L. A. M., van Looy, H. M. and Gabant, J. A. (1966). *J. Polym. Sci A-1,* **4,** 1905

60. van Looy, H. M., Rodriguez, L. A. M. and Gabant, J. A. (1966). *J. Polym. Sci. A-1*, **4,** 1927
61. Rodriguez, L. A. M. and van Looy, H. M. (1966). *J. Polym. Sci. A-1,* **4,** 1951; *ibid.,* 1971
62. Natta, G. and Pasquon, I. (1969). *Advan. Catal.,* **11,** 1
63. Kollar, L., Simon, A. and Kallo, D. (1968). *Magy. Kem. Foly,* **74,** (7) 289; [*Chem. Abstr.* (1968), **69,** 70208]
64. Kissin, Yu. V. and Chirkov, N. M. (1970). *Eur. Polym. J.,* **6,** 525
65. Buls, V. W. and Higgins, T. L. (1970). *J. Polym. Sci. A-1,* **8,** 1037
66. Simon, A., Jarivitzky, P. A. and Overberger, C. B. (1966). *J. Polym. Sci. A-1,* **4,** 2513
67. Overberger, C. G., Jarivitsky, P. A. and Mukamol, H. (1967). *J. Polym. Sci. A-1,* **5,** 2487
68. Gumboldt, A., Helberg, J. and Schlestzer, G. (1967). *Makromol. Chem.,* **101,** 229
69. Kollar, L., Simon, A., Osvath, J. (1968). *J. Polym. Sci. A-1,* **6,** 919
70. Adema, E. H. (1968). *J. Polym. Sci. C,* **16,** 3643
71. Schindler, A. (1966). *Makromol. Chem.,* **90.** 284; *ibid.* (1967), **102,** 263; *ibid.* (1968), **114,** 77; *ibid.* (1968), **118,** 1; ibid. (1966), *J. Polym. Sci. B,* **4,** 193
72. Schindler, A. and Strong, R. (1966). *Makromol. Chem.,* **90,** 284
73. Henrici-Olive, G. and Olive, S. (1969). *J. Polym. Sci. C,* **22,** 965; *ibid.* (1969). *Makromol. Chem.,* **121,** 70
74. Henrici-Olive, G. and Olive, S. (1969). *J. Organometal. Chem.,* **16,** 339
75. Lehr, M. H. (1968). *Macromolecules,* **1,** 178
76. Takada, M., Iimura, K., Nozawa, Y., Hizatome, M. and Koide, N. (1968). *J. Polym. Sci. C,* **23,** 741
77. Begley, J. W. and Pennella, F. (1967). *J. Catal.,* **8,** 203
78. Zambelli, A., Pasquon, I., Marinangeli, A., Lanzi, G. and Mognaschi, E. R. (1964). *Chim. Ind. (Milan),* **46,** 1464
79. Henrici-Olive, G. and Olive, S. (1968). *Angew. Chem. Int. Ed. Engl.,* **7,** 822
80. Reichert, K. H. and Malmann, M. (1969). *Angew. Chem. Int. Ed. Engl.,* **8,** 217
81. Henrici-Olive, G. and Olive, S. (1970). *J. Polym. Sci. B,* **8,** 205
82. D'yachkovskii, F. S., Shilova, A. R. and Shilov, A. E. (1966). *Polym. Sci. USSR,* **8,** 336; *ibid.* (1967). *J. Polym. Sci. C,* **16,** 2333
83. D'yachkovskii, F. S., Eritsyan, O., Kashireminov, Y., Matyska, B., Mach, K., Svestka, M. and Shilov, A. E. (1969). *Vysokomol. Soedin. Ser. A,* **11,** 543
84. Zambelli, A., Giongo, G. M. and Natta, G. (1968). *Makromol. Chem.,* **112,** 183
85. Miyazawa, T. and Ideguchi, T. (1963). *J. Polym. Sci. B,* **1,** 389
86. Zambelli, A. (1970). *Seventh Colloquium on N.M.R. Spectroscopy* (Aachen)
87. Cossee, P., Ros, R. and Schachtschneider, J. H. (1968). *4th International Symposium Catalysis (Moscow)*
88. Natta, G., Danusso, F. and Sianesi, D. (1959). *Makromol. Chem.,* **30,** 238
89. Natta, G., Pasquon, I. and Zambelli, A. (1962). *J. Amer. Chem. Soc.,* **84,** 1488
90. Zambelli, A., Natta, G. and Pasquon, I. (1963). *J. Polym. Sci. C,* **4,** 411
91. Zambelli, A., Natta, G., Pasquon, I. and Signorini, R. (1967). *J. Polym. Sci. C,* **16,** 2485
92. Boor, J. Jr. and Youngman, E. A. (1966). *J. Polym. Sci. A-1,* 1861
93. Youngman, E. A., Boor, J. Jr. (1967). *'Syndiotactic Polypropylene'* in *'Macromolecular Reviews',* vol. 2. (New York: Interscience)
94. Zambelli, A., Pasquon, I., Signorini, R. and Natta, G. (1968). *Makromol. Chem.,* **112,** 160
95. Natta, G., Zambelli, A., Lanzi, G., Pasquon, I., Mognaschi, E. R., Segre, A. L. and Centola, P. (1965). *Makromol. Chem.,* **81,** 161
96. Lehr, M. H. and Carman, C. J. (1969). *Macromolecules,* **2,** 217
97. Svab, J., Friml, K., Snupacek, J. Jr., Cermak, V. and Liska, V. (1968). *Chem. Prum.,* **18,** 398; [*Chem. Abstr.* (1969). **70,** 11740]
98. Svab, J., Cermak, V., Liska, V. and Hajek, Z. (1970). *Chem. Prum,* **20,** 472; [*Chem. Abstr.* (1971). **74,** 64507]
99. Zambelli, A., Lety, A., Tosi, C., Pasquon, I. (1968). *Makromol. Chem.,* **115,** 73
100. Suzuki, T. and Takaegami, Y. (1970). *Bull. Chem. Soc. Jap.,* **43,** 1484; [*Chem. Abstr.* (1970). **73,** 45917]
101. Tagaegami, Y. and Suzuki, T. (1969). *Bull. Chem. Soc. Jap.,* **42,** 848; [*Chem. Abstr.* (1969), **71,** 3680
102. Pino, P. (1965). *Advan. Polym. Sci.,* **4,** 393
103. Topchieva, I. N. (1966). *Russ. Chem. Rev.,* **35,** 741
104. Klabunovskii, E. I. (1968). *Russ. Chem. Rev.,* **37,** 969
105. Pino, P., Ciardelli, F. and Zandomeneghi, M. (1970). *Ann. Rev. Phys. Chem.,* **21,** 561

106. Pino, P., Montagnoli, G., Ciardelli, F. and Benedetti, E. (1966). *Makromol. Chem.*, **93**, 158
107. Pino, P., Ciardelli, F. and Montagnoli, G. (1968). *J. Polym. Sci. C*, **16**, 3265
108. Montagnoli, G., Pini, D., Lucherini, A., Ciardelli, F. and Pino, P. (1969). *Macromol.*, **2**, 684
109. Ciardelli, F., Carlini, C. and Montagnoli, G. (1969). *Macromol*, **2**, 296
110. Pieroni, O., Stigliani, G. and Ciardelli, F. (1970). *Chim. Ind. (Milan)*, **52**, 283
111. Lazzaroni, R., Salvadori, P. and Pino, P. (1970). *Chem. Commun.*, **18**, 1164
112. Pino, P., Lazzaroni, R., Salvadori, P. (1970). *Inorg. Chim. Acta*, Third Int. Symp, Venice (Italy)
113. Chiellini, E., Carlini, C., Botrini, C. and Ciardelli, F. (1969). *Chim. Ind. (Milan)*, **51**, 852
114. Ciardelli, F., Benedetti, E., Montagnoli, G., Luccherini, L. and Pino, P. (1965). *Chem. Commun.*, **13**, 255
115. Ciardelli, F., Carlini, C., Montagnoli, G., Lardicci, L. and Pino, P. (1968). *Chim. Ind. (Milan)*, **50**, 860
116. Ciardelli, F., Bano, H. and Carlini, C. (1970). *Chim. Ind. (Milan)*, **52**, 81
117. Marconi, W. (1967). *'The Polymerization of Dienes by Ziegler–Natta Catalysts'* in *'Stereochemistry of Macromolecules'*, 239 (New York: Marcel Dekker)
118. Natta, G. and Porri, L. (1969). *'Diene Elastomers'* in *'The Chemistry of Synthetic Elastomers', Part 2*, 597 (New York: Interscience)
119. Cooper, W. (1970). *Ind. Eng. Chem. Prod. Res. Develop.*, **9**, 457
120. Dolgoplosk, B. A., Makovetskii, K. L., Tiniakova, E. I. and Sharaev, O. K. (1968). *'Polymerization of Dienes by π-Allyl Complexes'* (SSSR: Nauka)
121. Timofeyeva, G. V., Kokorina, N. A. and Medvedev, S. S. (1969). *Vysokomol. Soedin.* **A11**, 596
122. Racanelli, P. and Porri, L. (1970). *Eur. Polym. J.*, **6**, 751
123. Porri, L., DiCorato, A. and Natta, G. (1969). *Eur. Polym. J.*, **5**, 1
124. Bujadoux, K., Jozefonvicz, J. and Neel, J. (1969). *IUPAC Internat. Symposium on Macromolecular Chemistry*, Budapest, Preprints vol. II, Paper 4/27
125. Iwamoto, M. and Yuguchi, S. (1967). *J. Polym. Sci. B*, **5**, 1007
126. Bresler, L. S., Grechanovsky, V. A., Muzsay, A. and Poddubnyi, I. Ya. (1970). *Makromol.,Chem.*, **133**, 111
127. Yoshimoto, T., Komatsku, K., Sakata, R., Yamamoto, K., Takeuchi, Y., Onishi, A. and Ueda, K. (1970). *Makromol Chem.*, **139**, 61
128. Smith, G. H. and Perry, D. C. (1969). *J. Polym. Sci. A-1*, **7**, 707
129. Cucinella, S., Mazzei, A., Marconi, W. and Busetto, C. (1969). *IUPAC Internat. Symposium on Macromolecular Chemistry*, Budapest, Preprints, vol. II, paper 4/26
130. Mazzei, A., Cucinella, S. and Marconi, W. (1969). *Makromol. Chem.*, **122**, 168
131. Porri, L., Natta, G. and Gallazzi, M. C. (1967). *J. Polym. Sci. C*, **16**, 2525
132. Sharaev, O. K., Alferov, A. V., Tinyakova, E. I. and Dolgoplosk, B. A. (1968). *Izv. Akad. Nauk SSSR, Ser. Khim.*, 2583
133. Babitskii, B. D., Kormer, V. A., Lobach, M. I., Poddubnuyi, I. Ya. and Sokolov, V. N. (1968). *Dokl. Akad. Nauk SSSR*, **180**, 420
134. Kormer, V. A., Lyashch, R. S., Babitskii, B. D., L'vova, G. V., Lobach, M. I. and Chepurnaya, B. A. (1968). *Dokl. Akad. Nauk SSSR*, **180**, 665
135. Mushina, E. A., Vydrina, T. K., Sakharova, E. V., Tinyakova, E. I. and Dolgoplosk, B. A. (1967). *Dokl. Akad. Nauk SSSR*, **177**, 361
136. Matsumoto, T. and Furukawa, J. (1967). *J. Polym. Sci. B*, **5**, 935
137. Dolgoplosk, B. A., Tinyakova, E. I., Vinogradov, P. A., Parengo, O. P. and Turov, B. S. (1968). *J. Polym. Sci. C*, **16**, 3685
138. Kormer, V. A., Babitskii, D. D., Lobach, M. I. and Chesnokova, N. N. (1969). *J. Polym. Sci. C*, **16**, 4351
139. Hirachi, K. and Hirai, H. (1970). *Macromolecules*, **3**, 382
140. Oreshkin, I. A., Ostrovskaya, I. Ya., Yakovlev, V. A., Tinyakova, E. I. and Dolgoplosk, B.A. (1967). *Dokl. Akad. Nauk SSSR*, **173**, 1349
141. Dolgoplosk, B. A., Oreshkin, I. A., Tinyakova, E. I. and Yakovlev, V. A. (1967). *Izv. Akad. Nauk SSSR, Ser. Khim.*, 2130
142. Oreshkin, I. A., Chernenko, G. M., Tinyakova, E. I. and Dolgoplosk, B. A. (1966). *Dolk. Akad. Nauk SSSR*, **169**, 1102
143. Durand, J. P., Dawans, F. and Teyssié, P. (1970). *J. Polym. Sci. A-1*, **8**, 979

144. Dolgoplosk, B. A. (1971). *Vysokomol. Soedin.* **13A,** 325
145. Dawans, F. and Teyssié, P. (1966). *Compt. Rend. Acad. Sci. Paris, Ser. C,* **263,** 1512
146. Durand, J. P., Dawans, F. and Teyssié, P. (1968). *J. Polym. Sci. B,* **6,** 757
147. Dawans, F. and Teyssié, P. (1969). *J. Polym. Sci. B,* **7,** 111
148. Lugli, G., Marconi, W., Mazzei, A. and Palladino, N. (1969). *Inorg. Chim. Acta,* **3,** 151
149. Azizov, A. G., Sharaev, O. K., Tinyakova, E. I. and Dolgoplosk, B. A. (1969). *Vysokomol. Soedin.,* **11B,** 746
150. Sharaev, O. K., Alferov, A. V., Tinyakova, E. I. and Dolgoplosk, A. B. (1967). *Izv. Akad. Nauk SSSR, Ser. Khim.,* **1170,** 2584
151. Azizov, A. G., Sharaev, O. K., Tinyakova, E. I. and Dolgoplosk, A. B. (1970). *Dokl. Akad. Nauk SSSR,* **190,** 582
152. Tinyakova, E. I., Alferov, A. V., Golenko, T. G., Dolgoplosk, A. B., Oreshkin, I. A., Sharaev, O. K., Chernenko, G. N. and Yakovlev, V. A. (1967). *J. Polym. Sci. C,* **16,** 2625
153. Mushina, E. A., Vydrina, T. K., Sakharova, E. V., Yakovlev, V. A., Tinyakova, E. I. and Dolgoplosk, B. A. (1967). *Vysokomol. Soedin,* **9B,** 784
154. Marechal, J. C., Dawans, F. and Teyssié, P. (1970). *J. Polym. Sci. A-1,* **8,** 1993
155. Babitskii, B. D., Kormer, V. A., Lapuk, I. M. and Skoblikova, V. I. (1968). *J. Polym. Sci. C,* **16,** 3219
156. Durand, J. P., Dawans, F. and Teyssié, P. (1970). *J. Polym. Sci. A-1,* **8,** 979
157. Matsumoto, T., Furukava, J. and Morimura, H. (1969). *J. Polym. Sci. B,* **7,** 541
158. Harrod, J. F. and Wallace, L. R. (1969). *Macromolecules,* **2,** 449
159. Anderson, W. S. (1967). *J. Polym. Sci. A-1,* **5,** 429
160. Porri, L., Gallazzi, M. C. and Vitulli, G. (1967). *J. Polym. Sci. B,* **5,** 629
161. Otsu, T. and Yamaguchi, M. (1969). *J. Polym. Sci. A-1,* **7,** 387
162. Morton, M. and Das, B. (1969). *J. Polym. Sci. C,* **27,** 1
163. Sokolov, V. N., Babitskii, B. D., Kormer, V. A., Poddubnyi, I. Ya. and Chesnokova, N. N. (1969). *J. Polym. Sci. C,* **16,** 4345
164. Ochiai, E., Hirai, H. and Makishima, S. (1966). *J. Polym. Sci. B,* **4,** 1003
165. Dauby, R., Dawans, F. and Teyssié, P. (1967). *J. Polym. Sci. C,***16,** 1989
166. Scott, H. (1966). *J. Polym. Sci. B,* **4,** 105
167. Jenkins, D. K., Timms, D. G. and Duck, E. W. (1966). *Polymer,* **7,** 419
168. Durand, J. B. and Dawans, F. (1970). *J. Polym. Sci. B,* **8,** 743
169. Dawans, F. and Teyssieé, P. (1967). *Makromol. Chem.,* **109,** 68
170. Teyssié, P., Dawans, F. and Durand, J. P. (1968). *J. Polym. Sci. C,* **22,** 221
171. Dawans, F. and Teysse, P. (1969). *Eur. Polym. J.,* **5,** 541
172. Cuzin, D., Chauvin, Y. and Lefebvre, G. (1969). *Eur. Polym. J.,* **5,** 283
173. Porri, L. and Gallazzi, M. C. (1966). *Eur. Polym. J.,* **2,** 189
174. Murahashi, S., Kamachi, M. and Wakabayashi, N. (1969). *J. Polym. Sci. B.,* **7,** 135
175. Wei, P. E. and Milliman, G. E. (1969). *J. Polym. Sci. A,* **7,** 2305
176. White, D. M. (1960). *J. Amer. Chem. Soc.,* **82,** 5678
177. Farina, M., Allegra, G. and Natta, G. (1964). *J. Amer. Chem. Soc.,* **86,** 516
178. Farina, M., Natta, G., Allegra, G. and Löffelholz, M. (1967). *J. Polym. Sci. C,* **16,** 2517
179. Löffelholz, M., Farina, M. and Rossi, U. (1968). *Makromol. Chem.,* **113,** 230
180. Farina, M., Pedretti, U., Gramegna, M. T. and Audisio, G. (1970). *Macromolecules,* **3,** 475
181. Farina, M., Audisio, G. and Natta, G. (1967). *J. Amer. Chem. Soc.,* **89,** 5071
182. Chatani, Y., Nakatami, S. and Tadokoro, H. (1970). *Macromolecules,* **3,** 481
183. Ter-Minasyan, R. I., Parenago, O. P., Frolov, V. M. and Dolgoplosk, B. A. (1970). *Dokl. Akad. Nauk SSSR,* **194,** 1372
184. Yakovlev, V. Ya., Dolgoplosk, B. A. and Tinyakova, E. I. (1969). *Vysokomol. Soedin.,* **11A,** 1645
185. Natta, G. and Dall'Asta, G. (1969). *'Elastomer from Cyclic Olefins',* in *'Polymer Chemistry of Synthetic Elastomers', Part 2,* 703. (New York: Interscience)
186. Rinehart, R. E. (1969). *'Catalysis by Transition Metal Salts in Water or Other Polar Solvents'* in *'Polymer Chemistry of Synthetic Elastomers' Part 2,* 867. (New York: Interscience)
187. Dall'Asta, G. (1968). *J. Polym. Sci. A-1,* **6,** 2397
188. Dall'Asta, G. and Manetti, R. (1966). *Rend. Accad. Naz. Lincei,* **41,** 351
189. Dall'Asta, G. and Motroni, G. (1968). *J. Polym. Sci. A-1,* **6,** 2405
190. Rinehart, R. E. and Smith, M. P. (1965). *J. Polym. Sci. A,* **3,** 1049

191. Michelotti, F. W. and Carter, J. H. (1965). *Amer. Chem. Soc. Div. Polymer Chem., Preprints*, **6** (1), 224
192. Schultz, R. G. (1966). *J. Polym. Sci. B,* **4,** 541
193. Oshika, T. and Tabuchi, H. (1968). *Bull. Chem. Soc. Japan,* **41,** 211
194. Rinehart, R. E. (1969). *J. Polym. Sci. C,* **27,** 7
195. Eleuterio, H. S. (1957). *U.S. Patent* 3 047 918
196. Natta, G., Dall'Asta, G., Bassi, I. W. and Carella, G. (1966). *Makromol. Chem.,* **91,** 87
197. Dall'Asta, G. and Manetti, R. (1968). *Eur. Polym. J.,* **4,** 145
198. Calderon, N., Ofstead, E. A. and Judy, W. A. (1967). *J. Polym. Sci. A-1,* **5,** 2209
199. Ofstead, E. A. (1969). *Synth. Rubber Symposium,* London
200. Dall'Asta, G. and Carella, G. (1966). *Italian Patent* 784 307
201. Günther, P., Haas, F., Marwede, G., Nützel, K., Oberkirch, W., Pampus, G., Schön, N. and Witte, J. (1970). *Angew. Makromol. Chem.,* **14,** 87
202. Haas, F., Nützel, K., Pampus, G. and Theisen, D. (1970). *Rubber Chem. Techn.,* **43,** 1116
203. Uraneck, C. A. and Trepka, W. J. (1966). *French patent* 1542040
204. Turner, L. and Bradshaw, C. P. C. (1966). *British patent* 1105565
205. Günther, F., Oberkirch, W. and Pampus, G. (1968). *French patent* 2 025 142
206. Dall'Asta, G. and Meneghini, P. (1969). *Italian patent* 859 384
207. Dall'Asta, G. *British Patent* 1 098 340 (Italian Priority 1965)
208. Dall'Asta, G., Bassi, I. W. and Fagherazzi, G. (1968). *Eur. Polym. J.,* **5,** 229
209. Fagherazzi, G. and Bassi, I. W. (1968). *Eur. Polym. J.,* **4,** 123, 151
210. Natta, G. and Bassi, I. W. (1967). *Eur. Polym. J.,* **3,** 33, 43
211. Dall'Asta, G. and Scaglione, P. (1969). *Rubber Chem. Techn.,* **42,** 1235
212. Scott, K. W., Calderon, N., Ofstead, E. A., Judy, W. A. and Ward, J. P. (1969). *Adv. Chem. Ser.,* **91,** 399 (Washington: Amer. Chem. Soc.)
213. For a review on this subject, see Bailey, G. C. *'Olefin Disproportionation',* in *'Catalysis Reviews',* vol. 3, p. 37 (1970). (New York: Marcel Dekker)
214. Wasserman, E., Ben-Efraim, D. A. and Volowski, R. (1968). *J. Amer. Chem. Soc.,* **90,** 3286
215. Hummel, K. and Ast, W. (1970). *Naturwissenschaften,* **57,** 245
216. Luskin, L. S. and Myers, R. J. (1964). *'Acrylic Ester Polymers', Encyclopedia Polymer Science and Technology,* vol. 1, (New York: Wiley)
217. Dixit, S. S., Deshpande, A. B., Anand, L. C. and Kapur, S. L. (1969). *Polym. Sci. A-1,* **7,** 1973
218. Abe, H., Imai, K. and Matsumoto, M. (1968). *J. Polym. Sci. C,* **23,** 469
219. Allen, P. E. M. and Casey, B. A. (1970). *Eur. Polym. J.,* **6,** 793
220. Murahashi, S., Yuki, H., Hatada, K. and Obotaka, T. (1967). *Kobunshi Kagaku,* **24,** 309
221. Tsuruta, T., Makimoto, T. and Kanai, H. (1966). *J. Macromol. Chem.,* **1,** 31
222. Blumstein, A., Malhotra, S. L. and Watterson, A. C. (1968). *Polymer Preprints Amer. Chem. Soc., Div. Polym. Chem.,* **9,** 167
223. Ballard, D. G. H. and Medinger, J. (1968). *J. Chem. Soc. B,* 1176
224. Yuki, H., Hatada, K., Niinomi, T. and Kikuchi, Y. (1970). *Polym. J.,* **1,** 36
225. Yuki, H., Hatada, K., Niinomi, T. and Miyajii, K. (1970). *Polym. J.,* **1,** 130
226. Ito, T., Aoshima, K., Todo, F., Uno, K. and Iwakuro, Y. (1970). *Polym. J.,* **1,** 278
227. Aylward, N. M. (1970). *J. Polym. Sci. A-1,* **8,** 319
228. Andreev, D. N., Krasulina, V. N., Mikailova, N. V., Nekrasova, T. I., Novoselova, A. V. and Smirnova, G. S. (1970). *Vysokomol. Soedin. Ser. B,* **12,** 123
229. Okuzava, S., Hirai, H. and Makishima, S. (1969). *J. Polym. Sci. A-1,* **7,** 1039
230. Hirai, H. and Ikegami, T. (1970). *J. Polym. Sci. A-1,* **8,** 2407
231. Otsu, T., Yamada, B. and Imoto, M. (1966). *J. Macromol. Chem.,* **1,** 61
232. Belopol'skaya, T. V. (1968). *Dokl. Akad. Nauk SSSR,* **180,** 1388
233. Frisch, H. L., Mallows, C. L., Heatley, F. and Bovey, F. A. (1968). *Macromolecules,* **1,** 533
234. Heatley, F. and Bovey, F. A. (1968). *Macromolecules,* **1,** 303
235. Bovey, F. A. (1969). *Appl. Polym. Symp.,* **10,** 99
236. Monjol, P. (1968). *Compt. Rend. Acad. Sci. Paris, Ser. C,* **267,** 1021
237. Ferguson, R. C. (1969). *Macromolecules,* **2,** 237
238. Klesser, E. and Grouski, W. (1969). *J. Polym. Sci. B,* **7,** 727
239. Yoshino, T. and Komiyama, J. (1966). *J. Amer. Chem. Soc.,* **88,** 176

240. Fowells, W., Schmerch, C., Bovey, F. A. and Hood, F. P. (1967). *J. Amer. Chem. Soc.*, **88**, 1396
241. Liquori, A. M., Anzuino, G., D'alagni, M., Vitagliano, V. and Costantino, L. (1968). *J. Polym. Sci. A-2*, **6**, 509
242. Liù, H. Z. and Liù, K. T. (1968). *Macromolecules*, **1**, 157
243. Miyamoto, T. and Inagaki, H. (1969). *Macromolecules*, **2**, 554; *ibid.* (1970). *Polymer J.*, **1**, 46
244. Dayants, J., Reiss, C. and Benoit, H. (1968). *Makromol. Chem.*, **120**, 113
245. Baca, J., Lochmann, L., Juzl, K., Coupek, J. and Lim, D. (1967). *J. Polym. Sci. C*, **16**, 3865
246. Lim, D., Coupek, J., Baca, J., Syroka, S. and Sneider, B. (1968). *J. Polym. Sci. C*, **23**,
247. Korotkov, A. A. and Krasulina, U. N. (1968). *Polymer Sci. USSR*, **10**, 1818
248. Azimov, Z. A., Korotkov, A. A. and Mitsengendler, S. P. (1968). *Vysokomol. Soedin, Ser. A*, **10**, 2145
249. Krasulina, V. N., Khachaturov, A. S., Mikhailova, N. V. and Erusalinski, B. L. (1970). *Vysokomol. Soedin. Ser. B*, **12**, 303
250. Marukaski, M. and Takida, H. (1969). *Makromol. Chem.*, **124**, 172
251. Berghmans, H. and Smets, G. (1968). *Makromol. Chem.*, **115**, 187
252. Guyot, A., Mordini, J. and Spitz, R. (1969). *Compt. Rend. Acad. Sci. Ser. C*, **269**, 483
253. Pham-Quang-To, Mordini, J. and Guyot, A. (1970). *Compt. Rend. Acad. Sci. Ser. C*, **271**, 1294
254. Osawa, Z., Kimura, T., Kasuga, T. (1969). *J. Polym. Sci. A-1*, **7**, 2007
255. Ando, I., Chujo, R. and Nishioka, A. (1970). *Polym. J.*, **1**, 609
256. Kotake, Y., Joh, Y. and Ide, F. (1970). *J. Polym. Sci. B*, **8**, 101
257. Joh, Y., Kotake, Y. (1970). *Macromolecules*, **3**, 337
258. Fujimoto, T., Kawabata, N. and Furukawa, J. (1968). *J. Polym. Sci. A-1*, **6**, 1209
259. Yoshino, T. and Iwanaga, H. (1968). *J. Amer. Chem. Soc.*, **90**, 2434
260. Yoshino, T., Komiyama, J. and Shinomiya, M. (1964). *J. Amer. Chem. Soc.*, **86**, 4482; *ibid.* (1965), **87**, 387
261. Segre, A. L., Ciampelli, F. and Dall'Asta, G. (1966). *J. Polym. Sci. B*, **4**, 633
262. Joh, Y., Yoshihara, T., Kurinara, S., Isukuma, I. and Imai, Y. (1968). *Makromol. Chem.*, **119**, 339
263. Joh, Y., Kurihara, S., Sakurai, T., Imai, Y., Yoshihara, T. and Tomita, T. (1970). *J. Polym. Sci. A-1*, **8**, 377
264. Joh, Y., Kurihara, S., Sakurai, T. and Tomita, T. (1970). *J. Polym. Sci. A-1*, **8**, 2383
265. Natta, G. and Dall'Asta, G. (1964). *Chim. Ind. (Milan)*, **46**, 1429
266. Hirai, H., Ikagami, T. and Makishima, S. (1969). *J. Polym. Sci. A-1*, **7**, 2059
267. Ketley, A. D. (1967). *'Stereospecific Polymerization of Vinyl Ethers'* in *'The Stereochemistry of Macromolecules'* vol. 2, (New York: Marcel Dekker)
268. Lal, J. (1968). *High Polym.*, **23**, 331
269. Yuki, H., Hatada, K., Ota, K., Kinoshita, I., Murahashi, S., Ono, K. and Ito, Y. (1969). *J. Polym. Sci. A-1*, **7**, 1517
270. Yuki, H., Hatada, K., Ota, K. and Sasaki, T. (1970). *Bull. Chem. Soc. Jap.*, **43**, 890
271. Shostakovskii, M. F., Annenkova, V. Z., Khaliullin, A. K., Inyutkin, A. I., Gaitseva, Z. A. and Salaurov, V. N. (1969). *Vysokomol. Soedin. Ser. A*, **11**, 1979
272. Joh, Y., Yuki, H. and Murahashi, S. (1970). *J. Polym. Sci. A-1*, **8**, 2775
273. Joh, Y., Yuki, H. and Murahashi, S. (1970). *J. Polym. Sci. A-1*, **8**, 3311
274. Okuyama, T., Fueno, T. and Furukawa, J. (1968). *J. Polym. Sci. A-1*, **6**, 993
275. Okuyama, T., Fueno, T., Furukawa, J. and Uyeo, K. (1968). *J. Polym. Sci. A-1*, **6**, 1001
276. Fueno, T., Okuyama, T., Matsumara, I. and Furukawa, J. (1969). *J. Polym. Sci. A-1*, **7**, 1447
277. Mizote, A., Higashimura, T. and Okamura, S. (1968). *Pure Appl. Chem.*, **16**, 457
278. Higashimura, T., Kusudo, S., Ohsumi, Y., Mizote, A. and Okamura, S. (1968). *J. Polym.*
279. Chsumi, Y., Higashimura, T. and Okamura, S. (1968). *J. Polym. Sci. A-1*, **6**, 3125
Sci. A-1, **6**, 2511
280. Higashimura, T., Okamura, S., Morishima, I. and Yonezawa, T. (1969). *J. Polym. Sci. B*, **7**, 23
281. Higashimura, T., Masuda, T., Okamura, S. and Yonezawa, T. (1969). *J. Polym. Sci. A-1*, **7**, 3129

282. Natta, G., Peraldo, M., Farina, M. and Bressan, G. (1962). *Makromol. Chem.*, **55,** 139
283. Fueno, T., Okuyama, T. and Furukawa, J. (1969). *J. Polym. Sci. A-1,* **7,** 3219
284. Liquori, A. M. and Pispisa, B. (1967). *J. Polym. Sci. B,* **5,** 375
285. Pino, P., Lorenzi, G. P. and Chiellini, E. (1968). *J. Polym. Sci. C,* **16,** 3279
286. Luisi, P. L., Chiellini, E., Franchini, P. F. and Orienti, M. (1968). *Makromol. Chem.*, **112,** 197
287. Chiellini, E., Montagnoli, G. and Pino, P. (1969). *J. Polym. Sci. B,* **7,** 121
288. Chiellini, E., Salvadori, P., Osgan, M. and Pino, P. (1970). *J. Polym. Sci. A-1,* **8,** 1589
289. Chiellini, E. (1970). *Macromolecules,* **3,** 527
290. Tsuruta, T. (1967). *'Stereospecific Polymerization of Epoxides'* in *'The Stereochemistry of Macromolecules',* vol. 2, (New York: Marcel Dekker)
291. Gurgiolo, A. E. (1967). *'Poly(Alkylene Oxides)'* in *'Reviews in Macromolecular Chemistry',* vol. 1. Ed. by Butler, G. B. and O'Driscoll (Marcel Dekker: New York)
292. Furukawa, J. and Saegusa, T. (1967). *'Epoxide Polymers'* in *'Encyclopedia of Polymer Science and Technology',* vol. 6 (New York: Wiley)
293. Ishii, Y. and Sakai, S. (1969). *'1,2-Epoxides'* in *'Ring-Opening Polymers',* vol. 2. (*Frisch-Kurt C.;* Marcel Dekker: New York)
294. Mazzei, A., Ghetti, G., DeChirico, A. and Marconi, W. (1969). *Chim. Ind. (Milan),* **51,** 1368
295. Strohmeier, W. and Hartmann, P. (1969). *Z. Naturforsch. B,* **24,** 777
296. Chang, E. Y. C. (1970). *J. Appl. Polym. Sci.,* **14,** 3137
297. Ishimori, M., Nakasugi, O., Takeda, N. and Tsuruta, T. (1968). *Makromol. Chem.,* **115,** 103
298. Ishimori, M., Hsiue, G. and Tsuruta, T. (1969). *Makromol. Chem.,* **128,** 52
299. Ishimori, M., Hsiue, G. and Tsuruta, T. (1969). *Makromol. Chem.,* **124,** 143
300. Nakaniwa, M., Ozaki, K. and Furukawa, J., (1970). *Makromol. Chem.,* **138,** 197
301. Vandenberg, E. J. (1969). *J. Polym. Sci. A-1,* **7,** 525
302. Kunz, I. and Kroll, W. R. (1970). *J. Polym. Sci. A-1,* **8,** 1601
303. Inoue, S., Isukuma, I., Kawaguchi, M. and Tsuruta, T. (1967). *Makromol. Chem.,* **103,** 151
304. Livshits, V. S., Krylov, O. V. and Klabunovskii, E. I. (1968). *Probl. Kinet. Katal. Akad. Nauk SSSR,* **12,** 263
305. Furukawa, J., Kumata, Y., Yamada, K. and Fueno, T. (1968). *J. Polym. Sci. C,* **23,** 711
306. Nakaniwa, M., Kameoka, I., Ozaki, K. and Furukawa, J. (1970). *Makromol. Chem.,* **138,** 209
307. Kumata, Y. and Furukawa, J. (1970). *Makromol. Chem.,* **134,** 317
308. Anderson, B. C., Hoover, C. L. and Vogl, O. (1969). *Macromolecules,* **2,** 686
309. Dall'Asta, G., Motroni, G., Manetti, R. and Tosi, C. (1969). *Makromol. Chem.,* **130,** 153
310. Kormer, V. A., Babitskii, B. D., Yufa, T. L. and Poletaeva, I. A. (1970). *Chem. Abstr.,* **73,** 121 392 v
311. Kormer, V. A., Yufa, T. L., Poletaeva, I. A., Babitskii, B. D. and Stepanova, Z. D. (1969). *Dokl. Akad. Nauk SSSR,* **185,** 873

5
Viscosity of Polymers

A. PETERLIN
Camille Dreyfus Laboratory, Research Triangle Institute, North Carolina

5.1 INTRODUCTION

In the last few years, five comprehensive reviews of the viscosity of polymers have been published[1–5]. The first two are mainly concerned with the zero-gradient steady-state viscosity of polymer solutions, the third one with that of concentrated solutions and melts and the last two with the molecular weight, temperature, gradient and frequency-dependence of the viscosity of polymer solutions and melts. All these reviews cover the field extensively up to about a year before publication, so that in this Review we may limit ourselves to the results published in the last 3 years. A short review article by Bloomfield[6] has surveyed hydrodynamic studies of biological macromolecules and their use in elucidating structural details of proteins, DNA, viruses and polysomes.

Models attempting to describe the hydrodynamic properties of polymer systems must connect molecular parameters like molecular weight, contour length, chain diameter and conformational properties, with the full range of observed rheological phenomena, i.e. with the stress tensor as a function of gradient and frequency or with the optical tensor which is proportional to, and at higher frequencies easier to measure than, the stress tensor. Such a requirement immediately eliminates rigid models, for example rigid ellipsoid, rigid rod, rigid ring or rigid dumb-bell models, and also all the variations of the elastic dumb-bell model which have only one relaxation time and a wrong type (for example, negative) hydrodynamic interaction[7]. The elastic *dumb-bell* is a very tempting model since it can be treated mathematically as a single-body problem in an apparently exact manner.

A much more realistic model is the *necklace* model with elastic links, links with finite maximum length or fixed length, and with finite internal viscosity. In tensor notation the hydrodynamic equation for the necklace is identical with the scalar equation for the dumb-bell. The hydrodynamic resistance of the bead is equally approximate as in the dumb-bell case. The main problem is in the correct expression for the hydrodynamic interaction which depends on the inverse distance between any two beads of the necklace. In order to keep the mathematics within manageable limits one averages such an inverse distance over all conformations of the undeformed (Zimm) or deformed (Fixman) coil. This is certainly an oversimplification which adversely affects the results. In the former case the hydrodynamic interaction is constant and independent of the gradient, as is the viscosity of a perfectly flexible coil, in striking contrast to experience. In the latter case, the viscosity drops as the square of the gradient, the decrease being more drastic with a better solvent. In a theta solvent the viscosity decreases to a minimum and then increases again. A decrease in the viscosity may also be a consequence of a finite resistance of the coil to rapid changes in shape (internal viscosity) even when a constant or zero hydrodynamic interaction occurs.

The majority of complications arising from an averaging of the inverse intramolecular distances at too early a stage in the calculations does not exist in the oscillating flow field where the amplitude of the gradient is usually so small that one can assume that the molecule is practically undeformed. In such a case a solution for the constitutive equation can be easily obtained and the *dynamic viscosity* may be calculated as a function of the frequency, ω, of the applied flow field, of polymer–solvent interaction, molecular weight and the internal viscosity of the macromolecule.

In concentrated solutions, and melts above the critical molecular weight M_c or cM_c/ρ, where c is the concentration and ρ is the density of polymer in g cm^{-3}, the viscous behaviour is determined by entanglements which act as temporary crosslinks. The basic concepts of the rheology of such a polymer network, as introduced by Lodge[8], have turned out to be excellently suited for the description of the dependence of the viscosity on gradient and frequency.

The relationship between steady flow and dynamic viscosity is still very puzzling however. Experimental data, in general, show a near identity between the steady-flow value $\eta(\dot{\gamma})$ and the absolute value of the complex dynamic viscosity $|\eta^*(\omega)|_{\omega=\dot{\gamma}}$. This relationship suggests that $\dot{\gamma}$ and ω are interchangeable. If this were so, one would expect that $\eta(\dot{\gamma}) = \eta'(\omega) < |\eta^*(\omega)|$ or, taking into consideration the above mentioned rotation of the stress tensor, $\eta(\dot{\gamma}) = |\eta^*(\omega)|^2/\eta'(\omega) > |\eta^*(\omega)|$. Both equations disagree with experiment. On the other hand, the correct mathematical treatment of the gradient and frequency-dependence of viscosity is completely different. Moreover, the streaming birefringence, which is proportional to the principal stress difference, increases at a more-than-linear rate with the gradient and decreases with frequency in perfect agreement with theoretical predictions and in disagreement with the assumed equivalence between $\dot{\gamma}$ and ω.

The above mentioned effects are significantly modified in polymers of such a low molecular weight that the coil is not yet fully formed so that the molecule more closely resembles a slightly bent rod. Such is the case with oligomers and with very rigid polymers as for example double-helix DNA or ladder molecules. With monomers, on the other hand, a rigid model (ellipsoid, dumb-bell) seems to be adequate. But with the monomer one must be prepared for a complete failure of the hydrodynamic approach which treats as a continuum either the solvent or the monomer medium in the case of melts. Negative values for intrinsic viscosity may indeed be observed if the solute acts as a plasticiser of the solvent[9].

5.2 DILUTE SOLUTIONS

5.2.1 Concentration dependence

From the basis of the principle of corresponding states, Utracki and Simha[10–14] have suggested that a plot of the reduced viscosity should be used, i.e.,

$$\bar{\eta} = (\eta/\eta_s - 1)/c[\eta] = \eta_{sp.}/c[\eta] \tag{5.1}$$

where η and η_s are the viscosity of solution and solvent, respectively, as functions of the reduced concentration $\bar{c} = c/\gamma$, with the concentration parameter γ depending on M and T but not on concentration. The method has been checked with polystyrene (PS), poly(isobutylene) (PIB), poly(vinyl chloride) (PVC), poly(vinyl alcohol) (PVA), cellulose derivatives, polyisoprene (PIP) and vinyl aromatic polymer solutions over the range of concentrations $c \leqslant 4/[\eta]$. Deviations occur at low molecular weight where the random coil is not yet fully developed. Ehrlich and Woodbury[15] investigated polyethylene (PE) in ethane, propane and ethylene in concentrations up to 0.2 g cm^{-3} between 150 and 250 °C and with pressures ranging between 15 and 30×10^4 lb in^{-2} and found that the relative viscosity was independent of pressure but increased as the solvent was changed from propane through ethane to ethylene. All data lay on a single linear plot of log $\eta_{rel.}$ *v.* $c[\eta]$.

The concentration dependence of the viscosity of dilute solutions may be described by a great many empirical formulae among which the linear Huggins–Kraemer equation

$$\eta_{sp}/c = [\eta]\,(1 + k'c[\eta]) \tag{5.2}$$

with Huggins' constant k', and Martin equation

$$\ln(\eta/\eta_s) = \log \eta_{rel.} = [\eta]\,(1 + k''c[\eta]) \tag{5.3}$$

and the Schulz–Blaschke equation

$$\eta = \eta_s\,\{1 + c[\eta]/(1 - kc[\eta])\} \tag{5.4}$$

are most commonly used. Since the experiments are usually performed at too high a concentration, one observes deviations from the correlation $k = k' = k''+1/2$, deviations from a straight line and a difference in the ordinate intercepts, i.e. different values of $[\eta]$[16].

A substantial amount of effort has been spent in obtaining theoretical or empirical equations yielding straight-line plots over a wider concentration range. The Schulz–Blaschke equation is just one of the more successful attempts in this direction[17, 18], particularly in poor solvents. Solomon and Gottesman[19] have suggested a plot of $c/\eta_{sp.}$ *v.* c as an alternative. By explicitly considering the c^2 term of equations (5.4) and (5.5) Maron and Reznik[20] recently deduced the equation

$$\Delta/c^2 = [\eta]^2/2+(k'-1/3)\,[\eta]^3c = I+Sc \tag{5.5}$$

with

$$\Delta = \eta_{sp.}-\ln \eta_{rel.} \tag{5.6}$$

which yields $[\eta]^2/2$ as the ordinate intercept I, and $k' = 1/3+S/[\eta]^3$ from the slope of a plot of Δ/c^2 *v.* c. The slope S vanishes when $k' = 1/3$ or $k'' = -1/6$, and only under these circumstances do the linear equations (5.2) and (5.3) represent the experimental data in a satisfactory manner[21]. Equation (5.5) was checked by the authors on poly(methyl methacrylates) (PMMA) and PS in benzene and toluene where it was indeed found to yield unambiguous intrinsic viscosities and slopes k' and $k'' = k'-1/2$.

An empirical equation relating k' and Flory's viscosity expansion factor for a polymer coil which yielded $k'_\theta = 0.52$ and $k_{min.} = 0.31$ has been derived by Sakai[22] from data on PS, poly(α-methyl styrene) (PαMS) and PMMA solutions. The author suspects that at the theta point k' still depends on the molecular weight of the solute[23]. A correlation between k' and $[\eta]$ with polymer compatibility in polymer mixtures has been obtained by Vasile and Schneider[24].

The near constancy of k' for a polymer series in the same solvent allows the determination of the intrinsic viscosity and hence molecular weight from a single measurement of viscosity at a fixed concentration. The method has been applied extensively to high and low density PE of varying polydispersity (Quakenbos[25]) and to PE and polypropylene (PP) in decalin at 135 °C (Elliott *et al.*[26]). The former author has applied the correlation between intrinsic viscosity and melt index for the characterisation of PE – linear or branched, wide or narrow molecular weight distribution – and the latter group have presented tables and graphs for the rapid determination of $[\eta]$ from a single viscosity measurement. A similar attempt has been made by Maron[27] and Berlin[28] on the basis of equations (5.2) and (5.3) which according to Schroff[29] leads to a good value of intrinsic viscosity only when k' lies between 0.3 and 0.4 (see equation (5.5)).

The presence of incomplete molecular dissolution of the polymer, i.e. the presence of gel particles, diminishes the viscosity of the solution because the gel particles, which contain a great many macromolecules crosslinked by entanglements or crystals or even permanent chemical crosslinks, swell very much less than the single macromolecule and hence exhibit a smaller hydrodynamically effective volume. A general survey of this situation has been given recently by Watterson *et al.*[30, 31]. An analysis of the situation which exists with cellulose nitrate which was corroborated by ultrafiltration and light scattering measurements has been presented by Schurz[32]. The association of polyacrylonitrile (PAN) solutions in dimethyl formamide–benzene mixtures has been followed viscometrically by Beevers[33] and of polymer–lithium species in hydrocarbon solvents by Morton *et al.*[34]. The stereocomplex formation of isotactic and syndiotactic PMMA has been investigated by Borchard *et al.*[35] and by Lin and Lin[36]. Braun and Quesada-Lucas[37] have deduced from cross-linking experiments with naphthalene sodium on poly(vinyl diphenyl-ethylene)–styrene (S) copolymers in dilute solution that the interpenetration of coils leads to early gelation. On the other hand, the change of hydrodynamic volume, i.e., of viscosity, caused by changes in emulsion type and phase composition can be used for the viscometric monitoring of polymerisation as has been demonstrated by Molau *et al.*[38] in studies of polymerising solutions of rubber in styrene.

5.2.2 Intrinsic viscosity at zero shear

Experimental work on polymer solutions during the last 30 years has been mainly concerned with the development of relationships between intrinsic viscosity at zero shear $[\eta]_0$, molecular weight M and molecular dimensions[1]. The Kuhn–Mark–Houwink–Sakurada equation in Flory's formulation

$$[\eta]_0 = K_m M^a = \phi_0 R^3/M = \phi_0 R_\theta^3 \alpha_\eta^3/M = 2.5\, N V_h/M \qquad (5.7)$$

with a between 0.5 (theta solvent) and 1.0 (best solvent), has been obtained for new polymer–solvent systems or checked with better fractionated samples (Yamaguchi[39] on PP, Koleske and Lundberg[40] on polycaprolactam (PCL), Beachell and Peterson[41] on polyurethane, Wallach[42] on polyimides, and Timpa and Segål[43] on cellulose trinitrate). The data may be evaluated in terms of coil dimensions in conjunction with information from sedimentation or diffusion[44]. The usual procedure, however, is based on the Flory–Fox formulation where ϕ_0 is 4.20×10^{24}, R and R_θ are root-mean-square (r.m.s.) gyration radii in the solvent investigated and in a theta solvent (unperturbed dimensions), respectively, $\alpha_\eta = R/R_\theta$ is Flory's coil expansion coefficient and V_h is the hydrodynamic volume of the molecule.

The proportionality of $[\eta]_0$ to hydrodynamic volume V_h divided by M suggests the possibility of correlating the volume with M if an independent measurement of M or of V_h is available. Combination with gel permeation chromatography which yields V_h has been used by Cote and Shida[45] for the evaluation of V_h as a function of M, particularly in the study of branching which leads to a reduction in the value of V_h. The hydrodynamic radius, on the other hand, can be derived from the translational frictional coefficient (diffusion constant) and used for the calculation of $[\eta]_0$[46]. The variation of ϕ_0 with chain topology, hydrodynamic interaction and excluded volume has been recently reviewed by Bloomfield and Sharp[47].

The unperturbed dimensions may be obtained either from measurement in a theta solvent or by extrapolating the data in good solvents by plotting $[\eta]/M^{\frac{1}{2}}$ v. $M^{\frac{1}{2}}$

$$[\eta]/M^{\frac{1}{2}} = K_\theta + b_T M^{\frac{1}{2}} \tag{5.8}$$

as suggested by Stockmayer and Fixman[48]. The ordinate intercept $K_\theta = \phi_0(R_\theta^2/M)^{\frac{3}{2}}$ contains the unperturbed r.m.s. gyration radius. From data in theta solvents, Beech and Booth[49] have determined the unperturbed dimensions of poly(ethylene oxide) (PEO), Evans *et al.*[50] of poly(tetrahydrofuran), Buch *et al.*[51] of poly(methyl-3,3,3-trifluoropropyl siloxamer) and poly (methylphenyl siloxamer) and Penzel and Schulz[52] of cellulose trinitrate. Wunderlich[53] and Vasudevan and Santappa[54] have found that the change from acrylates (A) to methacrylates (MA) slightly increases the unperturbed chain dimensions. The effect of increasing the length of the ester chain is much more noticeable, however, so that poly(lauryl MA) has nearly twice the dimensions of PMMA. Knecht and Elias[55] have found that the ester groups in the main chain of the former polymer, where the sequence of valency angles is 108 and 114 degrees, give rise to a less extended conformation than the methylene groups in the latter polymer where the valency angle is 109.5 degrees because they introduce a finite bending into the chain which is already in the fully *trans* conformation. Banks *et al.*[56, 57] find that with amylose tricarbanilate in a pyridine–water theta solvent and in pure pyridine the coil exhibits a partial permeability up to $M = 2.5 \times 10^6$.

Through the use of equation (5.8), Takeda *et al.*[58] have determined the unperturbed dimensions of chlorinated stereoregular polybutadienes (PBD), Mattiussi *et al.*[59] of linear PCL in aqueous formic acid, Papazian[60] of PS, Mohite *et al.*[61] of poly(*p*-chlorostyrene)(P*p*ClS), Ceccorulli *et al.*[62] of poly(*p*-methoxystyrene) and Meyerhoff and Shimotsuma[63] of poly(ethylene terephthalate) (PET) in *o*-chlorophenol. Banks and Greenwood[64, 65] have found with amylose and amylose acetate fractions (and Cowie *et al.*[66] with PαMS) that with an increased temperature difference $T-\theta$ the intrinsic viscosity may best be represented by the Ptitsyn–Eisner[67–69] formula derived from Peterlin's coil model[70, 71], in which the r.m.s. distance between two beads m links apart is proportional to $m^{(1+\varepsilon)/2}$. The downward bending of the experimental curve for P*p*ClS and the difference between the linearly extrapolated and directly measured values for K_θ have been pointed out by Matsumara[72]. According to Ueda and Kajitani[73, 74] equation (5.7) utilising a constant value of a gives a poor fit with the data obtained for narrow molecular-weight distribution fractions of poly(vinyl acetate) (PVAc), and that in equation (5.8) the second term is proportional to $M^{0.4}$ rather than $M^{0.5}$. The unperturbed dimensions vary slightly with the solvent and the theta temperature in good agreement with other observations (Kurata and Stockmayer[75], Dondos and Benoit[76]). Patel and Patel[77] have observed a negative value of ε for amylose benzoate below the theta temperature.

According to Reiss and Benoit[78] the inflection in the intrinsic viscosity–temperature curve for PS is a consequence of the fact that the helical conformation ascribed to the crystalline isotactic PS is preserved in solution below 50 °C but vanishes beyond 80 °C. Atactic PS shows a similar behaviour and thus probably contains stereoregular sequences. A similar transition occurs with poly(2-vinylpyridine)(PV2P) at 25 °C in benzene and tetrahydrofuran (Dondos[79]).

Noda *et al.*[80] have investigated monodisperse PαMS with M_w values of 4–747 × 10^4 in theta and good solvents over the temperature range 4–100 °C. The viscosity expansion coefficient α_η agrees with Flory's theory[81] (α^5 type) if M_w is above 10^6 and with Stockmayer and Fixman's theory[48] (α^3 type) if M_w is smaller. This situation, according to the authors (who refer to the recent results of Monte Carlo calculations on the conformation of long chains) may be the consequence of the fact that the expansion coefficient for short chains cannot be expressed by the same equation as for long chains. The calculations assuming either the tetrahedral[82] or cubic[83] lattice indeed show a change from the α^3 to the α^5 type in almost the same region of Z despite the fact that the coefficients are very different from each other in that region.

Rudine *et al.*[84] have used measurements of the intrinsic viscosity of the same sample in solvents of widely varying values of a (in equation (5.7)) to obtain a-order moments of molecular weight, M_v^a, from the known K_m, and have extrapolated to $a = 1$ in order to obtain the weight average value M_w. This simple expedient is effective with PS although there are probably no molecular weight distributions for which the ath root of the ath moment is a linear function of a over the whole range of values from 0.5 to 1. The method was also checked rather successfully using literature data on PMMA.

Mixed solvents, in particular a mixture of a solvent with a precipitant (non-solvent), are quite often used for reaching theta conditions in a convenient temperature range. But the preferential adsorption of one component, e.g. the solvent, by the polymer has differing effects on the intermolecular and intramolecular attractions and hence may cause deviations from a simple linear relationship between coil dimensions in the mixture with those in the pure components and also from a simple dependence of 'unperturbed' dimensions on the chemical structure of the components. According to Bates and Irwin[85] poly(hexane 1-sulphone) in a dioxane–n-hexane theta mixture possesses linear dimensions 15% larger than in methyl ethyl ketone (MEK), isopropanol or MEK–n-hexane. Dondos and Benoit[86] have observed two separate theta temperatures for PS, PMMA and P2VP, the upper one characterised by the disappearance of the second virial coefficient and the lower one by the relationship $b_T = 0$. Moldovan and Strazielle[87] have reported a maximum in the intrinsic viscosity of PEO in a 75/25 mixture of tetrachloride carbons and methanol at 25 °C. Dondos and Patterson[88] have suggested that the excess free energy ΔF of the mixed solvent is responsible for the observed deviations of $[\eta]_0$ for PV2P and PMMA from the normal linear variation with solvent composition. According to Dondos and Benoit[89], with PS and PV2P the difference between K_θ in a solvent mixture and K_θ in a pure solvent is of similar sign to the ΔF value for the mixture, and its temperature coefficient has an opposite sign from the heat of mixing ΔH.

Random and block copolymers of S and MMA have been investigated by Dondos[90] and by Dondos *et al.*[91], Fischer and Mächtle[92], Kotaka *et al.*[93], Matsuda *et al.*[94] and Ohnuma *et al.*[95], of vinyl 2-pyridine (V2P) and MMA by Dondos *et al.*[91], of α-methylstyrene (αMS) and ethylene by Tanzawa *et al.*[96] and of S and acrylonitrile (AN) by Lange and Baumann[97]. This list is far from complete. In block copolymers at low temperature the heterocontact interactions do not affect the coil dimensions so that the mean-square radius of gyration is a linear function of the mole fractions of the copolymer components[91]. In an ABA copolymer the nearly total collapse of the A component in a theta solvent for that component severely restricts the allowable conformations of the central (B) component, thus yielding an intrinsic viscosity for the copolymer less than that for the central block alone[95]. In random and alternating copolymers the heterocontacts, as estimated by the deviation from the linear additivity of the mean-square radius of gyration, affect not only the expansion in good solvents but also the unperturbed dimensions in a theta solvent[91]. The long-range interaction parameter b_T of the Stockmayer–Fixman equation is a maximum at an equimolar composition for a random copolymer in a good solvent for both components, while in a solvent which is good for component A and poor for B the maximum occurs at a composition rich in A and vice versa. The Huggins constant k' lies between 0.5 and 0.6 in theta solvents and decreases to 0.25 for an increasingly good solvent, but as a function of the copolymer composition it may show large deviations from linearity.

The short-range interactions, as given by the ratio $R_\theta/M^{1/2}$, may be expressed for block copolymers as a composition average of the components, while for alternating and random copolymers they show positive deviations proportional to the population of dyads of unlike monomers in the chain, being a maximum, P, for the alternating copolymer.

Oligomers have been recently reviewed by Sotobayshi and Springer[2] and by Bianchi and Peterlin[98]. In a good solvent the log $[\eta]_0$ *v.* log M plot shows a 0.5 slope at low values of M. As the monomer is approached the curves either bend down, as in the case of PS, PMMA or PEO, or approach a horizontal line, as in the case of poly(propylene glycol) (Meyerhoff[99], Sandel and Goring[100, 101]), cellulose and its derivatives. In all these cases the monomer has a finite positive intrinsic viscosity. The section with the slope 0.5 is missing in PE, the viscosity curve in this case exhibiting a steadily increasing slope as the value of M decreases. A similar dependence on the molecular weight is also observed in those cases where the intrinsic viscosity becomes negative at sufficiently low values of M, i.e. in those cases where the oligomeric solute acts as plasticiser for the solvent[102]. Using data obtained with low molecular-weight PS, PEO and PE, Rossi and Perico[103] have checked their calculations[104] based on the hydrodynamic theory of Kirkwood and Riseman[105] extended to include chains with a small number of bonds. Excellent agreement was found with the first two systems down to the tetramer while with PE similar agreement was only observed with oligomers with between 40 and 50 chain elements or greater.

Star, comb and randomly branched polymers have been systematically investigated by Berry[106] (PS stars, PVAc combs), Noda *et al.*[107] (PS combs), Nagasubramanian *et al.*[108] (trifunctionally branched PVAc), Moore and Millns[109] (long-chain branched PE), Meunier and van Leemput[110] (four- and six-branch PS) and Moritz and Meyerhoff[111] (randomly branched poly-(dodecyl MA)). Solutions of PE in the theta solvent diphenyl at 118 °C yield values of K_m and a equal to 4.92×10^{-2} and 0.20 respectively, which, according to the authors[109], indicates that the number of random trifunctional branch points per molecule is proportional to the molecular weight. The ratio, g_η, of theta point intrinsic viscosities of branched and unbranched molecules of the same molecular weight has been compared with the ratio of the mean-square radii of gyration, g_R, as calculated or determined from light scattering. The relationship

$$g_\eta = g_R{}^m \qquad (5.9)$$

has been observed, the value of m being 3/2 according to Flory's theory. For star-shaped molecules a value of m equal to $\frac{1}{2}$ has been calculated by Zimm and Kilb[112], while Berry finds $m = 0.60$ for star-shaped and 1.06 and 1.03 for two comb-shaped molecules and has suggested that m tends to 0.5 in the former and to 1.5 in the latter case. A value of m of 3/2 has also been observed by Noda *et al.* for the latter case. Nagasubramanian *et al.* have compared the data obtained at $a = 0.71$ (good solvent MEK) with theoretical predictions and have concluded that their values of g_η approach $g_R^{1/2}$ or g_s^3, where g_s is the ratio between the sedimentation resistance of branched and linear molecules.

5.2.3 Gradient and frequency dependence of intrinsic viscosity

Very few new investigations of the *steady-state gradient* dependence of $[\eta]$ have been undertaken, notable exceptions being the work of Ashare[113], Quadrat and Bohdanecky[114], Suzuki *et al.*[115] and Yamaguchi *et al.*[116] on PS, Noda *et al.*[117] on PS and PαMS and Ueberreiter *et al.*[118] on poly(vinyl carbazol). It has been found that the slope of a plot of $[\eta]_{\dot{\gamma}}/[\eta]_0$ versus the dimensionless generalised gradient $\beta = M[\eta]_0\, \eta_s\dot{\gamma}/NkT$ is more pronounced with a better solvent in agreement with the theory of Fixman[119]. At low molecular weights the decrease in slope starts at lower values of β than predicted for perfectly flexible coil, an effect which may be attributed to chain rigidity (Cerf[120, 121]). Quadrat and Bohdanecky have described an anomalous concentration dependence of the viscosity of dilute PS solutions in low viscosity theta solvents at high shear stress which they have attributed to the formation of double molecules under shear. At finite concentration, and as a function of shear rate, the viscosity drops to a minimum and then rises again. On extrapolation to zero concentration the effect disappears, so that the intrinsic viscosity, when plotted as a function of the gradient, does not show a minimum. Such an effect may also occur with solutions of PS of extremely high molecular weight (10^7) in good solvents as investigated by Wolff[122].

In an *oscillating* flow field either the complex dynamic intrinsic viscosity $[\eta^*]$ or shear modulus $[G^*]$ is measured. These quantities are interconnected by the equation

$$[G' + i\omega G''] = i\omega[\eta' - i\eta''] \qquad (5.10)$$

More *et al.*[123] have reported data on PS in the very high frequency range between 23 and 300 MHz, while Tanaka and Sakamishi[124] have reported similar results on PαMS and S–butadiene (BD) copolymers. Schrag and co-workers[125–128] have studied solutions of PS in an extremely viscous solvent (Aroclor) and have thus observed the full transition from $[\eta^*]_0$ to $[\eta^*]_\infty$. The data are extremely well described by the theory of Peterlin[129–132] taking into account the finite coil permeability (Tschoegl[133]) and internal viscosity (Cerf[120]). The ratio between the coefficient of internal viscosity ψ and the friction resistance of the statistical segment $f\eta_s$ (single bead of the Zimm necklace model[134]) for PS turns out to be 2.0 and independent of the viscosity of Aroclor which was varied between 2.5 and 70 poise. This result demands a substantial revision of the concept of internal viscosity[135].

The superimposition of steady and oscillatory shear has been studied experimentally and theoretically by Booij[136–138]. His thesis[137] contains a comprehensive survey of the field. The principal experimental information obtained by the author and by previous investigators is that the steady and oscillating state viscosities decrease with increasing shear rate especially at frequency values that are small in comparison to $\dot{\gamma}$. The phase angle between oscillating shear stress and shear rate has a value of 90 degrees when $\omega = \dot{\gamma}/2$. It is possible to provide a description which is at least qualitative by employing a variety of theories. With the necklace model, for example, it is necessary to assume a given mode of motion which vanishes when the increase of free energy of the mode reaches a critical value, which in various polymer liquids is approximately 3 kcal mol^{-1}. It is not unreasonable to suppose that this value is related to the energy barrier for free rotation about the main chain bonds.

5.2.4 Polyelectrolytes

Noda *et al.*[139] have investigated values of $\eta_{sp.}$ for the Na salt of poly(acrylic acid) (PAA) as a function of M, of the ionisation i and of the NaBr concentration (c_{salt}). In agreement with former data[140–142], the electrostatic part of the expansion factor, $\alpha^3_{\eta e}$, defined as the ratio of intrinsic viscosity to its value at infinite ionic strength, can be expressed as

$$\alpha^3_{\eta e} = [\eta]_c \quad /[\eta]_{c\,\mathrm{salt}\,=\,\infty} = 1 + B'(i)\,(M/c_{salt})^{1/2} \tag{5.11}$$

at all degrees of neutralisation if $M < 10^6$. Such a dependence is only predicted by the theory of Fixman[143]. All other theories[69, 144–149] predict a linear dependence on $M^{1/2}/c_{salt}$, and hence do not agree with the experimental results even qualitatively. Alexandrowicz[150] refutes this argument, claiming that his theory also yields a direct proportionality between $\alpha^3_{\eta e}$ and $c_{salt}^{-1/2}$ for $c_{salt}^{-1/2} > 3$. Okamoto and Wada[151] find lower values of $\eta_{sp.}$ for poly(methacrylic acid) (PMAA) than for PAA which they ascribe to the hydrophobic action of the methyl group. The proportionality of $\eta_{sp.}$ to between $c^{0.2}$ and $c^{0.4}$ (at higher values of M) is significantly greater than that usually observed ($\propto c^{0.5}$). Solutions of PMAA in mixtures of 0.002 N HCl and aliphatic alcohols have been investigated by Priel and Silberberg[152]. Bruce and Schwarz[153] have studied solutions of the ionic and non-ionic forms of poly(acrylamide) and found that the dependence of the viscosity on the gradient is well represented by Fixman's theory[119]. Banks *et al.*[154] have found that in aqueous solutions of linear amylose the intrinsic viscosity, when studied as a function of pH and of alkali and salt present, is constant up to pH=11 and then rapidly increases to a maximum at 0.15 N KOH. Addition of salt depresses these values below those observed even in neutral solutions. The authors explain this decrease in the viscosity in terms of helical conformation stabilised by ionic bonding. Schurz and Bayer[155] have determined a value of a equal to 1.33 in pure aqueous solutions of carboxymethyl cellulose (CMC) but that this value falls to 1.28 and 0.90 by the addition of 2% NaCl and 16% NaOH, respectively. Sodium CMC in water with added NaCl has been studied between 25 and 62.5 °C (Patel[156]). Hand and Williams[157] in a study of high molecular-weight DNA have found a nearly constant low value for $[\eta]_0$ for pH values below 3.5 and above 11.5, and a higher constant value for intermediate pH. According to Holt and Nasrallah[158] the ratio $\eta_{sp.}/c$ for poly(4-vinylpyridine 1-oxide) is 25 $cm^3\ g^{-1}$ for pH values below 4 with a rapid increase to a maximum (~80) at pH = 9 and a subsequent drop to 60 at pH = 12. In poly(2-vinylpyridine 1-oxide), however, there is a very uneven minimum range of $\eta_{sp.}/c$ for pH values between 4 and 10 with a rapid increase at lower and a less rapid increase at higher pH values (Holt and Tamann[159]).

Vink[160] has confirmed the linear correlation between specific viscosity and shear stress $\eta_s\dot{\gamma}$, as initially suggested by Strauss and Fuoss[161], for polyelectrolyte solutions of carboxymethylated cellulose ethers in water, with and without added NaCl, i.e.,

$$\frac{\eta_{sp.,\dot{\gamma}}-\eta_{sp.,\infty}}{\eta_{sp.,0}-\eta_{sp.,\infty}}=\frac{1}{1+B\beta} \tag{5.12}$$

which is practically identical with the correlation on uncharged cellulosic polymers investigated by Claesson and co-workers[162, 163]. Within experimental error B was found to be independent of the concentration of the solution but dependent on the ionic strength, and to some extent on the molecular weight also. Lin[164] has investigated the non-Newtonian intrinsic viscosity of pneumococcal DNA at pH = 6.8 with 0.15 and 1.0 N NaCl. The marked salt effect, which reduces $[\eta]_0$ by 20% which is evident at very low shear rates ($\dot{\gamma}$ between 0 and 2 s^{-1}), can be interpreted by a higher chain extension at lower salt concentration. At higher shear rates ($\dot{\gamma} \sim 5\ s^{-1}$) the salt effect becomes negligible as a consequence of coil extension by flow, an effect which overshadows the extension caused by charges on the macromolecule and the gegenions. Schurz *et al.*[165] have found that with native, and particularly with denatured, calf thymus DNA at pH = 7.1 a relatively long linear decrease of the intrinsic viscosity with the gradient, which extends up to 250 s^{-1}, occurs and afterwards a very slow decrease to 1/3 and 2/3 of the zero gradient value (6200 and 1300 $cm^3\ g^{-1}$), respectively occurs at $\dot{\gamma}$ = 2000 s^{-1}. Gibbs *et al.*[166] have described a maximum elasticity in hyaluronic acid solutions at pH = 2.5 but otherwise have observed the normal dependence of dynamic viscosity on the frequency.

5.2.5 Theory of intrinsic viscosity

Yamakawa[167] has revised the concept of hydrodynamic interaction as formulated by Oseen's tensor by explicitly considering the finite volume of the beads of the necklace model, and from this has obtained a new term proportional to a_h^2 where a_h is the hydrodynamic radius of the bead. This term vanishes if the tensor is averaged in the solution at rest without any orientation of the molecule. This is almost correct for a large molecule where for any end-to-end vector the intramolecular vectors are nearly completely at random, but is not true for a short molecule where the end-to-end vector has nearly the same orientation as the intrachain vectors between any two beads. The impenetrability of the beads also limits the smallest bead-to-bead distance to the bead diameter, and hence avoids the singularities of the Oseen tensor. As far as intrinsic viscosity is concerned, the effects of the corrections are negligibly small in the case of flexible chains but need attention in the case of short chains.

Thurston and Morrison[168] have calculated the exact eigenvalues λ_p of the Zimm random-coil model with hydrodynamic interaction for short chains. They have shown that with polystyrene solutions the dependence of the intrinsic viscosity on the molecular weight at small molecular weight values can indeed be very well represented by the curve with $h^* = f/(12\pi^3)^{1/2}\eta_s b = 0.5$ where b is the length of the segment.

Another deficiency of the necklace model is the neglect of the contribution of the single bead to the viscosity. All formulae yield zero intrinsic viscosity for a molecule with only one bead because the link number Z of such a molecule is 0. Since a free sphere of density ρ yields an intrinsic viscosity $2.5/\rho$ and a sphere prevented from rotation $4.0/\rho$, according to Peterlin[98, 102] it is necessary to introduce such an additive term in the intrinsic viscosity of a freely-draining coil and a term gradually decreasing with Z in the case of an impermeable coil. Monomers and very short oligomers have an intrinsic viscosity which is not directly derivable from the geometry of the molecule without explicit consideration of the shape and geometry of packing of solvent molecules. Since this contribution may be even negative[9], the additive term can be either positive or negative. A positive term explains the nearly horizontal intrinsic viscosity–molecular weight curve of so many polymers, particularly cellulose derivatives, in the oligomer range. In the case of paraffins, the negative contribution reduces the intrinsic viscosity to zero at finite values of Z and below that to negative values.

Iwata[169] has calculated $[\eta]_0$ for free-draining and impermeable ring polymers, while Kamide[170] has evaluated literature data relating to the viscosity of cellulose nitrate solutions in acetone and ethyl acetate and has confirmed the generally accepted view that over the molecular weight range $4 \times 10^4 < M < 4 \times 10^6$ the draining effects are not negligible. Fixman's[177] approach has been applied by Imai for the description of the intrinsic viscosity of homopolymers in general[172], of the Huggins' constant k'[173] and of copolymers[174]. He cautions about attempts to determine the unperturbed dimensions of copolymers from intrinsic viscosity measurements because, in general, there may be no theta state for a copolymer even if a proportionality of the intrinsic viscosity with $M^{1/2}$ is observed experimentally. There is no temperature at which all intersegmental potentials vanish simultaneously. (See Dondos and Benoit[88].)

The role of finite chain extensibility, treated by Reinhold and Peterlin[175] for the necklace and by Peterlin[176] for the dumb-bell model has been re-evaluated recently for the dumb-bell model by Stevenson and Bird[177], and Tanner and Stehrenberger[179]. Noda and Hearst[179] consider a worm-like model with fixed contour length. Tsuda[180] has calculated values of $[\eta]_0$ for a rigid model with N frictional elements and Ullman[181] for rod-like molecules of finite diameter. The calculated gradient dependence of intrinsic viscosity occurs at much higher values of β than observed experimentally with high molecular-weight polymers. It follows, therefore, that the finite chain extensibility cannot be the main reason for non-Newtonian viscosity, but it plays an important role as the fully-extended chain is approached, i.e. in oligomers and in the second Newtonian viscosity $[\eta]_\infty$, which drops to zero in contrast to a finite non-vanishing value for an infinitely extensible chain.

The expansion of the macromolecular coil during flow has been investigated by light scattering by Cottrell *et àl.*[182] using high molecular-weight ($M = 10^7$) PIB in decalin and by Champion and Davis[183] using PS in cyclohexanone. From the angular dependence of the scattered intensity in the flow plane as function of the gradient, the former authors suggest an extremely small deformation, as if the molecules were almost rigid, and the latter authors suggest an approximate agreement with the light scattering calculations of Peterlin and Reinhold[184] which were based on the perfectly flexible free-draining necklace model.

The internal viscosity ϕ of the macromolecule reduces the coil deformation in flow. For small values of ϕ, the coil is still sufficiently flexible to rotate with the volume element, i.e. with an angular velocity $\Omega = \dot{\gamma}/2$. Within this range it is possible to obtain exact solutions for the frequency dependence of the intrinsic viscosity provided that the amplitude of the generalised gradient $\beta = M[\eta]_0\eta_s\dot{\gamma}/NkT$ is so small that one can neglect the change of hydrodynamic interaction as a consequence of coil deformation. This is nearly always the case, and indeed for PS in viscous solvents[125–128] the dynamic intrinsic viscosity agrees completely with theoretical predictions[129–132].

If the internal viscosity is high, the rotational velocity of the coil differs from $\dot{\gamma}/2$. The difference depends on the conformation and orientation of the macromolecule. Cerf[185] has calculated the average value $<\Omega>$ and derived the gradient dependence of coil deformation and intrinsic viscosity for a wide range of $\psi/Z^{1/2}f\eta_s$ assuming constant hydrodynamic interaction. The results, however, are sensible only for a completely rigid coil without any change of intramolecular distances. Any deformation in flow changes these distances, and hence the hydrodynamic interaction, and this is characterised by a change of intrinsic viscosity[171, 186] even if the coil is free from internal viscosity.

The concept of internal viscosity has, to a certain extent, been generalised by Budtov and Gotlib[187] and by Booij and van Wiechen[138] who have introduced into each segment of the necklace model a force which is proportional to the rate at which the segment is deformed, and acts in the direction of the line between the end points of the segment. On the basis of the experimental data of Schrag and co-workers[125–128] which indicate that the ratio $\psi/f\eta_s$ is practically independent of η_s, Peterlin[135] has derived the internal viscosity from two effects, namely, the energy barriers between t, g and g' conformations and the need for chain displacement in a direction perpendicular to the applied force. The former contribution is independent of, and the latter is proportional to, the viscosity of the solvent. In low viscosity solvents ($\sim$1 cP) the former effect is important but in high viscosity solvents ($>$1 P) the latter effect predominates.

The often observed linear decrease of the intrinsic viscosity with the gradient (cellulosic polymers[162, 163], polyelectrolytes[160, 161, 164, 165]) has until now defied a theoretical explanation. All theories demand a decrease proportional to the square of the gradient.

5.3 CONCENTRATED SOLUTIONS AND MELTS

5.3.1 General considerations

The first Newtonian viscosity η_0 of melts is nearly proportional to the weight average molecular weight up to a critical molecular weight M_c, and proportional to a rather high power, as a rule 3.4, above M_c. In the former region there is very little, if any, gradient dependence of viscosity, and in the latter region the viscosity drops rapidly with shear stress to the second Newtonian viscosity η_∞ which is approximately equal to η_0 at low molecular weights. In concentrated solutions, the product ϕM_w^a rather than M_w determines the location of the system in the η–M plot. Here $\phi = c/\rho$ is the volume fraction of the polymer. Hayahara and Takao[188] have found that in AN copolymers the value of a depends on the excluded volume effect. Daum and Wales[189] have determined a value of $a = 2$ in PS solutions while Gupta and Forsman[190] have found that $a = 1$. A consequence of the similarity between the dependence of the melt and concentrated solution viscosity on M or ϕM^a is that the data for different molecular weight values may be plotted on a single master curve of log η *v.* log c (Simha and coworkers[14, 191], Vinogradov and Titkova[192]).

Since the intrinsic viscosity in a given solvent is a function of the molecular weight, a straightforward correlation might be expected between the intrinsic viscosity and the melt index which is inversely proportional to the viscosity at the temperature, pressure and flow regime of the melt flow experiment. Such a dependence has been observed by Manaresi *et al.*[193] on PET, Quackenbos[194] on high and low density PE, Baijal and Sturm[195] on PP, Keskkula and Taylor[196] on a 10% solution of PS in toluene and Catsiff[197] on poly(ethylene sulphide).

From an examination of the literature data on poly(decamethylene adipate) at 109 °C, on PS at 217 °C and on PVAc at 40 °C, Cross[198] has concluded that a single equation of the form

$$\eta_0 = K_1 M_w + K_2 M_w^{3.4} \tag{5.13}$$

gives a much better representation than two separate equations, i.e., $K_1 M_w$ for the lower and $K_2 M_w^{3.4}$ for the higher molecular-weight range. In particular there is a gradual transition instead of a sharp break at M_c.

According to an analysis performed by Allen and Fox[199, 200], the viscosity dependence on M, T and hydrostatic pressure p can be written as

$$\eta = F(\phi M_w)\zeta(T,p,c) \tag{5.14}$$

where the entropy factor F depends on molecular weight and concentration as expressed in equation (5.13) and the friction factor ζ reflects the influence of temperature, pressure, and concentration on the motion of an isolated chain element. Bartenev[211] has gone one step further, and claims that four independent factors describing independently the role of ϕ, M, T and

P are necessary. The factor ζ is expressed either in terms of the rate theory of flow (activation energy) or of the Williams–Landel–Ferry equation depending on the value of the glass transition temperature $T_g = T_0 + c_2$ with $c_2 \sim 50°C$.

The temperature dependence is reproduced quite well by the Vogel equation[202]

$$\log \eta = \log A + B/(T - T_0) \tag{5.15}$$

The kinetic parameter B, which is equal to the product $c_1 c_2$ of the constants in the Williams–Landel–Ferry expression for the temperature shift factor a_T, may be related to the internal barriers to rotation of the main-chain bonds in the isolated polymer molecule, which is V_0 for the t–g and $V_0 - \varepsilon_0/2$ for the g–t transition (see Table 5.1).

Table 5.1 Intramolecular energy barriers for bond rotation of vinyl polymers as calculated from the Vogel equation
(From Miller, A. A.[203], by courtesy of the American Chemical Society)

	T_g/K	$\varepsilon_0 = 2.5RT_g$ (kcal mol^{-1})	V_0 (kcal mol^{-1})
PE (linear)	160	0.80	3.5
PP	206	1.03	4.0
PIB	123	0.62	6.0
PS	323	1.61	3.9
PαMS	376	1.88	5.04
PMA		1.07	5.1
PMMA		1.50	6.15
PVAc		1.24	5.0
PDMS	81	0.40	2.1

A theoretical expression for the pressure-dependence of viscosity can be derived from the rate theory of viscosity[204], i.e.,

$$\eta_p = \eta_0 \, e^{bp} \tag{5.16}$$

For low and high molecular-weight PS at 140°C the experimental values[205], $b = 5.51 \times 10^{-3}$ and 2.90×10^{-3} bar^{-1}, respectively, agree quite well with the theoretical value, 4.43×10^{-3} bar^{-1}, derived from the free volume of the sample[206].

The non-Newtonian viscosity in the upper molecular-weight range is such an important effect that most studies on the viscosity of melts and highly concentrated solutions involve it. Cross[207] has correlated the literature data on bulk polymers in terms of a four-parameter equation

$$(\eta - \eta_\infty)/(\eta_0 - \eta_\infty) = 1/[1 + (\tau_0 \dot{\gamma})^b] \tag{5.17}$$

with the relaxation time (Graessley *et al.*[208])

$$\tau_0 = 12\pi M \eta_0 / cNkT(1 + \kappa c M) \tag{5.18}$$

where κ is an adjustable constant. The exponent b is connected with the polydispersity through the empirical relation $b = M_n/M_w$ yielding a value of unity for a monodisperse and a fractional value for a polydisperse sample[207]. The empirical equation of Bueche and Harding[209] has a value of $b = 0.75$. For a monodisperse polymer, Graessley[210] has derived from his model of entanglement formation that $b = 9/11 = 0.808$ together with a factor of 1.916 in front of the $(\tau_0\dot{\gamma})^b$ term. Bueche[211] has obtained $b = 6/7 = 0.857$ by consideration of entanglements of various complexity. Powell[212] finds that in poly(dimethyl siloxane) (PDMS) melts b lies between 1.06 and 1.98.

Another four constant formula in wide use is that of Sabia[213]

$$\log \eta/\eta_0 = (\eta/\eta_0 - a) \log [1 + (\tau_0\dot{\gamma})^b] \tag{5.19}$$

with $a = 2$ and $b = \frac{1}{3}$ for linear PE.

In viscosity measurements the effects of elastic energy[214–216] and of hydrostatic pressure[205, 206, 216, 217] are also important. The proportion of the pressure drop needed for elastic deformation depends on the flow rate but not on the length of the capillary. Hence the data from two capillaries of the same diameter but of different length can be used for the elimination of the elastic energy contribution to the capillary end-correction. It is also possible to measure the pressure difference between two points in the capillary sufficiently far away from the entrance. By measuring the pressure distribution along the capillary Han *et al.*[218] have found that a fully developed flow in molten polyethylene and polypropylene in the shear rate region 100–500 s^{-1} is attained within a length equivalent to one tube diameter. The extrapolated residual value at the exit to the capillary is the normal stress at the shear rate considered.

A more serious effect may be introduced by the large pressure gradient needed in the case of a very viscous liquid. The hydrostatic pressure at the entrance to the capillary can be so large that the viscosity of the melt increases appreciably. This phenomenon generally appears as an unexpected increase in the driving pressure with increasing shear rate as if the viscosity were increasing in those liquids which are either Newtonian or exhibit the usual decrease of viscosity with increasing gradient (Penwell and Porter[219]).

5.3.2 Experimental data

The variation of the first Newtonian viscosity as a function of the concentration and/or the molecular weight has been investigated, amongst others, by Frank[220, 221] (PP), Mieras and Van Rijn[222] (PP), Vinogradov *et al.*[223, 224] (PBD melt), Klein and Woernle[225] (PIB and PS in various solvents) Einaga *et al.*[226] (PS in Aroclor), Wallach[42] (polyimides), Blyler and Haas[227] (melts of copolymers of ethylene with AA and MAA), Narkis and Miltz[228] (crosslinked low-density PE) and Eisenberg[229] (liquid sulphur above 160 °C). In PBD the viscous flow activation energy and M_c increase with increasing ratio of *trans*-1,4 configurations. As a consequence, above M_c the zero-shear viscosity of samples with the same molecular weight but different *cis* to *trans* ratio may vary over more than a tenfold range. PIB and PS exhibit the largest increase of viscosity with concentration in a good solvent and the largest decrease of activation energy in a theta solvent. The intermolecular hydrogen bonds in copolymers of ethylene and AA or MAA act as temporary crosslinks and enhance the activation energy of flow, η_0, and its gradient dependence but do not influence the onset of liquid fracture. The actually observed value of η_0 for sulphur turned out to be 10^4 times smaller than the value calculated from the known molecular weight ($\sim$220 000). Consequently, it has proved necessary to include other mechanisms, e.g. bond interchange or chain-end interchange, in order to explain the much smaller value of the experimental measurement.

The gradient dependence of concentrated solutions and of plasticised melts has been reported by Schott[230] for PE, by Bueche and Tokcan[231] for P*p*ClS in various solvents and the pure melt, by Behrens and Folt[232], Collins and Metzger[233] and Schreiber[234] for PVC melts with a finite fraction of plasticiser, by Tager and Dreval[235] for PIB, PS and acetyl cellulose in various solvents, by Trizno *et al.*[236] for PVAc in methanol and methyl acetate, by Uy and Graessley[237] for diethyl phthalate, by Schurz *et al.*[238] for cellulose nitrate in butyl acetate, by Biddle and Pardhan[239] for hydroxyethyl cellulose in water, by Schutz[240] for amylose in water and by Paul *et al.*[241] for S–BD–S copolymers in good and poor solvents. Melts have been investigated by Johnson *et al.*[242] (branched PE from above to below the melting point), Mendelson[243] (PE, PP, poly(butene-1) (PB)), Nakajima[244] (PE), Saeda *et al.*[245] (linear PE), Bontinck[246] (PE, PB, polypentene, PP), Petraglia and Coen[247] (PP), Lee *et al.*[248] (narrow and wide molecular-weight distribution PDMS), Fujiki *et al.*[249] (copolymers of ethylene and vinyl acetate), Peyser *et al.*[250] (PB, copolymers of BD and S, BD and AN), Hyun and Karam[251], Ramsteiner[252], Toelcke *et al.*[253], Thomas and Hagan[254], Graessley and Segal[255,256] (PS), Elliott[257] (hydroxypropyl cellulose), Kraus and Gruver[258] (narrow molecular weight fractions of random copolymers of BD and S), Shih[259] (ethylene–propylene–diene copolymers) and by Scalco *et al.*[260] (ABS copolymers between 90 and 140 °C). A comprehensive survey of the flow properties of PS melts and their dependence on M, T and $\dot{\gamma}$ has been given by Casale *et al.*[261].

If the polydispersity of samples with different molecular weight values is sufficiently similar, the data can, in general, be plotted on master curves. With narrow fractions the agreement with equation (5.17) with $b = 0.8$ (Graessley[262]) is very satisfactory. Knowing the polydispersity, it is possible to construct theoretical master curves by superimposing curves corresponding to the components present which within reasonable accuracy agree with experimental master curves[254,256].

The flow of melts and concentrated solutions of branched polymers has been investigated by Bontinck[246], Mendelson *et al.*[263], Porter *et al.*[264], Schroff and Shida[265], Shida and Concio[266] (PE), Combs *et al.*[267], Shirayama *et al.*[268] (polyolefins), Graessley and Prentice[269] (PVAc in diethyl phthalate), Pritchard and Wissburn[270] (PVAc), Kraus and Gruver[271] (PBD) and Wang *et al.*[272] (poly-1-olefins). Branching sharpens the gradient dependence of the master curve. With PE, for example, the decrease in the value of η/η_0 from 1 to 0.01 occurs over six decades of the gradient for linear polymers and over four decades for the branched material. Increasing the number of long chain-branches at constant molecular weight increases the viscosity η_0, the value of the factor a above 3.4 and the activation energy, but decreases the ratio

η_0/η. Increasing the branch length in poly-1-olefins leads to a sharp maximum in the activation energy when the side-chain contains between four and six carbon atoms. The effects are extremely sensitive to the presence of plasticisers, i.e. to the presence of small molecules. Thus in concentrated solutions, the viscosity of a branched polymer is smaller than that of a linear polymer of the same molecular weight. The ratio of these values tends towards $g^{7/2}$ where g is the ratio of the mean-square radius of gyration of branched and unbranched molecules. Short chain-branches have a much smaller effect. As the length of short branches increases, the ratio η_0/η_∞ increases, but the value of η_0, of the activation energy, and of the critical shear rate for liquid fracture decreases. Fujimoto *et al.*[273] have found that with comb-shaped PS η_0 equals the value for the unbranched polymer at a branch molecular weight M_b of $3.5–4.0 \times 10^4$. If M_b is smaller, there may be no effective entanglement of the branches.

Ito[274] has studied the effect of hydrostatic pressure on the flow of polyoxymethylene, while Porter and co-workers[205, 206, 219] have studied PE. Hermann and Knappe[275] have investigated the gradient dependence of the viscosity of PMMA as a function of M ($M_n = 8–145 \times 10^3$), T(132–190 °C) and p (50–1050 atm). When $M_n > 27\,000$, a master curve may be drawn for each molecular weight, whereas when $M_n = 8000$ no such plot is possible. The master curves for different molecular weight may be superimposed over the downward curvature range up to the inflection point which occurs earlier as the molecular weight increases.

Hirata *et al.*[276] have shown that when linear PE melt is mixed in a screw extruder at an early stage in the process crosslinking predominates over chain scission but as the mixing proceeds branching develops. As a consequence, the viscosity drops and the relaxation spectrum broadens. Sieglaff and O'Leary[277] have found that with isotactic PP, in propylene copolymers and PVC[278] one or even two inflections occur in the log $\eta(\dot{\gamma})$ *v.* $1/T$ plots (γ = const.). Enhanced inflections are observed if the data are plotted at constant shear stress. The authors suggest that chain orientation and alignment leading to crystallisation occurs in the high stress profile at the capillary entry. Pritchard and Wissburn[270] report a reversibly large rise in the flow rate on continued shearing which is caused by a large effect on the entrance correction for capillary flow without a conspicuous effect on the melt viscosity. Plots of the time-dependence of η or $\tau = \dot{\gamma}\eta$ of PE melts over $\gamma = \dot{\gamma}t$ by Cooper and Pollet[279] are very similar to data of Matsuo *et al.*[280] on dilute solutions of high molecular-weight polymers.

Amongst other properties, the dynamic viscosity has been investigated by Weeks and Reid[281] (PE melts of varying polydispersity), Dunlop and Williams[282] (PP), Endo and Nagasawa[283] (concentrated PS solutions), Maxwell and co-workers[284–286] (branched PE, blends of PE/PS, PE/PMMA, PS/PMMA) and Wolkowicz and Forsman[287] (PE melts over a wide frequency range). Polyether from bisphenol-A and 4,4′-dichlorodiphenylsulphone has been studied by Mills *et al.*[288] in a steady and oscillatory shear flow. The plot of log η_0 *v.* log M_w has an initial slope of 2.5 at $M = 425$ which increases steadily to 5 at the highest value of $M (= 8.5 \times 10^4)$. To explain these results, the authors suggest a low M_c value of ~ 5000 corresponding to 15 monomer units. Such a small entanglement length ($M = M_c/2$), may be caused by strong polar interactions. Sakamoto *et al.*[289] have studied the melt rheology of an ethylene–MAA copolymer and its sodium and calcium salts between 100 and 160 °C. With the acid, $\eta(\dot{\gamma})$equals $|\eta^*(\omega)|_{\omega=\dot{\gamma}}$, whereas the complex viscosity of the salts is below that at steady shear. To explain this effect, the authors have suggested the existence of ionic domains affecting the long- but not the short-range segmental motion. The dynamic viscosity reflects long- and short-range segmental motions at the low and high frequency range of measurement, respectively. The steady flow viscosity, however, is concerned with long-range motions. Therefore, in the non-Newtonian range, the response of the charged system is different for the steady and oscillating flow.

Of particular interest are very extensive measurement of shear modulus G^* for narrow fractions of PS and PMMA by Onogi and co-workers[290–293] over a wide range of molecular weight, frequency and temperature. A set of data of $G'' = \omega\eta'(\omega)$ plotted versus $a_T\omega$ is shown in Figure 5.1. At the lowest

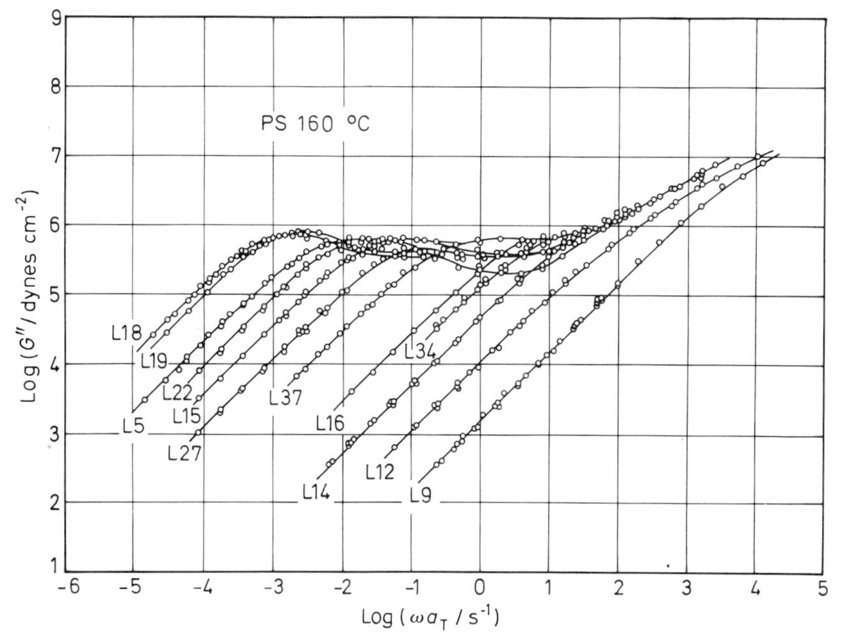

Figure 5.1 Master curve of G'' for narrow distribution PS between M_w = 580 000 (L18) and 8900 (L9).
(From Onogi, S. *et al.*, by courtesy of the American Chemical Society)

molecular weight investigated the slight inflection from the initial slope η'_0 to the final slope η'_∞ corresponds to the usual non-Newtonian viscosity in the molecular-weight range just above M_c. With higher molecular weights, the lower frequency branch is shifted to the left in accordance with the increase of η_0 proportional to M^a with $a \sim 3.4$. Initially the inflection to the η'_∞ straight line follows the usual conventional pattern extending over 2–3 decades of frequency, but with increasing molecular weight a horizontal plateau, and even a minimum in G'', develops which has very little in common with the usual transition from η'_0 to η'_∞.

If one considers the approximately valid empirical identity of ω and $\dot{\gamma}$, the abscissa can be considered as a generalised frequency or gradient and the ordinate as $\omega\eta'$ or $\dot{\gamma}\eta' = s$. The long plateau means that the shear stress remains constant over a wide range of gradient. During the initial rise of shear stress or applied pressure the gradient, i.e., the flow, increases at a rate slightly greater than proportional to the stress as expected for a normal non-Newtonian liquid. As soon as the stress reaches the plateau, however, the gradient jumps discontinuously to the right end of the plateau where the flow is again stabilised according to the ascending s–$\dot{\gamma}$ curve, i.e., to the second Newtonian viscosity η_{∞}.

Such a transition has been studied theoretically and experimentally by Vinogradov and co-workers[294, 295] and interpreted as plug flow yielding the liquid fracture phenomena. As seen in Figure 5.1, the molecular weight of the sample must be substantially above the critical molecular weight for chain entanglement. The critical shear stress for plug flow, however, is independent of the molecular weight and equal to about 10^6 dynes cm^{-2} for PS at 160 °C and PMMA at 220 °C. The extension of the flow jump at critical shear stress is proportional to the change in the inverse viscosity from η_0 to η_{∞}, i.e. proportional to $M^{2.4}$. These conclusions are to some extent corroborated by the viscosity dependence on shear stress for narrow distribution PS melts ($M_w = 1.8 \times 10^5$) investigated by Rudd[296], Stratton[297], and Thomas and Hagen[298]. These workers have found that at $s \sim 6 \times 10^6$ dynes cm^{-2} an abrupt drop in the value of η by nearly two decades occurs, which is very similar to the observations of Vinogradov.

Kataoka and Ueda[299–301] and Simmons[302] have measured the dynamic viscosity of solutions of PIB in decalin[299, 302], and of PDMS[300] and PE[301] melts using an oscillatory shear superimposed on a steady shear. The phase difference between the shear stress and the strain rate increases with increasing steady shear rate in general agreement with Booij's data[136–138].

5.3.3 Theory of melt flow

Below the M_c value, the flow of melts and concentrated polymer solutions is adequately described by the Rouse free-draining model with a characteristic relaxation time spectrum viscosity dependent on the frequency but not on the gradient.

Above M_c the polymer has a network structure with temporary crosslinks of finite lifetime. The rheology of such an entanglement network has been described and systematically studied by Lodge[8]. If the lifetime of an entanglement crosslink is independent of the gradient then the viscosity is also independent of the gradient. Considering the formation and destruction of entanglements during flow, however, leads to results which agree quite well with the empirical equation (5.17) (Graessley[210], Bueche[211]). Recently Graessley[262] has stressed the fact that if the macromolecules are densely packed the velocity field governing the frictional forces $v_{rel.}\zeta$ equation (5.14) is not a smoothly varying function of the coordinates. In an attempt to represent an extreme case of uncorrelated drag interaction, Graessley has modified the Rouse model by assuming independence of motion of the beads in the necklace or alternatively pair-wise coupling with retention of the spacial distribution. In this way a narrower relaxation time spectrum may be obtained together with a reasonably good description of the polydispersity effect on viscosity provided that the polydispersity is not too high.

A very elegant mathematical treatment of the viscosity contribution of four functional branch points has been derived by Chömpff[303–305], who has calculated the eigenvalue spectrum for a symmetrical crosslink between two freely-draining chains. In this model the crosslink P is permanent but mobile. As suggested by Duiser and Staverman[306], the two crosslinked chains may be transformed into a hypothetical, but mechanically equivalent situation by decoupling the chains, and allowing one chain, AC, to move with almost complete freedom. Once a configuration is chosen the coupling point P appears to the second chain BD as if it were fixed, and the sections BP and DP then chose their own configurations independently of each other. This method has been used for the calculation of the intrinsic viscosity of symmetric and asymmetric star-shaped, regularly branched and cyclic chain molecules in a theta solvent without hydrodynamic interaction. The results are applicable to the viscoelastic spectrum of micro-networks containing four functional branch points, entanglements or crosslinks.

Equations have been obtained and mathematical procedures developed to determine from the experimental data the average number of entanglements $(m-1)$ and the slip parameter δ. The results have been tested using a high molecular-weight ($M \sim 3.36 \times 10^6$) sample of poly(n-octyl MA). With $m = 81$ and $\delta = 10^{-5}$, the calculations allow the determination of the relaxation spectrum over nine decades of frequency. In particular the model predicts the existence of an extremely steep drop in $\eta'(\omega)$ from the low to high frequency plateau at ω between 10^{-6} and 10^{-4} s^{-1} (*see* Figure 5.1).

Of particular interest is the correlation between the frequency and the gradient dependence of the viscosity. Empirical evidence on PS (Cox and Merz[307], Ajroldi *et al.*[308], Akers and Williams[309] and Wales and den Otter[310]), PE (Onogi *et al.*[311] and Shida and Shroff[312]), and PMMA[308] melts support the view that $\eta(\dot{\gamma})$ is equal to the absolute value of $\eta^*(\omega)$ at $\dot{\gamma} = \omega$, at least over the entanglement region. Zapas and Phillips[313] have very convincingly shown that with a 10% solution of PIB in cetane at 25 °C in the frequency range 0.06–100 s^{-1} $\eta'(\omega)$ is below $\eta(\dot{\gamma})$ and that the two curves cannot be superposed by a shift along the $\dot{\gamma}$ or ω axis. The situation is even more extreme below M_c where the viscosity depends on the frequency but not on the gradient.

Zapas[314] has used the theory of Bernstein *et al.*[315, 316] concerning melt viscosity together with a heuristic potential function and derived, over a limited molecular-weight and gradient range, a quantitative relationship between the steady flow and dynamic viscosity. No arbitrary shift is required to fit the curves. Shida and Shroff[312] have derived the relaxation spectrum $H(\tau)$ from the loss modulus $G''(\omega)$ for high density PE of molecular weight $M_w = 0.86$, 1.1 and 1.28×10^5 respectively at 190 °C using the iterative procedure of Roesler and Twyman[317]. The calculated values of $\eta(\dot{\gamma})$ were found to be practically identical with $|\eta(\omega)^*|$ in perfect agreement with experiment. Similar calculations have been performed by Maruyama, Tanaka *et al.*[318, 319]. In spite of the similarity between $\eta(\dot{\gamma})$ and $|\eta^*(\omega)|$ described above and the partial success of the theoretical derivation of such a similarity, it is still not really clear why such a similarity should occur. The theoretical explanation for both effects is, in fact, quite different.

References

1. Berry, G. C. and Casassa, E. F. (1970). *Macromol. Rev.*, **4,** 1
2. Sotobayashi, H. and Springer, J. (1969). *Advan. Polymer Sci.*, **6,** 473
3. Berry, G. C. and Fox, T. G. (1968). *Advan. Polymer Sci.*, **5,** 261
4. Peterlin, A. (1968). *Advan. Macromol. Chem.*, **1,** 225
5. Mendelson, R. A. (1968). *Encyclopedia of Polymer Science and Technology,* Vol. 8, 587 (New York: Interscience)
6. Bloomfield, V. A. (1968). *Science,* **161,** 1212
7. Bird, R. B., Warner, H. R. Jr. and Evans, D. C. (1971). *Advan. Polymer Sci.*, **8,** 1
8. Lodge, A. S. (1964). *Elastic Liquids,* (New York: Academic Press)
9. Kuss, E. and Stuart, H. A. (1948). *Z. Naturforsch.*, **3a,** 204
10. Utracki, L. and Simha, R. (1963). *J. Polymer Sci., A,* **1,** 1089; *J. Phys. Chem.*, **67,** 1052
11. Utracki, L. (1964). *Polimery (Warsaw),* **9,** 144; *J. Polymer Sci. A-1,* **4,** 717
12. Simha, R. and Utracki, L. (1967). *J. Polymer Sci. A-2,* **5,** 853
13. Utracki, L., Simha, R. and Eliezer, N. (1969). *Polymer,* **10,** 43
14. Simha, R. and Chan, F. S. (1971), *J. Phys. Chem.*, **75,** 256
15. Ehrlich, P. and Woodbury, J. C. (1969). *J. Appl. Polymer Sci.*, **13,** 117
16. Melsheimer, J. (1971). *Kolloid-Z. und Z. Polymere,* **246,** 571
17. Grassie, N. and Roche, R. S. (1968). *J. Polymer Sci. C,* **16,** 4207
18. Sakai, T. (1968). *J. Polymer Sci. A-2,* **6,** 1659
19. Solomon, O. F. and Gottesman, B. S. (1968). *J. Appl. Polymer Sci.*, **12,** 971
20. Maron, S. H. and Reznik, R. B. (1969). *J. Polymer Sci. A-2,* **7,** 309
21. Ibrahim, F. and Elias, H. G. (1964). *Makromol. Chem.*, **76,** 1
22. Sakai, T. (1970). *Rep. Progr. Polymer Phys. Jap.*, **13,** 69
23. Sakai, T. (1970). *Macromolecules,* **3,** 96
24. Vasile, C. and Schneider, I. A. (1971). *Makromol. Chem.*, **141,** 127
25. Quakenbos, H. M. (1969). *J. Appl. Polymer Sci.*, **13,** 341
26. Elliott, J. H., Horowitz, K. H. and Hoodcock, T. (1970). *J. Appl. Polymer Sci.*, **14,** 2947
27. Maron, S. H. (1961). *J. Appl. Polymer Sci.*, **5,** 282
28. Berlin, A. A. (1966). *Vysokomol. Soedin.*, **8,** 1336
29. Schroff, R. N. (1968). *J. Appl. Polymer Sci.* **12,** 2741
30. Watterson, J. C., Lasser, H. R. and Elias, H. G. *Kolloid-Z. und Z. Polymere,* in the press
31. Watterson, J. G. and Elias, H. G. *Makromol. Chem.*, in the press
32. Schurz, J. (1969). *Faserforsch and Textiltechn.*, **20,** 481
33. Beevers, R. B. (1967). *Polymer,* **8,** 419, 463
34. Morton, M., Fetters, L. J., Pett, R. A. and Meier, J. F. (1970). *Macromol.*, **3,** 327
35. Borchard, W., Pyrlik, M. and Rehage, G. (1971). *Makromol. Chem.*, **145,** 169
36. Lin, H. Z. and Lin, K. J. (1969). *Macromolecules,* **1,** 157
37. Braun, D. and Quesada-Lucas, F. J. (1971). *Makromol. Chem.*, **142,** 313
38. Molau, G. E., Wittbrodt, W. M. and Meyer, V. E. (1969). *J. Appl. Polymer Sci.*, **13,** 2735
39. Yamaguchi, K. (1969). *Makromol. Chem.*, **128,** 19
40. Koleske, J. V. and Lundberg, R. D. (1969). *J. Polymer Sci. A-2,* **7,** 897
41. Beachell, H. C. and Peterson, J. C. (1969). *J. Polymer Sci. A-1,* **7,** 2021
42. Wallach, M. L. (1969). *J. Polymer Sci. A-2,* **7,** 1995
43. Timpa, J. D. and Segal, L. (1971). *J. Polymer Sci., A-1,* **9,** 2099
44. Peterlin, A. (1963). *Chimia,* **16,** 65
45. Cote, J. A. and Shida, M. (1971). *J. Polymer Sci. A-2,* **9,** 421
46. Rudin, A. and Johnson, H. K. (1971). *J. Polymer Sci. B,* **9,** 55
47. Bloomfield, V. A. and Sharp, P. A. (1968). *Macromolecules,* **1,** 380
48. Stockmayer, W. H. and Fixman, M. (1963). *J. Polymer Sci. C,* **1,** 137
49. Beech, D. R. and Booth, C. (1969). *J. Polymer Sci. A-2,* **7,** 575
50. Evans, J. M., Huglin, M. B. and Stepto, R. F. T. (1971). *Makromol. Chem.*, **146,** 91
51. Buch, R. B., Klimisch, H. M. and Johannson, O. K. (1969). *J. Polymer Sci. A-2,* **7,** 563; (1970). ibid., **8,** 541

52. Penzel, E. and Schulz, G. V. (1968). *Makromol. Chem.*, **113,** 64
53. Wunderlich, W. (1970). *Angew. Makromol. Chem.*, **11,** 189, 201
54. Vasudevan, P. and Santappa, M. (1971). *J. Polymer Sci. A-2,* **9,** 483
55. Knecht, M. R. and Elias, H. G. (1971). *Makromol. Chem.*, **150**
56. Banks, W., Greenwood, C. T. and Sloss, J. (1970). *Makromol. Chem.*, **140,** 109, 119
57. Banks, W., Greenwood, C. T. and Sloss, J. (1971). *Europ. Polymer J.*, **7,** 263
58. Takeda, M., Endo, R. and Matsuura, Y. (1968). *J. Polymer Sci. C,* **23,** 487
59. Mattiussi, A., Gechele, G. B. and Francesconi, R. (1969). *J. Polymer Sci. A-2,* **7,** 411
60. Papazian, K. A. (1969). *Polymer,* **10,** 399
61. Mohite, R. B., Gundiar, S. and Kapur, S. L. (1968). *Makromol. Chem.*, **116,** 280
62. Ceccorulli, G., Pizzoli, M. and Stea, G. (1971). *Makromol. Chem.*, **142,** 153
63. Meyerhoff, G. and Shimotsuma, S. (1970). *Makromol. Chem.*, **135,** 195
64. Banks, W. and Greenwood, C. T. (1969). *Polymer,* **10,** 257
65. Banks, W. and Greenwood, C. T. (1968). *Makromol. Chem.*, **114,** 245
66. Cowie, J. M. G., Bywater, S. and Worsfold, D. J. (1967). *Polymer,* **8,** 105
67. Ptitsyn, O. B. and Eisner, Yu. E. (1959). *J. Techn. Phys. USSR,* **29,** 1117
68. Ptitsyn, O. B. and Eisner, Y. Y. (1959). *Vysokomol. Soedin.*, **1,** 1200
69. Ptitsyn, O. B. (1961). *Vysokomol. Soedin.*, **3,** 1084, 1251
70. Peterlin, A. (1955). *J. Chem. Phys.*, **23,** 2464; (1964), *J. Polymer Sci. B,* **2,** 359
71. Ullman, R. (1964). *J. Chem. Phys.*, **40,** 2193
72. Matsumura, K. (1970). *Polymer J. Jap,* **1,** 322
73. Ueda, M. and Kajitani, K. (1967). *Makromol. Chem.*, **108,** 138
74. Ueda, M. and Kajitani, K. (1967). *Makromol. Chem.*, **109,** 22
75. Kurata, M. and Stockmayer, W. H. (1963). *Advan. Polymer Sci.*, **3,** 196
76. Dondos, A. and Benoit, H. (1971). *Macromolecules,* **4,** 279
77. Patel, C. K. and Patel, R. D. (1970). *Makromol. Chem.*, **131,** 281
78. Reiss, C. and Benoit, H. (1968). *J. Polymer Sci. C,* **16,** 3079
79. Dondos, A. (1970). *Makromol. Chem.*, **135,** 181
80. Noda, I., Mizutani, K., Kato, T., Fujimoto, T. and Nagasawa, M. (1970). *Macromol.*, **3,** 787
81. Flory, P. J. (1949). *J. Chem. Phys.*, **17,** 303
82. Suzuki, H. (1968). *Bull. Chem. Soc. Jap,* **41,** 538
83. Alexandrowicz, Z. (1969). *J. Chem. Phys.*, **51,** 561
84. Rudin, A., Bennett, G. W. and McLaren, J. R. (1969). *J. Appl. Polymer Sci.*, **13,** 2371
85. Bates, T. W. and Ivin, K. J. (1967). *Polymer,* **8,** 263
86. Dondos, A. and Benoit, H. (1969). *J. Polymer Sci. B,* **7,** 335
87. Moldovan, L. and Strazielle, C. (1970). *Makromol. Chem.*, **140,** 201
88. Dondos, A. and Patterson, D. (1969). *J. Polymer Sci. A-2,* **7,** 209
89. Dondos, A. and Benoit, H. (1970). *Europ. Polymer J.*, **6,** 1439
90. Dondos, A. (1971). *Makromol. Chem.*, **147,** 123
91. Dondos, A., Rempp, P. and Benoit, H. (1969). *Makromol. Chem.*, **130,** 233
92. Fischer, H. and Mächtle, W. (1969). *Kolloid-Z. und Z. Polymere,* **230,** 221
93. Kotaka, T., Tanaka, T., Ohnuma, H., Murakami, Y. and Inagaki, H. (1970). *Polymer J.*, **1,** 245
94. Matsuda, H., Yamano, K. and Inagaki, H. (1969). *J. Polymer Sci. A-2,* **7,** 609
95. Ohnuma, H., Kotaka, T. and Inagaki, H. (1970). *Polymer J.*, **1,** 716
96. Tanzawa, H., Tanaka, T. and Soda, A. (1969). *J. Polymer Sci. A-2,* **7,** 929
97. Lange, H. and Baumann, H. (1969). *Angew. Makromol. Chem.*, **9,** 16; 1970 ibid., **14,** 25
98. Bianchi, U. and Peterlin, A. (1968). *J. Polymer Sci. A-2,* **6,** 1759
99. Meyerhoff, G. (1971). *Makromol. Chem.*, **145,** 189
100. Sandell, K. S. and Goring, D. A. I. (1970). *Makromol. Chem.*, **138,** 77
101. Sandell, L. S. and Goring, D. A. I. (1970). *Macromolecules,* **3,** 50, 54
102. Peterlin, A. (1968). *Amer. Chem. Soc. Polymer Preprint,* **9,** 323
103. Rossi, C. and Perico, A. (1970). *J. Chem. Phys.*, **53,** 1223
104. Perico, A. and Rossi, C. (1970). *J. Chem. Phys.*, **53,** 1217
105. Kirkwood, J. G. and Riseman, J. (1948). *J. Chem. Phys.*, **16,** 565
106. Berry, G. C. (1971). *J. Polymer Sci. A-2,* **9,** 687
107. Noda, I., Horikawa, T., Kato, T., Fujimoto, T. and Nagasawa, M. (1970). *Macromolecules,* **3,** 795
108. Nagasubramanian, K., Saito, O. and Graessley, W. W. (1969). *J. Polymer Sci. A-2,* **7,** 1955

109. Moore, W. R. A. D. and Millns, W. (1969). *Brit. Polymer J.*, **1,** 81
110. Meunier, J. C. and van Leemput, R. (1971). *Makromol. Chem.*, **47,** 191
111. Moritz, U. and Meyerhoff, G. (1970). *Makromol. Chem.*, **139,** 23
112. Zimm, B. H. and Kilb, R. W. (1959). *J. Polymer Sci.*, **37,** 19
113. Ashare, E. (1968). *Trans. Soc. Rheol.*, **12,** 535
114. Quadrat, O. and Bohdanecky, M. (1968). *J. Polymer Sci. B*, **6,** 769
115. Suzuki, H., Kotaka, T. and Inagaki, H. (1969). *J. Chem. Phys.*, **51,** 1279
116. Yamaguchi, N., Sugiura, Y., Okano, K. and Wada, E. (1969). *Rep. Progr. Polymer Phys. Jap.*, **12,** 59
117. Noda, I., Yamada, Y. and Nagasawa, M. (1968). *J. Phys. Chem.*, **72,** 2890
118. Ueberreiter, K., Melsheimer, J. and Springer, J. (1969). *Kolloid-Z. und Z. Polymere*, **234,** 989
119. Fixman, M. (1966). *J. Chem. Phys.*, **45,** 793
120. Cerf, R. (1958). *J. Phys. Radium*, **19,** 122
121. Cerf, R. (1958). *Advan. Polymer Sci.*, **1,** 382
122. Wolff, C. (1968). *J. Chim. Phys.*, **65,** 1569
123. Moore, R. S., McSkimin, H. J., Gieniewski, C. and Andreatch, P., Jr. (1969). *J. Chem. Phys.*, **50,** 5088
124. Tanaka, H. and Sakanishi, A. (1969). *Rep. Progr. Polymer Phys. Jap.*, **12,** 72, 73
125. Thurston, G. B. and Schrag, J. L. (1968). *J. Polymer Sci. A-2*, **6,** 1331
126. Johnson, R. M., Schrag, J. L. and Ferry, J. D. (1970). *Polymer J. Jap.*, **1,** 742
127. Massa, D. J., Schrag, J. L. and Ferry, J. D. (1971). *Macromolecules*, **4,** 210
128. Osaki, K. and Schrag, J. L. (1971). *Polymer J. Jap.*, **2,** 541
129. Peterlin, A. (1966). *Kolloid-Z. und Z. Polymere*, **209,** 181
130. Peterlin, A. (1967). *J. Polymer Sci. A-2*, **5,** 179
131. Peterlin, A. and Reinhold, C. (1967). *Trans. Soc. Rheol.*, **11,** 15
132. Thurston, G. B. and Peterlin, A. (1967). *J. Chem. Phys.*, **46,** 4881
133. Tschoegl, N. W. (1964). *J. Chem. Phys.*, **39,** 149; (1964). ibid, **40,** 473
134. Zimm, B. H. (1956). *J. Chem. Phys.*, **24,** 269
135. Peterlin, A. (1972). *J. Polymer Sci. B*, **10,** 101
136. Booij, H. C. (1966). *Rheol. Acta*, **5,** 215, 222; (1968). ibid, **7,** 202
137. Booij, H. C. (1970). *Thesis*, Leiden
138. Booij, H. C. and van Wiechen, P. H. (1970). *J. Chem. Phys.*, **52,** 5056
139. Noda, I., Tsuge, T. and Nagasawa, M. (1970). *J. Phys. Chem.*, **74,** 710
140. Takahashi, A. and Nagasawa, M. (1964). *J. Amer. Chem. Soc.*, **86,** 543
141. Nagasawa, M. and Eguchi, Y. (1967). *J. Phys. Chem.*, **71,** 880
142. Lapanje, S. and Kovač, S. (1967). *J Macromol. Sci. A*, **1,** 707
143. Fixman, M. (1964). *J. Chem. Phys.*, **41,** 3772
144. Hermans, J. J. and Overbeek, J. T. G. (1948). *Rec. Trav. Chim.*, **67,** 761
145. Flory, P. J. (1953). *J. Chem. Phys.*, **21,** 162
146. Katchalsky, A. and Lifson, S. (1953). *J. Polymer Sci.*, **11,** 409
147. Kurata, M. (1966). *J. Polymer Sci. C*, **15,** 347
148. Alexandrowicz, A. (1967). *J. Phys. Chem.*, **46,** 3789, 3800; (1968). ibid., **47,** 4377
149. Quadrat, O. and Bohdanecky, M. (1964). *Collect. Czech. Chem. Commun.*, **29,** 2449
150. Alexandrowicz, A. (1971). *J. Phys. Chem.*, **75,** 442
151. Okamoto, H. and Wada, Y. (1970). *Rep. Progr. Polymer Phys. Jap*, **13,** 55
152. Priel, Z. and Silberberg, A. (1970). *J. Polymer Sci. A-2*, **8,** 689
153. Bruce, C. and Schwarz, W. H. (1969). *J. Polymer Sci. A-2*, **7,** 909
154. Banks, W., Greenwood, C. T., Hourston, D. J. and Proctor, A. R. (1971). *Polymer*, **12,** 452
155. Schurz, J. and Bayer, E. (1970). *Das Papier*, **24,** 384
156. Patel, J. R. (1970). *Makromol. Chem.*, **134,** 263
157. Hand, J. H. and Williams, M. C. (1970). *Nature (London)*, **277,** 369
158. Holt, P. F. and Nasrallah, E. (1968). *J. Chem. Soc.*, 233
159. Holt, P. F. and Tamami, B. (1970). *Polymer*, **11,** 553
160. Vink, H. (1970). *Makromol. Chem.*, **131,** 133
161. Strauss, U. P. and Fuoss, R. M. (1952). *J. Polymer Sci.*, **8,** 593
162. Claesson, A. and Lohmander, U. (1961). *Makromol. Chem.*, **44–46,** 461
163. Lohmander, U. and Stromberg, R. (1964). *Makromol. Chem.*, **72,** 143
164. Lin, O. C. C. (1970). *Macromolecules*, **3,** 80

165. Schurz, J., Uragg, H., Belegratis, M. and Gruber, E. (1970). *Z. Physiol. Chem.*, **351,** 843
166. Gibbs, D. A., Merrill, E. W. and Smith, K. A. (1968). *Biopolymers,* **60,** 777
167. Yamakawa, H. (1970). *J. Chem. Phys.*, **53,** 436
168. Thurston, G. B. and Morrison, J. D. (1969). *Polymer,* **10,** 421
169. Iwata, K. (1971). *J. Chem. Phys.*, **54,** 157
170. Kamide, K. (1969). *Makromol. Chem.*, **128,** 197
171. Fixman, M. (1965). *J. Chem. Phys.*, **42,** 3831
172. Imai, S. (1969). *J. Chem. Phys.*, **50,** 2107
173. Imai, S. (1969). *Proc. Roy. Soc. A.*, **308,** 497
174. Imai, S. (1969). *Brit. Polymer J.*, **1,** 161
175. Reinhold, C. and Peterlin, A. (1966). *J. Chem. Phys.*, **44,** 4338
176. Peterlin, A. (1961). *Makromol. Chem.*, **44–46,** 338
177. Stevenson, J. F. and Bird, R. B. (1971). *Trans. Soc. Rheol.*, **15,** 135
178. Tanner, R. I. and Stehrenberger, W. (1971). *J. Chem. Phys.*, **55,** 1950
179. Noda, I. and Hearst, J. E. (1971). *J. Chem. Phys.*, **54,** 2342
180. Tsuda, K. (1970). *Rheol. Acta,* **9,** 509
181. Ullman, R. (1969). *Macromolecules,* **2,** 27
182. Cottrell, F. R., Merrill, E. W. and Smith, K. A. (1969). *J. Polymer Sci. A-2,* **7,** 1415; (1970). ibid., **8,** 289
183. Champion, J. V. and Davis, I. D. (1970). *J. Chem. Phys.*, **52,** 381
184. Peterlin, A. and Reinhold, C. (1964). *J. Chem. Phys.*, **40,** 1029; (1965). ibid., **42,** 2172
185. Cerf, R. (1969). *J. Chim. Phys.*, **66,** 479
186. Peterlin, A. (1960). *J. Chem. Phys.*, **33,** 1799
187. Budtov, V. P. and Gotlib, Ya. Yu. (1965). *Vysokomol. Soedin.*, **7,** 478
188. Hayahara, T. and Takao, S. (1968). *Kolloid-Z. und Z. Polymere,* **225,** 100
189. Daum, U. and Wales, J. L. S. (1969). *J. Polymer Sci. B,* **7,** 459
190. Gupta, D. and Forsman, W. C. (1969). *Macromolecules,* **2,** 304
191. Simha, R. (1971). *J. Macromol. Sci. B,* **5,** 425
192. Vinogradov, G. V. and Titkova, L. V. (1970). *Kolloid-Z. und Z. Polymere,* **239,** 655
193. Manaresi, P., Giachetti, E. and DeFornasari, E. (1968). *J. Polymer Sci. C,* **16,** 3133
194. Quakenbos, H. M. (1969). *J. Appl. Polymer Sci.*, **13,** 341
195. Baijal, M. D. and Sturnn, C. L. (1970). *J. Appl. Polymer Sci.*, **14,** 1651
196. Keskkula, H. and Taylor, W. C. (1970). *J. Polymer Sci. B,* **8,** 867
197. Catsiff, E. H. (1971). *J. Appl. Polymer Sci.*, **15,** 1641
198. Cross, M. M. (1970). *Polymer,* **11,** 238
199. Allen, V. R. and Fox, T. G. (1964). *J. Chem. Phys.*, **41,** 337
200. Fox, T. G. and Allen, V. R. (1964). *J. Chem. Phys.*, **41,** 344
201. Bartenev, G. M. (1970). *J. Polymer Sci. A-1,* **8,** 3417
202. Vogel, H. (1921). *Physik. Z.*, **22,** 645
203. Miller, A. A. (1969). *Macromolecules,* **2,** 355
204. Hirai, N. and Eyring, H. (1959). *J. Polymer Sci.*, **37,** 51
205. Penwell, R. C. and Porter, R. S. (1971). *J. Polymer Sci. A-2,* **9,** 463
206. Penwell, R. C., Porter, R. S. and Middleman, S. (1971). *J. Polymer Sci. A-2,* **9,** 731
207. Cross, M. M. (1969). *J. Appl. Polymer Sci.*, **13,** 765
208. Graessley, W. W., Hazleton, R. L. and Lindeman, R. L. (1967). *Trans. Soc. Rheol,* **11,** 267
209. Bueche, F. and Harding, S. W. (1958). *J. Polymer Sci.*, **32,** 177
210. Graessley, W. W. (1967). *J. Chem. Phys.*, **47,** 1942
211. Bueche, F. (1968). *J. Chem. Phys.*, **48,** 4781
212. Powell, A. (1968). *Polymer,* **9,** 513
213. Sabia, R. (1963). *J. Appl. Polymer Sci.*, **7,** 347
214. Klein, J. and Fusser, H. (1968). *Rheol. Acta,* **7,** 118
215. Brenschede, E. and Klein, J. (1969). *Rheol. Acta,* **8,** 71; (1970). ibid, **9,** 130
216. McLuckie, C. and Rogers, M. G. (1969). *J. Appl. Polymer Sci.*, **13,** 1049
217. Duvdevani, I. J. and Klein, J. (1967). *SPE J.*, **23,** 41
218. Han, C. D., Charles, M. and Philippoff, W. (1969). *Trans. Soc. Rheol.*, **13,** 455
219. Penwell, R. C. and Porter, R. S. (1969). *J. Appl. Polymer Sci.*, **13,** 2427
220. Frank, H. P. (1966). *Rheol. Acta,* **5,** 89; (1968). ibid., **7,** 222, 344
221. Frank, H. P. (1969). *Angew. Makromol. Chem.*, **9,** 106
222. Mieras, H. J. M. A. and van Rijn, C. F. H. (1969). *J. Appl. Polymer Sci.*, **13,** 309

223. Vinogradov, G. V., Malkin, A. Ya. and Kulichikhin, V. G. (1968). *Vysokomol. Soedin,* **10,** 2522
224. Vinogradov, G. V., Malkin, A. Ya. and Kulichikhin, V. G. (1970). *J. Polymer Sci. A-2,* **8,** 333
225. Klein, J. and Woernle, R. (1970). *Kolloid-Z. und Z. Polymere,* **237,** 209
226. Einaga, Y., Osaki, K., Kurata, M. and Tamura, M. (1971). *Macromolecules.,* **4,** 87
227. Blyler, L. L. and Haras, T. W. (1969). *J. Appl. Polymer Sci.,* **13,** 2721
228. Narkis, M. and Miltz, J. (1969). *Polymer Eng. Sci.,* **9,** 153
229. Eisenberg, A. (1969). *Macromolecules,* **2,** 44
230. Schott, H. (1968). *Rheol. Acta,* **7,** 179
231. Bueche, F. (1969), and Tokcan, G. (1969). *J. Polymer Sci A-2,* **7,** 1385
232. Behrens, A. R. and Folt, V. L. (1969). *Polymer Eng. Sci.,* **9,** 27
233. Collins, E. A. and Metzger, A. P. (1970). *Polymer Eng. Sci.,* **10,** 57
234. Schreiber, H. P. (1969). *Polymer Eng. Sci.,* **9,** 311
235. Tager, A. A. and Dreval, V. E. (1970). *Rheol. Acta,* **9,** 517
236. Trizno, V. L., Konsetov, V. V., Mnatsakanov, S. S., Rozenberg, M. E. and Nikdaev, A. F. (1969). *J. Appl. Chem. USSR,* **42,** 1716
237. Uy, W. C. and Graessley, W. W. (1971). *Macromolecules,* **4,** 458
238. Schurz, J., Lederer, K. and Schmidt, K. H. (1969). *Das Papier,* **23,** 125
239. Biddle, D. and Pardhan, S. (1970). *Arkiv. Kemi.,* **32,** 43
240. Schutz, R. A. (1969). *Rheol. Acta,* **8,** 349
241. Paul, D. R., Lawrence, J. E. St. and Troell, J. H. (1970). *Polymer Eng. Sci.,* **10,** 70
242. Johnson, J. F., Barall, E. M., II and Porter, R. S. (1968). *Trans. Soc. Rheol.,* **12,** 1, 133
243. Mendelson, R. A. (1968). *Polymer Eng. Sci.,* **8,** 235; (1969). ibid., **9,** 350
244. Nakajima, N. (1970). *J. Appl. Polymer Sci.,* **14,** 2643, 2661
245. Saeda, S., Yotsuyanagi, J. and Yamaguchi, K. (1971). *J. Appl. Polymer Sci.,* **15,** 277
246. Bontinck, W. J. (1969). *Rheol. Acta,* **8,** 328
247. Petraglia, G. and Coen, A. (1970). *Polymer Eng. Sci.,* **10,** 79
248. Lee, C. L., Polmanteer, K. E. and King, E. G. (1970). *J. Polymer Sci. A-2,* **8,** 1909
249. Fujiki, T., Uemura, M. and Kosaka, Y. (1968). *J. Appl. Polymer Sci.,* **12,** 267
250. Peyser, Y., Dealy, J. M. and Kamal, M. R. (1971). *J. Appl. Polymer Sci.,* **15,** 1963
251. Hyun, K. S. and Karam, H. J. (1969). *Trans. Soc. Rheol.,* **13,** 335
252. Ramsteiner, F. (1970). *Rheol. Acta,* **9,** 374
253. Toelcke, G. A., Madonia, K. J. and Biesenberger, J. A. (1967). *Polymer Eng. Sci.,* **7,** 318
254. Thomas, D. P. and Hagan, R. S. (1969). *Polymer Eng. Sci.,* **9,** 164
255. Graessley, W. W. and Segal, L. (1970). *Amer. Inst. Chem. Eng. J.,* **16,** 261
256. Graessley, W. W. and Segal, L. (1969). *Macromolecules,* **2,** 49
257. Elliott, J. H. (1969). *J. Appl. Polymer Sci.,* **13,** 755
258. Kraus, G. and Gruver, J. T. (1969). *Trans. Soc. Rheol.,* **13,** 315
259. Shih, C. K. (1970). *Trans. Soc. Rheol.,* **14,** 83
260. Scalco, E., Huseby, T. W. and Blyler, L. L., Jr. (1968). *J. Appl. Polymer Sci.,* **12,** 1343
261. Casale, A., Porter, R. S. and Johnson, J. F. (1971). *J. Macromol. Sci. C,* **5,** 387
262. Graessley, W. W. (1971). *J. Chem. Phys.,* **54,** 5143
263. Mendelson, R. A., Bowles, W. A. and Finger, F. L. (1970). *J. Polymer Sci.,* **8,** 105, 127
264. Porter, R. S., Knox, J. P. and Johnson, J. F. (1968). *Trans. Soc. Rheol.,* **12,** 409
265. Schroff, R. N. and Shida, M. (1970). *J. Polymer Sci. A-2,* **8,** 1917
266. Shida, M. and Caniro, L. V. (1970). *J. Appl. Polymer Sci.,* **14,** 3083
267. Combs, R. L., Slonaker, D. F. and Coover, H. W. (1969). *J. Appl. Polymer Sci.,* **13,** 519
268. Shiratama, K., Matsuda, T. and Kita, S. I. (1971). *Makromol. Chem.,* **147,** 155
269. Graessley, W. W. and Prentice, J. S. (1968). *J. Polymer Sci: A-2,* **6,** 1887
270. Pritchard, J. H. and Wissburn, K. F. (1969). *J. Appl. Polymer Sci.,* **13,** 233
271. Kraus, G. and Gruver, J. T. (1970). *J. Polymer Sci. A-2,* **8,** 305
272. Wang, J. S., Porter, R. S. and Knox, J. R. (1970). *J. Polymer Sci. B,* **8,** 671
273. Fujimoto, T., Narukawa, H. and Nagasawa, M. (1970). *Macromolecules,* **3,** 57
274. Ito, K. (1969). *Rept. Progr. Polymer Phys. Jap.,* **12,** 131
275. Hermann, H. D. and Knappe, W. (1969). *Rheol. Acta,* **8,** 384
276. Hirata, S., Hasegawa, H. and Kishimoto, A. (1970). *J. Appl. Polymer Sci.,* **14,** 2025
277. Sieglaff, C. L. and O'Leary, K. J. (1970). *Trans. Soc. Rheol.,* **14,** 49
278. Sieglaff, C. L. (1969). *Polymer Eng. Sci.,* **9,** 81
279. Cooper, W. M. and Pollett, W. F. O. (1969). *J. Appl. Polymer Sci.,* **13,** 2312

280. Matsuo, T., Pavan, A., Peterlin, A. and Turner, D. T. (1967). *J. Colloid and Interface Sci.*, **24**, 241
281. Weeks, J. C. and Reid, G. C. (1970). *Rheol. Acta*, **9**, 69
282. Dunlop, A. N. and Williams, H. L. (1969). *J. Appl. Polymer Sci.*, **14**, 2753
283. Endo, M. and Nagazawa, M. (1970). *J. Polymer Sci. A-2*, **8**, 371
284. Chartoff, R. P. and Maxwell, B. (1969). *Polymer Eng. Sci.*, **9**, 159
285. Chartoff, R. P. and Maxwell, B. (1970). *J. Polymer Sci. A-2*, **8**, 455
286. Hill, A. S. and Maxwell, B. (1970). *Polymer Eng. Sci.*, **10**, 289
287. Wolkowicz, R. I. and Forsmann, W. C. (1971). *Macromolecules*, **4**, 184
288. Mills, N. J., Nevin, A. and McAinsh, J. (1970). *J. Macromol. Sci. B*, **4**, 863
289. Sakamoto, K., MacKnight, W. J. and Porter, R. S. (1970). *J. Polymer Sci. A-2*, **8**, 227
290. Onogi, S., Masuda, T. and Kitagawa, K. (1970). *Macromolecules*, **3**, 109
291. Masuda, T., Kitagawa, K., Inoue, T. and Onogi, S. (1970). *Macromolecules*, **3**, 116
292. Masuda, T., Kitagawa, K. and Onogi, S. (1970). *Polymer J. Jap.*, **1**, 418
293. Onogi, S., Masuda, T., Toda, N. and Koga, K. *Polymer J. Jap.*, 197
294. Vinogradov, G. V. and Ivanova, L. I. (1968). *Rheol. Acta*, **7**, 243
295. Vinogradov, G. V. Private communication
296. Rudd, J. F. (1960). *J. Polymer. Sci.*, **44**, 459; (1962). ibid., **60**, S7
297. Stratton, R. A. (1966). *J. Colloid and Interface Sci.*, **22**, 517
298. Thomas, D. P. and Hagan, R. S. (1966). *Polymer Eng. Sci.*, **6**, 473
299. Kataoka, T., Ueda, S. and Kurihara, H. (1971). *J. Polymer Sci. B*, **9**, 449
300. Kataoka, T. and Ueda, S. (1970). *Rep. Progr. Polymer Phys. Jap.*, **13**, 105
301. Kataoka, T. and Ueda, S. (1969). *J. Polymer Sci. A-2*, **7**, 475
302. Simmons, J. M. (1968). *Rheol. Acta*, **7**, 184
303. Chömpff, A. J. and Duiser, J. A. (1966). *J. Chem. Phys.*, **45**, 1505
304. Chömpff, A. J. and Prins, W. (1968). *J. Chem. Phys.*, **48**, 235
305. Chömpff, A. J. (1970). *J. Chem. Phys.*, **53**, 1566, 1577
306. Duiser, J. A. and Staverman, A. J. (1965). In *Physics of Non-Crystalline Solids*, Ed. by Prins, J. A., 376 (Amsterdam: North-Holland Publ. Co.)
307. Cox, W. P. and Merz, E. H. (1958). *J. Polymer Sci.*, **28**, 619
308. Ajroldi, G., Garbuglio, C. and Pezzin, G. (1967). *J. Polymer Sci.*, **5**, 289
309. Akers, L. C. and Williams, M. C. (1969). *J. Chem. Phys.*, **51**, 3834
310. Wales, J. L. S. and den Otter, J. L. (1970). *Rheol. Acta*, **9**, 115
311. Onogi, S., Fujii, T., Koto, H. and Ogihara, S. (1964). *J. Phys. Chem.*, **68**, 1598
312. Shida, M. and Shroff, R. N. (1970). *Trans. Soc. Rheol.*, **14**, 605
313. Zapas, L. J. and Phillips, J. C. (1971). *J. Res. Nat. Bur. Stand.*, **75A**, 33
314. Zapas, L. J. (1966). *J. Res. Nat. Bur. Stand.*, **70A**, 525
315. Bernstein, B., Kearsley, E. A. and Zapas, L. J. (1964). *J. Res. Nat. Bur. Stand.*, **60B**, 103
316. Bernstein, B., Kearsley, E. A. and Zapas, L. J. (1963). *Trans. Soc. Rheol.*, **7**, 391
317. Roesler, F. C. and Twyman, W. A. (1955). *Proc. Phys. Soc. B.*, **68**, 97
318. Maruyama, T., Takano, Y. J. and Yamamoto, M. (1968). *Rept. Progr. Phys. Jap.*, **11**, 99
319. Tanaka, T., Yamamoto, M. and Takano, Y. J. (1970). *Macromol. Sci. B*, **4**, 931

6
Fracture of Polymers

E. H. ANDREWS
Queen Mary College, University of London

6.1 INTRODUCTION

Although several extensive texts and reviews have been devoted to the fracture of polymeric solids during the past decade[1–3], the subject has advanced so rapidly that a fresh survey is highly appropriate. One of the reasons for the intense research activity in the deformation and fracture of polymers is, of course, their increasing importance as engineering materials.

Load-bearing properties and resistance to stress, corrosion, creep and fatigue have thus acquired an importance which they did not possess before. The rapidity with which plastics are replacing more familiar engineering materials has made the matter urgent. The long evolutionary years of metallurgical engineering have been denied to polymeric materials (with the exception, perhaps, of elastomers) and plastics are still frequently used in the absence satisfactory design criteria. For example, no collected design data exists for plastics in respect of their fatigue properties, and environmental stress cracking is still not sufficiently well understood for the engineer to design its avoidance.

Turning from these important practical aspects for a moment, the deformation and fracture of polymers constitute topics of immense scientific interest in their own right. The highly anisotropic atomic bonding in solid polymers results in a unique distribution of applied stress on a molecular scale with consequent implications for fracture viewed as a molecular event. Electron spin resonance and other techniques of mechanochemistry come into their own at this point and allow us to explore fracture in chemical terms.

A physical–chemical approach is necessary in considering environmental stress cracking and crazing, where again the mechanical factors of stress and elastic energy operate synergistically with chemical reaction and the thermodynamics of polymer–solvent interaction to produce these interesting and, in practice, serious phenomena.

Finally, the morphologist knows that polymer solids are not simply the continuous media of the engineer or the molecular aggregates of the synthetic chemist, but that they possess in general a physical microstructure or morphology which can profoundly modify their physical and mechanical behaviour. Such factors as molecular orientation, crystalline–amorphous structure, phase separation, blending and the use of fillers need to be considered here. A rigid, brittle polymer may be transformed into a resilient high-strength fibre by orientation and recrystallisation, whilst polyethylene can be transformed from the familiar tough material we know to a friable powdery solid simply by recrystallising at high pressure. In neither example does any change occur in the molecular structure of the polymer, but only in its physical morphology.

In order to provide a completely balanced picture of fracture in polymers, we would need to devote time to each of these three approaches to the subject. In the space available, however, we shall only be able to consider the first two, leaving aside a discussion of morphological aspects. These have, however, been recently reviewed by the present writer elsewhere[4, 5]. Firstly, we shall consider what we might call the engineering approach which considers solid materials as continuous media with certain bulk properties such as elastic moduli, yield stresses, fracture strengths and so on. The well-known time-dependence of such quantities in polymers can be included in such a treatment. This approach is particularly valuable because it avoids the need to know anything about the precise microscopical nature and structure of the solid concerned but only requires us to characterise the material by bulk parameters. The approach is also vital, of course, if we are to provide the engineer with design data since such data must always be expressed in terms of measurable continuum quantities such as stress and strain.

The continuum approach is embodied in the theory of fracture mechanics which has been developed very fully for linear elastic solids[6] (approximated by many metallic alloys). Unfortunately, polymers are always visco-elastic rather than elastic and many important materials (e.g. elastomers, low-density polyethylene, plasticised PVC and film) exhibit finite strain and non-linearity. It is often necessary therefore to employ a more general form of fracture mechanics than the linear theory and this is discussed in Section 6.2 of this chapter. The theory is followed in Section 6.3 by examples of the power of fracture mechanics both to describe fracture phenomena in a consistent manner and to provide insight into the *mechanisms* of fracture. Among these examples are tearing, fatigue, environmental fracture and the failure of adhesive joints.

The chemical, or molecular, approach to fracture stands in contrast to (but not in contradiction of) continuum theory. It is concerned with fracture as a kinetic process of bond breakage and especially with the early stages of fracture initiation. Section 6.4 of this chapter is devoted to this fairly recent and exciting field of study.

6.2 MECHANICS OF FRACTURE

6.2.1 Flaw theory of strength

In an atomically-perfect body placed under load, all interatomic bonds in the line of force will be equally stressed. Since the force–displacement curve for an atomic bond passes through a maximum it follows that, at some critical applied load, all such bonds will 'break' simultaneously, resulting in the complete dissociation of the solid in much the same way as a liquid dissociates at the critical thermodynamic temperature. Although no such failure has ever been observed in a solid, some near-perfect 'whisker' crystals and glass fibres do disintegrate into powder when they break. The vast majority of fracture events in solids, however, cause separation of the loaded body into two or several pieces, indicating at once that fracture is usually an heterogeneous process occurring at selected points within the material.

This heterogeneity is further underlined when we consider the magnitude of the forces necessary to produce fracture. The theoretical value of the fracture stress (i.e. the force per unit area necessary to cause fracture) can be calculated from the maximum in the interatomic force–displacement curve and the number of bonds per unit area. The laws of force between atoms vary according to the chemical nature of the bond but are all such that the maximum theoretical fracture stress σ_f turns out to be of the order

$$\sigma_f = 0.1\, E$$

where E is the Young's modulus of the solid. A corresponding maximum shear stress to cause the atoms to slide past one another on a slip plane is

$$\tau_f = 0.1\, G$$

where G is the shear modulus of the material. Although some microscopic 'whisker' crystals and specially-prepared glass fibres or rods approach these theoretical strengths within a factor of less than 10, the majority of materials are 100–1000 times weaker. This weakness is attributed to the existence of flaws or imperfections on a very small scale, which cause local stress concentrations where the stresses are very much higher than the nominal stress in the body. Thus the stress necessary to break atomic bonds is achieved locally when the overall stress in the body is relatively low. When this occurs the flaw propagates as a crack. Since concentration at its tip is self-sustaining it continues to grow, eventually causing macroscopic fracture.

These considerations indicate that the strength of most real solids is governed by the presence of flaws and the manner in which they propagate under stress. This is the basic tenet of continuum fracture theory (or 'fracture mechanics') and enables us to develop the theory mathematically.

6.2.2 Stresses at a crack; linear fracture mechanics

The distribution of stress around a crack or narrow elliptical hole in a stressed lamina has been the subject of many studies[7], from the time of Inglis (1913)[8] to the present day. All analytical solutions of this problem are based upon linear elasticity theory and infinitesimal strain conditions, though solutions have also been obtained for an ideal elastic–plastic material.

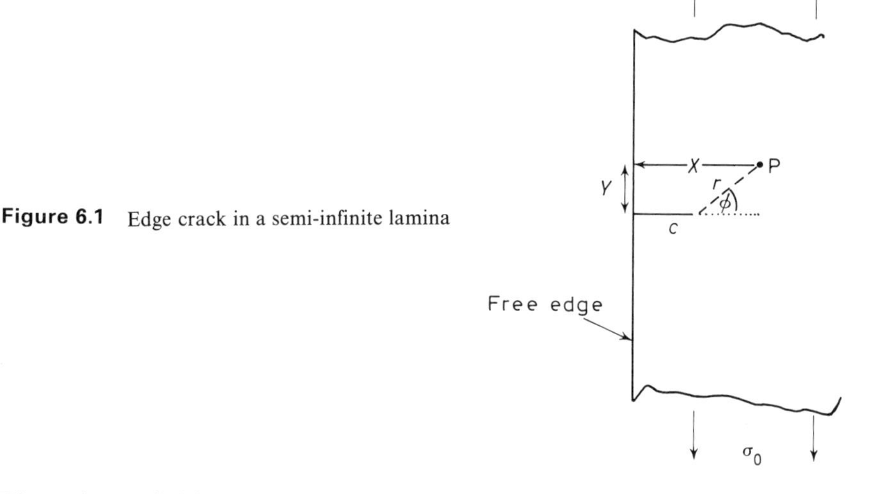

Figure 6.1 Edge crack in a semi-infinite lamina

The only available results for highly elastic or visco-elastic materials are empirical[9], but these do reveal that the general nature of stress concentrations around a crack are similar, even in these cases, to the results of elasticity theory. That is, the major principal stress has its maximum value at the crack tip and decays more or less radially with distance from the tip. The maximum value of the stress and the rate of decay with distance both increase with decreasing tip radius. At some point a little removed from the extreme tip the state of stress (in a tensile field) approaches that of biaxial tension for very thin sheets and of hydrostatic tension in the median plane of thick sheets.

The general form of the stresses around a crack can be deduced from dimensional considerations without appeal to any theory of elasticity. Consider the simple but important case of a uniformly stressed, semi-infinite sheet, containing an infinitely sharp edge crack at right angles to the applied stress (Figure 6.1). The stresses at a point P with coordinates (r, ϕ) referred to the crack tip as origin must be of the form

$$\sigma_{ij}(\mathrm{P}) = \sigma_0 f\left(\frac{c}{r}, \phi\right) \tag{6.1}$$

where σ_{ij} are the components of the stress tensor, σ_0 is the uniform applied stress and c the crack length. Equation (6.1) is useful, as will be seen later, in the application of fracture mechanics to materials not obeying classical elasticity theory and which cannot therefore be treated by linear fracture mechanics.

Equation (6.1) can, of course, be made explicit for a linear elastic solid and Irwin[6] has given the following approximate solution for the infinitely sharp crack, based upon Westergaard's stress-function solution[10].

$$\sigma_{ij} = \frac{K}{(2r)^{\frac{1}{2}}} f_{ij}(\phi) - p_{ij} \tag{6.2}$$

where f_{ij} are known and p_{ij} is a small correction term when $i = j = 1$ (1 representing the crack axis in a cartesian coordinate system) and is zero otherwise.

Linear fracture mechanics is based upon equation (6.2) and is concerned with the quantity K which is called the 'stress intensity factor' since it completely governs the stress at any given point P near the crack tip. Comparison of equations (6.1) and (6.2) shows that, for the edge crack in a tensile field (a case for which K is written as K_I),

$$K_\mathrm{I} = \alpha \sigma_0 c^{\frac{1}{2}} \tag{6.3}$$

where the constant α is, in fact, equal to $\sqrt{\pi}$ for the semi-infinite sheet. If, as is usually the case in practice, the sheet cannot be considered infinite, α becomes a function of (c/b) where b is the plate width and various explicit forms for α have been proposed.

The basic idea of fracture mechanics is that the process of fracture, i.e. the propagation of an existing flaw, is governed entirely by the intensity of the stress field around the crack tip, i.e. by the parameter K. For given environmental conditions, K should be independent of the specimen shape and the method of loading (although equation (6.3) will change according to the method of loading and specimen shape). This means that K should have a characteristic value for fracture in a given material and this is largely borne out in practice. Exceptions include the effect of a transition at the crack tip from plane strain (centre of thick sheets) to plane stress (surface of thick sheets and whole of thin sheets) and certain rather special effects encountered which will be discussed later. The characteristic value of K_I for catastrophic fracture is denoted K_IC and is known as the 'fracture toughness'. Once it is known for a given material it enables us to predict the fracture stress for a specimen containing cracks of a known size, using equation (6.3) or its appropriate equivalent for the shape and loading regime in question. Thus, the fracture stress from equation (6.3) is

$$\sigma_f = K_{IC}/\alpha c^{\frac{1}{2}} \tag{6.4}$$

6.2.3 Energy-balance approach; the surface work

Linear fracture mechanics are inapplicable to materials which are highly elastic or visco-elastic and a more general method is therefore required which does not require explicit knowledge of the stress distribution around a crack. As long ago as 1921, Griffith[11] proposed an energy-balance criterion. He maintained that pre-existent cracks would propagate catastrophically if the energy thereby released from the stress field in the body was greater than the surface energy of the newly created crack interfaces. This criterion can be written

$$-\partial\mathscr{E}/\partial A > S \tag{6.5}$$

where $\mathscr{E}$ is the total elastic stored energy in the stressed body, A, the interfacial area of the crack and S the surface energy.

Although Griffith used linear elasticity theory to evaluate the left-hand side of equation (6.5) and thus limited his solution to ideal materials, the equation remains completely general, being an expression of the basic law of energy conservation. Rivlin and Thomas (1953)[12] showed how equation (6.5) could be applied to highly elastic materials, and their evaluation of $(-\partial\mathscr{E}/\partial A)$ for one form of specimen will illustrate the method.

Consider the test piece illustrated in Figure 6.2. It consists of a long sheet of material, of width l_0 containing a crack which terminates as shown. The

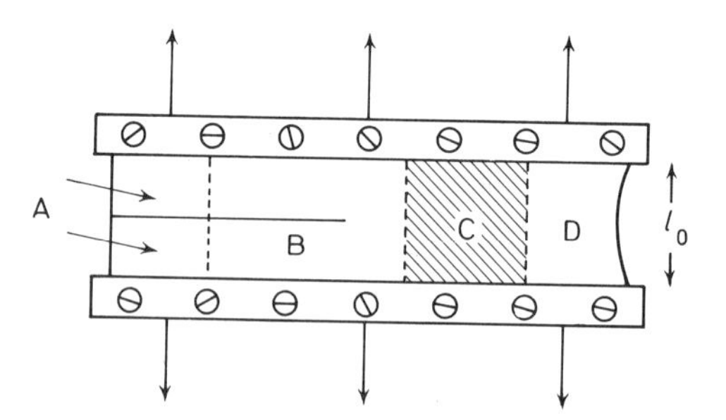

Figure 6.2 Pure shear test piece

sheet is gripped along its edges and stressed in a direction perpendicular to its length. The region A is stress-free because of the crack, the region B has a complex state of stress because of the crack tip, as does region D because of the free edge. Region C, however, is in a well-defined state of pure shear (strains $\varepsilon_1 = -\varepsilon_3$; $\varepsilon_2 = 0$). If the crack grows by a length Δc the net effect is to decrease the size of region C by that length (i.e. by a volume $l_0 h_0 \Delta c$ where h_0 is the sheet thickness before deformation) and to increase region A by the same amount. The loss of stored energy is thus

$$-\Delta\mathscr{E} = (l_0 h_0 \Delta c) W_p$$

where W_p is the recoverable elastic energy density in pure shear for the material which can be evaluated empirically in a separate experiment. Thus, since $\Delta A = 2h_0\Delta c$.

$$-(\Delta\mathscr{E}/\Delta A)_{\text{limit}} = -\partial\mathscr{E}/\partial A = \tfrac{1}{2}W_p l_0 \qquad (6.6)$$

The corresponding solution for the edge crack was found[12] by similar reasoning to be

$$-\partial\mathscr{E}/\partial A = kcW_0 \qquad (6.7)$$

where k was an unknown constant and W_0 the recoverable energy density in simple extension in the bulk of the test piece. Applying equation (6.7) to a linear elastic material for which $W = \sigma^2/2E$ (where E is Young's modulus) we obtain

$$-\partial\mathscr{E}/\partial A = kc\sigma^2/2E \qquad (6.8)$$

Comparing this with equation (6.3) it is seen that

$$(-\partial\mathscr{E}/\partial A) = K_{\text{I}}^2\left(\frac{k}{2E\alpha^2}\right)$$

i.e. the energy supply per area of crack is proportional to K_{I}^2. This emphasises the essential equivalence of the stress-intensity and energy-balance approaches. The latter, however, is not limited to linear materials and infinitesimal strains as is the former and must therefore be regarded as altogether more general and, indeed, necessary for most polymeric systems.

We now return to Griffith's equation (6.5). Using the analyses described above it was possible to evaluate $-\partial\mathscr{E}/\partial A$ at the critical loading conditions required to propagate cracks and thus to test the Griffith criterion. It was found, for many materials including metals, glassy plastics and elastomers, that $(-\partial\mathscr{E}/\partial A)$ does indeed have a characteristic value for propagation as suggested by Griffith but that this value was usually larger by orders of magnitude than the surface energy S. Orowan[13], Benbow and Roessler[14], Berry[15] and others therefore proposed that S be replaced by what we shall call the 'surface work', denoted here $\mathscr{T}$. Then, for an edge-crack of length c we have that the specimen fracture stress σ_f is, from equation (6.8),

$$\sigma_f = \sqrt{(2E\mathscr{T}/kc)} \qquad (6.9)$$

The surface work includes the thermodynamic surface energy, but additionally includes energy dissipated in the highly stressed region around the tip by plastic flow (P) and visco-elastic response or other forms of mechanical hysteresis (H). It may also include energy 'lost' dynamically (D) by radiation as sound or travelling stress waves although these factors are normally neglected.

Thus we have the surface work

$$\mathscr{T} = S+P+H+D \qquad (6.10)$$

where all energies are referred to unit area of crack surface. Normally the components $P+H$ predominate because, at the high strains usually developed at a crack tip, most real materials are inelastic. There are important exceptions however. If atomic bonds can be broken at low stresses as, e.g. when chemical attack occurs simultaneously with mechanical loading or when secondary bonds are weakened by plasticisation, cracks can grow without the tip being subject to high stresses and $\mathscr{T}$ can approach S in value. Examples are ozone-cracking in rubbers[16] and solvent-cracking in polymer glasses[17]. Alternatively, if S is very low, i.e. the atomic bonding across the crack plane is intrinsically weak, the tip region will again be subject to low stresses and plastic or visco-elastic responses will be minimised. This is the case for, e.g. the cleavage of mica or the separation of very weak adhesive joints. Thirdly, in a material where plasticity does not occur but which is simply visco-elastic, reduction of the rate of application of load and elevation of the temperature can progressively reduce H until $\mathscr{T} \rightarrow S$. This appears to be the case for the tearing of non-crystallising cross-linked gum rubbers[18].

On the other hand, deliberate attempts can be made to increase $\mathscr{T}$ (and thus the resistance to fracture) by modifying materials. The addition of particulate fillers, such as carbon black, to rubber is a successful example of this, the dispersed particles or aggregates causing enhanced local stresses at points throughout the already highly stressed tip region. The hysteresis loss is greatly increased both by relaxation of the cross-linked molecular network and by filler–matrix breakdown, and extremely tough materials result. Rubber-modified plastics (high-impact polystyrenes, ABS etc.) operate by the same principle, except that here the local stress raisers are soft particles in a hard matrix and the local energy dissipation mechanism is multiple crazing around the particles. Finally, fibre reinforcement is a further method of increasing $\mathscr{T}$ for cracks transverse to the fibre axis. Here the energy necessary to pull out fibres which otherwise bridge the growing crack must be included in $\mathscr{T}$. Cracks running *parallel* to closely packed and aligned fibres may propagate at *reduced* $\mathscr{T}$ because any plastic zone at the tip may be limited to a size commensurate with the distance between the fibres instead of being free to enlarge to its natural dimensions.

6.2.4 Visco-elastic solids

In the foregoing section it has been shown that the energy-balance method enables fracture mechanics to be applied to non-linear solids subject to finite strains. All that has been said, however, applies to materials whose behaviour in bulk (i.e. in regions remote from the immediate crack tip) is elastic. All polymers are, however, visco-elastic to some extent and some, like low-density polyethylene or plasticised PVC, are strikingly so. Recently the energy-balance method has been extended to cover even these materials[19].

It might be thought at first that all we need to do to take account of mechanical hysteresis (illustrated in Figure 6.3 for polyethylene) is to replace the energy density W in equations (6.6) and (6.7) by the recoverable part of that density. This is, however, an over-simplification since in passing, e.g. from the pure shear region C in Figure 6.2 to the unstressed region A, an element of material is taken first to a higher stress than that with which it started, before the stress relaxes. The energy loss, being a function of the maximum stress in the cycle, is not therefore simply related to the energy density of the bulk, but depends for each element upon its total stress history.

This problem can be overcome by a more rigorous derivation of $(-\partial\mathscr{E}/\partial A)$ and the full details are given elsewhere[19]. The derivation will, however be outlined here.

Consider the edge crack in an infinite visco-elastic sheet loaded by a uniform tensile stress (Figure 6.1). The coordinates of a point P in the sheet

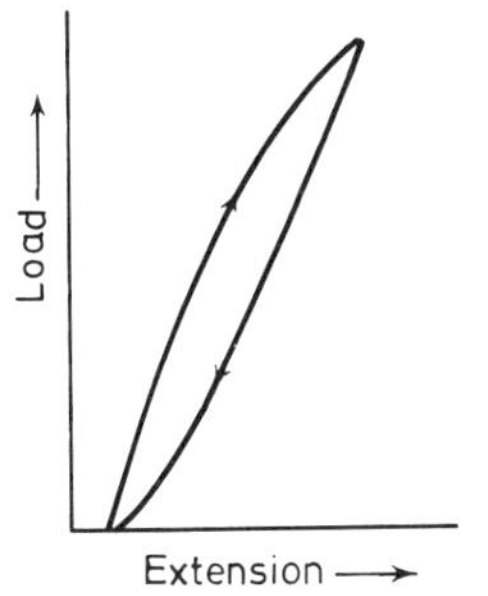

Figure 6.3 Mechanical hysteresis in polyethylene

are conveniently defined by the rectangular coordinates X, Y, as shown, rather than r, ϕ as used in equation (6.1), but by similar dimensional arguments we have,

$$\sigma_{ij}(\mathrm{P}) = \sigma_0 f\left(\frac{X}{c}, \frac{Y}{c}\right) \equiv \sigma_0 f(x, y) \tag{6.11}$$

where

$$x = X/c,\ y = Y/c \tag{6.12}$$

and

$$W_\mathrm{I}(\mathrm{P}) = W_0 f(x, y) \tag{6.13}$$

where W_I is now the work of deformation per unit volume (input energy density) at P and W_0 the input energy density at points remote from the crack. Notice that we no longer refer to stored energy.

If the crack now propagates by a small amount Δc at constant applied stress σ_0, the stress field may be divided into two regions. In one region the stresses increase, and in the other they decrease. (Strictly speaking, we must consider each component of the stress tensor separately, since, e.g. in a given element σ_{11} may increase whilst σ_{22} decreases, i.e. the regions must be separately designated for each stress component. The result, however, is unaffected since the work contributions from different stress components can be treated separately and then superimposed. We shall continue the analysis therefore in terms of a single stress component referred to, for simplicity, as 'the stress'.)

For an element subject to increasing stress, the material continues on its monotonic stress–strain curve and the increase in W_I occasioned by the crack growth is given simply by differentiation of equation (6.13).

$$\frac{\mathrm{d}W_{\mathrm{I}}}{\mathrm{d}c}(\mathrm{P}) = W_0\left\{\frac{\partial f}{\partial x}\frac{\partial x}{\partial c}+\frac{\partial f}{\partial y}\frac{\partial y}{\partial c}\right\}$$

$$= -\frac{W_0}{c}\left\{x\frac{\partial f}{\partial x}+y\frac{\partial f}{\partial y}\right\}$$

$$= \frac{W_0}{c}f_1(x, y) \tag{6.14}$$

where f_1 is another function.

For elements in the relaxing region the situation is less simple because the retraction stress–strain curve is different from the extension curve. Figure 6.4 helps to clarify this, where increments of work ΔW are plotted against

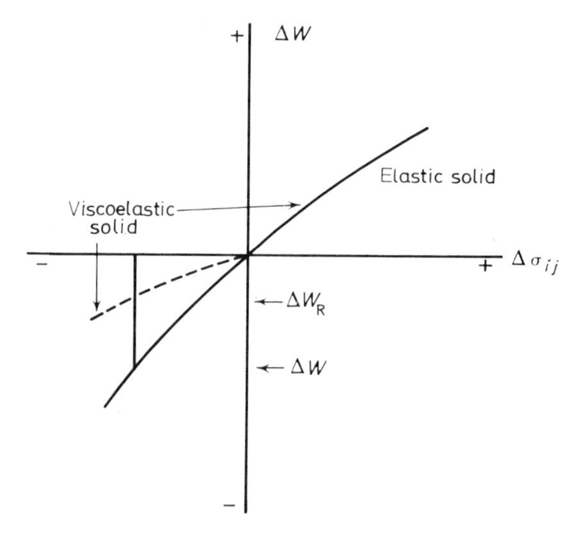

Figure 6.4 Stress and work increments for elastic and hysteresial cases

increments of stress $\Delta\sigma$ both for positive and negative changes. The origin refers to the state of stress before incremental changes. Clearly the incremental work of retraction is some fraction β of the corresponding value for an elastic solid, β of course being a function of the origin stress σ. For an element which relaxes, therefore, we can write,

$$\frac{\mathrm{d}W_{\mathrm{R}}}{\mathrm{d}c}(\mathrm{P}) = -\beta(\sigma)\frac{W_0}{c}f_1(x, y) \tag{6.15}$$

Summing over the stress field we therefore have,

$$\partial\mathscr{E}/\partial A = \partial\mathscr{E}/2h\partial c = \frac{W_0}{2hc}\left\{\sum_{\mathrm{PI}} f_1(x, y)\delta v_{\mathrm{I}} - \sum_{\mathrm{PR}} f_1(x, y)\beta(\sigma)\delta v_{\mathrm{R}}\right\} \tag{6.16}$$

where h is the sheet thickness and δv_{I} and δv_{R} are volume elements in loading (I) and retractive (R) regions respectively.

Now a volume element

$$\delta v = h\mathrm{d}X\mathrm{d}Y = c^2 h\mathrm{d}x\mathrm{d}y \tag{6.17}$$

so that

$$-\partial\mathscr{E}/\partial A = \frac{W_0 c}{2}\left\{-\sum_{\mathrm{PI}} f_1(x, y)\mathrm{d}x\mathrm{d}y + \sum_{\mathrm{PR}} f_1(x, y)\beta(\sigma)\mathrm{d}x\mathrm{d}y\right\}$$

But since, from equation (6.11), $\sigma(\mathrm{P})$ is a function of x, y and σ_0 only, we find that both summations are functions of x, y and σ_0 only. These summations are therefore independent of the crack length c being constant in x, y space in which the crack length c transforms to c/c, i.e. to unity. We therefore have

$$-\partial\mathscr{E}/\partial A = k_1 c W_0 \tag{6.19}$$

where k_1 is a function of σ_0. If the slopes of the extension stress–strain curve and the retraction curve at a given stress σ_0 are in a ratio independent of σ_0 (as is nearly the case for some materials), k would become constant. Experience indicates that it varies relatively slowly with σ_0 for elastomers[20] and for various polyethylenes[19], with extreme values of π for $\sigma_0 \to 0$ and about unity for large σ_0.

Equation (6.19) is almost identical to the elastic solution of equation (6.7) except that W_0 here represents the work of deformation per unit volume (the input energy) whether or not this energy is recoverable. Clearly, if $\beta = 1$, i.e. the material is elastic, equations (6.19) and (6.7) are identical, so that the derivation covers elastic materials as a limiting case of the visco-elastic solid.

If the summation in the above equations is carried out over the whole stress field, right up to the crack boundaries, it must be remembered that the 'available energy' $-\partial\mathscr{E}/\partial A$ given by equation (6.19) is no longer required to provide the component H of $\mathscr{T}$ in equation (6.10) since this has already been compensated for in the calculations. Thus equation (6.19) will, for propagation conditions, give

$$-\partial\mathscr{E}/\partial A = k_1 c W_0 = \mathscr{T} - H \tag{6.20}$$

6.3 APPLICATIONS OF FRACTURE MECHANICS

6.3.1 Brittle fracture

Brittle fracture is taken to mean fracture at low strain. Low, of course, is a relative term and since the room-temperature tensile yield strains of many rigid plastics are between 2 and 4% this will provide an upper limit to brittle fracture strains. In metals, of course, yield strains can be factors of ten smaller than this and the 'low-strain' range must be defined accordingly.

In the low-strain region the force–deflection curve for polymeric solids can be adequately represented by a straight line (though the departures from linearity may be important for some purposes) and inelastic behaviour is minimised. Even polymers, therefore, can be treated as approximate linear solids in this deformation range and linear fracture mechanics can be applied. The energy-balance criterion is, of course, also valid (it will be remembered that Griffith's original fracture criterion was for brittle fracture in glass).

Although fracture mechanics has its simplest application in brittle fracture, the subject is far from straightforward. For example, although the surface work $\mathscr{T}$ often decreases with decreasing temperature in the brittle range due to a decrease in the plasticity contribution (see equation (6.10)), the tensile strength of polymers usually *increases* with decreasing temperature.

We shall consider first of all the surface work for brittle crack propagation and secondly its relation to tensile strength via the concept of the 'intrinsic' flaw.

The only systematic work on brittle fracture has been carried out on polymeric glasses such as polymethylmethacrylate (PMMA), polystyrene (PS) and polyvinyl chloride (PVC). Measurements of $\mathscr{T}$, deduced using equation (6.9), have been made by several workers at 20 °C. The results are collected in Table 6.1.

Table 6.1

Material	$\mathscr{T}$/J m^{-2}	*Workers*
PS	2.5×10^3 slow cracks	Benbow and Roesler[14]
PS	3.0×10^2 fast cracks	Benbow and Roesler[14]
PS	8.8×10^2	Svennson[21]
PS	7.1×10^2	Berry[15]
PS	4.0×10^2	Broutman and McGarry[22]
PMMA	4.9×10^2	Benbow and Roesler[14]
PMMA	4.4×10^2	Svennson[21]
PMMA	1.4×10^2	Berry[15]
PMMA	1.3×10^2	Broutman and McGarry[22]
Polyester	1.2×10^4	Broutman and McGarry[22]

The variation of up to a factor of three or more for one material must be attributed to different test methods and differently prepared specimens (variations in molecular weight, thermal history and so on). It is clear, however, that in all cases the surface work for isotropic glassy plastics at room temperature is greatly in excess of the true surface energy (~ 0.5 J m^{-2}) and involves plastic deformation components as discussed earlier.

Table 6.2 Values of the product $E\mathscr{T}$ for PMMA at 35 °C for different temperatures of introduction of the crack
(From Cessna and Sternstein[23] by courtesy of Interscience)

Temperature/°C	$E\mathscr{T}$/(MN^{2}m^{-3})
−70	0.475
30	0.504
55	0.625

In all the cases quoted above, propagation was from an initial crack introduced (in various ways) at room temperature. Cessna and Sternstein[23] showed that $\mathscr{T}$ is reduced if the original crack is introduced at lower temperatures as seen in Table 6.2.

It is to be concluded that the plastic-work contribution to $\mathscr{T}$ is determined, at least in part, by the plastic zone created unavoidably at the tip of the starter crack during its formation.

On the other hand, if $\mathscr{T}$ is measured at different temperatures using starter cracks introduced at room temperature, $\mathscr{T}$ rises with reducing

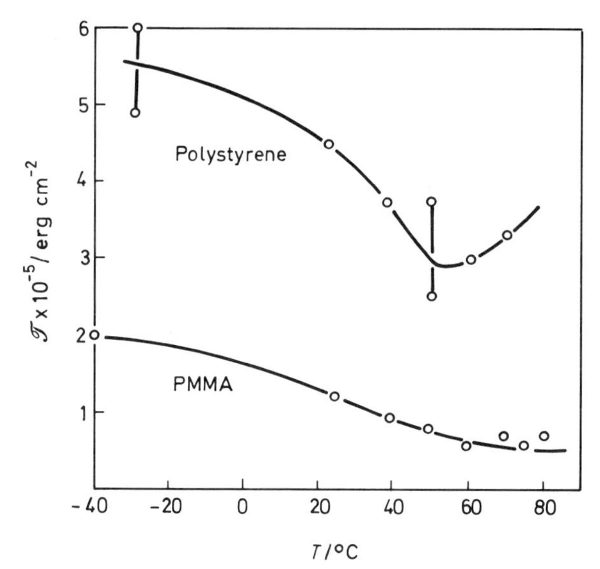

Figure 6.5 Dependence of surface work on temperature for polystyrene (upper curve) and PMMA (lower curve)
(From Broutman and McGarry[25], by permission of J. Wiley)

temperature for both PMMA and PS (Figure 6.5)[22], and this undoubtedly reflects the rising yield stress of the material at the tip whose volume is fixed during introduction of the starter crack at 20 °C.

The contribution of P to $\mathscr{T}$ (equation (6.10)) can be almost totally eliminated under some conditions. Beardmore and Johnson[24] found that surface steps were sometimes formed during the tensile testing of PMMA at 78 K; the steps mark the intersection of low-temperature crazes (see later) with the specimen surface. It is known that such steps are efficient stress raisers and that the step height plays effectively the same role as the crack length in an edge crack. Using the maximum measured step height of 150 nm and the observed fracture stress of 140 MN m^{-2} in equation (6.9), these workers deduced a value for $\mathscr{T}$ of between 0.4 and 0.7 J m^{-2}. This is two orders of magnitude smaller than the values listed above and agrees with the likely value of the true surface energy S.

Finally, any other factors which tend to minimise plastic flow at the starter-crack tip will give reduced values for $\mathscr{T}$. One such factor is orientation, and hot-drawn glasses fractured *along* the drawing direction can have surface work values reduced by two orders of magnitude compared with the isotropic values[25]. For fracture perpendicular to the draw direction, $\mathscr{T}$ is, of course, increased.

It might be thought that increasing the speed of fracture initiation would give smaller $\mathscr{T}$ because less plastic flow would occur in the shorter time and this appears to be the case for PS as cited in Table 6.1. However, Williams, Randon and Turner[27] found that the fracture toughness K_{IC} for PMMA increased by a factor of about two over six decades of increasing crack speed. The decrease of plastic strain at higher speeds appears to be offset by the increase in plastic flow stress also to be expected at higher deformation rates, so that it is not possible to predict whether the plastic work will rise or fall in magnitude as speed increases. The fracture toughness for mild steel actually passes through a minimum when plotted as a function of crack speed[27].

Having examined the fracture toughness or surface work and its dependence upon temperature and rate, what can be said about the tensile strength of brittle materials from a fracture mechanics view-point? Reference to equation (6.9) will show that σ_f and $\mathscr{T}$ are related through Young's modulus E and the crack length c. In a tensile test, of course, there is no intentional crack in the specimen but, as discussed at the outset, most real materials contain stress-raising flaws which can be regarded as 'intrinsic cracks' whose length we label c_0. A knowledge of σ_f and $\mathscr{T}$ for the same material under given conditions (of rate, temperature and orientation for example) thus leads to a value for the parameter c_0.

Berry[28] investigated the dependence of c_0 on temperature and other variables for PMMA. Figure 6.6 shows that c_0 was of the order of 10^{-4} m

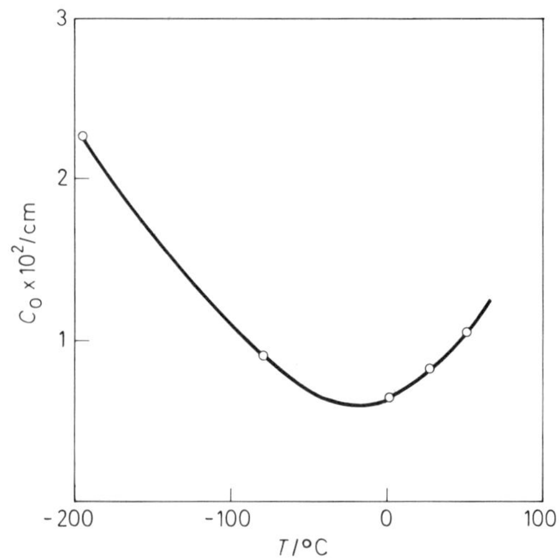

Figure 6.6 'Intrinsic' flaw size in PMMA as a function of temperature
(From Berry[28], by permission of J. Wiley)

varying by a factor of c. 3 over the temperature range 77–340 K. Variation by a further factor of 2 was obtained by varying the molecular weight of the polymer from 10^5 to 6×10^6; the large c_0 was for the higher molecular weight.

Much higher c_0 values than these can be obtained. A sheet of PMMA stretched biaxially by 65% had a value around 10^{-3} m and this is also the figure for isotropic polystyrene. It is fairly clear that such large 'intrinsic' flaws do not exist in the material before testing and that they are actually produced during loading of the specimen. Their length at the instant of catastrophic failure then serves as an effective 'intrinsic' flaw.

These test-induced flaws are associated, in isotropic glassy polymers at least, with craze growth. At higher temperatures catastrophic tensile failure appears to initiate within a growing craze and runs initially through the craze[29]. The parameter c_0 in such cases is not the craze length but a considerably shorter dimension associated with the structure of the craze. It is not accidental, however, that the most readily crazed polymer, polystyrene, has the largest 'intrinsic' flaw size observed for isotropic materials.

At liquid nitrogen temperatures, the work of Beardmore and Johnson[24] shows that catastrophic fracture can be initiated at surface steps caused by shear on the craze planes, rather than from within the craze.

In solids where brittle fracture occurs in the absence of prior crazing (such as glass), other mechanisms[30] exist for the growth of flaws smaller than the instantaneous critical size for catastrophic fracture (at any stress there will always be a value of c_0 which satisfied equation (6.9)), but this mechanism will be discussed in Section 6.4.

6.3.2 Tearing of rubber

When the strain at fracture is large, as in elastomers, a propagating fissure does not retain the small tip radius we associate with the word 'crack', and we speak instead of a 'tear'. Tearing is therefore a phenomenon associated with materials whose fracture strains, for whatever reason, are large. Enlargement of the tip radius in elastomers is strictly a finite-strain effect; the undeformed tip radius is as small as that of a brittle crack. Moreover, the stress intensity at the tip is governed by the undeformed tip radius[9], and is therefore still very high. In fact deformation at the tip modifies the stress distribution in such a way as to spread these high stresses over a much larger volume of material, increasing the energy losses and thus $\mathscr{T}$. Tearing is therefore often associated with very high $\mathscr{T}$ values.

A further difference between brittle crack propagation and tearing is that the latter is more controllable. That is, we can arrest a propagating tear more easily than a brittle crack by changing the constraints on the specimen. This effect is due to two causes. Firstly, we have greater control over the bulk stored-energy density W_0 (equations (6.6) and (6.7)) in high-compliance materials like rubber or paper, since quite large movements of the specimen boundaries correspond to given changes in W_0. It is therefore easier to make fine adjustments to W_0 and thus to the energy available for propagation than in a rigid solid. Secondly, tearing in some materials is predominantly a visco-elastic process as $\mathscr{T}$ is strongly dependent on the rate of tearing[18]. Any tendency for the tear to accelerate is therefore offset by a rise in $\mathscr{T}$ which can only be accommodated by increasing W_0. At constant W_0, therefore, catastrophic failure does not occur. These generalisations are not universally true and it is common knowledge that brittle cracks *can* be propagated slowly and that tearing *can* be catastrophic.

It will not be necessary here to review the tearing of rubbers exhaustively since this has been done previously[2,31]. We shall simply summarise the main features of the fracture mechanics analysis of the phenomenon, considering again the surface work and the parameter c_0.

The surface work in an unfilled non-crystallising rubber (called the 'tearing energy, T' by the authors mainly responsible for these studies) is

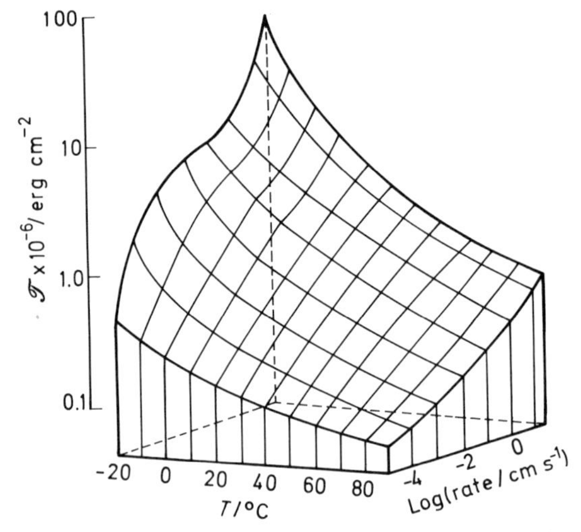

Figure 6.7 Surface work for SBR as a function of temperature and rate of tearing
(From Thomas[32a], by courtesy of the Institution of the Rubber Industry)

predominantly visco-elastic in origin. That is $\mathcal{T} \sim H$ in equation (6.10). This is evident[18] both from the strong rate and temperature dependence of $\mathcal{T}$ shown in Figure 6.7 and from the fact that this data transforms by the WLF* equation into a single master curve[32], Figure 6.8. The value of $\mathcal{T}$ ranges from 10^5 J m^{-2}, compared with $\sim 10^3$ J m^{-2} for glassy plastics, to 10^2 J m^{-2} and lower for low-rate, high-temperature data. The variation is thus very much larger than in materials for which $\mathcal{T} \approx P$ (equation (6.10)).

The introduction of 'reinforcing agents', whether natural ones like strain-induced crystallisation or artificial ones like particulate fillers or fibres, superimposes upon the basic visco-elastic energy losses new losses of a different kind. This is illustrated in Figure 6.9 for a fine thermal carbon black[33] added to the elastomer whose basic behaviour was shown in Figure 6.7. The new losses form a high plateau on the energy-loss surface which persists, however, only over limited ranges of rate and temperature. In these regions, of course, the material is particularly resistant to tear and tensile failure. The actual mechanism of energy dissipation for fillers is a combination of enhanced relaxation in the molecular network[34], due to very high strain concentrations around the particles, and actual filler–rubber breakdown[35]. In strain-crystallising elastomers, notably natural rubber, the heat of crystallisation is lost to the strain field during cyclic deformation and represents a large non-visco-elastic energy loss which completely masks the

*Williams, Landel and Ferry.

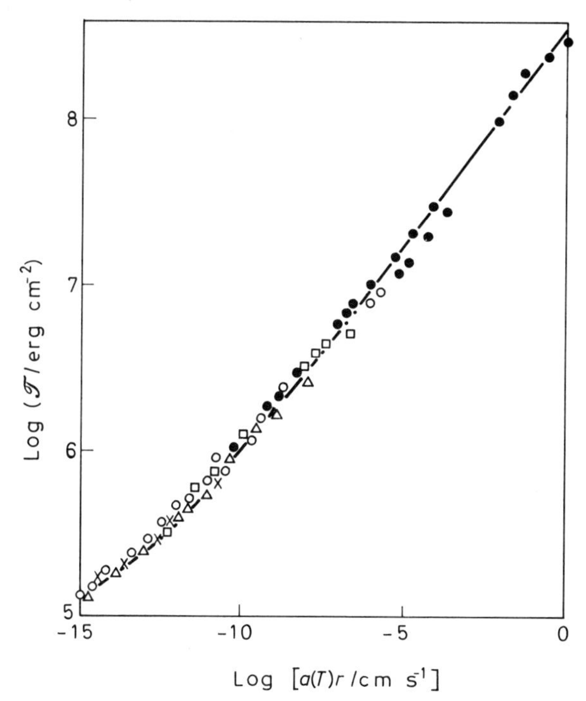

Figure 6.8 WLF master curve for surface work in gum SBR
(From Mullins[32], by courtesy of the Institution of the Rubber Industry)

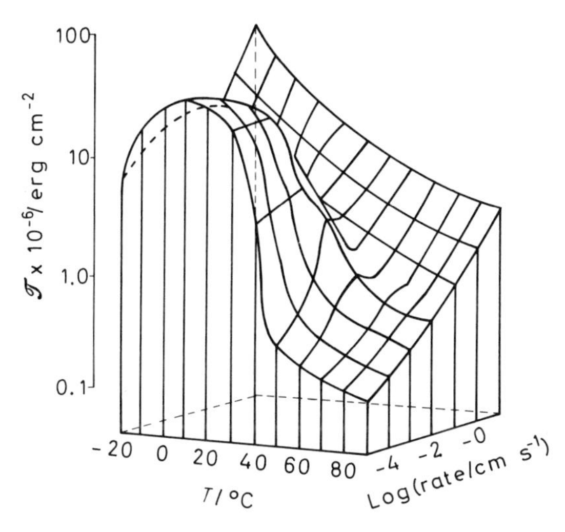

Figure 6.9 Surface work for SBR containing fine thermal carbon black
(From Greensmith[33a], by permission of J. Wiley)

visco-elastic contribution except at very high strain rates and low temperatures.

Knowing the rate and temperature dependence of $\mathscr{T}$ it is possible to predict the time dependence of fracture stress on the assumption that failure occurs from intrinsic flaws c_0. Greensmith[36] showed that the time to fracture as a function of the energy density W_0 could be predicted accurately both for constant strain and constant rate-of-strain tests assuming a c_0 of *c*. 10^{-5} m.

So far we have viewed the surface work for tearing as an empirically determined quantity. The earlier analysis (Section 6.2.4) enables us to consider its actual character in terms of the hysteresial properties of the material. We have from equations (6.18)–(6.20) that

$$\mathscr{T} - H = cW_0 \left\{ \sum_{\text{PR}} g(x, y)\beta(\sigma_p) - \sum_{\text{PI}} g(x, y) \right\} \tag{6.21}$$

where g is a function. Now in a cross-linked elastomer we can set $\mathscr{T} = H + S$ only so that

$$S = cW_0 \left\{ \sum_{\text{PR}} g(x, y)\beta(\sigma_p) - \sum_{\text{PI}} g(x, y) \right\} \tag{6.22}$$

This gives the actual surface energy, i.e. the work required to break atomic bonds, a quantity which will be virtually independent of rate and temperature. Now $\beta(\sigma_p)$ is the ratio of recoverable work to input work increments around the stress σ. To simplify, let us assume that for our cross-linked material the relaxing stress field can be divided into two parts, a low-strain part in which hysteresis is negligible (i.e. $\beta \rightarrow 1$) and a high-strain part in which hysteresis is approximately independent of σ_p and has a maximum value, $(1-\beta)$ of h_f, where the suffix indicates that this value is attained at strains approaching the fracture strain. Then equation (6.22) is simplified to

$$S = cW_0(q_1 - q_2 h_f) \tag{6.23}$$

where q_1, q_2 are constants.

We may now apply this equation to tensile fracture by setting $W_0 = W_f$, the fracture energy density, and $c = c_0$, the 'intrinsic' flaw size. We obtain

$$W_f = \frac{S}{c_0}(q_1 - q_2 h_f) \tag{6.24}$$

This equation suggests a unique relationship between the energy density to fracture and the hysteresis ratio h_f at fracture, and such a relationship has indeed been established for visco-elastic elastomers by Grosch, Harwood and Payne[37]. Their empirical relation was

$$W_f = K^3 h_f^2 \tag{6.25}$$

where K is a constant and h_f is the total energy loss: input ratio for a complete stress cycle. It therefore differs somewhat from the incremental definition used here, and this may account for the difference between equations (6.24) and (6.25). Over certain ranges of the product $q_2 h_f$, however, equations (6.24) and (6.25) behave in a very similar fashion, a plot of log W_f against log $(1 - q_2 h_f/q_1)^{-1}$ having a slope of almost exactly 2 for $(q_2 h_f/q_1)$ between

0.5 and 0.8. It is possible therefore that equation (6.25) is an empirical approximation to equation (6.24). In any case the increase of W_f with h_f is clearly predicted by the analysis.

To summarise, therefore, we can say that in a cross-linked elastomer the surface work consists of a surface-energy term required for actual bond fracture plus a visco-elastic dissipation term which is operative in a limited high-stress region around the tip and which effectively diminished the amount of the input energy which is available for bond fracture. Although S is small compared with H under most conditions, equation (6.24) shows that the value of S has a direct influence upon the fracture strength, which would reduce to zero if $S \rightarrow 0$. This is self-evident physically, of course, since if $S \rightarrow 0$ all stresses in the tip region would also tend to zero and hysteresis would be eliminated, giving $H \rightarrow 0$ also. We shall return to this argument again when we consider adhesion in Section 6.3.5.

6.3.3 Fatigue fracture of polymers

It is only quite recently that the fatigue properties of polymers have become important, i.e. since the rapid growth of engineering applications. An exception to this is the fatigue behaviour of rubber which has long been an object of interest to those concerned with tyre failure. Fatigue in polymers must be considered under two headings, only one of which can be considered here. Firstly, at high frequencies and loads, cyclic failure can be caused by 'heat build-up' or, more strictly, by the rise in specimen temperature due to visco-elastic or plastic energy dissipation and aggravated by the low thermal conductivity of the materials. This matter has been considered elsewhere by the author[38] and in recent papers by Constable, Williams and Burns[39] and by Zilvar[40]. Since it does not involve the application of fracture mechanics it will not be considered further here.

At low frequencies, or at low loads or under conditions where heat can readily escape from the specimen, fatigue failure occurs by the initiation and propagation of fatigue cracks and is susceptible to fracture mechanics treatment.

When a body containing a crack is loaded appropriately, the crack tends to propagate. For it to do so, we have already seen that a certain amount of energy must be provided per area of the propagating crack. We have already seen that this critical energy is a function of temperature and speed of propagation but a new factor must now be introduced. If $\mathscr{T}_c$ is the surface work required for catastrophic or steady propagation of the crack (i.e. the quantity we have previously termed the surface work $\mathscr{T}$), it is usually found that for

$$-\partial\mathscr{E}/\partial A < \mathscr{T}_c \tag{6.26}$$

some small-scale propagation has occurred but that the crack arrests before macroscopic propagation is observed. It is further found that reduction of the stress to zero followed by its renewed application (keeping $(-\partial\mathscr{E}/\partial A)$ below the critical value) results in a further incremental growth and so on.

The amount of growth per cycle is found to obey an equation of the form

$$\frac{\mathrm{d}c}{\mathrm{d}N} = f(-\partial\mathscr{E}/\partial A) \tag{6.27}$$

for $(-\partial\mathscr{E}/\partial A) < \mathscr{T}_c$, and this applies to such varied materials as metals, rigid plastics, soft plastics and elastomers. Equation (6.27) is, of course, in harmony with the philosophy of fracture mechanics that all crack propagation events are governed by the stress field at the crack tip. Equation (6.27) can, for linear elastic materials, be equally well written in terms of the stress intensity factor K. Empirically, equation (6.27) is found to assume the simple form

$$\mathrm{d}c/\mathrm{d}N = B(kcW_0)^n \tag{6.28}$$

The 'constants', B and n, vary from material to material, and may also vary in a given material with temperature, cycling rate and, even, with the stress level. Some values for n are given in Table 6.3.

Table 6.3

Material	*n, high* $(-\partial\mathscr{E}/\partial A)$	*n, low* $(-\partial\mathscr{E}/\partial A)$
Metals	2	2
PMMA [41]	2.5	2.5
Polethylene [19]	1.7	4.0
Plasticised PVC	3.0	3.0
SB rubber [42]	4.0	4.0
Natural rubber [42]	2.0	1.0

Apart from natural rubber, where strain-induced crystallisation occurs at the crack tip during each cycle, there is a general tendency for n to increase as the material becomes more compliant, and n is probably related therefore to the stress and strain distribution at the crack tip. This has been established [43] experimentally for natural rubber where the value of n was shown to be a direct consequence of the geometry of stress decay with distance from the tip.

In the light of equation (6.28), a plot of log $\mathrm{d}c/\mathrm{d}N$ *v.* log cW_0 provides a characteristic plot defining fatigue crack-growth behaviour and plots of this kind have been obtained for many elastomers, PMMA, polyethylene, plasticised PVC, polycarbonate, epoxy resin, ABS resin and nylon [44].

Some of these results are suspect because the stress intensity factor method has been used for materials like polyethylene and plasticised PVC which are decidedly non-linear and have finite strains, and this may account for the absence of a truly linear plot in such cases. This is illustrated for polyethylene in Figure 6.10, which shows some results of Andrews and Walker [19] analysed by the energy-balance method and Figure 6.11 which shows data of Hertzberg *et al.* [44] on similar material analysed in terms of the stress intensity factor.

A knowledge of B and n enables the fatigue lifetime to be predicted as follows. Assuming growth from an edge crack of original (intrinsic) length c_0, we have, from equation (6.28),

$$\int_{c_0}^{\infty} \frac{\mathrm{d}c}{c^n} = B(kW_0)^n \int_0^{N_f} \mathrm{d}N$$

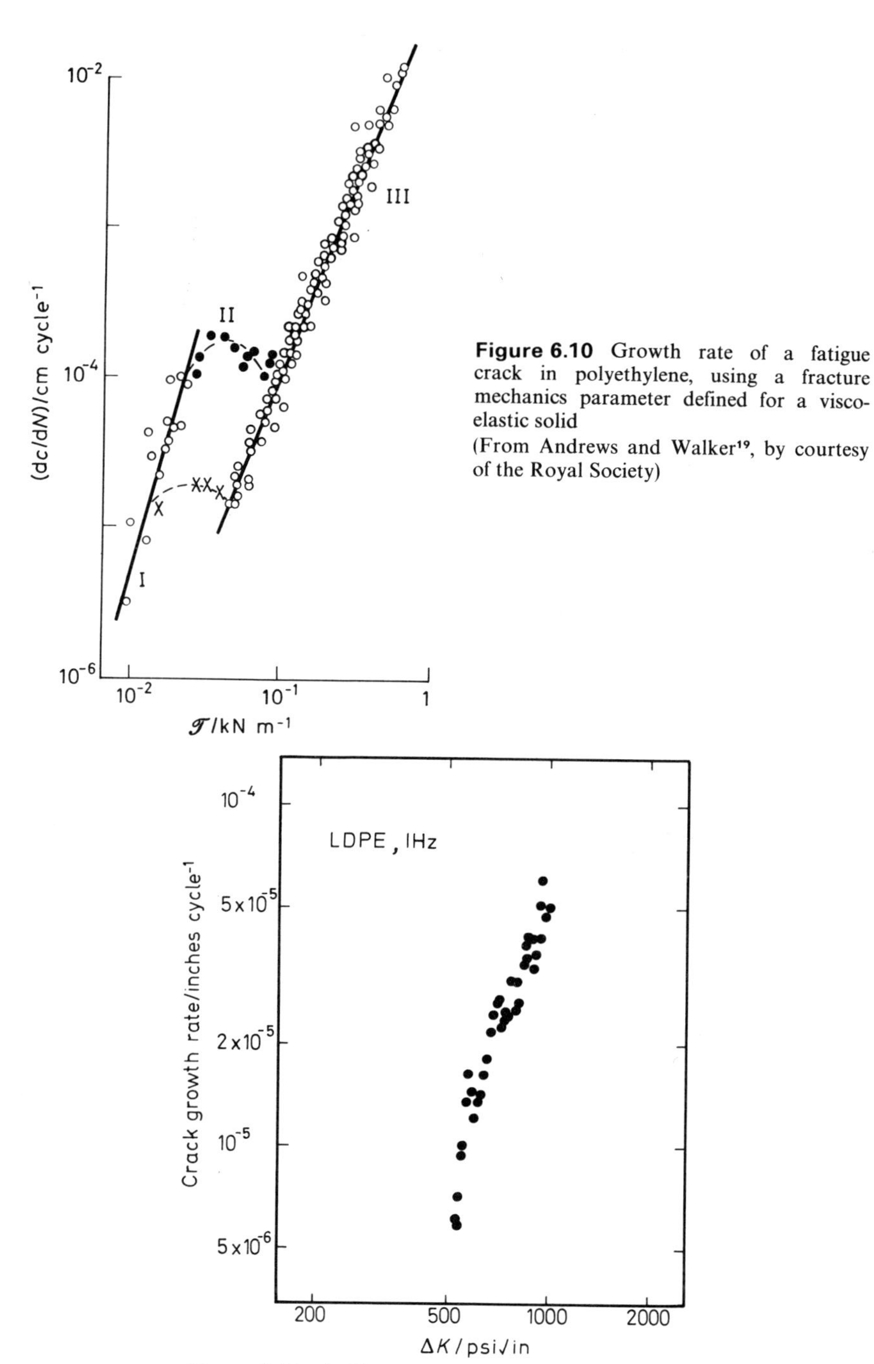

Figure 6.10 Growth rate of a fatigue crack in polyethylene, using a fracture mechanics parameter defined for a visco-elastic solid
(From Andrews and Walker[19], by courtesy of the Royal Society)

Figure 6.11 As Figure 6.10 but using a linear fracture mechanics parameter as abcissa
(From Hertzberg *et al.*[44], by courtesy of *J. Mat. Sci.* and Chapman and Hall)

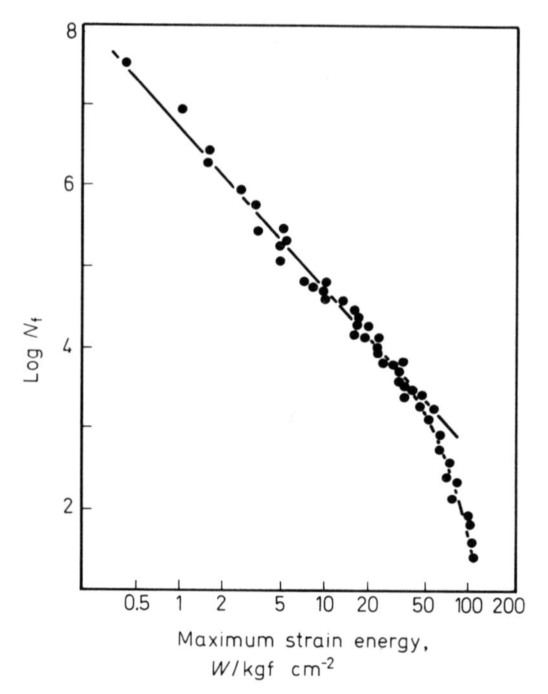

Figure 6.12 Fatigue life of NR gum rubber as a function of strain energy density. Solid straight line is a theoretical curve
(From Mullins in *Chemistry and Physics of Rubberlike Substances*, 294, by courtesy of The Natural Rubber Producers' Research Association)

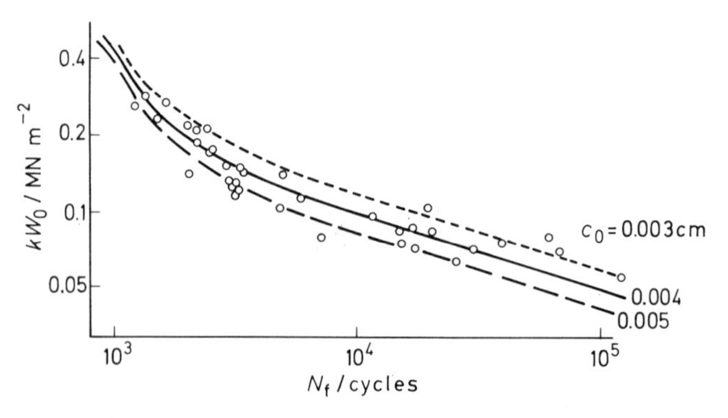

Figure 6.13 Fatigue life of low-density polyethylene as a function of strain energy density. Lines are theoretical curves for different choices of the 'intrinsic' flaw size
(From Andrews and Walker[19], by courtesy of the Royal Society)

where W_0 is the (constant) maximum value for the cycle and N_f is the lifetime in cycles. The upper limit for crack length is conveniently set to infinity for failure, but the specimen width can equally well be used. Providing $n \neq 1$ the integration gives, finally

$$N_f = \{Bk^n W_0^n c_0^{n-1}\}^{-1} \tag{6.29}$$

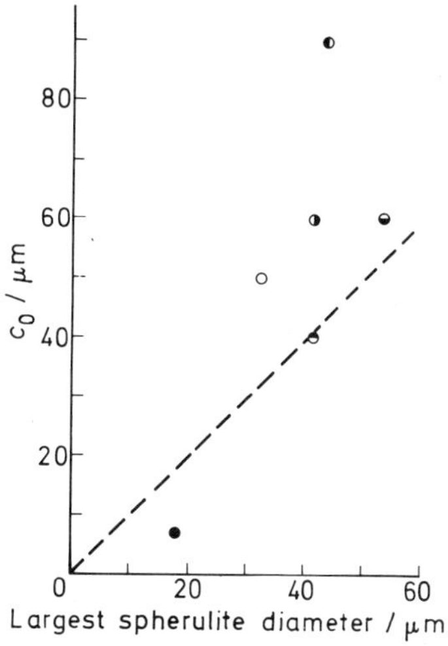

Figure 6.14 'Intrinsic' flaw size for fatigue in various polyethylenes as a function of largest spherulite diameter. The line is the 1:1 relationship
(From Andrews and Walker[19], by courtesy of the Royal Society)

In cases where n and B vary with $(-\partial\mathscr{E}/\partial A)$ as, e.g. for polyethylene in Figure 6.10, the integration may be carried out in stages for which B and n can be assigned constant values.

A comparison between theoretical and actual fatigue lives has not been made by many of the workers who studied growth behaviour, presumably because parallel 'true fatigue' tests were not done. Such comparisons exist

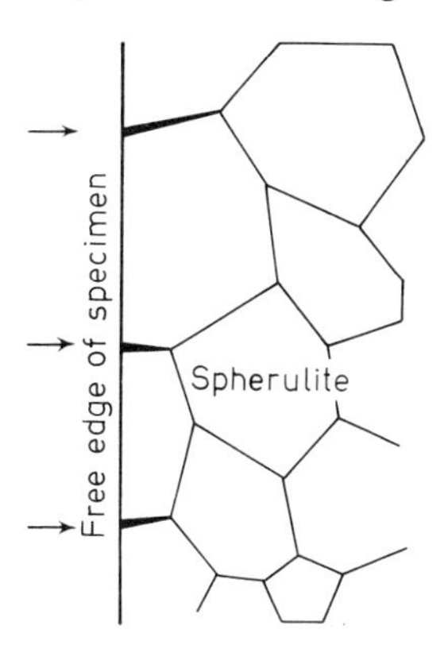

Figure 6.15 Formation of inter-spherulitic boundary cracks (schematic)
(From Andrews and Walker[19], by courtesy of the Royal Society)

however for elastomers[42] (Figure 6.12) and for polethylene[19] (Figure 6.13) and are found to show excellent agreement. Only a single fitting constant is required to adjust the theoretical curves to the experimental points, namely the 'intrinsic flaw' size c_0. Figure 6.13 shows that c_0 is very precisely defined by the fitting procedure. The values of c_0 in the cases quoted were 2.5×10^{-5} m for natural rubber and $0.7–9 \times 10^{-5}$ m for various polyethylenes. In the latter materials there exists some correlation between c_0 and maximum spherulite diameters (Figure 6.14) and it has been suggested[19] that the 'intrinsic' flaws in this case are spherulite boundary cracks opened during cutting of the specimen or else during the first few cycles of the fatigue test (Figure 6.15). In the case of elastomers, c_0 may be either a manifestation of surface scratching or else of ozone crack growth in the surface. Unlike the c_0 values sometimes required to explain brittle fracture in rigid polymers, those for fatigue initiation are sufficiently small to be real characteristics of the material structure.

6.3.4 Environmental stress cracking and crazing

The first application of fracture mechanics to environmentally-assisted fracture was in the ozone cracking of cross-linked unsaturated elastomers. Gent and Braden[16] observed that there existed a critical value $\mathscr{T}_c$ of the parameter $(-\partial\mathscr{E}/\partial A)$, independent of stress, crack length and elastic modulus, below which single cracks did not propagate and above which they propagated at a speed depending only on the ozone concentration and the temperature (especially near T_g). The critical value of $(-\partial\mathscr{E}/\partial A)$ was low, at about 0.05 J m^{-2}, approaching a true surface energy value. This result can now be understood in terms of the very low stresses acting in the tip region. Molecular scission at the tip occurs mainly by chemical attack and stress is only needed to fracture bonds in a liquid-like degradation product[45]. This stress is so low that surrounding material is not strained sufficiently for significant hysteresis to develop, giving the low $\mathscr{T}_c$ values observed.

The work on ozone cracking has been reviewed in detail elsewhere[46] and will not be elaborated here. Its importance as a foundation for other environmental failure studies merits emphasis, however, since it suggested to Andrews and Bevan[47] that a similar approach might be possible to crack and craze growth in thermoplastics. In preliminary studies, they showed that a critical value of $(-\partial\mathscr{E}/\partial A)$ was necessary for continued propagation of environmental stress cracks in PMMA immersed in methylated spirit. It had been recognised previously that a critical surface stress or strain was necessary for crazes or cracks to develop in glassy polymers, but this was the first demonstration that the controlling parameter is, in fact, the surface work (or stress intensity factor). The value $\mathscr{T}_c$ for crack arrest was found to vary with temperature, falling rapidly as temperature increased up to a characteristic temperature T_c and staying constant thereafter.

Subsequently[48] the same authors investigated the critical value $\mathscr{T}_c$ for craze or crack propagation in a variety of solvents.

According to equation (6.7) the critical condition is

$$-\partial\mathscr{E}/\partial A = \mathscr{T}_c = kcW_{0,c}$$

and

$$W_{0,c} \propto c^{-1} \tag{6.30}$$

It was found that a plot of $W_{0,c}$ *v.* c^{-1} (the latter corrected for finite width of specimen) gave reasonable linearity for some solvents such as water and methylated spirit, but gave increasing scatter for an ascending series of aliphatic alcohols from methanol to butanol. Alcohol–water mixtures

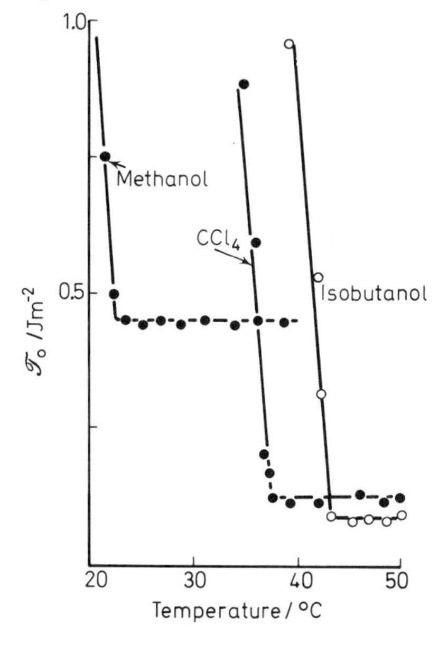

Figure 6.16 Minimum surface work for craze formation, for PMMA in various solvents, as a function of temperature

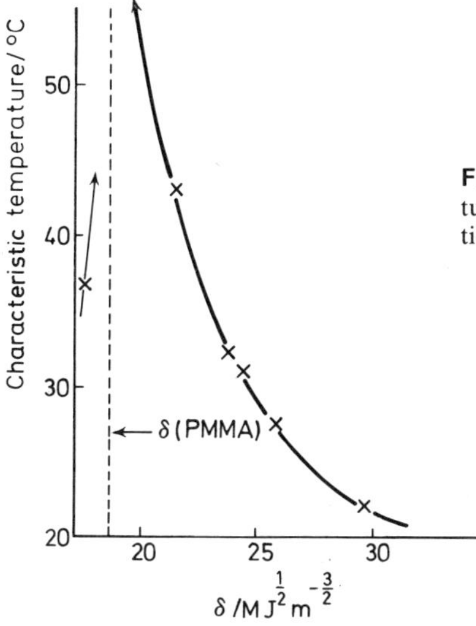

Figure 6.17 The characteristic temperature for PMMA–alcohol systems as a function of solvent solubility parameter.

gave worsening scatter as the alcohol increased. The scatter was attributed to the hysteresis of the 'craze matter' (the orientated and voided polymer which fills the craze) which supported different levels of stress across the craze according to its stress history. A lower boundary to the experimental points gave a minimum for $\mathscr{T}_c$ for each solvent and temperature and the behaviour of this value was investigated. Figure 6.16 shows the lower boundary $\mathscr{T}_c$ values for different solvents and PMMA as functions of temperature. The same behaviour previously noted for methylated spirit is found in all solvents but the characteristic temperature T_c varies with solubility parameter in the manner shown in Figure 6.17. At temperatures above T_c, $\mathscr{T}_c$ is constant at a value denoted $\mathscr{T}_c^*$ which, like T_c, varies with δ (solvent). However, whereas T_c rises to a maximum at δ (solvent) = δ (polymer), $\mathscr{T}_c^*$ goes through a minimum at this point (Figure 6.18).

Experiments on PMMA swollen in alcohols[49] show that the glass transition temperature is depressed significantly and a plot of T_c *v.* T_g (swollen) is a

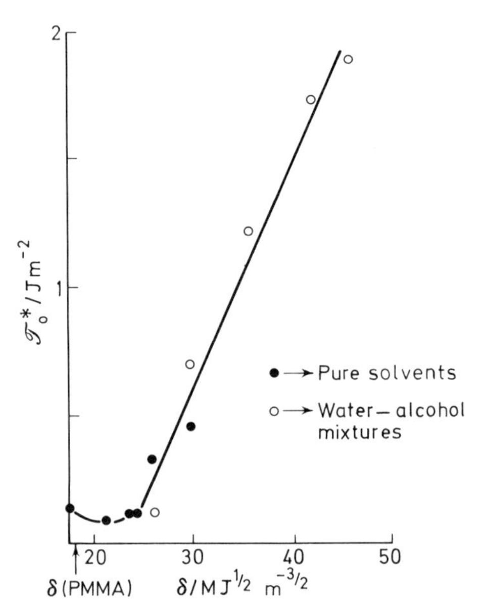

Figure 6.18 Surface work for crazing (above the characteristic temperature) as a function of solvent solubility parameter

1:1 relation for a constant polymer fraction of *c.* 0.7 in the swollen mixture. It was concluded therefore that T_c must be the glass transition temperature of a swollen region at the crack or craze tip; the polymer content was approximately constant at 0.7.

Andrews and Bevan[48], and Gent[50] proposed cavitation criteria for craze growth. It is supposed that cavities are created at the tip of a propagating craze by the action of the hydrostatic component of the local stress field. If the swollen material is below T_g it will behave approximately as an elastic–plastic solid and for cavity growth will require a hydrostatic stress

$$p = \frac{2Y}{3}\psi + \frac{2\gamma}{r} \tag{6.31}$$

where Y is the yield stress of the swollen region, ψ a nearly constant term, γ the cavity surface energy and r the cavity radius. This can be integrated to give the work of craze formation giving

$$\mathscr{T}_c = 2.42\,(h\gamma/\rho)f^{\frac{2}{3}} + 0.33\,Y\psi hf \qquad (6.32)$$

where h is the craze thickness, ρ the mean distance between voids and f the void fraction. The second term is proportional to the yield stress Y of the swollen tip material which decreases linearly with temperature and becomes zero at T_c. This linear decrease of Y is thought to account for the dependence of $\mathscr{T}_c$ upon temperature below T_c. The first term in equation (6.32) is relatively insensitive to temperature and depends chiefly on the void surface energy γ. Since this term equals $\mathscr{T}_c$ above T_c, the behaviour of $\mathscr{T}_c^*$ with respect to solubility parameter (Figure 6.18) is plausibly due to the variation of γ. The latter is, of course, the interfacial energy between swollen polymer and solvent and is not determined easily by experiment. Its decrease towards zero as δ (solvent) $\rightarrow \delta$ (polymer) is to be expected, however, and supports the above interpretation of $\mathscr{T}_c^*$.

Once the critical condition for crazing is achieved, the kinetics of craze propagation must be considered. Marshall, Culver and Williams[51] studied the velocity of craze growth for the system PMMA and methanol at 20 °C. They found that the constant craze velocity achieved under steady load was a function not of the applied stress but of the stress intensity factor K_0 for the 'starter' edge-crack from which the craze was grown. They explained their results in terms of the diffusion of solvent through the craze as follows.

Assuming that the craze propagates at constant thickness, the craze-opening displacement (COD) is always equal to the crack-opening displacement of the starter crack, given by linear fracture mechanics as

$$\text{COD} = K_0^2/EY \qquad (6.33)$$

where Y is the yield stress of the tip material and E the bulk Young's modulus. At the growing tip the void volume created per unit area of craze must equal the craze-opening displacement since the state of plane strain prohibits sideways contraction of the deforming zone. If we further suppose that the void volume consists effectively of tubular voids of constant bore running continuously through the craze and conducting the solvent to the tip, we obtain the void volume/area of craze

$$v = K_0^2/EY = \pi nlD^2/4 \qquad (6.34)$$

where n is the number of tubes, l their length and D their diameter. If n and l are considered constant, we have

$$D^2 = \alpha K_0^2$$

From the Navier–Stokes channel-flow theory, the velocity of flow through a tube of diameter D is

$$V = \frac{D^2}{12\mu}\frac{\mathrm{d}P}{\mathrm{d}X}$$

where μ is the viscosity and $\mathrm{d}P/\mathrm{d}X$ the pressure gradient. If solvent flows in through the side of the test piece

$$\mathrm{d}P/\mathrm{d}X = 2P/L$$

where P is atmospheric pressure and L the sheet thickness, so that

$$V = \frac{\alpha' K_0^2 P}{L\mu} \tag{6.35}$$

where α' is a constant. It is now assumed that craze velocity is governed by the rate of delivery of solvent to the tip, i.e. that it is given by V, so that $V(\text{craze}) = V(\text{solvent})$.

Experimental results agree well with the predicted dependence of craze velocity upon K_0^2 and also inversely upon L (Figure 6.19).

Andrews and Levy[52] extended kinetic measurements to a range of solvents and temperatures and also investigated the effects of varying the applied

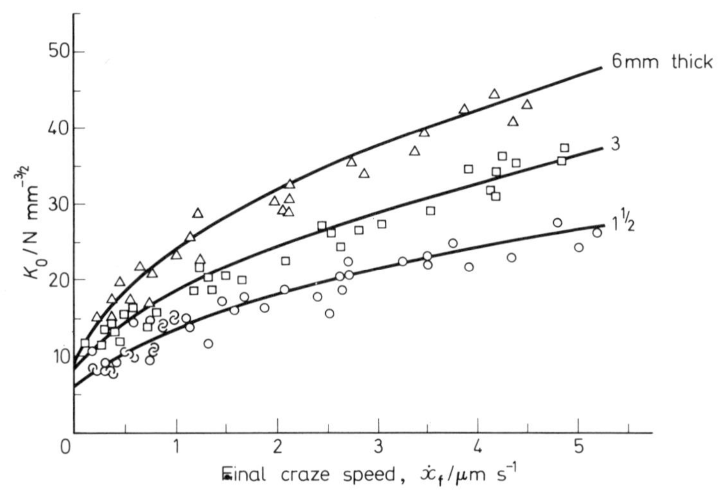

Figure 6.19 Craze velocity as a function of initial stress intensity factor for PMMA in methanol and three sheet thicknesses. Lines give theory (From Marshall *et al.*[51], by courtesy of The Royal Society)

stress and of removing the starter crack (to eliminate its stress field) after production of craze. The fuller picture which emerges from this work is outlined by the data of Figures 6.20 and 6.21. Craze velocities *v.* applied stresses for different c_0/b ratios (where b is the specimen width) are shown in Figure 6.20. The velocities are independent of the craze length, as found by Marshall *et al.*[51], and variation of the applied stress during craze propagation causes the velocity to vary reversibly, along the velocity–stress curve. This is entirely in harmony with the work of Marshall *et al.*, except that K_0 is no longer calculated simply from the *first* applied stress but from the instantaneous stress and c_0. Difficulty arises, however, when the starter crack is removed. On the theory outlined above, K_0 is reduced to zero by this operation and velocity should also become zero. The craze kinetics are found, however, to be identical whether the starter crack is present or absent (providing the net section stress is employed when the starter crack is present)! Furthermore, the influence of c_0 upon craze velocity at a given stress, whether or not the starter-crack has been removed, is minimal for temperatures below T_c but

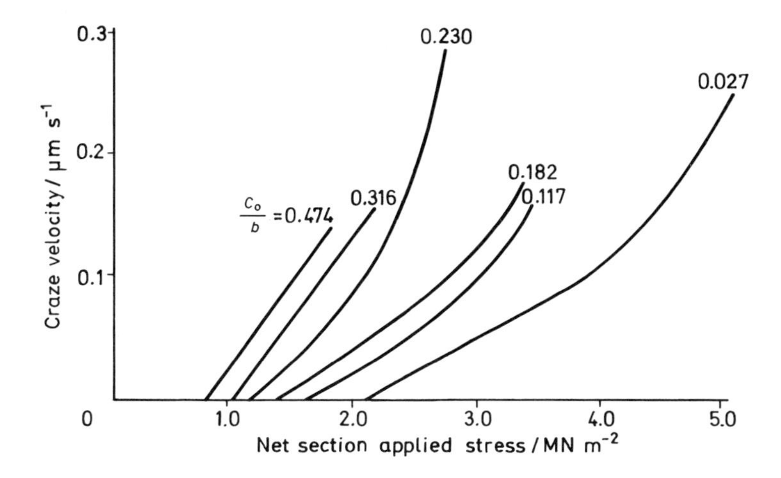

Figure 6.20 Craze velocity as a function of stress for different starter crack lengths c_0 (b is the specimen width). PMMA in ethyl alcohol at 20 °C

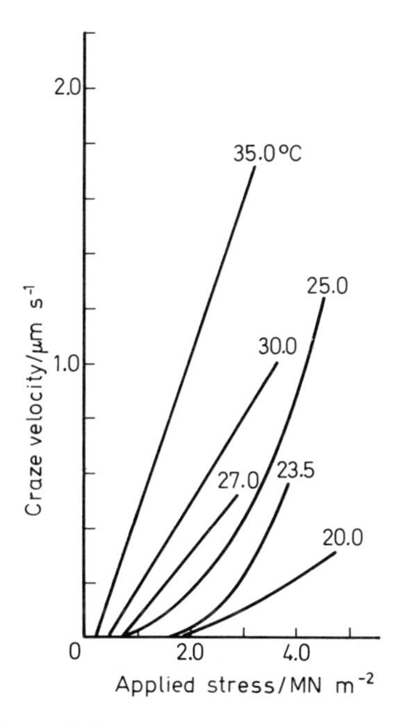

Figure 6.21 Craze velocity as a function of stress for different temperatures. PMMA in ethyl alcohol.

very strong for temperatures higher than T_c. It has not yet proved possible to find a convincing explanation for these effects, but they are likely to arise from the inelastic response of the craze matter to deformation.

Secondly, a strong temperature dependence of velocity is found (Figure 6.21). This effect is too great to be explained in terms of the temperature dependence of the parameters in equation (6.35) unless Y is the yield stress of the swollen material at the craze tip rather than the bulk material.

These studies are being continued and a comprehensive theory can be expected to emerge in due course.

6.3.5 Failure of adhesive joints

Although adhesion has been the subject of much study from both the mechanical and chemical viewpoints, the application of fracture mechanics ideas is relatively recent[53–55]. In a series of investigations on the bonding of elastomeric adhesives to various substrates, Gent and others have shown how powerful the fracture mechanics approach can be both in providing meaningful strength data and in unravelling the underlying physics of adhesive failure.

Gent and Petrich[56] and Gent[57] determined the resistance to separation of an adhesive bond between a visco-elastic adhesive (an SBR elastomer) and a rigid substrate (Mylar film), both for peeling and for tensile separation normal to the bonded plane. They found that both measures of adhesion varied with rate of deformation in the adhesive and with temperature in accordance with the WLF equation indicating that the observed strength reflects a visco-elastic property of the adhesive rather than a thermodynamic property of the interface.

Results from the two tests, however, showed marked differences at equivalent rates of deformation and, moreover, reacted differently to changes in the thickness of the adhesive layer. The resistance to peel increased with layer thickness, but the tensile strength was reduced. These differences were completely explicable in terms of fracture mechanics; the bond failure was viewed as the propagation of a crack along the interface requiring a given work of crack formation, denoted θ, per area of newly created surface. The symbol θ is wholly equivalent to $\mathscr{T}$, but serves to distinguish between adhesive and cohesive failure. Just like $\mathscr{T}$, the failure energy θ is a function of rate and temperature.

The differences between peel and tensile behaviour are explicable because, for test specimens of different geometry, the (constant) quantity θ is given by different functions of the external constraints and physical dimensions. For example, a thin layer of adhesive stressed normal to its plane approximates to the pure-shear test piece discussed in Section 6.2.3. and

$$\theta = h_0 W_p \tag{6.36}$$

where h_0 is the layer thickness. Clearly the stored energy density W_p will decrease as h_0 increases to keep θ constant. In contrast, for a peel test

$$\theta \simeq P \tag{6.37}$$

where P is the peel force per unit width of adhesive layer, and is to a first approximation independent of layer thickness (bending forces in the layer actually cause P to rise with increasing h_0 as observed).

Gent and Kinloch[58] extended these studies to the model systems illustrated in Figures 6.22 and 6.23. The first is analogous to the centre-crack arrange-

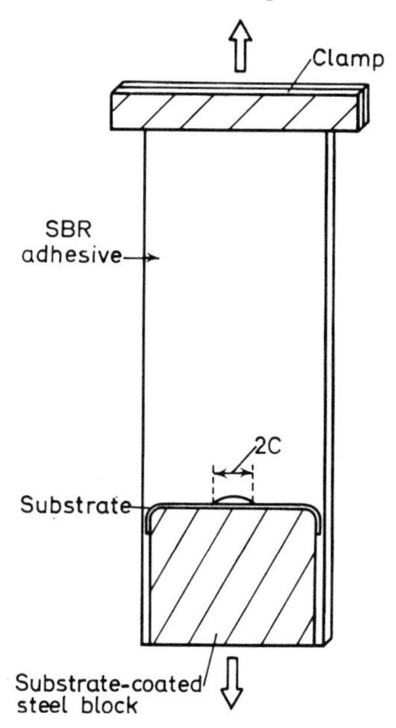

Figure 6.22 Simple extension test piece for adhesion studies
(From Gent and Kinloch[58], by permission of J. Wiley)

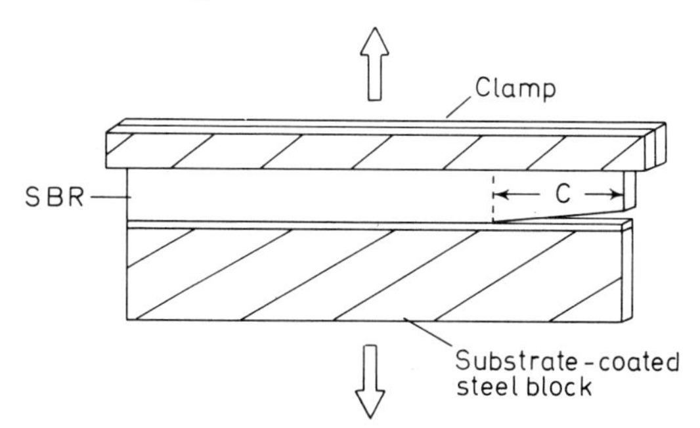

Figure 6.23 Pure shear test piece for adhesion studies
(From Gent and Kinloch[58], by permission of J. Wiley)

ment often used for cohesive-failure measurements, and is called the simple-extension test piece. The adhesive is separated from the substrate over a small central crack length and is effectively of semi-infinite dimensions, avoiding layer thickness effects. In contrast, the pure shear test piece (Figure 6.23) is designed deliberately to emphasise the role of layer thickness (equation (6.36)).

For the simple extension test piece, θ is given by the equivalent to equation (6.7)

$$\theta = kcW_0 \tag{6.38}$$

The overall results from these studies are shown in Figure 6.24 which is a reduced plot of log θ against log $\dot{c}$ where $\dot{c}$ is the rate of crack propagation. The reduction to a single master curve of data gathered over a wide range of rates and temperatures is by application of the WLF equation.

Results for peel, pure shear and simple extension test pieces agree closely, confirming the validity of the fracture mechanics approach.

Values of θ at the lowest rates and highest temperatures used were of the order of 300 erg cm^{-2} (0.3 J m^{-2}). Under these conditions visco-elastic losses

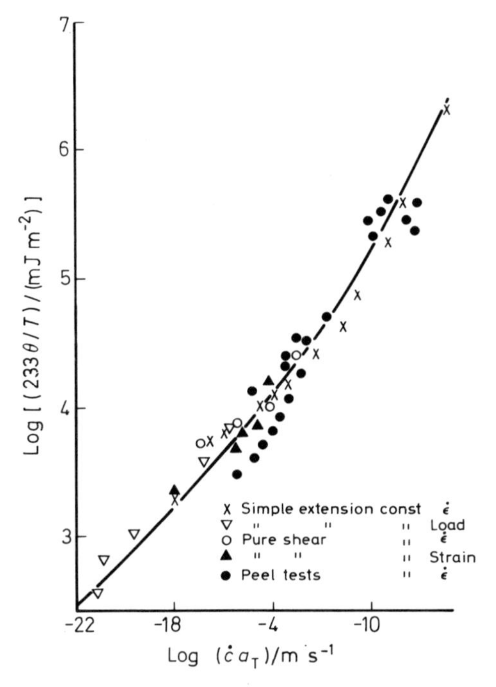

Figure 6.24 Adhesive failure energy as a function of reduced fracture rate showing that points from different test pieces define a single curve (From Gent and Kinloch[58], by permission of J. Wiley)

are minimised and θ should tend to values characteristic of the simple interfacial energy, say 10^{-2}–10^{-1} J m^{-2}, and the trend seems to bear this out.

Andrews and Kinloch[59] continued these studies by varying the nature of the substrate, whilst Gent and Schultz[60] used the original materials but varied the interfacial energy term by debonding specimens immersed in different organic solvents. In both cases a most interesting effect was observed, and is illustrated in Figures 6.25 and 6.26 for the two types of experiment.

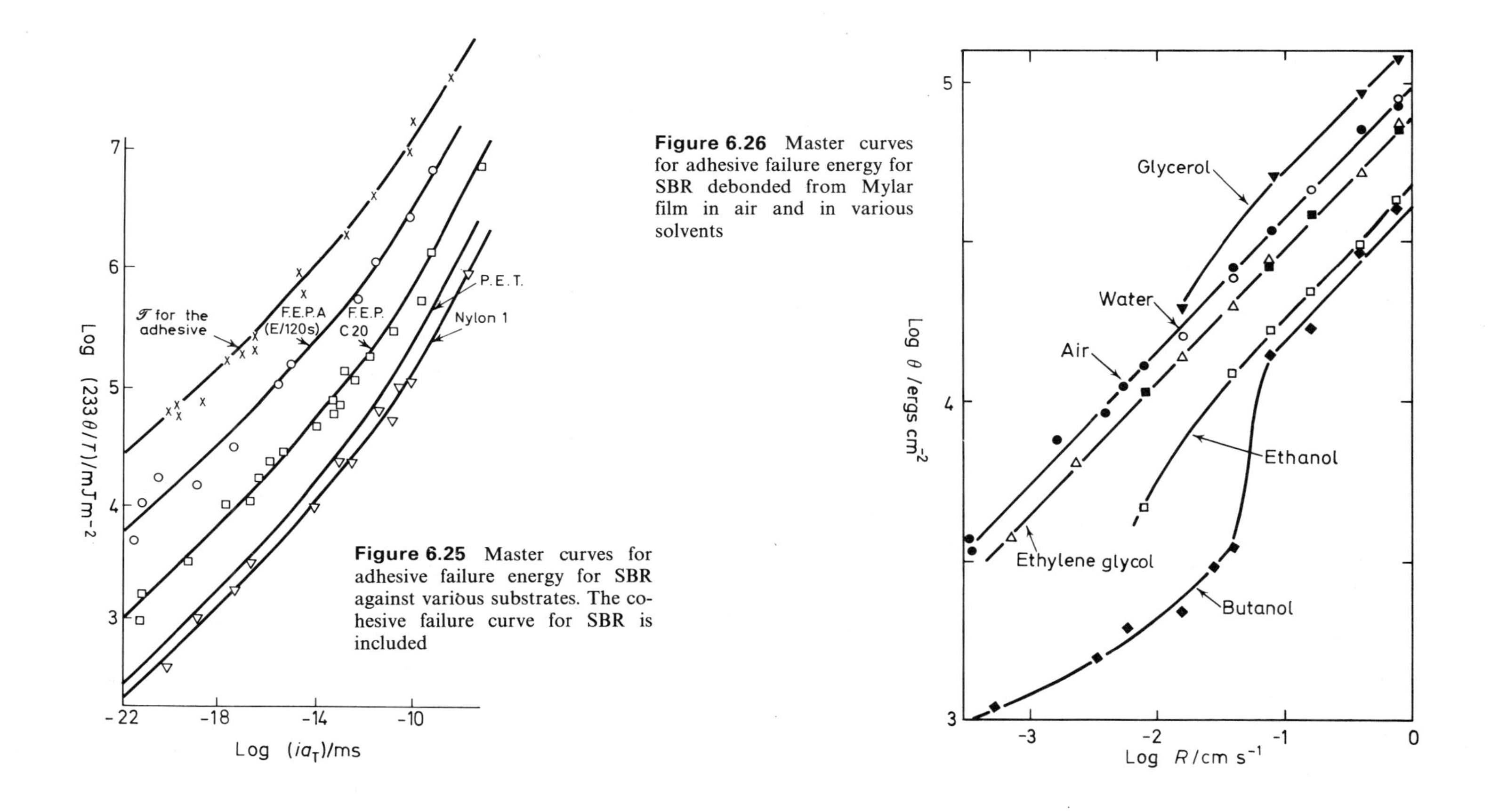

Figure 6.25 Master curves for adhesive failure energy for SBR against various substrates. The cohesive failure curve for SBR is included

Figure 6.26 Master curves for adhesive failure energy for SBR debonded from Mylar film in air and in various solvents

The failure energy θ was found to give parallel curves when plotted logarithmically against reduced rate, each curve correspondingly to a different interfacial energy. In Andrews and Kinloch's data the curves for different substrates were also parallel to the cohesive failure curve for the elastomer.

These results can be summarised by the expression

$$\theta = \theta_R h' \tag{6.39}$$

where θ_R is the reversible or thermodynamic work of adhesion and h' is a number reflecting the hysteresis or irreversible work of formation of surface. At first sight equation (6.39) is a very strange expression, since, following equation (6.10) we would expect the total failure energy to be given by a *sum* of interfacial and dissipative terms,

$$\theta = \theta_R + H \tag{6.40}$$

Equations (6.39) and (6.40) can be reconciled if H is proportional to θ_R, so that

$$H = h\theta_R$$

and

$$\theta = \theta_R(1+h) \equiv \theta_R h' \tag{6.41}$$

This equation has the required property that $\theta \to \theta_R$ as hysteresis $\to 0$.

Why should the hysteresis losses be proportional to θ_R, however? The answer can best be made in terms of the stress distribution at the crack tip.

Let σ_b be the stress necessary to fracture the atomic bonds across the adhesive–substrate interface. This stress will be related to the reversible work of adhesion,

$$\theta_R \approx \xi\sigma_b^2/2E \tag{6.42}$$

where E is the slope of the stress–strain relationship for the bonds in question (equivalent to the Young's modulus of a hypothetical solid, bonded exclusively by such bonds), and ξ is the interatomic distance. (Equation (6.42) is derived by dividing the energy density at bond fracture by the number of atomic layers in unit volume normal to the applied stress, to give an energy per layer.)

Since the stress distribution at an edge or centre crack is given by equation (6.1), it follows that the stress at any point P will be related to the stress σ_b at the extreme tip by

$$\begin{aligned}\sigma_{ij}(\mathrm{P}) &= \sigma_b f(\tfrac{c}{r},\phi)_p / f(\tfrac{c}{r},\phi)_{\mathrm{tip}} \\ &= \alpha\sigma_b f(\tfrac{c}{r},\phi)_p\end{aligned}$$

where α is a function of c only. Thus

$$\sigma_{ij}(\mathrm{P}) = \alpha f(\tfrac{c}{r},\phi)_p (2E\theta_R/\xi)^{\frac{1}{2}} \tag{6.43}$$

Now the energy lost by hysteresis in a stress cycle is directly related to the maximum stress developed in that cycle. Thus the hysteresis loss at the point P is directly related to θ_R by equation (6.43). The total hysteresis loss will be obtained by summing over the stress field, but this summation does not affect the term in θ_R. We obtain, therefore, that

$$H = g(\theta_R) \tag{6.44}$$

where g is a direct-relation function. Exact proportionality, as required by equation (6.41) and observed experimentally, requires that

$$H \propto \sum_{P} \sigma_{ij}^{2}(P)/2E \tag{6.45}$$

This is exactly the case for a Hookean material in which a constant fraction of the work of deformation is lost hysteresially, and more detailed argument shows that equation (6.45) is justifiable also for non-Hookean materials such as used in the studies reported.

6.4 FRACTURE AS A MOLECULAR PROCESS

6.4.1 Introduction

All fracture must ultimately occur by the rupture of interatomic bonds and this will be true whether the chain molecules in a polymer are degraded by rupture of their covalently bonded 'backbone' or separate by sliding when the cumulative secondary bonding forces along their length are overcome by the applied stresses. Not surprisingly, main-chain fracture is predominant in cross-linked or network polymers, in highly crystalline materials where the crystalline zone act as anchor points for the stress-bearing molecules, at low temperatures where molecular mobility is minimised and at high molecular weights where entanglements act as effective physical 'cross-links'.

The occurrence of main-chain fracture under conditions of severe mechanical working is well established and, indeed, provides the basis of the subject we call mechano-chemistry. Thus high molecular weight elastomers like natural rubber are degraded by mastication, the molecular weight distribution becoming sharper because the longer molecules are preferentially degraded. Other methods that have been used to produce main-chain fracture mechanically include grinding, ball-milling, calendaring and shearing in various kinds of rheometer. Studies have been made on solids, melts and solutions and upon most of the common thermoplastics and elastomers.

An extensive review of the mechano-chemistry of polymers, with 164 references, has recently been published by Casale, Porter and Johnson[61] and it is therefore unnecessary for us to dwell upon the subject here. It is important to note, however, that the molecular mechanism operative in all the degradation processes cited is the formation of macro-radicals by direct stress rupture of primary bonds. These radicals may then stabilise or decay in a variety of ways including recombination, reaction with neighbouring molecules, reaction with radical acceptors (especially atmospheric oxygen), disproportionation or the formation of stable secondary radicals. It is the decay or stabilisation processes which command the greatest interest since they determine whether the mechanically-worked polymer will degrade, gel, emit volatile products etc., and the importance of this for polymer processing is self-evident.

Our concern in this review is not with mechano-chemistry in general, but with molecular rupture occurring under conditions which also lead to the macroscopic fracture of solids. Thus we are interested in the same molecular mechanisms of primary-bond rupture, radical formation and their consequences, but only in relation to solid polymers under stresses which tend to produce fracture of the bulk material. An example will help to clarify the distinction. Free radicals can be produced and identified at ambient temperature by grinding glassy thermoplastics like polymethylmethacrylate, but the same materials subject to tensile load up to the point of fracture yield no measurable radical concentration by e.s.r. spectroscopy. If main-chain fracture is to be viewed as a cause or a concommitant of macroscopic fracture, and if the relation between the two is to be established, results from grinding or mastication are clearly of only secondary significance compared with data from, say, tensile tests.

In general it is found that main-chain fracture in solids under tensile load occurs to a significant extent only when the molecules are 'anchored' within the structure. Thus thermoplastics do not evidence molecular fracture except possibly at very low temperatures, but orientated, crystalline fibres, containing short lengths of molecule between crystallites, display a strong effect. Similarly, cross-linked elastomers, well below T_g, give strong e.s.r. spectra when deformed, the radical concentration increasing with the cross-link density[62].

These phenomena will be considered further after a brief discussion of the theoretical considerations applying to molecular fracture in polymer solids under tensile load. An excellent review of some of the subject matter covered in this section has already been published by Kausch von Schmeling and should be consulted[3].

6.4.2 Kinetic theory and fracture mechanics

The direct observation of molecular fracture by e.s.r.[63] and other spectrochemical techniques is a relatively recent development, but the idea of fracture as an interatomic bond-rupture process is by no means new. It is implicit in Griffith's theory, since the surface energy of the crack is nothing more than the dissociation energy of the atomic bonds broken in its creation. However, the time dependence of fracture phenomena which could not be explained by Griffith's theory, demanded a more careful analysis. Models of fracture on an atomic scale were set up and attempts made to predict the macroscopic fracture behaviour for comparison with experimental data.

The most successful of these models was a simple 'thermofluctuation' or kinetic model of the fracture process advanced by Taylor[64] in 1947 and developed by Stuart and Anderson[65] and by Zhurkov[66] and co-workers. The fracture of interatomic bonds is viewed as an activated process, i.e. one involving the surmounting of a potential energy barrier. This is shown in Figure 6.27(a) where states A and B represent unbroken and broken bonds, respectively. The energy barrier can be surmounted because random thermal fluctuations ensure that the thermal or kinetic energy of an atom varies in time, with a finite possibility of exceeding the potential barrier and producing a transition A → B or B → A. The frequency with which such transitions occur is given by chemical rate theory as

$$\nu = \nu_0 \exp(U/kT)$$

where ν_0 is the frequency of atomic thermal vibration and has a value of 10^{12}–$10^{13}\,s^{-1}$, U is the height of the energy barrier, i.e. the 'activation energy', k is Boltzmann's constant and T the absolute temperature.

In the unstrained state, clearly, the potential energy of the unbroken bond configuration is lower than for a broken bond, so that the transition B → A

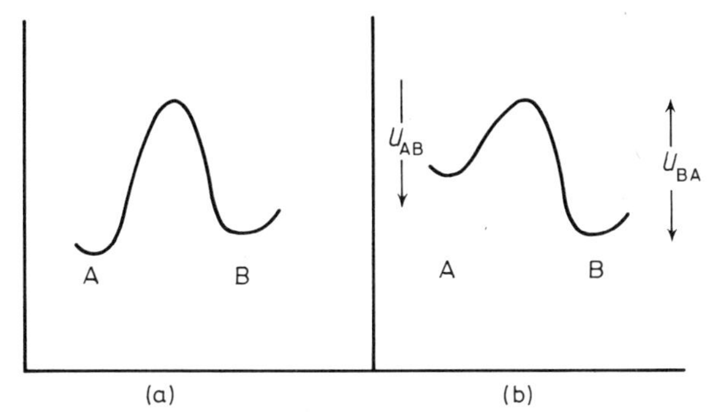

Figure 6.27 Potential energy diagram (schematic) for bond fracture (a) unstressed (b) under tensile stress

occurs more readily than A → B. As a result, no net bond fracture occurs. If, however, a stress is applied to the specimen, the strain energy stored in each molecule favours the process A → B since this energy is relieved by bond fracture. The stress-distorted energy barrier is shown in Figure 6.27(b). Now the net frequency of bond fracture is

$$\nu^* = \nu_0 \exp(-U_{AB}/kT) - \exp(-U_{BA}/kT) \tag{6.46}$$

and is positive. As soon as U_{BA} exceeds U_{AB} by a factor of 2 or more, the reverse transition B → A becomes negligible and we have

$$\nu^* \approx \nu_0 \exp(-U_{AB}/kT) \tag{6.47}$$

Since U_{AB} is a stress-modified energy barrier, it can be written

$$U_{AB} = U_0 - f(\sigma) \tag{6.48}$$

where U_0 is the value of U_{AB} in the unstrained state, σ is the macroscopic stress and f is a function. The simplest functionality is usually adopted, namely,

$$U_{AB} = U_0 - \beta\sigma$$

where β is a constant with the dimensions of volume and is thus referred to as the 'activation volume'.

If we now stipulate, as a fracture criterion, that a certain number, N, of bonds must be broken for the remaining intact bonds to be unable to carry the load, we obtain the time to fracture t_f as

$$t_f = \frac{N}{\nu^*} = \frac{N}{\nu_0} \exp(U_0 - \beta\sigma)/kT \tag{6.49}$$

or,

$$\ln t_f = C + (U_0 - \beta\sigma)/kT \tag{6.50}$$

If this is true, the logarithm of the lifetime under stress should plot linearly against the stress and equation (6.50) is fully substantiated by experimental data (see Figure 6.28).

When the constant U_0 is evaluated from such experiments using equation (6.50), its magnitude is found to agree almost precisely with the values

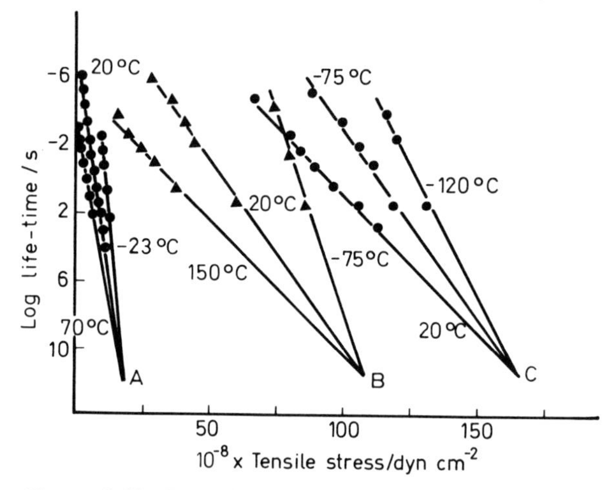

Figure 6.28 Dependence of lifetime on tensile stress for various materials and temperatures. (A) Unoriented PMMA; (B) viscose yarn; (C) nylon 6 yarn
(From Zhurkov and Tomashevsky[66], by courtesy of the Institute of Physics)

obtained for chemical bond rupture by thermal degradation. Zhurkov and Tomashevsky[66] give the comparisons shown in Table 6.4. This impressive agreement can be taken as strong support that bond breakage in fracture does occur by a thermofluctuation mechanism just as in thermal degradation.

Before proceeding to the direct experimental study of molecular fracture it is helpful to pause to ask whether the theory of thermofluctuations is

Table 6.4

Polymer	U_0/kcal mol^{-1}	U(*thermal*)/kcal mol^{-1}
PVC	35	32
Polystyrene	54	55
PMMA	54	52–53
Polypropylene	56	55–58
Teflon	75	76–80
Nylon 6	45	43

compatible with continuum theory. It is frequently said that the two theories are not reconcilable and this statement must be examined. For simplicity, we shall take the case of a crack and consider the kinetic theory applied to the highly stressed tip zone. The question then becomes 'Do the two theories give irreconcilable accounts of its propagation?'.

Reference to Figure 6.27 reminds us that an unbroken bond is in a lower energy state than a broken bond. This is often forgotten; if states A and B had been equi-energy states the application of *any* tensile or shear stress would favour bond breakage A → B. However, since the energy barrier to A → B is initially higher than to B → A, some non-zero threshold stress, σ_0, is required to redress the balance. Under this stress, given by

$$\beta\sigma_0 = [U_{AB} - U_{BA}]_{\sigma = 0}, \tag{6.51}$$

the processes A → B and B → A are equi-probable and only *additional* stress will provide a net increase in the number of broken bonds.

The right-hand side of equation (6.51) is simply the energy difference between associated and dissociated atomic bonds and this is precisely the nature of a surface energy. The surface energy S of Griffith's theory, when expressed in molar terms, is thus of the same nature as the term $\beta\sigma$ in equation (6.50).

This argument is slightly complicated by the fact that the high-energy state B may decay by a variety of processes other than the simple reverse reaction B → A. Three possibilities exist, (i) the free radicals may decay into alternative, more stable free radicals which remain trapped, (ii) the free radicals may react with a neighbouring molecule to form a cross-link or (iii) the free radicals may react with oxygen or other foreign species and become inactive.

In (i) and (iii) we have a situation in which B → A is inhibited. Only if it is completely prevented does $\sigma_0 \to 0$, and this would only occur if the *final* state C (i.e. after (i) or (iii) has occurred) possessed a potential energy equal to or lower than that of A. Such a situation is unlikely except under conditions of thermal degradation. Case (ii) preserves the integrity of the molecular network and is thus akin to a recombination reaction.

Thus we see that the threshold stress (or equivalent surface energy) may be modified or decreased, e.g. by reaction with oxygen, but that it is unlikely to become zero at normal temperatures.

6.4.2.1 The energy requirement $\mathscr{T}$

Having seen how the surface energy requirement S has its analogue in kinetic theory we must now turn to the total-energy requirement $\mathscr{T}$, which has a minimum value of S but is usually much greater in magnitude. The additional energy, it will be remembered, is dissipated because the stresses required to fracture bonds at the crack tip have to be transmitted to those bonds through the surrounding material where flow and fracture processes of various kinds occur under the influence of this stress. The magnitude of H (i.e. $\mathscr{T} - S$) is thus governed by the volume of material under high stress which, in turn, is controlled by the crack geometry. An infinitely sharp tip would have zero volume under infinite stress. This cannot occur, of course, since sharpness cannot increase beyond atomic dimensions, but it illustrates the fact that crack-tip geometry (or more precisely, the geometry of the stress field there) controls the quantity $\mathscr{T}$.

In kinetic theory one is concerned solely with the stress σ on an atomic bond at the extreme tip, but of course the magnitude of this stress in relation to the applied stress is a function of the crack geometry. The sharper the tip, the larger will be the 'stress-concentration factor' and the faster will bonds fracture under a given external stress. It is, therefore, the efficiency with which stress is concentrated that constitutes the kinetic theory equivalent to the $\mathscr{T}$ of continuum theory. When we observe in a visco-elastic rubber that the energy requirement at a given temperature increases with crack velocity we are simply finding that the higher tip stress necessary (by kinetic theory) to maintain the faster growth can only be obtained by subjecting a larger volume of 'loss-prone' material to higher stresses.

It should, finally, be pointed out that both the kinetic theory and continuum theory appeal to heterogeneity, the latter in terms of the 'intrinsic' flaw and the former in recognising that crack growth (and possibly the coalescence of many microcracks) is dominant in the fracture process (Zhurkov *et al.*[67]).

6.4.3 Stress on chemical bonds

Before we consider the use of e.s.r. for the direct observation of molecular fracture, it is relevant to consider the way in which stress is carried, on a molecular scale, within the solid. The complex micro-morphology of most polymers gives rise to an inhomogeneous stress field in which some molecular segments (and thus some interatomic bonds) are more highly stressed than others. This situation was investigated experimentally by Zhurkov *et al.*[68] using infrared spectroscopy.

They observed that absorption bands associated with C—C bond vibration in various polymers (nylon 6, polypropylene, polycaproamide) were shifted

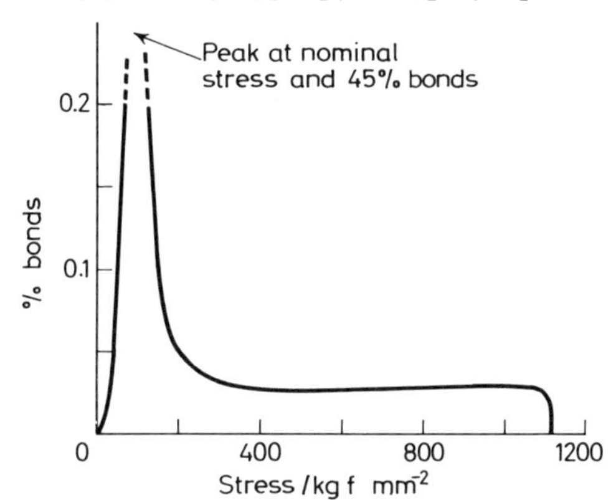

Figure 6.29 Stress distribution among atomic bonds in stressed polypropylene
(From Zhurkov *et al.*[69] in *Fracture* by courtesy of Chapman and Hall)

to lower frequencies and that the extent of this shift was a function of the magnitude of stress. Distortion of the absorption peak also occurred and this could be interpreted in terms of a *distribution* of stresses over the bonds of a loaded specimen. The authors were able to produce the kind of distribution curve shown in Figure 6.29, showing that a small but significant number of bonds were stressed to a much higher degree than the nominal, achieving maximum stresses nearly ten times the applied stress. The actual maximum stress observed was found to be a decreasing function of temperature and, if it is assumed that this maximum stress is the actual stress under which the molecule breaks, this evidence supports the idea of a thermally assisted (i.e. kinetic) mechanism for bond fracture.

6.4.4 Molecular fracture; study by electron spin resonance

It has already been stated that the free radicals (or, more strictly, the unpaired electrons contained therein) produced by molecular fracture can be observed directly using e.s.r. spectroscopy. The spectrum is obtained by placing the specimen in a resonant cavity, irradiated by microwaves of a suitable fixed frequency ν, and subjecting the system to a strong magnetic field H_0 which splits the energy of unpaired electrons into their Zeeman levels. Transitions between the levels are excited by the microwave radiation and energy is therefore absorbed; the resonant condition is $h\nu = g\mu_B H_0$, where h is Planck's constant, μ_B is Bohr's magneton and g the spectroscopic splitting factor. The value of H_0 is swept through the resonance position and the first derivative of the energy absorption is recorded as the e.s.r. spectrum. It consists of a series of peaks because electrons in different atomic environments have slightly different resonance conditions (g factors). The resonances in solids are broad because the magnetic field is modified locally by the magnetic fields of neighbouring spins.

In principle, the e.s.r. spectrum provides both a measure of the *number* of unpaired electrons in unit volume (from its intensity) and an identification of the macro-radical involved. In practice, the former is readily deduced, but identification presents considerable difficulties. Nevertheless, some spectra have been assigned with a fair degree of certainty as indicated in Figure 6.30.

Molecular fracture was first observed in crushed specimens of polymers such as polyethylene, polycaprolactam and polymethylmethacrylate (Zhurkov and Tomashevsky[66]) and the spectra of Figure 6.30 were obtained in this way. It is more difficult to detect free radicals produced by tensile stressing and so far none have been observed in glassy thermoplastics such as polymethylmethacrylate and polystyrene under these conditions. In semicrystalline polymers (e.g. nylon 6, polyethylene terephthalate), however, radicals are observed once the stress exceeds *c.* 60% of the breaking stress and radical concentrations up to 10^{16} cm^{-3} have been measured. Free radicals are also observed in abundance during the tensile testing of cross-linked elastomers such as natural rubber and polychloroprene, at temperatures below *c.* -120 °C[62, 69].

The radicals are stable at these temperatures, but decay immediately on warming the material through the glass transition temperature, i.e. as the molecules regain mobility. In the elastomer networks the loss of the e.p.r. signal is accompanied by increased cross-linking and the evolution of hydrogen as the free radicals attack neighbouring molecules (see Figure 6.31). Some of the radicals decay in a different manner, combining with dissolved oxygen to produce a peroxy radical (Figure 6.30).

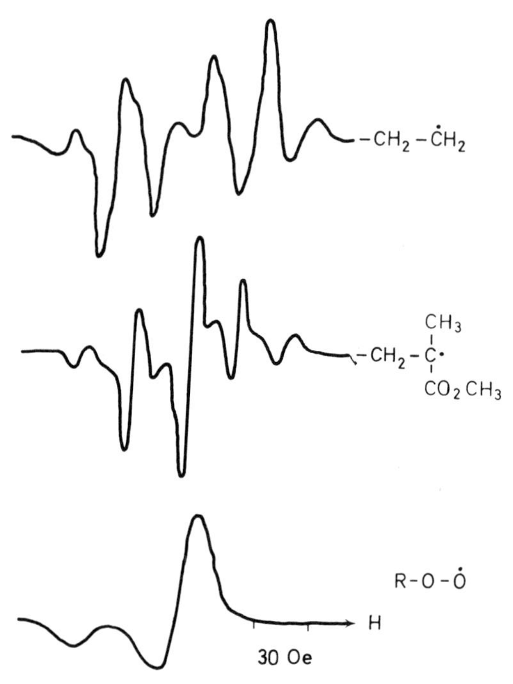

Figure 6.30 Some e.s.r. spectra obtained from mechanically crushed polymers
(From Zhurkov and Tomashevsky[66], by courtesy of the Institute of Physics)

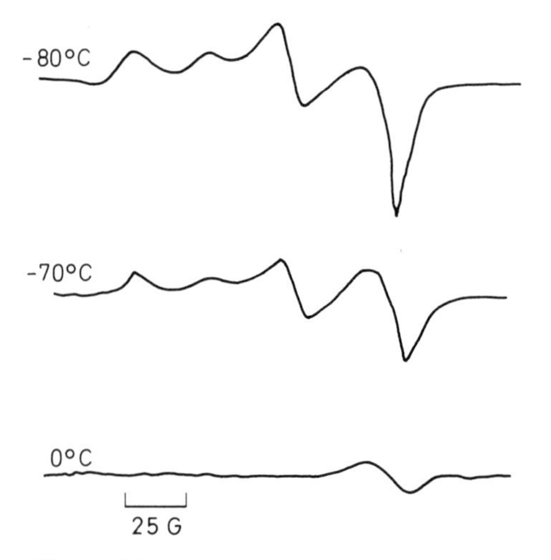

Figure 6.31 Decay of the e.s.r. spectrum for strained *cis*-polyisoprene as the temperature is raised through T_g (*c.* −70 °C)
(From Reed and Natarajan, by permission of J. Wiley, to be published)

The concentration of free radicals produced depends upon the stress level and attempts have been made to verify the kinetic theory equation (6.47) directly using the e.p.r. results. Zhurkov and Tomashevsky[66], using stepwise loading of their nylon fibres, concluded that the rate of radical formation was exponential in the stress, i.e.

$$\mathrm{d}C/\mathrm{d}t = B\exp(\alpha\sigma) \tag{6.52}$$

where C is the radical concentration and B and α are constants.

Comparing this with the kinetic theory equation for lifetime under load they found that the 'activation volume' β of equation (6.49), was identical in magnitude with the constant α of equation (6.52).

Roylance *et al*[70], whilst agreeing with Zhurkov's results for constantly increasing load, found that the rate of radical production in *constant-stress* tests was not constant, but that the radical concentration reached an equilibrium value at a given stress (see Figure 6.32). They showed, for nylon 6, that this 'steady-state' concentration C_s was given by

$$C_s = A\exp(\beta\sigma-\gamma)/kT \tag{6.53}$$

or, at constant temperature,

$$C_s = A'\exp(\alpha\sigma) \tag{6.54}$$

where α, β, γ, A, A' are constants.

The discrepancy between constant load and constant rate-of-loading tests can probably be explained in terms of the distribution of stresses amongst the atomic bonds discussed earlier. In a constant load test, the more highly stressed bonds break, until only normally stressed bonds remain. As the

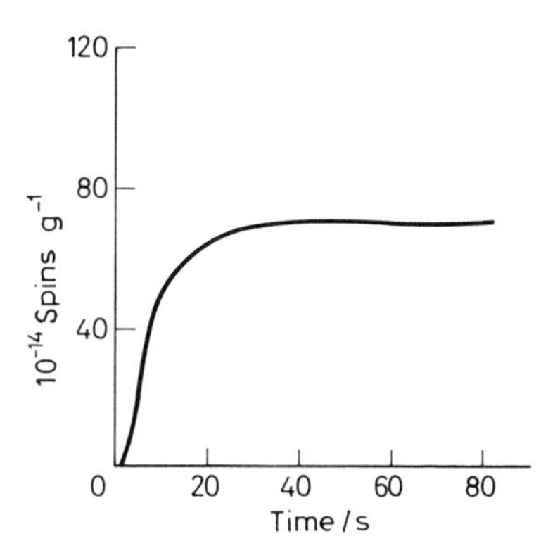

Figure 6.32 Spin concentration as a function of time under steady applied load (From Roylance, DeVries and Williams[70] in *Fracture*, by courtesy of Chapman and Hall)

'over-stressed' bonds are eliminated, the rate of radical formation naturally falls until the radical concentration reaches a plateau value. In an increasing load test, however, the 'over-stressed' bond population is continually replenished as the overall stress level rises. It is then reasonable to expect to observe a constant rate of formation of radicals as found by both Russian and American workers.

The e.p.r. results show that molecular fracture occurs at bonds subject to above-average stresses. They do not tell us whether the 'over-stressed' bonds are distributed more or less uniformly through the solid or are located in regions of stress concentration in the continuum sense, e.g. at a flaw or crack tip. Although, therefore, there is no basic conflict between continuum and kinetic theory, there remain many questions as to the precise way in which molecular fracture events relate to, and govern, the initiation and propagation of the cracks which are known to cause eventual macroscopic fracture. This matter is considered further in the following section.

6.4.5 Mechanical and chemical consequences of molecular fracture

The chemical consequences of molecular fracture under tensile loading are not different in kind from those of other mechano-chemical tests such as grinding or milling. Some of these consequences, however, are relevant to the mechanical (and thus the fracture) behaviour of the material under test and will therefore be briefly reviewed.

Regel *et al.*[71] deformed PMMA in the vacuum of a mass spectrograph and observed the release of molecular fragments similar to those obtained by thermal degradation of the polymer. They concluded that mechanically induced fracture was producing wholesale molecular degradation and cited this evidence in support of the kinetic theory of fracture. These results must be treated with some caution, however, since the release of low molecular weight polymer, and any residual monomer, into the vacuum of the spectrometer would be followed by thermal degradation at the ion source and the observation of thermal degradation products in the spectrum. The role of applied stress could simply be to accelerate the outward diffusion of pre-existing low molecular weight components of the polymer (the enhancement of diffusion rates by tensile stress is a well known effect in polymers).

Evidence from e.s.r. studies is perhaps more reliable, and these show that a variety of chemical reactions may follow initial rupture of primary bonds. E.S.R. spectra are relatively stable at temperatures below T_g, as shown earlier in Figure 6.31, but the prominence in such spectra of peroxy-radical signals indicates that chemical reaction is taking place and that the stable spectra observed are probably secondary rather than primary in nature. This is strikingly illustrated by some recent experiments by Reeve[69].

Polychloroprene vulcanisates deformed below about $-150\,^{\circ}$C exhibit wholesale craze formation (whitening) if their pre-extension at room temperature is $\sim 100\,\%$. The e.s.r. spectra from these tests are dominated by the peroxy radical. If the pre-orientation is $300\,\%$, however, further deformation below $-150\,^{\circ}$C produces no crazes. A strong e.s.r. signal is still obtained, however, but this time without any trace of peroxy-radical formation (this spectrum is broad and also symmetrical in contrast to the other). It is concluded that the formation of crazes in $100\,\%$ pre-oriented specimens allows atmospheric oxygen to diffuse in rapidly to react with radicals during, or shortly after, their formation. If crazes do not form, oxygen is excluded and different stable radicals are observed.

Reed and Natarajan[62] also found that the peroxy radical dominated e.s.r. spectra from natural rubber peroxide vulcanisates deformed below T_g, but obtained a quite different spectrum from sulphur-cured materials. Either the locus of initial fracture in the latter case is different (polysulphide cross-links break instead of main chains) or else the primary radicals react with sulphur in preference to oxygen to give a different spectrum.

As the temperature rises above T_g, the e.s.r. spectrum decays, indicating that further chemical reaction occurs as soon as the molecules gain mobility. This decay is associated with two effects, namely the evolution of hydrogen and an increase in cross-link density in the vulcanisates studied by Andrews and Reed[72], and Reed and Natarajan[62]. Hydrogen evolution was found in both sulphur- and peroxide-cured natural rubber and also in polychloroprene[69]. It could, however, be suppressed by changing the accelerator used in curing natural rubber, suggesting that chemical additives can greatly modify the mode of decay of the radicals as T rises through T_g.

The increase in cross-link density is probably associated with hydrogen release, although the exact mechanism is still in doubt. The change in cross-link density is associated with the initial cross-link density and the spin concentration as shown in Table 6.5 for one series of experiments.

Table 6.5 Natural rubber–sulphur vulcanisates tested at 120 K

M_N = initial inter-crosslink molecular weight
M_F = final inter-crosslink molecular weight

%Sulphur	M_N	M_F	$\Delta M/M_N$	*Spins/g*
1	11 158	10 668	4.5%	4.0×10^{17}
2.5	6 082	5 370	13.4%	8.9×10^{17}
4	4 385	3 850	12.2%	9.5×10^{17}
6	3 364	2 934	12.8%	11.1×10^{17}

Except for very lightly cured materials, the fractional increase in cross-link density (or decrease in M_c) is roughly constant at *c.* 13%. The spin concentration rises with decreasing molecular weight between cross-links. Both of these effects point to an increase in the number of molecular fracture events with increasing cross-link density; the spin concentration shows a linear plot *v.* M_c^{-1} (Figure 6.33). This suggests that the number of fracture events should reduce toward zero for un-crosslinked glasses, as is observed, in fact.

To summarise the relevant chemical effects of molecular fracture, therefore, we can say that the initial fracture event can produce reaction in the glassy state below T_g and/or as the material warms up through T_g. The result can be either degradation of the molecules or cross-linking, or both effects may occur simultaneously. The release of volatile reaction products such as hydrogen often occurs. A close analogy with the results of high-energy radiation might be deduced from these comments, but it has been shown that the e.s.r. signal from stretched and irradiated specimens of the same polymer are not identical[62].

We turn finally to the mechanical consequences of molecular fracture.

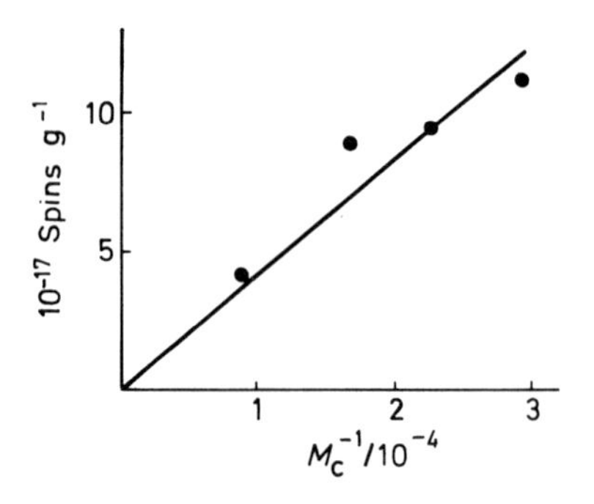

Figure 6.33 Spin concentration in *cis*-polyisoprene deformed below T_g as a function of cross-link density (reciprocal of M_c).

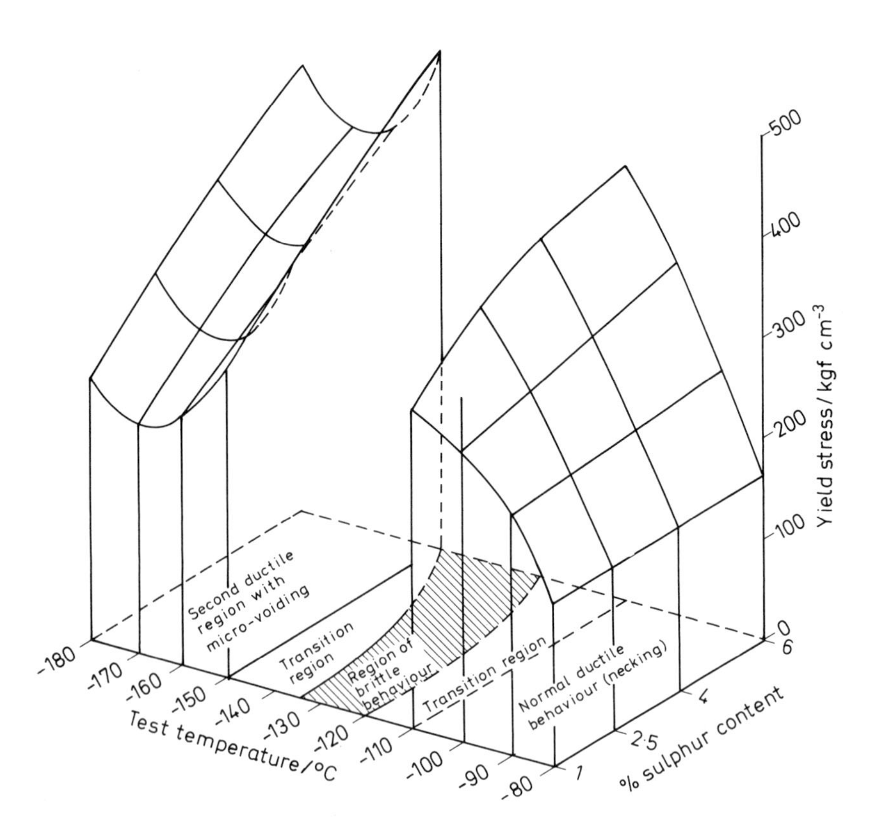

Figure 6.34 Yield stresses for pre-oriented *cis*-polyisoprene below T_g, showing regions of necking, brittle facture and yield by micro-void formation
(From Reed and Natarajan, by permission of J. Wiley, to be published)

It has already been noted that the appearance of e.s.r. spectra is frequently associated with whitening of the specimen. Yield and flow in pre-strained elastomeric vulcanisates (both natural rubber and polychloroprene) seems to occur in the glassy state by two quite distinct mechanisms.

Below T_g, but above some temperature $T_{c,1}$, plastic deformation occurs by the familiar 'cold drawing' process in which a neck is formed and propagated along the specimen. No whitening of the material is observed, no volatile gases are produced and no e.s.r. spectra are generated. Below $T_{c,1}$ (which is a rate-dependent quantity but is roughly −120 °C in natural rubber) there is a range of brittle behaviour, i.e. no plastic deformation occurs. This is eventually followed, below a second temperature $T_{c,2}$ (*c.* −140 °C for natural rubber) by a renewal of ductility which, however, occurs without neck formation. Narrow white bands appear in the specimen and broaden until they occupy the whole gauge length. Since no diminution of cross-section takes place during elongation three regions must be void-containing, i.e. they are generalised crazes. In this region gases are evolved and e.s.r. spectra produced.

Figure 6.34 shows the behaviour of the yield stresses for natural rubber (pre-oriented 100%) over these three regions.

Polychloroprene exhibits similar effects except that no brittle range has been observed, i.e. $T_{c,1} = T_{c,2}$. This is probably a function of pre-orientation and morphology. The transition from cold-drawing to crazing behaviour occurs at *c.* −150 °C, and, again, e.s.r. spectra are only found at $T < T_c$. In polychloroprene there is a secondary mechanical transition at *c.* −150 °C and this may be involved in determining T_c.

At higher pre-extension (300%) in polychloroprene, molecular orientation precludes cold drawing, and at $T < T_c$ no striations or whitening are observed either. However, strong e.s.r. spectra are still obtained. This means that although spectra only occur in conjunction with crazing in 100% pre-oriented samples, visible crazing is not *necessary* for the production of free radicals. Again, polychloroprene tested under a liquid (isopentane) at temperatures of −160 to −180 °C, where strong e.s.r. spectra are found and where crazing would occur in a gaseous environment, necked and cold-drew. It appears, therefore, that visible crazing is a frequent, but not essential, concomitant of molecular fracture.

We say *visible* crazing, because low-angle x-ray scattering (Zhurkov *et al.*[67]) has shown that micro-voids ranging in size from 1–100 nm are formed in many polymers on loading. Their materials included nylon 6, polypropylene, PVC and polyvinylbutyral, and the concentration of 'micro-cracks' ranged up to *c.* 10^{15} cm^{-3}. The authors considered that a critical concentration of micro-cracks must exist to produce macroscopic fracture by coalescence. They also found a correlation between total micro-crack surface area and free-radical concentration from e.s.r. spectra.

The evidence is, therefore, that molecular fracture under tensile stress frequently gives rise to visible cavitation or craze formation, and that it may possibly always produce micro-voids on some scale or another. When craze formation is extensive, it provides a mechanism for plastic deformation, but when less evident it may lead directly to fracture by coalescence of the micro-voids. The data of Reed and Natarajan suggest that molecular fracture and cavitation occur when the bulk yield stress rises above some critical value with decreasing temperature.

6.5 CONCLUSION

There are now three well-established approaches to the study of fracture in polymers, namely fracture mechanics, the molecular and the morphological. Only the first two have been covered in this review, but even here there are numerous advances to be reported and a great deal of current research is in progress. It is difficult to suggest which approach is the most fruitful, although fracture mechanics is the only aspect of the subject which lends itself directly to engineering applications and design studies. If fracture is to be regarded as a topic of engineering interest, rather than one of pure scientific concern only, it seems highly desirable to relate all studies eventually to the mechanics of the process.

Independently of this, however, the microscopical approaches (both molecular and morphological) hold out the promise of improved materials. In the development of metallic alloys, detailed understanding of the microscopical processes of deformation and fracture have enabled metallurgists to optimise properties by control of the microstructure. It is not unreasonable to expect that analogous benefits will, in due course, flow from the much younger science of polymer physics.

Continuing and important advances across the whole field of fracture in polymeric solids can be confidently expected during the coming years, and it is hoped that this review will serve both to stimulate and to maintain the already thriving interest in the subject.

References

1. Rosen, B. (editor) (1964). *Fracture Processes in Polymeric Solid,* (New York: John Wiley & Sons)
2. Andrews, E. H. (1968). *Fracture in Polymers,* (London: Oliver & Boyd)
3. Kausch, H. H. (1970). *Reviews in Macromol. Chem.,* **5,** 97
4. Andrews, E. H. (1972). *'Fracture'* in *Materials Science of Polymers.* (Amsterdam: North-Holland Pub. Co.)
5. Andrews, E. H. (1972). *Pure and Appl. Chem.,* (to be published)
6. Tiffany, C. F. and Masters, J. N. (1964). *Amer. Soc. Testing Materials, STP381,* 249
7. Andrews, E. H. (1968). *Fracture in Polymers,* 143 (London: Oliver & Boyd)
8. Inglis, C. E. (1913). *Trans. Inst. Naval Architects, (London),* **60,** 219
9. Andrews, E. H. (1961). *Proc. Phys. Soc.,* **77,** 483
10. Westergaard, H. M. (1939). *J. Appl. Mech.,* **6(2),** A49
11. Griffith, A. A. (1921). *Phil. Trans. Roy. Soc.,* **A221,** 163
12. Rivlin, R. S. and Thomas, A. G. (1953). *J. Polymer Sci.,* **10,** 291
13. Orowan, E. (1950). *Proc. Symp. on Fatique and Fracture of Metals,* 139 (New York: John Wiley & Sons)
14. Benbow, J. J. and Roesler, F. C. (1957). *Proc. Phys. Soc. B,* **70,** 201
15. Berry, J. P. (1963). *J. Appl. Phys.,* **34,** 62
16. Braden, M. and Gent, A. N. (1960). *J. Appl. Polymer Sci.,* **3,** 90
17. Andrews, E. H. and Bevan, L. (1966). in *Physical Basis of Yield and Fracture,* 209 (London: Institute of Physics)
18. Greensmith, H. W. and Thomas, A. G. (1955). *J. Polymer Sci.,* **18,** 189
19. Andrews, E. H. and Walker, B. J. (1971). *Proc. Roy. Soc. A,* **325,** 57

20. Greensmith, H. W. (1963). *J. Appl. Polymer Sci.,* **7,** 993
21. Svennson, N. L. (1961). *Proc. Phys. Soc.,* **77,** 876
22. Broutman, L. J. and McGarry, F. J. (1965). *J. Appl. Polymer Sci.,* **9,** 589
23. Cessna, L. C. and Sternstein, S. S. (1965). *J. Polymer Sci. B,* **3,** 825
24. Beardmore, P. and Johnson, T. L. (1971). *Phil. Mag.,* **23,** 1119
25. Broutman, L. J. and McGarry, F. J. (1965). *J. Appl. Polymer Sci.,* **9,** 609
26. Marshall, G. P., Culver, L. E. and Williams, J. G. (1969). *Plastics and Polymers,* **37,** 75
27. Williams, J. G., Radon, J. and Turner, C. E. (1968). *Polymer Eng. and Sci.,* (April), 130
28. Berry, J. P. (1963). *J. Polymer Sci. A,* **1,** 993
29. Murray, J. and Hull, D. (1970). *J. Polymer Sci. A2,* **8,** 1521
30. Charles, R. J. (1959). in *Fracture,* (B. Averbach, editor), 225, (New York: John Wiley & Sons)
31. Lake, G. J., Lindley, P. B. and Thomas, A. G. (1969). in *Fracture,* 1969, 493 (London: Chapman and Hall)
32. Mullins, L. (1959). *Trans. I. R. I.,* **35,** 213
32a. Thomas, A. G. (1956). *Trans. I. R. I.,* **32,** 236
33. Greensmith, H. W. (1956). *J. Polymer Sci.,* **21,** 175
33a. Greensmith, H. W. (1960). *Trans. Soc. Rheul.,* **4,** 184
34. Harwood, J. A. C. and Payne, A. R. (1966). *J. Appl. Polymer Sci.,* **10,** 315
35. Andrews, E. H. and Walsh, A. (1958). *Proc. Phys. Soc.,* **72,** 42
36. Greensmith, H. W. (1964). *J. Appl. Polymer Sci.,* **8,** 1113
37. Grosch, K. A., Harwood, J. A. C. and Payne, A. R. (1966). in *Physical Basis of Yield and Fracture,* 144 (London: Institute of Physics)
38. Andrews, E. H. (1969). in *Testing of Polymers,* Vol. 4 (W. E. Brown, editor), (New York: John Wiley & Sons)
39. Constable, I., Williams, J. G. and Burns, D. J. (1970). *J. Mech. Eng. Sci.,* **12,** 20
40. Zilvar, V. (1971). *Plastics and Polymers,* **39,** 328
41. Burns, D. J., Borduas, H. F. and Culver, L. E. (1967). *Proc. 23rd S.P.E. Ann. Tech. Conf., Detroit,* 233
42. Lake, G. J. and Lindley, P. B. (1966). in *Physical Basis of Yield and Fracture,* 176 (London: Institute of Physics)
43. Andrews, E. H. (1961). *J. Appl. Phys.,* **32,** 542
44. Hertzberg, R. W., Nordberg, H. and Manson, J. A. (1970). *J. Mat. Sci.,* **5,** 521
45. Andrews, E. H. and Braden, M. (1961). *J. Polymer Sci.,* **55,** 787
46. Andrews, E. H., Barnard, D., Braden, M. and Gent, A. N. (1963). in *Chemistry and Physics of Rubber-like Substances,* 329 (London: Maclaren and Sons)
47. Andrews, E. H. and Bevan, L. (1966). in *Physical Basis of Yield and Fracture,* 209 (London: Institute of Physics)
48. Andrews, E. H. and Bevan, L. (1972). *Polymer,* to be published
49. Andrews, E. H., Levy, G. M. and Willis, J. (1972). *Polymer,* (to be published)
50. Gent, A. N. (1970). *J. Mat. Sci.,* **5,** 925
51. Marshall, G. P., Culver, L. E. and Williams, J. G. (1970). *Proc. Roy. Soc. (London), A,* **319,** 165
52. Andrews, E. H. and Levy, G. M. (1972). (to be published)
53. Williams, M. L. (1969). *J. Appl. Polymer Sci.,* **13,** 29
54. Ripling, E. J., Mostovy, S. and Patrick, R. L. (1964). *Mat. Res. and Stand.,* **4,** 129
55. Malyshev, B. M. and Salganik, R. L. (1965). *Int. J. Frac. Mech.,* **1,** 114
56. Gent, A. N. and Petrich, R. P. (1969). *Proc. Roy. Soc. (London), A,* **310,** 433
57. Gent, A. N. (1971). *J. Polymer Sci. A2,* **9,** 283
58. Gent, A. N. and Kinloch, A. J. (1971). *J. Polymer Sci. A2,* **9,** 659
59. Andrews, E. H. and Kinloch, A. J. (1972). to be published
60. Gent, A. N. and Schultz, J. (1971). *Amer. Chem. Soc. Preprints, Washington Conf. (Sept.),* **31,** No. 2, 113
61. Casale, A., Porter, R. S. and Johnson, J. F. (1971). *Rubber Chem. and Technol.,* **44,** 534
62. Reed, P. E. and Natarajan, R. (1972). *J. Polymer Sci.,* (to be published)
63. Ayscough, P. B. (1967). *Electron Spin Resonance in Chemistry,* (London: Methuen and Co.)
64. Taylor, N. W. (1947). *J. Appl. Phys.,* **15,** 943
65. Stuart, H. A. and Anderson, D. L. (1953). *J. Amer. Ceram. Soc.,* **36,** 416
66. Zhurkov, S. N. and Tomashevsky, E. E. (1966). in *Physical Basis of Yield and Fracture,* 200 (London: Institute of Physics)

67. Zhurkov, S. N., Kuksenko, V. S. and Slutsker, A. I. (1969). in *Fracture,* 531, (London: Chapman and Hall)
68. Zhurkov, S. N., Vettegren, V. I., Korsukov, V. E. and Novak, I. I. (1969). in *Fracture,* 545, (London: Chapman and Hall)
69. Reeve, B. (1972). *Ph.D. thesis,* Univ. of London
70. Roylance, D. K., DeVries, K. L. and Williams, M. L. (1969). in *Fracture,* 551, (London: Chapman and Hall)
71. Regel, V. R., Muinov, T. M. and Pozdnyakov, O. F. (1966). in *Physical Basis of Yield and Fracture,* 194. (London: Institute of Physics)
72. Andrews, E. H. and Reed, P. E. (1967). *J. Polymer Sci. B,* **5,** 317

7
Polymer Degradation

N. GRASSIE
University of Glasgow

7.1 INTRODUCTION

During the period 1966–1970 there have been substantial advances in most of the important aspects of polymer degradation including that induced by heat, light, oxygen, high-energy radiation and mechanically, and accounts have been given at various times of the current position in some of these[1–4]. Progress in all these directions has been profoundly influenced by the many new analytical techniques which have become freely available in the form of commercial instruments during the past decade or which are currently being developed[5–8]. Spectroscopic techniques in the form of n.m.r. and e.s.r., thermal analysis in the form of t.g.a., d.t.a., d.s.c. and t.v.a., and gas–liquid and gel-permeation chromatography have been particularly effective. These have allowed a very much closer analysis, firstly, of the changes in polymer structure which occur during degradation and which have been particularly difficult to characterise by the classical analytical methods and, secondly, of the volatile products, the detection and analysis of trace amounts of which are often of great significance in allowing progress towards a clearer understanding of reaction mechanism.

There has been great preoccupation during the period under review with the production of more highly stable polymers. To this end it has been found advantageous, as a general principle, to build cyclic structures into the polymer chain backbone, at the same time retaining that degree of flexibility necessary to obtain optimum physical properties. Thus there has developed a vast literature relating to the synthesis of these materials and the assessment of their general stability, usually by thermal analysis techniques[1, 9–11]. High-temperature polymers have been omitted from this review, however, because accounts of fundamental investigations of degradation mechanisms are only beginning to appear and it is not yet possible to discuss them adequately in terms of general principles.

7.2 POLY(VINYL ESTERS)

7.2.1 Poly(vinyl chloride)

7.2.1.1 Introduction

During the period under review earlier discussion has continued unabated about three fundamental factors in the degradation of poly(vinyl chloride), namely, whether the reaction is ionic or free radical, the chemical nature of the points in the molecule at which reaction is initiated, and whether or not it is catalysed by the hydrogen chloride formed. The position at the end of 1966 has been comprehensively reviewed[12]. These questions are intimately associated with the mode of action of stabilisers, upon which the commercial application of this polymer in particular is so dependent. They have not been finally and unequivocally answered but a great deal of progress has been made. This has been assisted by the recognition that much of the earlier confusion about even the essential fundamental features of the degradation of poly(vinyl chloride) has arisen because of inattention to the importance of the precise conditions under which investigations were carried out, and clear distinctions are now being made in particular between thermal, photo and oxidative effects.

7.2.1.2 Thermal

Claims that the purely thermal elimination of hydrogen chloride proceeds by an ionic or radical mechanism have been almost equally divided. Palma and Carenza's[13] proposal of a radical mechanism is based only upon a resemblance between kinetic data for the thermal reaction and that induced by high-energy radiation. Geddes[14] has demonstrated that typical free-radical initiators increase the rate of the reaction although this is not necessarily proof that the mechanism in their absence is radical. Others appear to be on much surer ground. Thus it has been demonstrated[15, 16] that degrading poly(vinyl chloride) can initiate the radical depolymerisation of poly(methyl methacrylate) and the radical oxidation of polybutadiene[17] and that the build up of radicals, measured by e.s.r. in thermally degrading poly(vinyl chloride) is depressed by the presence of poly(methyl methacrylate), polystyrene and poly(α-methyl styrene)[18] due to the fact that chlorine atoms, which propagate the reaction in poly(vinyl chloride) are lost to the second polymer, initiating its radical depolymerisation. It has also been shown[19] that tritium and ^{14}C-labelled toluene become incorporated into thermally degrading poly(vinyl chloride). Comparing the behaviours of poly(vinylidene chloride) and poly(vinyl) chloride), Hay[20] concludes, however, that while the former undoubtedly degrades by a radical mechanism it certainly cannot be exclusive in poly(vinyl chloride) since radicals are not produced in sufficient concentration to be detected by e.s.r.; the singlet spectrum which appears is consistent with the polyene residue. Pyrolysed anthracene, a source of free radicals, has also been shown to be a very much less effective inhibitor of the thermal, compared with the photo-oxidative, degradation[21]. Nor is the reaction inhibited by nitric oxide[22], although the subsequent breakdown of the residual polyene chain to give aromatic compounds is thus inhibited.

The acceleration of the reaction by additives with mobile protons such as acids, alcohols and phenols[23, 24] and the variability of the rate in different solvents[25] provides strong evidence for an ionic mechanism and it has been pointed out[26] that catalysis by hydrogen chloride can be accounted for by the strongly basic nature of the chloride ion formed by dissociation of the hydrogen chloride in the reaction medium. Whether the reaction is ionic or radical is clearly a function of additives and the reaction medium but one is inclined to agree with Braun[27] that the purely thermal non-oxidative degradation of pure poly(vinyl chloride) is non-radical.

It is quite clear that the thermal instability of poly(vinyl chloride) cannot be accounted for by the normal poly(vinyl chloride) structure but that labile structural abnormalities must be responsible. Both small-molecule and polymeric models have been studied. A comparative investigation[28] of the thermal stabilities of 2-chlorobutane, 2,4-dichloropentane, 2,4,6-trichloroheptane, 4-chloropent-2-ene, 6-chlorohepta-2,4-diene and 3-chloropent-1-ene, which represent normal and abnormal structures, demonstrated that the stability of saturated models is not greatly affected by chain length but diminishes rapidly as the number of double bonds increases. Double bonds within the chain reduce stability very much more than they do at the ends[28]. Thus terminal unsaturation is not sufficient to explain the instability of commercial polymers. As polymeric models, radical-initiated poly(vinyl chloride), linear syndiotactic poly(vinyl chloride) and copolymers of vinyl chloride with 1-chloropropene and 2-chloropropene have been compared[29–31] from which it has been deduced that branched polymers are considerably less stable than linear polymers and that tertiary chlorine atoms at branch structures initiate the reaction more rapidly than tertiary hydrogen atoms. In this connection also head-to-head polymer is more stable than head-to-tail material[32].

The conjugated polyene structures which result from loss of hydrogen chloride and cause discoloration of the residual polymer are most amenable to study by ultraviolet spectrometry. Rates of dehydrochlorination and the length distribution of polyene sequences vary widely with the origin of the polymer[33] but, for the pure thermal degradation reaction, have been reported to lie within the range 13–30 units[25, 34]. Sequences in degraded poly(vinyl bromide) are appreciably longer[34]. It has been suggested that the lengths of sequences may be limited by the presence of structural abnormalities[25] and it has indeed been demonstrated that branch points in the form of 2-chloropropene units[30] as well as groups such as diethylfumarate and isobutene[34] reduce the length of polyene sequences, the former by providing an increased number of points of initiation and the latter by the blocking influence on the dehydrochlorination process. However, the concentration of abnormalities in a normal poly(vinyl chloride) is not sufficient to account for the relatively short sequence lengths which are probably rather due to transfer processes during dehydrochlorination[13]. With increasing extent of degradation, as measured by dehydrochlorination, a growing polyene deficit is found[35]; polyene structures are consumed in subsequent intra- and inter-chain processes. This kind of behaviour is related to the action of certain colour stabilisers for poly(vinyl chloride) and to the decoloration of the discoloured polymer[36, 37]. Decoloration occurs in air in the solvents tetrahydrofuran, methyl ethyl ketone and dioxane, all of which peroxidise readily. The number of double bonds in the polymer decreases, the peroxide is consumed and solvent fragments become incorporated into the polymer as decoloration proceeds.

The question whether or not hydrogen chloride catalyses the reaction was previously one of the major points of controversy. The issue was complicated by inattention to experimental detail in a great many investigations. There now seems to be a fairly general concensus of opinion in favour of catalysis. At temperatures below 200 °C autocatalytic features only appear if hydrogen chloride is allowed to accumulate in the system[23]. Rate constants have been deduced for catalysed and uncatalysed processes[38] and energies of activation of 21 and 28 kcal mol^{-1} respectively have been reported. A mechanism has been proposed[26] in which the basic chloride ion, formed by dissociation of hydrogen chloride, is the active species. Above 200 °C autocatalytic features can exist due to the exothermicity of the reaction such that it becomes explosive around 236 °C[39]. It has been deduced that the explosive decomposition temperature should increase with pressure and it has been shown to increase to 313 °C at 38.1 kbar.

Kelen and his co-workers[40] have derived theoretical relationships concerning extent of dehydrochlorination and concentration and length distribution of polyene sequences, assuming random initiation, propagation by a series of allyl-activated decomposition steps and termination either at chain ends or in an abnormal propagation step. They claim good agreement with the experimental data. Further papers[40, 41] refine the theoretical treatment.

7.2.1.3 Thermal oxidation

The rate of degradation, measured by hydrogen chloride evolution, is accelerated when poly(vinyl chloride) is thermally degraded in oxygen, but, in apparent contradiction, discoloration is retarded[42]. This is due to the fact that a very much higher proportion of shorter polyene sequences is formed than in the purely thermal reaction[43]. It has also been demonstrated that compounds, such as naphthyl sulphide, which destroy peroxides, are effective retarders[42] and the radical nature of the reaction is confirmed by the fact that in blends (mixtures) of poly(vinyl chloride) and poly(butadiene) the rate of oxygen absorption increases with the poly(vinyl chloride)/ poly(butadiene) ratio[17]. Radicals formed in the poly(vinyl chloride) phase apparently diffuse into the polybutadiene phase and accelerate its oxidation. The fact that the stability is inversely proportional to the intrinsic viscosity[44] demonstrates that chain ends are not particularly susceptible and it has been suggested that there may be a greater concentration of chain branches at which thermal oxidation may be initiated in higher molecular weight material. Thus, in contrast to the purely thermal reaction, thermal oxidation is a radical reaction initiated at abnormalities in the polymer chain and involving peroxide. It is not clear whether the short polyene sequences are formed in the primary degradation process or whether they are formed by the subsequent oxidation of longer primary sequences.

7.2.1.4 Photo-oxidation

The dehydrochlorination chain process which leads to polyene sequences and discoloration plays very little part in the photo-oxidation which occurs in poly(vinyl chloride) below 150 °C and which can seriously affect the mechanical properties of the polymer[45]. It is suggested that chlorine atoms are eliminated followed by hydroperoxidation of the polymer radical by the Bolland mechanism. Subsequent decomposition of the hydroperoxide leads to scission and cross-linking. Scission and cross-linking are accelerated by the covalent attachment of the chelate substituents, copper salicylate, ferrocene and copper phthalocyanine, to the polymer molecules[46]. Degradation has been followed by stress relaxation of plasticised film[45]. An energy of activation of 12 kcal mol^{-1} has been deduced for the rate-controlling step which is believed to be hydrogen abstraction from the substrate by peroxy radicals. The radical nature of the reaction is emphasised by the very effective inhibiting power of pyrolysed anthracene, which is a source of stable free radicals[21].

Kwei[47] has studied the photo-oxidation of *dl*- and *meso*-dichloropentanes as models for syndiotactic and isotactic sequences in poly(vinyl chloride). The rate of oxidation of the model compounds is similar to that of poly(vinyl chloride), the volatile products are similar, the CH_2/CHCl ratio increases linearly with time, the structure $\sim CHCl{-}CH_2{-}\underset{\underset{O}{\|}}{C}{-}CH_2\sim$ is found in poly(vinyl chloride) after photo-oxidation, and the *dl* model is oxidised 1.5 times faster than the *meso* model. It is concluded that the behaviour of the model is closely similar to that of poly(vinyl chloride), that syndiotactic sequences in poly(vinyl chloride) are probably more susceptible to photo-oxidation than isotactic sequences and that attack occurs much more readily at the chloromethylene rather than the methylene groups.

The production of hydrogen chloride which may occur during photo-oxidation, particularly at the higher temperatures, is associated with a quite separate and purely thermal reaction.

7.2.1.5 *High-energy radiation*

Study of the effects of high-energy irradiation has assisted understanding of the degradation of poly(vinyl chloride) as of other polymers, principally because the initiation process, usually brought about by irradiation at 77 K, can be observed separately from the propagation processes which occur during post-irradiation storage at higher temperatures. These investigations are almost invariably associated with e.s.r. spectroscopy which can give direct information about the intermediate radical species. Degradation reactions in poly(vinyl chloride) have the additional advantage that they are associated with the development of conjugation which can be readily analysed by use of u.v. spectroscopy.

The radical concentration in poly(vinyl chloride) irradiated by 1.6 MeV electrons *in vacuo* increases to a limiting value which is higher the lower the radiation temperature[48] within the range 100–320 K. Under the influence of irradiation the radicals decay in a first-order process whose rate constant is proportional to the dose rate. During post-irradiation storage at ordinary temperatures the concentration of radicals is constant although there is an increase in the concentration of conjugated double bonds[49]. Ultraviolet spectral measurements show[50–51] that irradiation below 200 K gives trapped alkyl radicals generated by C—Cl bond scission. These dehydrochlorinate during post-irradiation storage at 250 K to give allyl and polyenyl radicals. Radical chain-transfer processes result in polyenes and regenerate alkyl radicals. Thus allyl and dienyl radicals attain steady-state concentrations while polyene absorption increases. The radical and polyene portions of the u.v. spectrum can be distinguished by oxygen scavenging[50].

Polymer irradiated *in vacuo* subsequently exhibits enhanced rates of thermal degradation[50, 52], from which it is presumed that irradiation produces thermally-labile structures. It has also been demonstrated that the thermal and γ-ray initiated reactions at 80–130 °C are similar kinetically. Radical formation and decay curves on irradiation by γ-rays at 20–90 °C have been obtained and the energy of activation for radical decay is found to rise quite sharply from 5 kcal mol^{-1} at 40–60 °C to 53 kcal mol^{-1} at 70–90 °C. This has been explained in terms of a 'cage' model for radical interaction.

7.2.1.6 *Dechlorination*

Perhaps the earliest quantitative investigation of the degradation of poly (vinyl chloride) was described in the now classical paper by Flory[53] in which he demonstrated the essential head-to-tail structure of the polymer from the proportion of residual chlorine left after dechlorination by zinc. This reaction has been applied recently to the investigation of microstructural differences in poly(vinyl chloride). Milan and Smets[54] have confirmed that cyclopropane units are formed in the reaction and have also shown that some olefinic structures are revealed by i.r. and n.m.r. spectroscopy. The concentration of olefin is greater in polymer prepared in butyraldehyde solution than in bulk and emulsion polymers. Guyot *et al.*[55] believe the reaction in dioxane solution is catalysed by $ZnCl_2$, that water acts as a co-catalyst and that the reaction is a little faster when the content of syndiotactic units is high. They also find olefinic structures using i.r. and n.m.r. spectroscopy and believe that these structures are subsequently hydrogenated by the combined action of zinc and hydrochloric acid. They also suggest that dehydrochlorinated structures increase the reactivity of the neighbouring chlorine atom to dechlorination. Gel-permeation chromatography demonstrates that there is no chain scission during the reaction.

7.2.1.7 Related polymers

The thermal degradation of pure poly(vinylidene chloride) *in vacuo* is clearly analogous to that of poly(vinyl chloride) in the sense that hydrogen chloride is liberated leaving a conjugated polyene residue. E.S.R. gives clear evidence of a radical mechanism[20] and this is confirmed by retardation by triphenylmethane. The fact that the radical inhibitors anthraquinone and diphenylpicrylhydrazyl accelerate the reaction demonstrates the confusion which may arise in the application of inhibitors of this kind at elevated temperatures. In solution in *N*-methyl-2-pyrrolidone and hexamethylphosphoramide, decomposition rates are solvent-dependent[56] and substantially higher than in the solid although polymer preparation has no significant effect. Initiation is thus random rather than chain terminal and there is evidence that the reaction is base catalysed by the solvent. The reaction proceeds well beyond 50% elimination of chlorine which, because insolubility does not develop until a very late stage and aromatic compounds are ultimately produced as in degrading poly(vinyl chloride), has been explained in terms of intra-chain aromatisation with elimination of hydrogen chloride. Intramolecular elimination of hydrogen chloride from the intermediate polyene structure has been confirmed in a degradation study of poly(vinylene chloride) obtained by dehydrochlorination of poly(vinylidene chloride) by sodium amylate[57].

Copolymers of vinylidene chloride with methyl methacrylate, methacrylonitrile and styrene are all less stable than the homopolymer. The methyl methacrylate copolymer eliminates methyl chloride as well as hydrogen chloride leaving $\beta\gamma$-unsaturated γ-lactone structures in the residue. In the methacrylonitrile copolymer it is believed that the increased rate is due to increased chain mobility while the styrene units catalyse hydrogen chloride loss[58, 59]. On the other hand, the rate of decomposition of vinylidene chloride grafted to ethylene–propylene copolymers and polybutadiene is depressed[60] and head-to-head poly(vinylidene chloride) is much more stable than the normal head-to-tail material[32]. The pattern of chlorobenzenes obtained on pyrolysis at 430 °C and analysed by gas chromatography has been used to determine the sequence distribution in triads in copolymers of vinylidene chloride and vinyl chloride. The same technique has been used to determine the structure of chlorinated poly(vinyl chloride)[61].

Poly(α-chloroacrylonitrile), which may be regarded as a hybrid of poly-

(vinyl chloride) and polyacrylonitrile, also liberates hydrogen chloride thermally[62] and under the influence of basic materials[63] to form polyene structures although it is at least 80 °C less stable than poly(vinyl chloride). Comparison of the u.v. spectra of polyacrylonitrile and poly(α-chloroacrylonitrile) demonstrates that there is no significant nitrile group condensation as in polyacrylonitrile[64]. Copolymerised α-chloroacrylonitrile units introduce instability into poly(methyl methacrylate) and polystyrene[65]. Random scission occurs at α-chloroacrylonitrile units followed by depolymerisation to monomer in the methyl methacrylate copolymer. Since the threshold temperature of the degradations of homopolymer and copolymers of α-chloroacrylonitrile are similar (*c.* 150 °C) it is concluded that they are all radical processes initiated by scission of the particularly labile C—Cl bonds.

Chlorinated 1,4-polybutadiene has been described as head-to-head, tail-to-tail poly(vinyl chloride). The steric structure of this material has been elucidated by analysis of such characteristic degradation products as *o*-, *m*- and *p*-dichlorobenzenes and vinylidene chloride[66].

Chlorination of polyisobutene occurs at both methyl and methylene groups although predominantly at the latter and the stability decreases progressively in the composition range 0–0.5 chlorine atoms per monomer unit and remains at a minimum from 0.5–1.0 chlorine atoms per monomer unit[67]. In the temperature range 190–240 °C hydrogen chloride is the principal volatile product of thermal degradation but the mechanism is quite clearly different from that in poly(vinyl chloride) because removal of hydrogen and chlorine atoms from adjacent carbon atoms is impossible.

The thermal degradation of chlorinated polystyrene[68] and the photo-oxidation of chloroprene[69] may also be associated with similar reactions in poly(vinyl chloride).

7.2.2 Poly(vinyl acetate)

On thermal treatment, poly(vinyl acetate) and poly(vinyl alcohol) liberate acetic acid and water respectively leaving a conjugated unsaturated residue in strict analogy to the behaviour of poly(vinyl chloride). The earlier suggestion that in poly(vinyl acetate) this is a non-radical chain reaction initiated at chain terminal structures and propagated from unit to unit along the chain has recently been disputed[70].

Relatively small doses of both 2537 Å and γ-rays cause both scission and cross-linking in poly(vinyl acetate) and for the same amount of energy dissipated u.v. is about four times more efficient than the high-energy radiation[71]. Although the rate of cross-linking is greater than the rate of chain scission so that the polymer becomes insoluble, it has been shown that under γ-irradiation there is preferential scission at branch points early in the reaction[72, 73]. Simulated branches (methyl groups) were produced by copolymerisation with traces of isopropenyl acetate but these cause cross-linking rather than scission. High yields of acetic acid, methane, carbon monoxide and hydrogen and traces of carbon dioxide are obtained with large doses of high-energy radiation[74].

Under alkaline conditions poly(vinyl acetate) is hydrolysed to poly(vinyl alcohol) and acetic acid. In copolymers with vinyl and acrylic esters the rate of hydrolysis is reduced with increasing proportions of comonomer and increasing length and branching of the side chain while small amounts (1%) of acid comonomers greatly increase the rate of hydrolysis. It is believed that these effects are mainly steric and environmental[75].

7.2.3 Poly(vinyl alcohol)

Evolution of water in high yield in the thermal degradation of poly(vinyl alcohol) at 140°C is accompanied by some chain scission to give small amounts of aldehydes and ketones with the general formulae (1) and (2)[76].

$$\underset{\text{(1)}}{\mathrm{CH(=O)}\!-\!(\mathrm{CH}=\mathrm{CH})_n\mathrm{Me}} \qquad n = 0,1,2,3,4\ldots \qquad \underset{\text{(2)}}{\mathrm{Me}\!-\!\mathrm{C(=O)}\!-\!(\mathrm{CH}=\mathrm{CH})_n\mathrm{Me}}$$

Although poly(vinyl alcohol) is usually prepared by hydrolysis of poly(vinyl acetate), it is interesting to note that polyacetaldehyde with the poly(vinyl alcohol) structure degrades in a generally similar way[77]. No systematic study has been made of the degradation reactions of vinyl acetate–vinyl alcohol copolymers but temperatures quoted[78] do suggest that poly(vinyl alcohol) containing 20% of unhydrolysed acetate groups is much more thermally stable than the fully hydrolysed material. Hydrogen peroxide is perhaps an unexpected product of the photo-sensitised oxidation of poly(vinyl alcohol) in aqueous solution and it has been suggested that the primarily formed hydroperoxide can decompose by two routes[79].

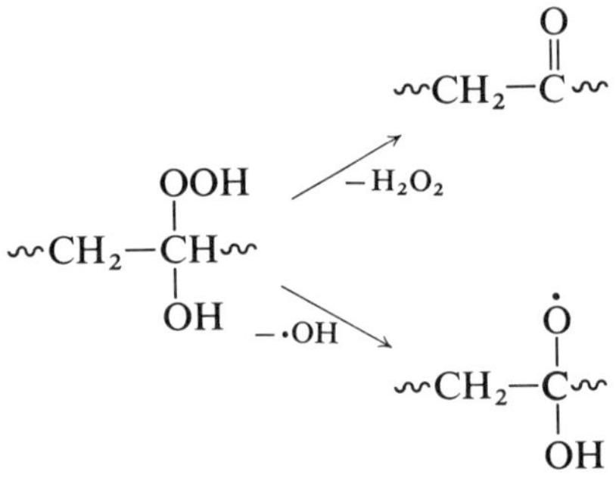

$$\sim\mathrm{CH_2}\!-\!\mathrm{C(OOH)(OH)}\!\sim \;\xrightarrow{-\mathrm{H_2O_2}}\; \sim\mathrm{CH_2}\!-\!\mathrm{C(=O)}\!\sim$$

$$\sim\mathrm{CH_2}\!-\!\mathrm{C(OOH)(OH)}\!\sim \;\xrightarrow{-\cdot\mathrm{OH}}\; \sim\mathrm{CH_2}\!-\!\mathrm{C(\dot{O})(OH)}\!\sim$$

The second reaction is followed by chain scission, hydroperoxidation of the resulting radical and decomposition of the hydroperoxide ultimately to give acidic products. A 1:1 ratio between chain scission and acid groups is indeed found.

7.2.4 Copolymers of vinyl chloride and vinyl acetate

The thermal degradation of copolymers of vinyl chloride and vinyl acetate covering the whole composition range have been studied in bulk[80] and in tritoluyl phosphate solution[81]. Hydrogen chloride and acetic acid are produced in the same molar ratio as the two monomers in the copolymer. At each end of the composition range, incorporation of the comonomer unit results in a copolymer less stable than the homopolymer. Minimum stability occurs for compositions of 40–50 mol % vinyl acetate in the bulk degradation and 30–40 mol % vinyl acetate for degradation in solution. Thus the acid-elimination process is facilitated by increasing heterogeneity in the molecular chain and it proceeds from unit to unit along the chain as in the homopolymers leaving a polyconjugated residue. Both of the acidic products catalyse the reaction but benzoic and other aromatic acids either retard it or have no effect. The accelerating effect of chain heterogeneity may be due to the inductive effect of the chlorine atoms weakening the C—H bonds of the adjacent methylene group, thus facilitating the elimination of acetic acid.

$$\sim CH_2-\underset{\underset{\underset{\underset{Me}{|}}{C=O}}{|}}{\underset{|}{O}}{CH}-\overset{\overset{H}{|}}{\underset{\underset{H^{\delta+}}{|}}{C}}-\underset{\underset{Cl^{\delta-}}{|}}{CH}\sim \longrightarrow \sim CH_2-CH=CH-\underset{\underset{Cl}{|}}{CH}\sim \quad + \ MeCOOH$$

or there may be neighbouring-group participation by the acetate group in the elimination of hydrogen chloride.

$$\sim CH_2-\underset{\underset{O-\underset{\underset{Me}{|}}{C}=O}{|}}{CH}-CH_2-\overset{\overset{Cl}{|}}{\underset{\underset{H}{|}}{C}}\sim \longrightarrow \sim CH_2-\underset{\underset{O-\underset{\underset{Me}{|}}{C}=O}{|}}{CH}-CH=CH\sim \quad + \ HCl$$

In either case, the unsaturation will provide allylic activation for elimination of further units in sequence.

7.2.5 Polyacrylonitrile

The application of acrylonitrile-based polymers in the synthetic fibre industry during the past 20 years has prompted many studies of the thermal degradation of polyacrylonitrile, and especially of the thermal coloration of the polymer which occurs above 200 °C. There has been a great deal of controversy about the chemical nature of the structures causing colour. Both conjugated polyene structures in the chain backbone and condensed-ring structures incorporating conjugated carbon–nitrogen sequences have been proposed. During the past 5 years the latter explanation has come to be fairly generally accepted and Peebles and his colleagues[82] and Braun and El Sayed[83] in particular have offered convincing confirmatory evidence that the basic reaction consists of polymerisation of nitrile groups.

$$\begin{array}{ccc} CH_2 & CH_2 & CH_2 \\ CH & CH & CH \\ | & | & | \\ CN & CN & CN \end{array} \longrightarrow \begin{array}{ccc} CH_2 & CH_2 & CH_2 \\ CH & CH & CH \\ | & | & | \\ C{=}N & C{=}N & C{=}N \end{array}$$

Earlier workers[84] suggested fairly complete cyclisation with long conjugated sequences. Watt[85] expresses doubt concerning long sequence length but retains the concept of a ladder structure and explains the thermal stability in terms of increased resistance of cyclised units to chain scission. Other investigators[82] conclude that the amount of cyclisation is very small and that most of the acrylonitrile units remain unchanged.

The manufacture of carbon fibres by pyrolysing acrylonitrile-based fibres at high temperatures has recently had the effect of intensifying interest in the thermal degradation of polyacrylonitrile, particularly under temperature-programmed conditions and over wider temperature ranges and the cyclisation mechanism is fundamental to current concepts of how polyacrylonitrile is converted in high yield to carbon fibre[85–87].

The principal features of the pyrolysis of polyacrylonitrile at high temperatures in an inert atmosphere or *in vacuo* may be summarised as follows[88]. Up to 350 °C, chain fragments containing conjugated carbon–nitrogen sequences are produced. The more volatile of these are removed from the system, especially *in vacuo*, and considerable weight loss occurs. The gaseous products of the reaction are principally ammonia, which is probably formed from terminal imine structures in the coloured polymer, and hydrogen cyanide which is eliminated from units which have not undergone cyclisation, possibly in conjunction with the chain-scission reaction. Above 350 °C, the cyclised structures lose hydrogen and become aromatic. This process continues to at least 700 °C, above which further hydrogen formation may result from intermolecular condensation of aromatic structures. Finally, beginning at *c.* 900 °C, the char begins to lose nitrogen, indicating breakdown of the heterocyclic rings and rearrangement to give pure carbon. Pyrolysis of polyacrylonitrile in air has not been studied in as much detail as in nitrogen but an important effect of oxygen is to inhibit polymerisation of nitrile groups. When the reaction does occur at higher temperatures, the characteristics are altered and it is probable that oxidation reactions are contributing.

DTA studies[89, 90] have shown that an intense exotherm occurs in the temperature range 250–300 °C, and it has been suggested[86, 91] that it is associated with nitrile group polymerisation. An alternative explanation is that it is associated with formation of ammonia[92]. However, the former explanation has recently been confirmed by the demonstration that pyrolysis conditions can be chosen such that the exotherm can be induced without weight loss[88].

Under temperature-programmed conditions the autocatalytic appearance of volatile products[92] is caused by the rapid increase in the temperature of the polymer during the exotherm. The large proportion of chain fragments formed, which is greater in pyrolysis carried out *in vacuo* than in nitrogen[88], suggests that chain scission is fairly competitive with propagation of the conjugated sequences. Thus it seems that initiation of cyclisation occurs at frequent intervals along the acrylonitrile chain and is terminated when it reaches an adjacent conjugated segment. Scission of the non-conjugated connecting units results in chain fragments volatile at pyrolysis temperatures. The conclusion that the cyclised segments are relatively short is in agreement with the model advanced by Noh and Yu[93] to explain the kinetics of the coloration and the changes in the infrared spectrum of the polymer. The size of the exotherm is very sensitive to conditions, increasing with both sample size and heating rate[88], and the degree of fragmentation, in turn, depends upon the size of the exotherm.

Early work on both polyacrylonitrile and polymethacrylonitrile, which has been reviewed in the series of papers by Peebles and his colleagues[82], has shown that the coloration reaction may be initiated through cyclic intermediates or at abnormal structures. The importance of chain-terminal structures is demonstrated by the observation that the exotherm becomes broader and less intense with decreasing molecular weight[90] and that the rates of coloration of redox-initiated polymers depends upon the concentration of acid end groups[94]. It is clear from the current literature that there is intense interest in the effects of structure on the reactions at low temperature in view of their importance to the quality of the ultimate carbon fibre produced at higher temperatures.

In an attempt to gain information about initiation through cyclic structures the reaction has been studied in dimethylformamide solution at 153 °C [95]. However, the polar nature of dimethylformamide would seem to make it a probable initiator so that any conclusion deduced from the experimental results is likely to be invalidated. Nevertheless, it is clear that longer conjugated segments are produced in solution than in the solid and this has been associated with greater chain mobility in solution. A decrease in viscosity in the initial stages has been attributed to intramolecular cross-linking and a later increase in viscosity due to straightening of the molecules resulting from the formation of the more rigid, co-planar, cyclic segments with conjugated bonds.

It has also been demonstrated that cross-linking by the formation of metal chelates improves thermal stability[96], while acrylonitrile segments grafted on to polytetrafluorethylene are less thermally stable than pure polyacrylonitrile[97].

7.3 HYDROCARBON POLYMERS

7.3.1 Polystyrene

7.3.1.1 Thermal

Interest in the more fundamental aspects of the thermal degradation of polystyrene can be traced back for nearly 40 years and there is general agreement about the principal features of the reaction. Thus the reaction proceeds by a radical chain mechanism, the main volatile products are monomer, dimer, trimer and tetramer, and in radical-initiated polymers the molecular weight of the residue decreases sharply in the initial stages of the reaction and later more slowly.

Although monomer comprises 40–50% of the products at 300–350 °C, it has recently been shown that the monomer/oligomer ratio increases with temperature[98] and it has been claimed that under flash pyrolysis conditions at very high temperatures quantitative yields of monomer are obtained[99]. Substitution by methyl groups in the ring also increases the relative yield of monomer[98] and the stabilities of polystyrenes substituted in the *para*-position are directly related to the Hammett σ values of the substituents[100]. Degradation of polystyrene in solution is faster than in the solid[101] but the difference decreases with increasing temperature. It has been suggested that solvation decreases with increasing temperature so that the environment of the polymer molecules in solution becomes more and more like that in the solid.

For at least 30 years there has been controversy over the reasons for the sharp initial decrease in the molecular weight of degrading polystyrene. In general, those who have fitted theoretical calculations to molecular weight and extent of volatilisation data are satisfied that it can be accounted for by the intermolecular transfer step in the radical process[102,103]. On the other hand, those who have taken a more direct chemical approach in general favour the idea that weak links are present in polystyrene. The position was elegantly reviewed in 1967 by Cameron and MacCallum[104].

In an attempt to identify weak links in polystyrene, Cameron and Kerr[105] isolated chain scission from volatilisation by working at lower temperatures than hitherto. They found, plotting chain scissions against time, that a straight line is obtained which passes through the origin for anionically-prepared polymers but makes a positive intercept on the chain scissions axis for radical-initiated polymer. Thus they concluded that the latter incorporates weak links while the former does not. Wegner and Patat[101] also found no sharp initial decrease in molecular weight in anionic polymers. It has also been shown[106] that these weak links are not head-to-head linkages, chain branches or unsaturated structures although unsaturated and head-to-head structures[107] do reduce the pyrolytic stability of polystyrene by lowering the energy of activation for the bond rupture which occurs after the weak link scission phase. McNeill and his co-workers[108,109] confirmed that weak links are not associated with unsaturated structures by demonstrating that the decrease in molecular weight in the weak-link phase is the same in cationic and free-radical polymers, although unsaturation is exclusively terminal in the former and distributed at random throughout the molecules of the latter. There continue to be suggestions that the weak links may be oxygenated structures introduced inadvertently into the polystyrene[101,106] but there is no positive evidence of this.

There is, of course, no denial by those workers who favour weak links that intermolecular transfer plays an important part in the overall degradation process and this has been proved experimentally by showing that the degradation of polystyrene containing thermally labile structures can induce degradation in normal polystyrene at temperatures at which it does not normally degrade[110].

Apart from the controversy for and against weak-link scission in the early stages, there is general agreement that the overall reaction consists of initiation in which radicals are formed; those radicals then compete in depropagation, intramolecular and intermolecular transfer, which result in monomer production, dimer, trimer, etc. production, and chain scission, respectively. Initiation apparently occurs preferentially at chain ends[111]; radicals produced by main-chain scission probably recombine without escaping from the 'cage' in which they are formed[112]. It is generally agreed that the zip length for propagation is short, of the order of 2–3 [102, 103, 113]. In the past, theoretical treatments have normally assumed second-order termination by the mutual destruction of pairs of radicals but Cameron[111] has accounted for his evidence of first-order termination in terms of intramolecular transfer followed by elimination of a small radical (7.1) rather than the normal process in which a neutral molecule is eliminated and a chain-terminal radical remains in the polymer residue to continue the degradation process, (7.2)

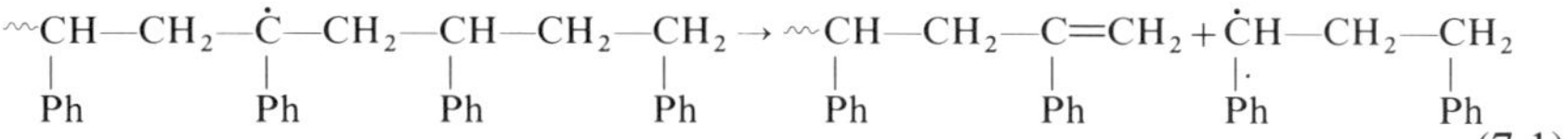

$$\sim\underset{\mathrm{Ph}}{\mathrm{CH}}-\mathrm{CH_2}-\underset{\mathrm{Ph}}{\dot{\mathrm{C}}}-\mathrm{CH_2}-\underset{\mathrm{Ph}}{\mathrm{CH}}-\mathrm{CH_2}-\underset{\mathrm{Ph}}{\mathrm{CH_2}} \rightarrow \sim\underset{\mathrm{Ph}}{\mathrm{CH}}-\mathrm{CH_2}-\underset{\mathrm{Ph}}{\mathrm{C}}=\mathrm{CH_2}+\underset{\mathrm{Ph}}{\dot{\mathrm{C}}\mathrm{H}}-\mathrm{CH_2}-\underset{\mathrm{Ph}}{\mathrm{CH_2}} \quad (7.1)$$

$$\sim\underset{\mathrm{Ph}}{\dot{\mathrm{C}}\mathrm{H}}+\mathrm{CH_2}=\underset{\mathrm{Ph}}{\mathrm{C}}-\mathrm{CH_2}-\underset{\mathrm{Ph}}{\mathrm{CH}}-\mathrm{CH_2}-\underset{\mathrm{Ph}}{\mathrm{CH_2}} \quad (7.2)$$

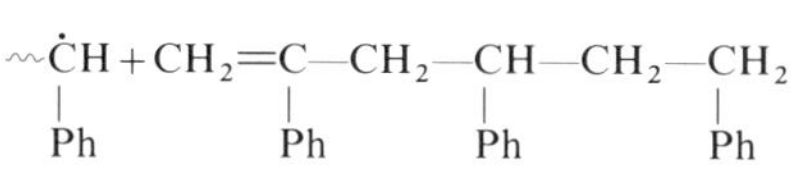

There seems nothing to choose energetically between these two processes.

7.3.1.2 Thermal and photo-oxidation

The primary product of oxidation of polystyrene is a hydroperoxide formed in a type of Bolland mechanism[114]. At the high temperatures of thermal oxidation (200–300 °C) its decomposition is instantaneous, although in carbon tetrachloride solution under u.v. radiation at low temperatures concentrations of peroxide as high as four peroxide groups per 1000 styrene units have been observed[115]. In both photo- and thermal-oxidation, aromatic carbonyl structures accumulate as a result of hydroperoxide decomposition and a combination of n.m.r., i.r. and u.v. spectral measurements have confirmed acetophenone (3) rather than benzal acetophenone (4) type structures[115] as well as conjugated unsaturation in the chain backbone (5)[116] which causes the yellow coloration.

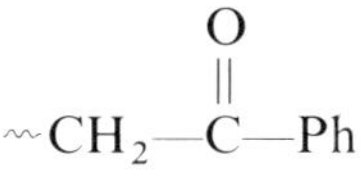

$$\sim\mathrm{CH_2}-\overset{\mathrm{O}}{\overset{\|}{\mathrm{C}}}-\mathrm{Ph} \quad (3) \qquad \sim\overset{\mathrm{Ph}}{\overset{|}{\mathrm{C}}}=\mathrm{CH}-\overset{\mathrm{O}}{\overset{\|}{\mathrm{C}}}-\mathrm{Ph} \quad (4) \qquad \sim\mathrm{CH}=\overset{\mathrm{Ph}}{\overset{|}{\mathrm{C}}}-\mathrm{CH}=\overset{\mathrm{Ph}}{\overset{|}{\mathrm{C}}}\sim \quad (5)$$

$$\sim\mathrm{CH_2}-\overset{\mathrm{Ph}}{\overset{|}{\underset{\bullet}{\mathrm{C}}}}-\mathrm{CH_2}\sim \quad (6)$$

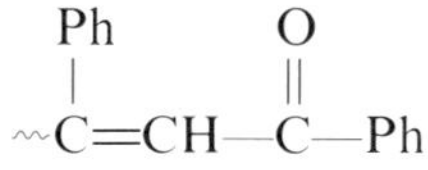

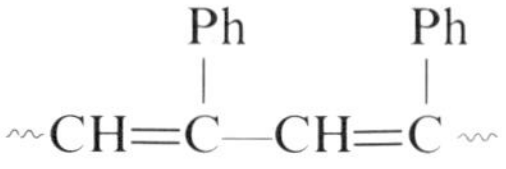

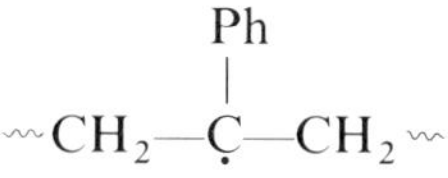

In the absence of oxygen or even in film at ordinary temperatures in an oxygen atmosphere, where the rate of oxidation is restricted by diffusion, u.v. radiation causes insolubility[115]. This is presumably due to the fact that a large proportion of the primarily formed tertiary radicals (6) combine to form cross-links or eliminate hydrogen to form (5). Under conditions in which the primary radicals can react freely with oxygen, as in the photo-oxidation in solution at low temperatures[117] or during thermal oxidation of molten polymer at higher temperatures[115], chain scission due to decomposition of hydroperoxides predominates and increased solubility is observed. Gel-permeation chromatography and sedimentation-velocity measurements have been shown to be particularly valuable techniques for following this kind of degradation because unzipping, cleavage, cross-linking and dimerisation are distinguishable[118, 119].

The relative inertness and the short kinetic chain-length of the thermal oxidation of polystyrene compared with the model compound cumene, has been attributed to steric hindrance to propagation in the Bolland mechanism. The reaction is catalysed by a mixture of cobalt acetate and sodium bromide in chlorobenzene–acetic acid solution. Since the mobile chain carrier $Co^{2+}(CH_3COO)Br$ is now involved, oxidation occurs readily at temperatures as low as 40 °C and benzaldehyde, acetophenone and phenol are observed as final products[120].

The influence of the wavelength of radiation on the rate of photo-oxidation of polystyrene has been observed[121]. From 2483 to 2537 Å the rate is independent of wavelength. Thereafter it falls sharply till at 2800 Å it is 20% of that at 2537 Å. Radiation at 3130 Å is ineffective. These effects seem to be associated as much with the ability of the polystyrene to absorb as with the energy content of the radiation, since the wavelength limit of absorption of polystyrene lies just above 2800 Å.

A start has also been made to studying the reactions of the atmospheric pollutants, nitrogen peroxide and sulphur dioxide, with polystyrene and some other polymers[122–124].

7.3.1.3 Mechanical

The mechanical degradation of polystyrene has been studied in solution by high-speed stirring[125] and by flow through a capillary viscometer[126], in the melt during extrusion[127] and in the solid during ball milling[128]. During high-speed stirring, degradation occurs in the boundary layer region extending to a distance equal to approximately the length of the polymer molecules from the leading edge of the blade tip, so it seems that the molecules subjected to the critical shearing stress are completely extended in the stream lines of flow[125]. Degradation during extrusion is consistent with the mechanical forces on the molecules simply reducing the activation necessary for thermal degradation so that it occurs from *c.* 50 °C lower than normal[127]. It has also been suggested that weak links in polystyrene may have a substantial effect on mechanical rupture and the strength properties of the polymer[129]. In mechanically degraded solid polymer, radicals have been detected by their reaction with nitric oxide[128] and have been identified as (7) and (8)[130].

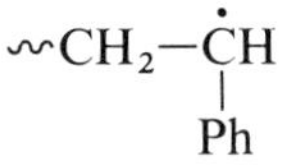

$$\sim CH_2-\dot{C}H(Ph) \quad (7) \qquad \sim CH_2-\dot{C}(Ph)-CH_2\sim \quad (8) \qquad \sim CH_2-CH(C_6H_6^{\bullet})-CH_2\sim \quad (9)$$

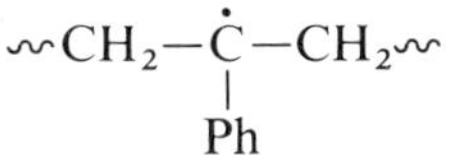

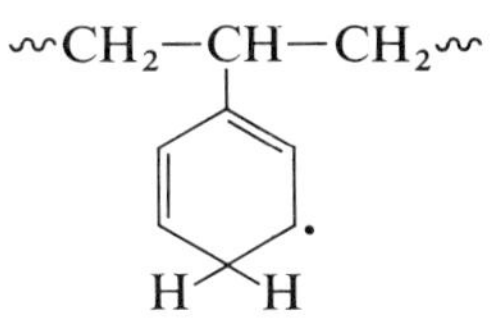

This is different from the radical found during high-energy irradiation which is of the cyclohexadienyl type (9).

7.3.1.4 *Styrene copolymers and polymer mixtures*

When acrylonitrile units are copolymerised into polystyrene molecules, the rate of volatilisation increases in direct proportion to the acrylonitrile content[131]. The primary effect of the acrylonitrile units on stability is to cause an increased rate of chain scission but the unzipping process which follows chain scission is not greatly affected by the acrylonitrile units even in copolymers containing 24.9 mol % of acrylonitrile. Thus acrylonitrile appears among the products and the proportion of chain fragments (dimer, trimer and tetramer) increases, these fragments also incorporating nitrile units[132]. Yellow coloration develops in the residue from high-acrylonitrile copolymers at advanced stages of degradation and spectral measurements suggest that this may be due to conjugated unsaturation in the polymer-chain backbone which may be associated with the liberation of hydrogen cyanide from the acrylonitrile units. The rate constant of the chain-scission process which is associated with acrylonitrile units is *c*. 30 times that for 'normal' scission in styrene segments of the polymer chains[133].

In copolymers of styrene and methyl methacrylate the products, namely both monomers and the oligomers of styrene, are more or less as expected from the known behaviours of the two homopolymers[134]. The concentration of weak links is proportional to the styrene content which suggests that they are associated with single styrene units rather than with adjacent pairs. Sequences of at least ten styrene units in the copolymer molecules are necessary for dimer, trimer and tetramer to be produced. The presence of styrene in the copolymer has a disproportionately large effect in increasing its thermal stability. This is due to the fact that in the copolymerising mixture of monomers the disproportionation termination reaction which occurs in pure poly(methyl methacrylate) is almost completely replaced by combination, so that the thermally labile, unsaturated chain-terminal structures which exist in poly(methyl methacrylate) are present in extremely low concentrations even in a 4:1, styrene–methyl methacrylate copolymer.

Poly(methyl α-phenyl acrylate) is an interesting link between poly(methyl methacrylate) and polystyrene. In the early stages of thermal degradation at 210–280 °C, random scission is followed by unzipping to monomer[135]. Chain-end initiation becomes important beyond 45 % conversion but there is no evidence for transfer throughout degradation.

There is no interaction between poly(methyl methacrylate) and polystyrene when they are degraded as a mixture[136] so that mixtures and copolymers of the same composition may be readily distinguished by thermal analysis techniques. On the other hand, poly(α-methyl styrene), which, like poly(methyl methacrylate), degrades quantitatively to monomer at temperatures below those at which polystyrene degrades, can induce degradation in polystyrene. Although the molten mixture of polystyrene and poly(α-methyl styrene) is heterogeneous, consisting of micelles of poly(α-methyl styrene) embedded in polystyrene, the monomer radicals produced by complete unzipping of the poly(α-methyl styrene) can apparently diffuse into the polystyrene and initiate decomposition[137].

The pyrolysis-gas chromatography technique has been applied to the characterisation of both styrene–acrylonitrile and styrene–methyl methacrylate copolymer systems[138,139]. An unexpected product, 2,4-diphenyl thiophene (10), is found in the thermal degradation of the copolymer of styrene

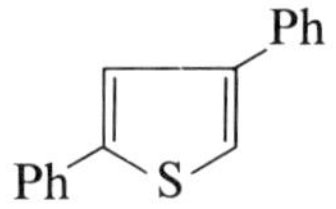

Ph

Ph S

(10)

and sulphur dioxide[140]. This could be formed by the simultaneous loss of two molecules of water from a repeat unit but it seems more likely to be the final product of a complex degradation sequence.

7.3.1.5 Related polymers

A number of papers have appeared which deal with various aspects of the degradation of a number of polymers related to polystyrene in the sense that they incorporate aromatic structures attached to a saturated chain.

The greater part of the thermal degradation of poly(α-methyl styrene) consists of random scission followed by depropagation of the radicals to monomer. There is evidence of inhibition by catalyst fragments in the early stages of the reaction, however[141]. In solution, in mixtures of cyclohexanol and cyclohexane, rate constants for chain scission and monomer production are inversely dependent upon solvent viscosity, showing that the overall reaction is diffusion-controlled. The rate determining step is probably diffusion out of the 'cage', of the radicals formed by chain scission[142].

Poly(*m*-aminostyrene) and poly(*m*-acetamidostyrene) are more stable than polystyrene in air and the same or less stable in nitrogen[143]. The antioxidant characteristics of the substituent groupings play the major role in stabilisation. The products of degradation, both major and minor, are closely analogous to those of polystyrene[144,145], except that the residue is insoluble, probably due to a cross-linking reaction involving the amino substituent.

The thermal degradation of poly(2-vinyl pyridine) is also very similar to that of polystyrene but chelation by Cu^{2+} ions drastically decreases the stability, fast degradation occurring at 100 °C. The Cu^{2+} ions make possible a new redox initiation process[146]. Both poly(2-vinyl pyridine) and poly(4-vinyl pyridine) are very much less stable than polystyrene to high-energy radiation[147].

The photolysis and radiolysis of poly(phenyl vinyl ketone)[148,149], poly(vinyl benzophenone[150] and poly(vinyl pyrrolidone)[151] have also been studied and the mechanical degradation of the latter by repeated freezing and thawing of its aqueous solution has many similarities to the degradation induced by ultrasonic irradiation[152]. Similar cryodegradation has been observed in sulphonated polystyrene cross-linked with divinylbenzene and in the quaternary ammonium derivative of this polymer[153].

7.3.2 Polyethylene

The purely thermal degradation of polyethylene has for long been known to result in a complex series of saturated and unsaturated hydrocarbons with molecular weights limited only by their volatility under the conditions in which the degradation is carried out. This is known to occur by a free-radical mechanism dominated by intra- and inter-molecular transfer processes. Very much better analysis of these products has recently been possible using g.l.c. so that a more quantitative picture of the various constituent reactions is beginning to emerge[154]. There has always been very much greater interest, however, in thermal oxidation and photo-oxidation, especially the latter, because of their direct association with the deterioration in polymer properties which occurs during processing and the subsequent ageing and weathering of commercial articles fabricated from this material.

Reviews on various aspects of photo-degradation have appeared and two of the most recent by Cicchetti[155] and Winslow and his colleagues[156] give an authoritative and comprehensive account of the subject at least up to the end of 1968. Both primary and secondary initiation of photo-oxidation are possible, the former occurring when ultraviolet light directly activates the molecular oxygen or oxygen-substrate complexes, while the latter is operative when the radiation is absorbed by oxidation products such as peroxides or carbonyl structures or metallic or other impurities. Secondary initiation is by far the more important and carbonyl groups are the principal sensitisers. These may be derived from carbon monoxide impurity or catalyst fragments introduced during the preparation of the polymer, from autoxidation during high-temperature processing or by the action of ozone on out-door exposure.

The photo-degradation which occurs during weathering is principally a non-radical scission process, during the occurrence of which carbonyl and vinyl structures appear and carbon monoxide is liberated. Only minor concentrations of vinylene and vinylidene groups appear and no hydroperoxide is produced. Indeed hydroperoxide groups present in thermally oxidised polymer disappear after only brief exposure to 2537 Å radiation. The chain scission has been explained in terms of Norrish type 1 and 2 mechanisms[157].

$$\sim CH_2-C(=O)-CH_2-CH_2\sim \;\nearrow\; \sim CH_2-\dot{C}(=O) + \cdot CH_2-CH_2\sim \qquad \text{type 1}$$

$$\sim CH_2-C(=O)-CH_2-CH_2\sim \;\searrow\; \sim CH_2-C(=O)-CH_3 + CH_2=CH\sim \qquad \text{type 2}$$

Type 1 is temperature-dependent and accounts for the formation of carbon monoxide and gel while type 2 is independent of temperature and is not quenched by oxygen, although it is inhibited in the glass transition region probably due to restriction of freedom of internal motion of the polymer chain. At 120 °C the two processes make approximately equal contributions to a total quantum yield of *c.* 0.05, while at ambient temperature type 2 predominates.

Attention has been drawn to the fact that a structure is present in radical-polymerised polyethylene which leads to the production of a triene on u.v. radiation[158]. This occurs whether or not oxygen is present but oxidation is inhibited while the triene is being consumed. Although this review does not cover the more technological aspects of polymer degradation, it is important to mention that Chan and Hawkins[159], by the application of internal reflection spectroscopy, have been able to correlate the very early stages of degradation with the subsequent deterioration of bulk properties thus obviating the need for accelerated weathering tests which, in any case, may not produce true ageing conditions.

It is well established that the thermal oxidation of polyethylene proceeds by the Bolland hydroperoxidation mechanism and more quantitative investigations are now being pursued. By use of dicumyl peroxide as initiator it has been shown that in film at 116–130 °C the rate is proportional to the square root of the initiator concentration and independent of oxygen pressure and has a chain length in the range 65–140[160]. Four quite distinct chain-rupture reactions occur involving P· radicals, $PO_2\cdot$ radicals and the decomposition of hydroperoxide[161]. The decomposition of each molecule of hydroperoxide yields, on average, only 0.2 radicals[162] a fact which may contribute to the greater stability of polyethylene against autoxidation compared with polypropylene. The increase in density which occurs on ageing polyethylene has been associated with the increased carbonyl content rather than with changes in crystallinity[163].

The effect of high-energy radiation on polyethylene is to produce a variety of structures, principally vinylidene, vinyl and vinylene groups but also conjugated diene and triene, cyclic groups and allyl and higher polyenyl radicals[164]. The unusual effect of the simultaneous removal of four hydrogen atoms to form the diene has been commented upon[165] and shown to be favoured in crystalline regions where chains are more likely to be in the proper alignment. This seems to be confirmed by the fact that diene is absent in polymer irradiated in the molten state. Studying the distribution of radicals trapped in regions of different morphological structure, Bohm[166] has shown that reactions between oxygen and free radicals occur much more readily in the amorphous than in the crystalline zones. A reaction between hydrogen and vinyl end groups has also been observed during irradiation.

7.3.3 Polypropylene

7.3.3.1 Introduction

Among the polyolefins, the greatest activity has undoubtedly been centred upon polypropylene. The promotion of this material to be one of the large-tonnage polymers of commerce has made it urgent to understand and control the thermal oxidation and photo-oxidation reactions which occur during processing and ageing, and which are much more acute than were the problems of comparable processes in polyethylene.

7.3.3.2 *Thermal*

The purely thermal decomposition of polypropylene, like that of polyethylene, is dominated by intramolecular radical-transfer reactions which result in a continuous spectrum of saturated and unsaturated hydrocarbon products. In the degrading system the chain terminal radical (11) exists in greater concentration than (12)[167],

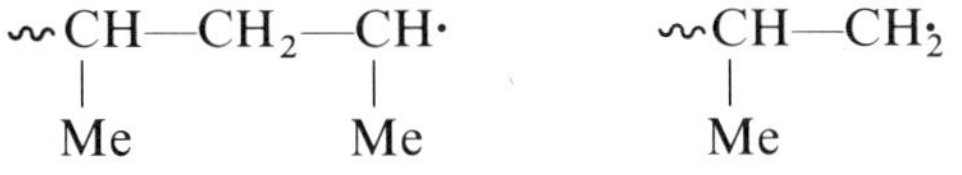

$$\sim\mathrm{CH(Me)-CH_2-CH(Me)}\cdot \qquad\qquad \sim\mathrm{CH(Me)-CH_2}\cdot$$

(11) (12)

while the transfer process involves reaction of these radicals with tertiary rather than methylene hydrogen atoms[168]. Thus, the three dominant volatile products, pentane, 2-methylpent-1-ene and 2,4-dimethylhept-1-ene can be accounted for by 1,5- and 1,7-backbiting followed by chain scission[167, 169].

7.3.3.3 *Thermal oxidation*

The thermal oxidation of solid polypropylene has been shown by Ross and Chien[170] to be limited by oxygen diffusion to a depth of 0.635 mm from the surface. The primary process is one of hydroperoxidation by the Bolland mechanism but the hydroperoxide groups tend to be formed in adjacent sequences rather than at random[171]. Hydroperoxidation is initiated by the products of hydroperoxide decomposition and terminated primarily by disproportionation of peroxy radicals[172] or by the reaction

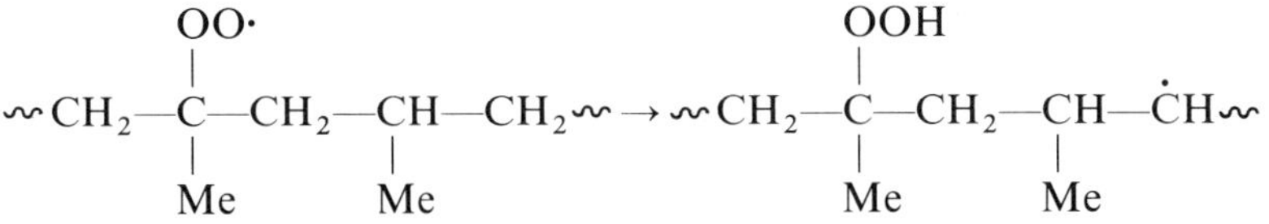

$$\sim\mathrm{CH_2-C(OO\cdot)(Me)-CH_2-CH(Me)-CH_2}\sim \rightarrow \sim\mathrm{CH_2-C(OOH)(Me)-CH_2-CH(Me)-\dot{C}H}\sim$$

Rate constants for initiation, propagation and termination have been measured at 110 °C[172, 173] ($k_i = 3\times10^{-4}\ \mathrm{s^{-1}}$, $k_p = 1.9\ \mathrm{l\ mol^{-1}\ s^{-1}}$, $k_t = 3\times10^{6}\ \mathrm{l\ mol^{-1}\ s^{-1}}$) and it has been found possible to rationalise the more important products of the reaction – water, formaldehyde, acetaldehyde, acetic acid, acetone and hexane-2,5-dione – with the reaction scheme. A very thorough investigation has been made of the non-volatile products of thermal oxidation[174]. Functional groups, two of which are formed per chain scission, comprise γ-lactone (17 %), aldehyde (21 %), ketone (21 %), acid (25 %) and ester (16 %). These are compared with the products from polyethylene and with the products of process degradation (mechanical oxidation)[175] and photo-oxidation[176] in both of which only one functional group is formed per chain scission and high proportions of vinyl groups are formed in addition to those listed above for thermal oxidation. The thermal decomposition of polypropylene hydroperoxide consists of two consecutive reactions[177]. The first consumes up to 90% of the total hydroperoxide and consists of homolytic decomposition of hydroperoxide groups. The second involves decomposition of β-hydroxyhydroperoxide structures formed as a by-product of the first reaction. The chemiluminescence observed during the oxidation of polypropylene has been shown to be a function of the hydroperoxide rather than the peroxy radical concentration[178].

In order to eliminate diffusion effects and thus study the kinetics of the reaction more effectively, investigations have been carried out in thin films[179] and in solution[180, 181]. Bawn and Chaudhri[180] report orders of reaction with respect to polymer and oxygen of 1.7 and 1 respectively, which cannot be readily explained by the Bolland mechanism. These authors add a number of additional steps, involving the intermediate radicals, to the Bolland mechanism and show that this can account qualitatively for the kinetic results.

The effects of both initiation and inhibiting additives on the mechanism of thermal oxidation of polypropylene are of considerable interest[182–185]. In catalysis by cobalt salts, CO^{III} is found to be the effective catalyst; Co^{II} is relatively inactive[186].

7.3.3.4 Photo-oxidation

Advances in our understanding of the photo-oxidation of polypropylene are principally due to a series of papers by Carlsson, Wiles and Kato[187–192]. Using polymer films and wavelengths greater than 3000 Å in air, they found that the rate of photo-oxidation increases with tacticity and suggest that this is due to a stereo-dependent step in the oxidation chain process. The major products of the reaction include hydroperoxide and carbonyl groups and since both of these are also formed inadvertently during thermal processing and will be present in most fabricated articles which are subjected to ageing, their photo-degradation *in vacuo* was studied[189, 190].

The effect of ketonic products was observed by studying polypropylene which had been extensively air-oxidised at 225 °C. Air-oxidation results principally in two polymeric ketones, (13) and (14).

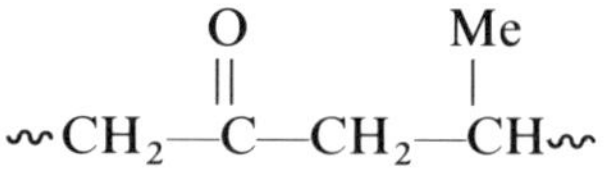

$$\sim CH_2-\overset{\overset{\displaystyle O}{\|}}{C}-CH_2-\overset{\overset{\displaystyle Me}{|}}{C}H\sim \qquad\qquad \sim\overset{\overset{\displaystyle Me}{|}}{C}H-CH_2-\overset{\overset{\displaystyle O}{\|}}{C}-Me_3$$

(13) 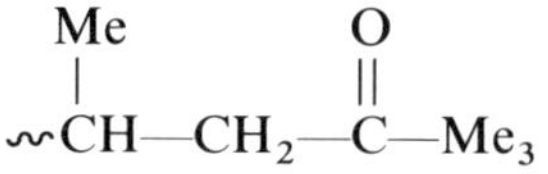 (14)

On subsequent photolysis *in vacuo* (13) decomposes by the Norrish type 1 mechanism to give carbon monoxide and two macro-radicals, while (14) decomposes by the Norrish type 2 mechanism to give acetone and an unsaturated chain end. Both processes have quantum yields of *c.* 0.08 and are believed to be the source of initiation in the early stages of the photo-oxidative deterioration of polypropylene. In the later stages, initiation of photo-oxidation is probably dominated by the photolysis of hydroperoxide. The mechanism of photolysis of polypropylene hydroperoxide under high vacuum was studied using 3650 Å radiation. The primary step involves cleavage to t-alkoxy and hydroxy radicals. The hydroxy radicals abstract hydrogen atoms to give water as the major volatile product. The main polymeric products are produced by reaction or scission of the alkoxy radicals.

By studying the surface reaction using attenuated total reflection, Carlsson and Wiles[191,192] believe that restriction of the photo-oxidation to a thin surface layer is not due to restrictions on diffusion of oxygen but to trace quantities of polymeric carbonyl and hydroperoxide groups present predominantly in the surface layers and formed during processing or the subsequent life of the polymer.

Initiation of photo-degradation in polypropylene has also been attributed to titanium and aluminium oxides from the polymerisation catalyst[193] and to iron particles introduced from production machines, especially during the pelleting process[194]. It is of interest to note that the greater rate of thermal oxidation of isotactic compared with atactic poly(but-1-ene) has also been attributed to the presence of 'ash' from the polymerisation catalyst[195].

7.3.3.5 High-energy radiation

Both scission and cross-linking occur when polypropylene is exposed to high-energy radiation. This is intermediate between the behaviours of polyethylene, which predominantly cross-links, and polyisobutene, which prodominantly scissions. Irradiation at 77 K results in alkyl radicals $\sim\dot{C}(CH_3)-CH_2\sim$ [196,197] and at higher temperatures these radicals undergo cross-linking and scission reactions. Under irradiation at room temperature intermolecular bonds are formed in isotactic polymer in a non-free-radical reaction of vinylidene double bonds formed by rupture of the polymer chain, while in atactic material the major proportion of cross-links is formed by recombination of free radicals[198]. The helical configuration of isotactic polypropylene with the methyl groups on the outside of the helix, combined with the low mobility of the molecules, apparently inhibits radical-recombination reactions. During radiolysis, oxygenated structures, present as impurity, are decomposed[198,199] and gel formation in polypropylene only begins when all these oxygenated structures have reacted and after an appreciable quantity of double bonds has accumulated.

7.3.3.6 Mechanical (ethylene–propylene copolymers)

During mill mastication, the rate of degradation decreases with increasing temperature up to 315 °F[200]. Free-radical acceptors demonstrate that this is due to mechanical rupture of carbon–carbon bonds. Above 350 °F, thermooxidative degradation becomes dominant and the higher the temperature the greater the extent of degradation. Mastication at 68 °F causes a decrease in molecular weight, a narrowing of the distribution, and the process is non-random[201]. In the region 182–315 °F there is little change in molecular weight while at 480 °F, mastication broadens the distribution and degradation is random.

7.3.4 Polyisobutene

A close analysis of the products of the thermal degradation of polyisobutene from C_1 to C_{24} has been made using gas chromatography[202]. Independence of rate upon molecular weight[203] supports initiation by random scission and the products can be accounted for in terms of unzipping to monomer and intramolecular transfer. A very much higher proportion of monomer is obtained from polyisobutene than from polyethylene so that intramolecular transfer is relatively more difficult than depropagation in the former. This is to be expected since, in polyisobutene, six out of eight hydrogen atoms per monomer unit are primary and require greater activation energy for transfer than the secondary hydrogen atoms. In addition, the secondary atoms in polyisobutene are sterically obstructed. The structure of the products demonstrates that they are formed by transfer of primary (~CH_2·) radicals rather than tertiary (~$\dot{C}$Me) radicals and this is explained in terms of the greater reactivity of primary radicals and their greater availability from scission processes following transfer.

Under high-energy radiation polyisobutene degrades faster in solution than in the solid. Two explanations have been given, namely that 'cage' recombination of radicals is reduced in solution or that free radicals produced primarily in the solvent initiate degradation in the polymer. The latter explanation is supported by the fact that in heptane solution the rate of degradation is independent of the molecular weight of the polymer, and of the viscosity of the solution[204]. A decrease in rate in solutions with concentrations less than 5% is attributed to a change in the conformation of polyisobutene radicals in dilute solution.

Mechanical degradation of polyisobutene during laminar flow in hexadecane, 1,2,4-trichlorobenzene and decalin has been studied[205–207]. A minimum shear of 60 dyn/cm^2 × T(K) is necessary to achieve shear degradation[206]. A negative temperature coefficient for degradation is to be expected since stresses and extending forces increase with decreasing temperature at fixed shear rate due to increased viscosity[205]. Degradation is likely to occur through rupture of chemical bonds at points of entanglement. Investigating the effect of initial molecular weight, concentration, temperature and shear stress on mechanical degradation in decalin solution, Ram and Kadim[207] have expressed degradability in terms of a single parameter, Y, which is characteristic of each polymer–solvent system, and given by,

$$Y = (\eta_f - \eta_s)\,\tau_\omega\, M_f/\eta_f C$$

in which C is the concentration of the solute, η_f and η_s are the final equilibrium viscosity and the viscosity of the solvent, τ_ω is the shear stress at the wall and M_f is the final equilibrium molecular weight. Determination of Y for a variety of systems could contribute to a systematisation of this branch of polymer degradation which is so difficult to quantify.

Although the molecular weight decreases progressively there is little change in the nature of the molecular-weight distribution during ultrasonic degradation of polyisobutene in 1,2,4-trichlorobenzene[208]. A reduction in rate with increasing viscosity has been correlated with a reduction in cavitation[209].

7.3.5 Poly(4-methylpent-1-ene)

The products of thermal degradation of poly(4-methylpent-1-ene) at 291–341 °C have been accounted for by the same intramolecular radical-transfer process as occurs in lower polyolefins, transfer to the tertiary hydrogen atoms predominating. Isobutene and propane are the main components volatile at room temperature and only very small amounts of monomer are formed by depropagation. The sharp initial fall in molecular weight has been attributed to weak links which are probably oxygenated groups formed during storage, because deliberate peroxidation of the polymer enhances the initial sharp decrease in molecular weight[210].

7.3.6 Polydienes

In view of its long commercial history and the inherent instability resulting from its unsaturation, it is not surprising that a great deal has been known for some considerable time about the degradation reactions of natural rubber (*cis*-polyisoprene). Our knowledge of the reactions of polybutadiene is less complete. There have been a number of reviews on the subject. The most recent and authoritative by Bevilacqua[217] appeared in 1970 although it only includes material published to the end of 1965. Although there has been considerable interest in a number of aspects of the degradation of polydienes since 1965, no clearly discernible pattern of progress emerges. This may be due, in part at least, to the structural variability of these polymers as well as the complexity of commercial formulations.

In the area of pure thermal degradation, the pyrolysis-gas chromatography technique has been applied to the characterisation of the polyisoprene structure[212]. The dimer fraction from several polyisoprenes has been examined in particular and dienes have been identified which may be associated with pairs of adjacent 1,4-units, 1,4-units adjacent to 1,2-units and pairs of adjacent 1,2-units in the parent polymer. The thermal degradation of vulcanised natural rubber may be activated by mechanical stress[213].

Profound structural changes are induced in polydienes by ultraviolet radiation *in vacuo*[214]. Thus in 1,4-polybutadiene and polyisoprene, extensive *cis–trans* isomerisation occurs as well as loss of chain unsaturation and the formation of vinyl, vinylidene and cyclopropyl groups. In buta-1,2-diene the pendant vinyl groups react in pairs, probably to form one or more of the structures (15)–(17).

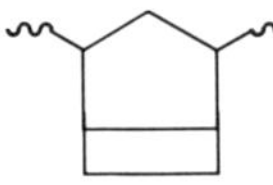

(15) 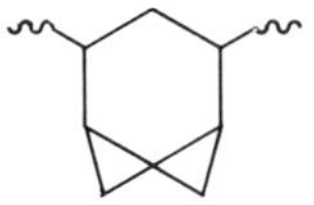 (16) 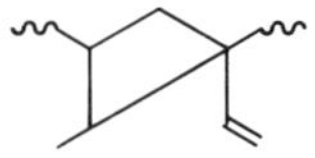 (17)

Similar reactions probably occur at the pendant vinylidene groups in high 3,4-polyisoprenes. In solution, nitroso compounds sensitise the photo-degradation of *cis*-polyisoprene by visible light[215]. Degradation, measured by a decrease in solution viscosity, is due to free radicals formed by decomposition of the sensitiser. Under irradiation by 4 MeV electrons in nitrogen, cross-linking predominates over chain scission in natural rubber[216]. The scission which occurs is principally associated with cross-links and radiation-induced and sulphur cross-links behave similarly.

Under thermal-oxidation conditions, butadiene polymers and copolymers undergo both scission and cross-linking, with scission slightly in excess[217–219]. A limiting solubility is rapidly achieved which allows a quantitative estimate of the scission/cross-linking ratio. Reaction with sulphur causes an excess of cross-linking over scission during oxidation. Formic acid is a major low-molecular weight product of oxidation and acetone and acetaldehyde are also found but these products do not provide such helpful clues to the oxidation mechanism as does levulinaldehyde in the case of polyisoprene. The formic acid was previously attributed to the oxidation of pendant vinyl groups but yields do not correlate with vinyl-group content so it is now believed to be a product of main-chain scission. An essential difference between the thermal oxidations of 1,4-polybutadiene and 1,4-polyisoprene is the frequency of intermolecular addition of peroxy radicals. In polyisoprene it is undetectable, while in polybutadiene it occurs at least one quarter as frequently as hydrogen transfer. In the thermal-oxidative and photo-oxidative degradation of ABS rubbers it is the butadiene units which are primarily attacked[220]. The inhibiting effect of thermolysed anthracene on the thermal- and photo-oxidative degradation of 1,4-polybutadiene is due in part to the presence of free radicals and partly to light absorption[221]. It is well known that the oxidation of rubber is accelerated by metallic impurities. For a wide range of rubbers those metals which undergo one-electron transfer (for example Co, Mn, Ce and Fe) are the most active accelerators, those which undergo two-electron transfer (for example Pb, Sn) are less active, while metals which cannot undergo electron transfer (for example Zn and Na) are inactive[222].

As expected, the molecular weight distribution in natural rubber is narrowed by mill mastication. However, an unexpected observation was the development of a secondary peak at the high molecular weight end of the distribution[223] which also moves to lower molecular weight and later disappears. It is believed to be due to recombination of a small proportion of radicals primarily formed in the shearing process.

Some interesting observations have also been made on some of the degradation reactions of butyl rubber. Thus, the influence of the atmospheric pollutant, nitrogen dioxide, is to cause chain scission[224]. Initially, a small concentration of particularly susceptible structures react. These 'weak links' are assumed to be associated with the isoprene content of the rubber. The ultrasonic degradation of butyl rubber has been studied in cyclohexene and toluene solutions[209]. The rate is reduced as the viscosity of the medium is increased and this may be due to the reduction of cavitation with increase in viscosity. The limiting degree of polymerisation after prolonged irradiation depends upon solvent, radiation intensity, solution viscosity and initial degree of polymerisation.

The rate of growth of ozone cracks in butyl rubbers as in butadiene–styrene rubbers is determined by the segmental mobility of the polymer. Reaction with ozone is so rapid that it does not constitute the rate-controlling step[225]. In highly unsaturated rubbers the distance of diffusion of the ozone is so small that the rate-controlling step is the motion of the polymer chains and the dependence on ozone concentration is correspondingly small. In butyl rubber, which has a low unsaturation, penetration distances are larger so that the rate exhibits a concentration dependence.

7.3.7 Polytetrafluorethylene and related polymers

The thermal degradation of polytetrafluorethylene is known to give high yields of monomer but increasing amounts of a wax fraction consisting of dimer, trimer, etc., is found due to secondary reactions if conditions are such that the monomer concentration builds up in the hot zone, as, for example, in thick layers of degrading material[226]. The influence of carbon black is to retard the evolution of volatile degradation products in the early stages of the reaction but this is followed by a small enhancement in rate[227]. The early retardation is believed to be due to termination of radical chains by the trapped radicals known to be present in carbon black, while later acceleration is due to re-initiation by scission of the composite molecules, $\sim CF_2—CH_2—C_x\sim$, thus formed.

During the period under review there has been rather more interest in degradation induced by high-energy radiation. Above 400 °C, monomer is the major product while at lower temperatures increasing amounts of a waxy fraction are formed as a result of secondary reactions of the monomer[226]. The dependence of rate on layer thickness cannot be entirely explained in terms of diffusion of monomer, however. It is believed also to be affected by the fact that appreciable proportions of small radicals diffuse from the surface thus modifying the steady-state concentration of radicals in the degrading polymer[228, 229]. Degradation by high-energy radiation in air occurs in a four-step process[230]. Thus formation of a chain side fluorocarbon radical, $\sim CF_2-\dot{C}F\sim$, is followed by reaction with oxygen to give a peroxy radical, ROO·, which is ruptured by the radiation; the chain-end radicals thus produced are converted to chain-side radicals by fluorine abstraction. Owing to the high stability of these fluorinated radicals only a very small proportion are lost during the cycle, which accounts for the great sensitivity of polytetrafluorethylene to radiation damage.

In the pyrolysis of poly(vinyl fluoride) poly(vinylidene fluoride) and polytrifluorethylene, the formation of hydrogen fluoride is accelerated by prior

γ-irradiation[231]. In each case cross-linking predominates over scission during γ-irradiation. The thermal production of hydrogen fluoride from unirradiated poly(vinyl fluoride) is almost stoicheiometric, although volatilisation proceeds to nearly 100% due to the subsequent production of benzene and other aromatics as from poly(vinyl chloride). After irradiation, however, volatilisation stops at the stoicheiometric production of hydrogen fluoride. With unirradiated poly(vinylidene fluoride) nearly the maximum residue for volatilisation of two hydrogen fluoride molecules is produced although the residue is increased by pre-irradiation.

Polychlorotrifluorethylene is *c.* 150 °C less stable than polytetrafluorethylene and also gives lower yields of monomer. Poly(1,1-dichloro-2,2-difluorethylene) is even less stable, monomer being the major product (80%) together with a black, involatile residue[232] which may be associated with the formation of cross-links during polymerisation and which is believed to have the structure, (18),

$$\overset{\wr}{\underset{\wr}{\mathrm{C}}}\mathrm{Cl{-}CF_2{-}CCl_2{-}}\overset{\wr}{\underset{\wr}{\mathrm{C}}}\mathrm{Cl}$$

(18)

7.4 METHACRYLATE AND ACRYLATE POLYMERS

7.4.1 Polymethacrylates

7.4.1.1 Thermal

The mechanism of the thermal degradation of poly(methyl methacrylate) has been firmly established for some considerable time as being a two-stage process. The reaction is initiated at unsaturated chain-terminal structures at 200 °C, while at temperatures some 50–100 °C higher initiation is predominantly by random scission. In either case the radicals so formed 'unzip' to give quantitative yields of monomer. Work reported during the period under review has confirmed this picture[233–239] and further shows that termination is bimolecular at low temperatures and changes to unimolecular at high temperatures. The most obvious interpretation of this is that pairs of radicals mutually destroy each other at low temperatures while the ultimate small radical from complete unzipping escapes from the system at high temperatures. Undoubtedly the most graphic and convincing proof of this picture of the thermal degradation of poly(methyl methacrylate) is provided by McNeill's thermal volatilisation analysis (t.v.a.) studies[235, 240] in which the effects of variables such as molecular weight, polymer preparation, unsaturation and comonomers are clearly demonstrated. Although monomer production is a general reaction of the methacrylates, ester decomposition to yield methacrylic acid and the corresponding olefin becomes possible when the alcohol residue incorporates β-hydrogen atoms. This reaction was previously shown to predominate in poly(t-butyl methacrylate)[241] but it plays a significant part even in poly(ethyl methacrylate)[240].

While the mixing of poly(methyl methacrylate) with polystyrene has no effect on the thermal degradation properties of either[136] profound changes occur in mixtures of poly(methyl methacrylate) and poly(vinyl chloride)[15,16]. As well as methyl methacrylate and hydrogen chloride, carbon dioxide, methyl chloride and anhydride structures are formed and the volatilisation characteristics of the reaction are profoundly changed. These features have been explained in terms of attack on the poly(methyl methacrylate) by chlorine atoms produced during dehydrochlorination of poly(vinyl chloride) and by reaction between methyl methacrylate ester groups and hydrogen chloride. Very much higher yields of methyl chloride are obtained when the methyl methacrylate and vinyl chloride are copolymerised[242].

7.4.1.2 Ultraviolet and high-energy radiation

The u.v.-initiated degradation of poly(methyl methacrylate) was comprehensively reviewed by Fox in 1967[2]. Chain scission occurs and volatile products appear, including methyl formate, methanol, carbon monoxide, carbon dioxide, hydrogen and methane, most of which are derived from the ester group. As the temperature is raised, increasing amounts of monomer are formed and it becomes the predominant product in molten polymer. Thus the picture has emerged of primary elimination of ester groups followed by chain scission and unzipping.

$$\sim CH_2-\underset{COOMe}{\overset{Me}{C}}-CH_2-\underset{COOMe}{\overset{Me}{C}}\sim \;\rightarrow\; \sim CH_2-\overset{Me}{\dot{C}}-CH_2-\underset{COOMe}{\overset{Me}{C}}\sim \;\rightarrow$$

$$+\cdot COOMe \rightarrow HCOOMe + MeOH + CO + CO_2$$

$$\sim CH_2-\overset{Me}{C}=CH_2 + \cdot\underset{COOMe}{\overset{Me}{C}} \rightarrow \text{monomer}$$

During the period under review some further contributions have been added to the large volume of material dealt with by Fox. Thus solutes added to poly(methyl methacrylate) irradiated in dioxane and methylene chloride solution may act as sensitisers or inhibitors of chain scission[243]. Results may be explained in terms of an electronic energy-transfer mechanism involving the lowest excited triplet levels of the polymer and added solute.

Poly(methyl methacrylate) becomes thermally unstable as a result of either u.v. or γ-irradiation, monomer being liberated at 160 °C, which is well below the normal thermal degradation temperature[244]. It appears that thermal depolymerisation is initiated at unstable structures formed during irradiation. In the γ-ray-initiated degradation of poly(methyl methacrylate) the ratio of methyl-containing volatiles to chain scission is *c.* 1[245]. In addition, it is found that ethyl mercaptan protects against chain scission but that the amount of volatile products is not affected. It is clear that as in the u.v.-initiated reaction, side-chain and main-chain scission are closely related, the former preceding and inducing the latter.

The u.v.- and high-energy radiation-induced degradation of poly(methyl methacrylate) has been observed to be slower in mixtures than in either pure syndiotactic or isotactic polymers[246]. Since there are also differences in optical density and turbidity the degradation behaviour is interpreted in terms of the formation of a 'sterio-complex'.

7.4.1.3 Mechanical

As a result of degradation carried out on a ball mill, it is suggested that the energy required for chain rupture in poly(methyl methacrylate) is accumulated in the bonds of the main chain through conversion of elastic deformations to heat[247]. In methyl methacrylate–styrene copolymers, degraded by drilling at liquid nitrogen temperatures, poly(methyl methacrylate) radicals are primarily formed. These decay to produce styrene radicals[248]. This transfer reaction depends upon the environment in which the monomer units find themselves and especially upon the proximity of one to the other. Thus the rate of decay decreases in the order: random copolymer > block copolymer > homopolymer mixture.

7.4.1.4 Hydrolysis

Acid hydrolysis of isotactic poly(methyl methacrylate) proceeds smoothly to 100% conversion whereas in syndiotactic polymer the reaction is limited to 85%. In the former, the reaction tends to be autocatalytic and monomer units tend to react in blocks while in the latter, monomer units react at random[249, 250]. In the isotactic polymer it is clear that hydrolysis is activated by neighbouring acid groups while in syndiotactic polymer the residual 15% of unreacted units are deactivated by flanking acid residues.

There is also a strong neighbouring-group effect in alkaline hydrolysis[251]. Whereas methyl acrylate units are hydrolysed completely, only 9% of the ester groups react in poly(methyl methacrylate) although this proportion increases in copolymers with methyl acrylate or styrene. It is believed that only those methyl methacrylate units at the centre of isotactic triads are susceptible in the homopolymers but that any methyl methacrylate unit attached to a methyl acrylate or a styrene unit can be hydrolysed in the copolymer. Benzyl and ethyl methacrylate homopolymers and copolymers with methyl acrylate behave similarly[252], but in copolymers of methyl methacrylate and benzyl methacrylate the methyl methacrylate units are protected by benzyl methacrylate and the benzyl methacrylate units sensitised by methyl methacrylate.

7.4.2 Polyacrylates

In spite of their close similarity in structure, the thermal degradation behaviours of the polymethacrylates and polyacrylates are quite different. The acrylates yield insignificant amounts of monomer and large amounts of chain fragments. It is presumably because of the analytical difficulties that more detailed studies of the acrylates have been neglected until quite recently. In the thermal degradation of poly(methyl acrylate) at 286–310 °C, rate curves and changes in molecular weights both suggest that random scission of the polymer chain occurs. This is followed by a radical process involving intra- and inter-molecular transfer. Approximately 90% of the products comprise chain fragments but carbon dioxide and methanol are significant minor products, and some cross-linking occurs as well as chain scission[253]. The formation of carbon dioxide, methanol and chain fragments are all inhibited by radical inhibitors so these products must all be formed in competing reactions of the radical[254].

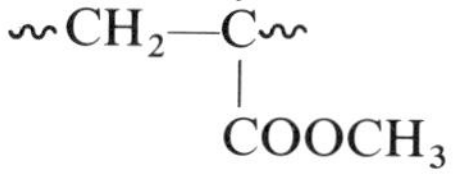

$$\sim CH_2-\dot{C}(COOCH_3)\sim$$

The cross-linking reaction which leads to gel formation is rather difficult to account for. The conditions of polymerisation affect the amount of gel formed and it has been proposed that it is associated with the degree of branching of the polymer so that degree of conversion and molecular weight are both significant variables[255]. The thermal degradation of poly(benzyl acrylate) at 260–300 °C follows essentially the same pattern[256].

The major decomposition reaction of poly(acrylic acid) in the temperature range 25–150 °C is elimination of water between pairs of adjacent units to form cyclic poly(acrylic anhydride). At high temperatures, carbon dioxide is liberated and unsaturation appears[257, 258]. This general behaviour is similar to that of poly(methacrylic acid) but poly(methacrylic anhydride) appears to be rather more stable. By heating copolymers of acrylic acid and ethylene, this reaction has been used to introduce unsaturation into a polyethylene structure which then becomes much more susceptible to thermal and oxidative degradation[259].

While poly(methyl acrylate) and poly(ethyl acrylate) give products rather similar to those from the corresponding methacrylates during u.v. irradiation[2], poly(t-butyl acrylate) liberates isobutene in an ester decomposition as in its thermal decomposition[260]. A photo-induced rotational equilibrium is established between *cis*- and *trans*-ester spacial configurations; isobutene is liberated from the *cis* form[261].

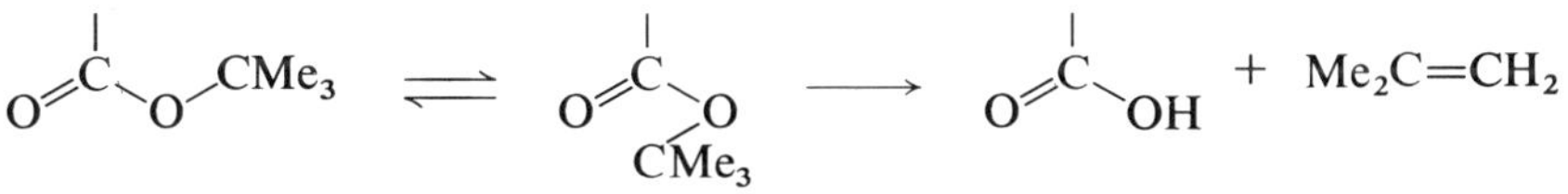

$$O{=}C(-)-O-CMe_3 \rightleftharpoons O{=}C(-)-O(CMe_3) \longrightarrow O{=}C(-)-OH + Me_2C{=}CH_2$$

In aqueous acetonitrile solution, α-cyanoacrylate polymers undergo hydrolytic scission of the polymer chain; the principal volatile product is formaldehyde[262]. The reaction is faster in alkaline solution and is believed to proceed as follows:

$$\sim CH_2-\underset{COOR}{\overset{CN}{C}}-CH_2-\underset{COOR}{\overset{CN}{C}}\sim + OH^- \rightarrow \sim CH_2-\underset{COOR}{\overset{CN}{C}}-CH_2OH + {}^-\underset{COOR}{\overset{CN}{C}}\sim$$

$$\downarrow OH^-$$

$$\sim CH_2-\underset{COOR}{\overset{CN}{C^-}} + CH_2(OH)_2$$

7.4.3 Copolymers of methyl methacrylate

7.4.3.1 Copolymers of methyl methacrylate and acrylonitrile

The superficial effects of copolymerised acrylonitrile on the thermal degradation of poly(methyl methacrylate) are quite different at 220 and 280 °C. At 220 °C, the unzipping process from chain ends cannot pass through acrylonitrile units which thus act as a very efficient inhibitor. Random scissions of methyl methacrylate segments can occur slowly but molecules rather than radicals are produced. At 280 °C, two fundamental changes occur. Firstly, random scission is effectively into radicals which depropagate immediately and, secondly, depropagation can pass through acrylonitrile units which are liberated as monomer. Thus acrylonitrile no longer acts as an inhibitor at the higher temperature[263]. In the u.v. degradation of methyl methacrylate–acrylonitrile copolymers at 160 °C chain scission occurs preferentially at acrylonitrile units[264]. The resultant radicals depropagate to, but cannot pass through, the first acrylonitrile unit. Thus, in the photo-reaction, acrylonitrile acts both as an accelerator (to chain scission) and an inhibitor (to unzipping).

7.4.3.2 Copolymers of methyl methacrylate and methyl acrylate

Stabilisation of poly(methyl methacrylate) by copolymerised methyl acrylate is due to partial blockage of the depropagation process by methyl acrylate units[265, 266]. Thus degradation only occurs at temperatures at which poly-(methyl methacrylate) devoid of terminal unsaturation degrades (260–330 °C). This involves random scission followed by depropagation. When depropagation reaches an isolated unit of methyl acrylate, there is competition between depropagation, intramolecular transfer and intermolecular transfer which result in methyl acrylate monomer, large-chain fragments and chain scission, respectively. Depropagation is unlikely to occur from a radical chain end which comprises a sequence of more than one methyl acrylate unit. Yellow coloration is associated with carbon–carbon conjugation in the chain backbone and it seems probable that the minor products, hydrogen and methane, are liberated from adjacent monomer units in a reaction of the polymer radicals which competes with chain scission. The complete absence of methanol is surprising in view of its abundance as a product of thermal degradation of poly(methyl acrylate).

The principal characteristics of the u.v.-initiated reaction at 170 °C are similar to those of the thermal reaction. Minor differences can be accounted for in terms of the mechanism proposed for the thermal reaction bearing in mind the differences in the temperatures at which the two investigations were carried out[267].

7.4.3.3 Copolymers of methyl methacrylate and α-chloroacrylonitrile

α-Chloroacrylonitrile introduces drastic instability into poly(methyl methacrylate) molecules[65]. Dehydrochlorination and chain scission to radicals occurs at α-chloroacrylonitrile units at 150 °C; the radicals unzip to give a mixture of the monomers although with proportionally less α-chloroacrylonitrile than in the copolymer.

7.5 POLYMERS WITH HETEROATOMS IN THE CHAIN BACKBONE

7.5.1 Polyethers

7.5.1.1 Introduction

The degradation reactions of the lowest member of the polyether series, poly(methylene oxide) (more commonly polyoxymethylene or polyformaldehyde) are characterised by high monomer (formaldehyde) yields which are clearly formed in unzipping reactions closely akin to those of the methacrylates. The degradation reactions of the higher members of the series, poly(ethylene oxide), poly(propylene oxide), poly(tetramethylene oxide), etc., are very much more closely akin to those of polyethylene in which chain fragments other than monomer are formed. It is for this reason that polyethers will be discussed under the two headings, polyoxymethylene and higher polyethers.

7.5.1.2 Polyoxymethylene

The thermal- and photo-oxidative degradation of polyoxymethylene, $\sim CH_2—O—CH_2—O\sim$, was adequately reviewed to the end of 1965 by Dudina and his colleagues[268]. The thermal depolymerisation process proceeds from the chain-terminal hydroxyl groups in the parent polymer at 150 °C but the polymer is a good deal more stable when the molecules are 'end capped' by acetate or other groups[269] and initiation then occurs by random chain scission. Thermal oxidation, which has been studied at 160–180 °C, is a random oxygen-initiated depolymerisation. There had been contradictory evidence as to whether these reactions proceed by ionic or radical mechanisms and Dudina and his colleagues[268] concluded that it 'seems justified to suggest that ions and radicals are simultaneous active centres'.

A comparative study of the thermal degradation of hydroxyl- and acetyl-terminated polymers has been made[270, 271]. Radical inhibitors have no effect on the reaction, minor products to be expected in a radical process could not be detected and the reaction is very sensitive to basic catalysis by methoxide, acetate and even chloride ions. Thus, while an ionic mechanism undoubtedly occurs in the presence of basic catalysts, a molecular mechanism must be preferred on energetic grounds in the uncatalysed reaction. Later workers[272], were unable to obtain any data which contradicted this.

The products of the thermal oxidation of polyoxymethylene are formaldehyde, water, formic acid, methyl formate, trioxane and other larger fragments of the macromolecule[273]. As a result of studies of both the polymer[274] and model compounds[275], it is clear that this is a radical reaction involving peroxide radicals. At 190 °C, a reduction in molecular weight occurs even at very low oxygen pressures (0–5 mmHg)[276]. However, no formaldehyde is formed in the oxygen-pressure range 0–25 mmHg. This means that with insufficient oxygen the chain scission results in new chain-terminal structures which have good heat resistance. It is thus assumed that a terminal radical resulting from chain scission, $\sim\dot{O}CH_2$, is converted to the stable $\sim OCH_3$ structure at low oxygen pressures but becomes oxidised and is ultimately converted to $\sim O—CH_2—OH$ at high oxygen pressures. Only when these terminal hydroxyl groups are formed does formaldehyde appear as a product. Sukhov and his colleagues[276] present a very plausible and comprehensive mechanism which accounts for all the principal experimental facts.

Under 2537 and 3650 Å radiation *in vacuo* at ambient temperatures both hydroxyl and 'end-capped' polyoxymethylenes evolve hydrogen and carbon monoxide in a molar ratio of 3.5–4.0[277]. The necessary chromophore for absorption of radiation is provided by the ester carbonyl group in the 'end-capped' material but there are no comparable absorbing structures in hydroxyl-terminated polymer. The production of carbon monoxide and hydrogen may be explained in similar terms to their production during vacuum photolysis of cellulose, namely by decomposition of terminal hydroxyl groups,

$$\sim CH_2—O—CH_2—OH \rightarrow H_2 + \sim CH_2—O—CH{=}O \rightarrow \sim CH_2—O\cdot + CO + 0.5\,H_2$$

Additional hydrogen may be produced by elimination between neighbouring molecules in a cross-linking reaction.

Irradiation in air produces well-defined changes in the hydroxyl and carbonyl regions of the infrared spectrum[277]. Radiation at 2537 Å is *c.* 100 times more effective than at 3650 Å. Carbonyl absorption is resolved into three components at 1750, 1785 and 1815 cm^{-1}. These are attributed speculatively to the structures, (19)–(21).

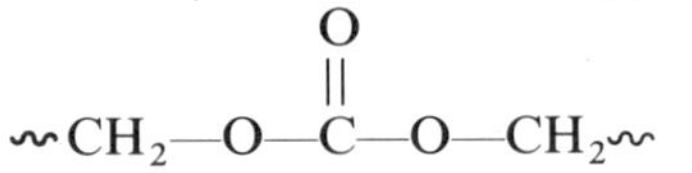

$$\sim CH_2—O—\overset{\overset{O}{\|}}{C}—O—CH_2\sim \quad (19)$$

$$\sim CH_2—O—\overset{\overset{O}{\|}}{C}—O—\overset{\overset{O}{\|}}{C}—O—CH_2\sim \quad (20)$$

$$\sim O—\overset{\overset{O}{\|}}{C}—O—\overset{\overset{O}{\|}}{C}—O—\overset{\overset{O}{\|}}{C}\sim \quad (21)$$

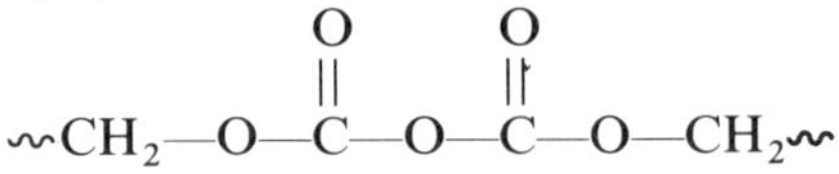

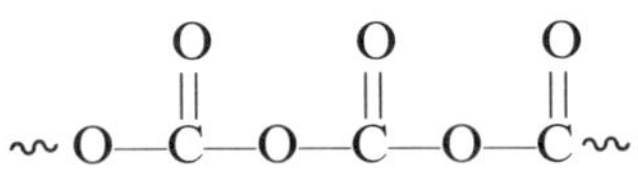

Under 3650 Å radiation, the concentration of (19) increases autocatalytically while the two anhydrides increase linearly. On the other hand, a steady-state concentration is rapidly built up under 2537 Å radiation. A mechanism is proposed in which hydroperoxides are formed at methylene groups. These hydroperoxides eliminate water giving the carbonyl groups which may be photolysed by 2537 Å but not 3650 Å radiation.

7.5.1.3 Higher polyethers

The thermal degradation of poly(tetramethylene oxide) (22), poly(hexamethylene oxide) (23) and polydioxolane (24) all proceed by a radical chain mechanism[278].

$$\sim O-(CH_2)_4-O-(CH_2)_4\sim \qquad \sim O-(CH_2)_6-O-(CH_2)_6\sim$$

(22) (23)

$$\sim CH_2-CH_2-O-CH_2-O\sim$$

(24)

Initiation occurs exclusively at the carbon atom in the α-position relative to the ether linkage and leads to the destruction of the ether bonds.

$$\sim CH_2-CH_2-CH_2-\dot{C}H-O-CH_2-CH_2\sim \rightarrow \sim CH_2-CH_2-CH_2-CHO+\cdot CH_2-CH_2\sim$$

The volatile products consist principally of C_1–C_4 alkanes and alkenes, acetaldehyde, propionaldehyde, butyraldehyde, tetrahydrofuran and hydrogen.

The thermal oxidations of poly(ethylene oxide) (25), poly(propylene oxide) (26) and poly(tetramethylene oxide) (22)[279, 280] are also radical chain processes.

$$\sim O-CH_2-CH_2\sim \qquad \sim O-CH_2-CHMe\sim$$

(25) (26)

In (25), hydroperoxides are the primary reaction products but in (26) decomposition of the peroxy radical occurs simultaneously. Oxidation mechanisms have been proposed which account for the main products and whose principal features are similar to those outlined above for polyoxymethylene.

Ozonisation of non-crystalline samples of (26) followed by reduction with lithium aluminium hydride gives dipropylene glycol but the di-primary (27), di-secondary (28) and primary–secondary (29) compounds have been separated by g.l.c.[281].

$$(HO-CH_2-\underset{\displaystyle Me}{\underset{|}{C}}H_2)_2O \qquad (HO-\underset{\displaystyle Me}{\underset{|}{C}}H-CH_2)_2O$$

(27) (28)

$$HO-CH_2-\underset{\displaystyle Me}{\underset{|}{C}}H-O-CH_2-\underset{\displaystyle Me}{\underset{|}{C}}H-OH$$

(29)

The presence of (27) and (28) demonstrates that some head-to-head, tail-to-tail structures must have been present in the original polymer.

In the mechanical degradation of poly(ethylene oxide) by high-speed stirring[282] at 25–40 °C in benzene solution with thiophenol as a radical acceptor, the ultimate number of bonds broken per chain is independent of concentration of polymer but increases with stirring rate and in poorer solvents.

7.5.2 Polysiloxanes

As the temperature is raised, the first observable change in polydimethylsiloxane *in vacuo* is an increase in molecular weight which occurs at 170–300 °C and which has been associated with intermolecular condensation of terminal hydroxyl groups[283–285]. The principal volatile products, which appear at higher temperatures, are methane and cyclic oligomers, principally trimer (30) and tetramer, which are also associated with reaction at the chain terminal hydroxyl groups[284–286].

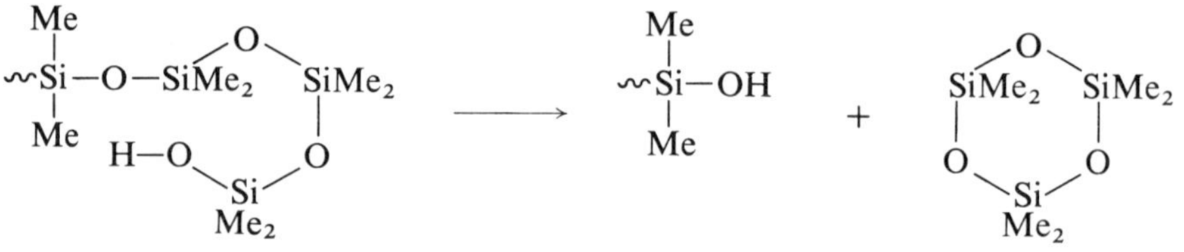

(30)

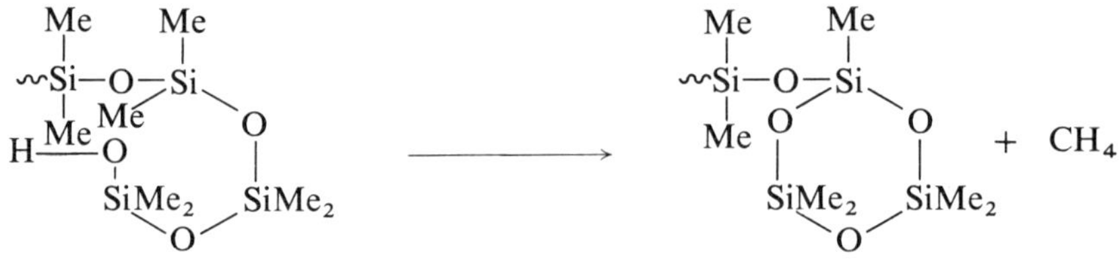

+ CH_4

The reduction in the rate of volatilisation which is observed as the reaction proceeds is associated with the decrease in concentration of hydroxyl groups so that these two processes may be regarded respectively as the propagation and termination steps of a molecular depolymerisation reaction. Depolymerisation to cyclic structures of this kind must be favoured by the spiral structure of polysiloxane molecules[284]. The importance of end groups is emphasised by the stabilising effect of terminal methyl groups[284, 285] and ONa, OK and SO_3H terminal structures have also been investigated[287]. Low molecular weight cyclic oligomers are also produced from polymers of 3,3,3-trifluoropropylmethylsiloxane (31)[288] but in addition the silicon–carbon bond is broken and fluorine is transferred from the γ-position to the silicon atom,

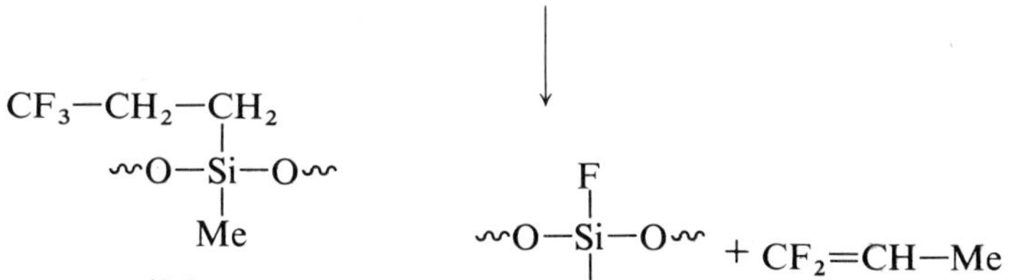

$$CF_3-CH_2-CH_2-Si(Me)(\sim O-)(-O\sim) \;(31) \longrightarrow \sim O-Si(F)(Me)-O\sim + CF_2{=}CH-Me$$

(31)

The thermal stability is also increased by the incorporation of heteroatoms such as B (32), Al, P, Ti, V, in the siloxane chain[289–291],

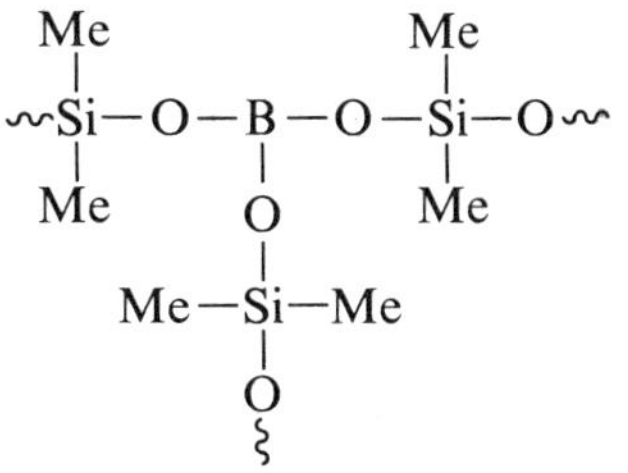

$$\sim Si(Me)_2-O-B(-O-Si(Me)_2-O\sim)-O-Si(Me)_2-O\sim$$

(32)

although the ultimate volatile degradation products are the same as for the parent material. A similar stabilising effect is observed with certain organometallic additives, for example, tetrabutoxytitanium and tributoxyaluminium[284, 285]. These compounds act as cross-linking agents so that heteroatoms become incorporated into the silicone structure. The stabilising effect is greater, the greater the metal–oxygen bond energy.

There have been a number of interesting isolated observations on degradation reactions of silicon-containing polymers. Thus, cyclo-linear-polyphenylsiloxane, (33),

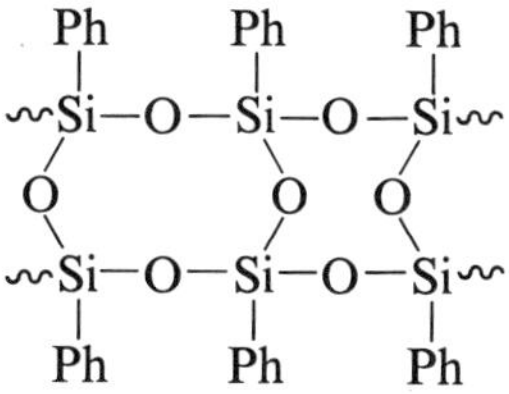

(33) 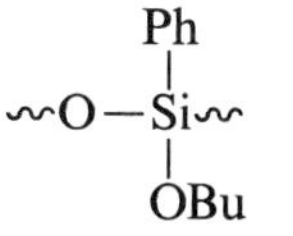 (34) 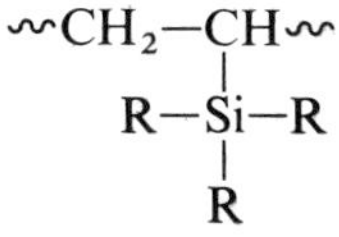 (35)

$$\sim O-Si(Ph)(OBu)\sim \qquad \sim CH_2-CH(SiR_3)\sim$$

which is relatively highly thermally stable liberates benzene as a principal volatile product[292] of thermal degradation. On the other hand, under thermal oxidative conditions, polyphenylbutoxysiloxane (34) degrades in three well-defined steps[293]. Up to 400 °C oxidation of the butoxy groups occurs in a radical chain reaction involving hydroperoxide and which ultimately results in hydroxyl groups attached to the silicon atoms. At 400–600 °C, cross-linking occurs by condensation of hydroxyl groups, and finally, above 600 °C, the phenyl groups are oxidised. Polymers have been prepared by the action of a glow discharge on octamethyltrisiloxane and other low molecular weight siloxanes[294]. On heating these polymers, cross-linking occurs at 150–300 °C but heating in vacuum to 500 °C does not cause appreciable degradation. In air at 475 °C the organic part of the structure is completely oxidised and a residue of silicon oxides remains. In the photolysis (2537 Å) of mixed phenyl–methyl polysiloxane the principal volatile product is hydrogen with smaller amounts of methane and benzene. Increased phenyl content results in an increase in stability[295].

In the ultrasonic degradation of polydimethylsiloxane in toluene solution the rate of bond scission increases with dilution and power input[296]. The thermal stabilities of some poly(vinyl trialkylsilanes) (35) have been measured. The stability increases in the order n-butyl < ethyl < methyl and a marked increase in stability occurs when alkyl groups are replaced by phenyl[297].

7.5.3 Cellulose and cellulose derivatives

There have been some rather isolated investigations of thermal-[298, 299] and thermal-oxidative[300] reactions of cellulose.

Under ultraviolet radiation, cleavage of the pyranose ring occurs at the 1,2-position in cellulose acetate[301]. In the presence of air and moisture the process of decomposition is increased due to photo-oxidation and hydrolysis. In oxygen, peroxides, carboxyl- and carbonyl-groups accumulate and the volatile products in order of abundance are carbon dioxide, carbon monoxide, methane, water, ethane, acetone as well as formaldehyde. The same products are formed by irradiation *in vacuo* although in different relative amounts. Reaction proceeds by a radical chain mechanism, the initially formed aceto- and acetoxy radicals decomposing to give the three major products, carbon dioxide, carbon monoxide and methane (from the methyl radicals).

$$\cdot COCH_3 \rightarrow CO + \cdot CH_3; \quad \cdot OCOCH_3 \rightarrow CO_2 + \cdot CH_3$$

Arabinose units are formed *in vacuo* and xylose units in oxygen.

Methyl cellulose undergoes random chain scission under ionising radiation[302]. Carboxy and carboalkoxy radicals are formed leading to carboxylic acids and esters, the formation of the latter being linearly related to chain cleavage.

Cyanoethyl cellulose proves to be more resistant to thermal oxidation than other cellulose ethers[303]. During reaction at 150 °C, peroxides are formed and decompose. These peroxides form on the second carbon atoms of the cyano-ethoxyl group and their decomposition facilitates the formation of nitrile radicals and thus of hydrogen cyanide which occurs as a trace product. Acrylonitrile, water, carbon dioxide and traces of ammonia and formaldehyde are also formed.

7.6 POLYMERS WITH AROMATIC GROUPS IN THE CHAIN BACKBONE

7.6.1 Introduction

The most obvious common feature of the degradation reactions of polymers incorporating aromatic groups in the chain backbone is that these aromatic groups are invariably the most stable parts of the molecule. Thus attention inevitably becomes focused upon the inter-aromatic groups. In order to be allowable in the structure of a commercial polymer these inter-aromatic groupings must have reasonable thermal stability as well as the ability to bring to the polymer chain that degree of flexibility which will allow the development of useful physical properties. The number of possible structures is thus fairly limited and although the various aromatic polymers to be considered in this section may appear rather diverse, they are brought together so that the importance of inter-aromatic groupings on polymer stability may be emphasised. Polyamides are included because a great deal of current interest is focused upon the aromatic polyamides.

7.6.2 Poly(ethylene terephthalate)

In the thermal oxidation of poly(ethylene terephalate) it is the methylene groups adjacent to the low concentrations of ether linkages, formed by the occasional condensation of pairs of ethylene glycol molecules during the preparation of the polymer, which are most susceptible to attack[304]. Hydroperoxidation occurs at 130 °C. It is only above 200 °C that oxidation of the true poly(ethylene terephthalate) structure occurs, initiated by hydroperoxidation at the methylene groups of the ethylene glycol residues. The ultimate effect is chain scission and the production of water and the expected variety of acidic materials which, in turn, contribute to the further degradation of the polyester.

An interesting extension to earlier work on the purely thermal degradation of poly(ethylene terephthalate) concerns polymers prepared by condensation of terephthalic acid with the bisphenols of the type (36), (37), (38) and (39)[305–307]

$HO-C_6H_4-CH_2-C_6H_4-OH$ (36)

$HO-C_6H_4-C(Me)(Ph)-C_6H_4-OH$ (37)

$HO-C_6H_4-C(Ph)(Ph)-C_6H_4-OH$ (38)

$HO-C_6H_4-C(Me)(Me)-C_6H_4-OH$ (39)

In the region 275–350 °C cross-linking is dominant while above 350 °C scission becomes increasingly important and volatile products appear. Carbon monoxide and carbon dioxide are produced as a result of homolytic scission at the ester group but a certain amount of decomposition in the bisphenol units is evidenced by small amounts of hydrogen and methane, depending upon the structure of the bisphenol. The aromatic constituents on the central carbon atom have a very positive stabilising influence.

The photolysis of poly(ethylene terephthalate) by 2537 and 3130 Å radiation can be explained in terms of the reactions,

$$\sim\!C_6H_4\!-\!\overset{O}{\overset{\|}{C}}\!-\!O\!-\!CH_2\!-\!CH_2\!-\!O\!-\!\overset{O}{\overset{\|}{C}}\!-\!C_6H_4\!\sim \;(36) \longrightarrow \sim\!C_6H_4\cdot\;(40) + CO + \cdot O\!-\!CH_2\!-\!CH_2\!\sim\;(41)$$

$$(36) \longrightarrow \sim\!C_6H_4\cdot + CO_2 + \cdot CH_2\!-\!CH_2\!\sim\;(42)$$

In addition to the carbon monoxide and carbon dioxide, the ultimate products can be accounted for by combination, disproportionation and hydrogen abstraction by the radicals (40), (41) and (42)[308]. That the radical (43) may be an important intermediate is indicated by the identification of structures (44) and (45) during the exposure of poly(ethylene terephthalate) to ultraviolet radiation in air.

$$\sim\!O\!-\!\overset{O}{\overset{\|}{C}}\!-\!\dot{C}_6H_3\!-\!\overset{O}{\overset{\|}{C}}\!-\!O\!\sim \quad (43) \qquad \sim\!O\!-\!\overset{O}{\overset{\|}{C}}\!-\!C_6H_3(OH)\!-\!\overset{O}{\overset{\|}{C}}\!-\!O\!\sim \quad (44)$$

$$\sim\!O\!-\!\overset{O}{\overset{\|}{C}}\!-\!C_6H_2(OH)_2\!-\!\overset{O}{\overset{\|}{C}}\!-\!O\!\sim \quad (45)$$

γ-Irradiation of poly(ethylene terephthalate) results in radicals (46), (47) and another observable in the e.s.r. spectrum and as yet unassigned[309, 310].

$$\sim\!C_6H_4\!-\!\overset{O}{\overset{\|}{C}}\!-\!O\!-\!\dot{C}H\!-\!CH_2\!\sim \quad (46) \qquad \sim\!\overset{O}{\overset{\|}{C}}\!-\!\dot{C}_6H_3\!-\!\overset{O}{\overset{\|}{C}}\!\sim \quad (47)$$

In the amorphous regions the decay of these radicals is accounted for by the physical movement and mutual destruction even below the glass transition temperature but in the crystalline regions decay occurs at 100 °C, which is so far below the melting point (260 °C), it is necessary to invoke 'radical-site hopping' by hydrogen atoms. Conversion of (47) to (46) on a 1:1 basis can be induced by the radiation from laboratory fluorescent lighting at ambient temperatures[312].

Campbell and Turner[311, 313] have attempted to separate the cross-linking and scission reactions which occur during γ-irradiation of poly(ethylene terephthalate) by using oxygen as a radical scavenger to suppress cross-linking. They were able to demonstrate a 1:1 relationship between chain scission and the formation of carboxyl groups but, as well as being a radical scavenger, it was clear that oxygen is also involved in a radiation-induced oxidation reaction.

In inert media poly(ethylene terephthalate) is mechanically degraded in a vibratory mill[314], scission to radicals occurring at the ∿O—CH_2∿ linkage. Traces of moisture favour a mechanochemically activated hydrolysis.

7.6.3 Polyamides

Current interests in the applications of polyamides are reflected by papers on the degradation of poly(caproamide) (nylon 6)[315,316], aromatic polyamides[317] and piperazine polyamides[318,319].

Poly(caproamide) is shown to give the monomer, caprolactam, as the principal product of thermal degradation[315] while wavelengths in the region of solar radiation cause scission at the amide bond although only bonds in the amorphous regions are susceptible[316].

Comparison has been made of the thermal stabilities of the aromatic polyamides made by condensation of *m*- and *p*-phenylene diamines with iso- and terephthalic acids[317]. The order of stability is *p*,tere- > *m*,tere- > *p*,iso- > *m*,iso- and this order is closely associated with rigidity and softening point. The principal products of degradation of all four polymers are hydrogen, carbon monoxide, carbon dioxide, water, benzene, toluene and benzonitrile, although the relative amounts differ from polymer to polymer. Carbon dioxide, water and benzene demonstrate intense hydrolytic reaction of the amide bond and subsequent decarboxylation of the resulting carboxyl end group.

The stabilities of a variety of piperazine aromatic polyamides have been compared[318,319]. Methyl substitution in the piperazine ring decreases the thermal stability while terephthaloyl units increase stability compared with isophthaloyl units. Thermal degradation is associated with random scission at amide groups resulting in large quantities of carbon monoxide but there is some competition by hydrolytic processes although to a much smaller extent than in aliphatic polyamides.

7.6.4 Polycarbonates

Earlier work on the degradation of polycarbonates has recently been reviewed by Davis and Golden[320]. Whether the commercial material, poly[2,2-propane-bis(4-phenyl carbonate)] (48) gels or undergoes chain scission during thermal treatment depends upon both the method of preparation[321] and the way in which degradation is carried out[322]. Thus, if the products of degradation are continuously removed then further condensation and gelation occurs, but if they are allowed to accumulate they lead to chain scission. The principal volatile products of thermal degradation, which has been studied in the range 300–389 °C, are carbon dioxide and bisphenol A. The carbonate group is the most susceptible part of the molecule and rearranges to form a pendent carboxyl group in the *ortho* position to an ether linkage.

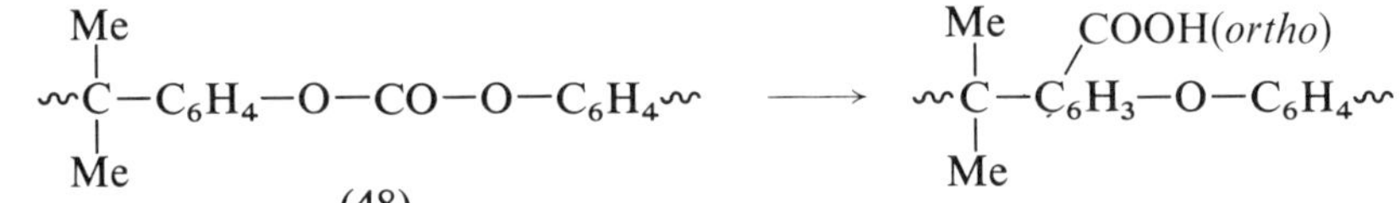

$$\sim\!\!\underset{\text{Me}}{\overset{\text{Me}}{\text{C}}}-C_6H_4-O-CO-O-C_6H_4\!\sim \longrightarrow \sim\!\!\underset{\text{Me}}{\overset{\text{Me}}{\text{C}}}-\overset{\text{COOH}(ortho)}{C_6H_3}-O-C_6H_4\!\sim \quad (48)$$

Subsequent reactions lead to the principal products. A small amount of methane is obviously associated with breakdown at the propyl group.

The thermal oxidation of polycarbonate occurs at a very much lower temperature and involves both the carbonate and propyl groups[321]. Oxidation of the methyl groups proceeds by a hydroperoxidation mechanism while water, one of the ultimate products of oxidation, leads to hydrolysis of the carbonate structures. The thermal oxidation process is accelerated by small amounts of sodium chloride which reacts with phenolic products or end groups to give hydrogen chloride which accelerates the hydrolysis of carbonate linkages[323]. The effect may be eliminated by the addition of stabilisers which react with hydrogen chloride. Substitution of a methyl group by phenyl markedly increases oxidation stability while introduction of methyl substituents on the aromatic rings *ortho* to the carbonate group increases the hydrolytic stability[324]. Infrared and n.m.r. spectral measurements demonstrate that high-temperature oxidation of these phenyl-substituted polycarbonates proceeds by scission of carbonate linkages, removal of phenyl substituents at the central carbon atom and the emergence of aldehyde, ketone and quinonoid carbonyl groups[324].

Under γ-irradiation, scission occurs at the carbonate group with liberation of carbon monoxide and carbon dioxide. Under u.v., however, radiation is absorbed in the surface skin which becomes insoluble as a result of cross-linking through the aromatic groups.

7.6.5 Phenol–formaldehyde resins and related polymers

The thermal degradation of phenol–formaldehyde resins stops after a certain degree of decomposition which increases with increasing degradation temperature in the range 250–400 °C[325]. This is due to additional structures being formed which increases the network rigidity. The principal volatile products of degradation are xylene (76 %), phenol (10 %), cresol (10 %), with smaller amounts of benzene and other aromatics. It has been suggested that the following reactions are involved.

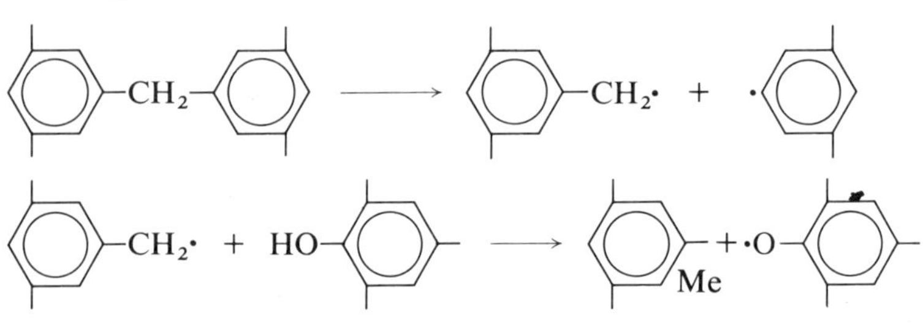

$$\text{Ar}-CH_2-\text{Ar} \longrightarrow \text{Ar}-CH_2\cdot + \cdot\text{Ar}$$

$$\text{Ar}-CH_2\cdot + HO-\text{Ar}- \longrightarrow \text{Ar}(\text{Me})- + \cdot O-\text{Ar}-$$

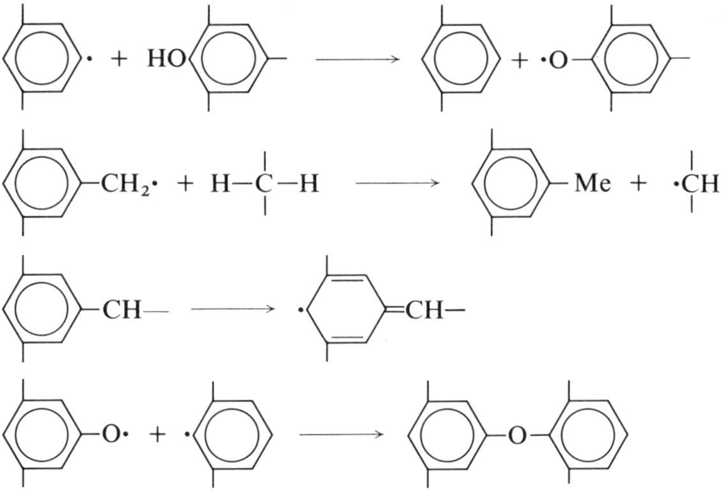

Under thermal oxidative conditions a good deal of purely thermal degradation products are still formed but methylene and methylol structures are also oxidised to carbonyl and carboxyl groups, respectively, and carbon monoxide, carbon dioxide and benzaldehyde appear among the volatile products[326, 327].

A resin prepared by condensation of cyanuric acid, phenol and formaldehyde[328] is more stable than phenol–formaldehyde resin below 330 °C but less stable above 330 °C. A sharp decrease in carbonyl absorption shows that the isocyanuric ring is rapidly destroyed at the higher temperatures.

Di(chloromethyl)benzene may be condensed with aromatics like benzene, naphthalene, phenanthrene[329] or a variety of heterocyclics[330] in the presence of Friedel–Crafts catalysts to form polymers which are structurally similar to phenol–formaldehyde resins in the sense that they comprise alternate aromatic and methylene groups. On thermal degradation in nitrogen, the carbocyclic polymers give a lower residual char than phenol–formaldehyde resins but the yield of char increases with the molecular weight of the aromatic unit[329]. Among the heterocyclic polymers there is evidence that thiophene gives a more stable polymer than benzene[330].

7.6.6 Poly-*p*-xylylene

Poly-*p*-xylylene (49) is also structurally related to the phenol–formaldehyde

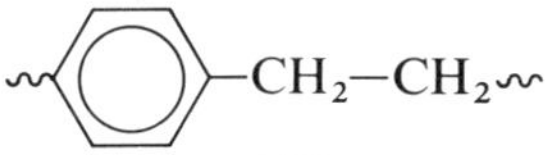

(49)

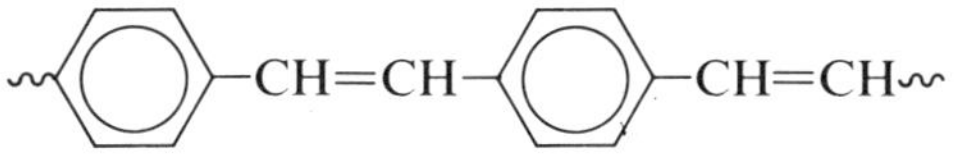

(50)

resins. In the thermal degradation, two parallel but independent processes occur[331]. Firstly, scission at inter-methylene carbon–carbon bonds is followed by depolymerisation to give dimer to pentamer with a predominance of tetramer but no monomer. Secondly, dehydrogenation of the methylene groups leads to a conjugated structure (50). The reaction seems to occur preferentially at abnormal structures[332] which may be branch points.

7.6.7 Miscellaneous aromatic polymers

The major volatile product of thermal degradation of polyphenylene (51) is hydrogen together with a carbon char. Polyphenylene oxide (52) gives

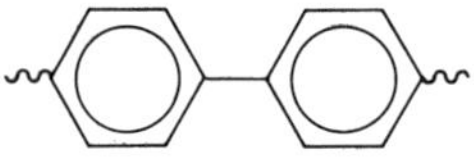

$$\sim C_6H_4-C_6H_4\sim \quad \sim C_6H_4-O\sim \quad \sim C_6H_4-S\sim \quad \sim C_6H_4-SO_2\sim$$

(51) 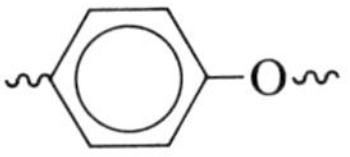 (52) 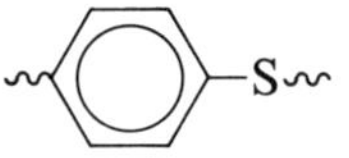 (53) 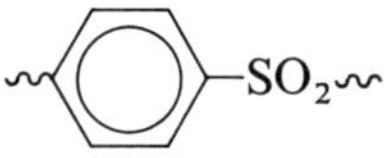 (54)

low molecular weight fragments and hydroxyl end groups while polyphenylene sulphide (53) gives mainly dimeric and trimeric chain fragments and hydrogen sulphide and there is almost complete removal of sulphur as sulphur dioxide from polyphenylene sulphone (54)[333]. Perhaps surprisingly (53) is more thermally stable in both inert and oxidising atmospheres than its fully fluorinated analogue[334].

In the photolysis of a phenoxy-resin with the structure (55)[335, 336] hydrogen,

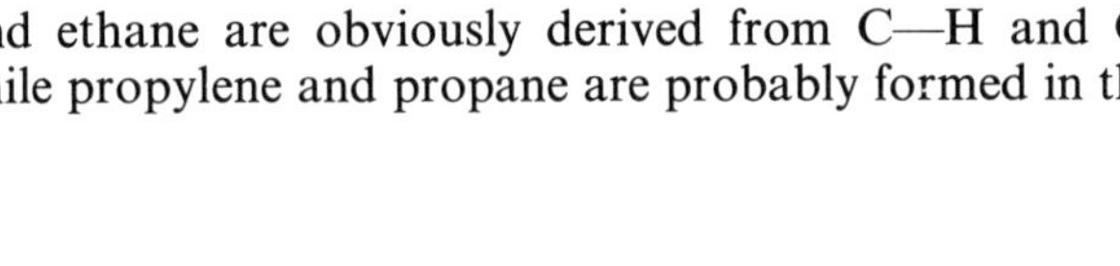

$$\sim C_6H_4-C(Me)_2-C_6H_4-O-CH_2-CH(OH)-CH_2-O\sim$$

(55)

methane and ethane are obviously derived from C—H and C—C bond cleavage while propylene and propane are probably formed in the sequence

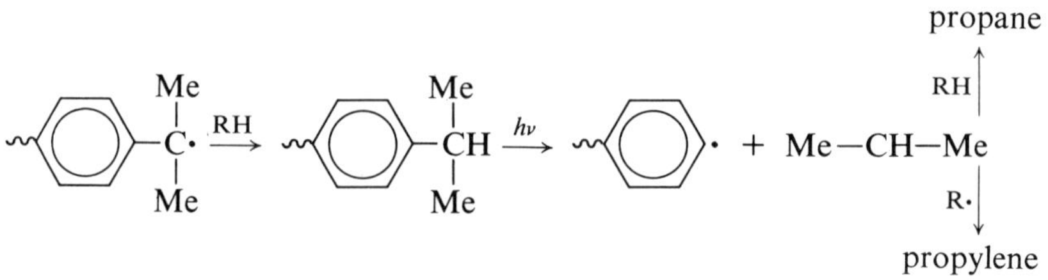

$$\sim C_6H_4-\dot{C}(Me)_2 \xrightarrow{RH} \sim C_6H_4-CH(Me)_2 \xrightarrow{h\nu} \sim C_6H_4\cdot + Me-\dot{C}H-Me$$

$$Me-\dot{C}H-Me \xrightarrow{RH} \text{propane}; \quad Me-\dot{C}H-Me \xrightarrow{R\cdot} \text{propylene}$$

These products are also formed during photo-oxidation but with acetone and oxides of carbon in addition.

$$\sim\!C_6H_4\!-\!\dot{C}Me_2 \xrightarrow{O_2} \sim\!C_6H_4\!-\!CMe_2\!-\!OO\cdot \xrightarrow{RH} \sim\!C_6H_4\!-\!CMe_2\!-\!OOH$$

$$\downarrow$$

$$\sim\!C_6H_4\!-\!OH + MeCOMe$$

$$R\!-\!CH_2\cdot \xrightarrow{O_2} RCH_2OO\cdot \xrightarrow{RH} R\!-\!CH_2OOH \longrightarrow R\!-\!CH_2O\cdot + \cdot OH$$

$$R\!-\!CH_2O\cdot \xrightarrow{R\cdot} R\!-\!CHO \xrightarrow{O_2} RCOOH$$

$$RCOOH \xrightarrow{h\nu} RH + CO_2 \qquad R\!-\!CHO \xrightarrow{h\nu} \dot{R}H + CO$$

In both the thermal- and high energy-induced degradation of the polysulphone, (56), at 380 °C sulphur dioxide, methane, diphenyl ethers and

$$\sim\!C_6H_4\!-\!O\!-\!C_6H_4\!-\!SO_2\!-\!C_6H_4\!-\!O\!-\!C_6H_4\!-\!CMe_2\!\sim$$

(56)

phenols are the principal products, demonstrating that all three interaromatic groups are involved in the degradation process[337,338]. As in the photo-oxidation of (55), oxides of carbon are formed in addition to the thermal degradation products[339]. In the degradation of poly(butene-1-sulphone) by high-energy radiation, C—S bond scission is highly specific and is followed by elimination of sulphur dioxide and depolymerisation, especially at higher temperatures[340].

The degradation properties of epoxy resins based on 2,2-bis(4-hydroxyphenyl) propane have much in common with those just described with the additional complication of reaction at the amino cross-links[341–343].

References

1. Conley, R. T. (editor) (1970). *Thermal Stability of Polymers*. (New York: Marcel Dekker)
2. Fox, R. B. (1967). *Progress in Polymer Science*, Vol. 1, 45–89 (A. D. Jenkins, editor) (London: Pergamon)
3. Grassie, N. (1966). *Encyclopaedia of Polymer Science*, Vol. 4, 647–716 (New York: Wiley)
4. Reich, L. and Stivala, S. S. (1969). *Autoxidation of Hydrocarbons and Polyolefins*, (New York: Marcel Dekker)

5. Ke, B. (editor) (1964). *Newer Methods of Polymer Characterisation.* (New York: Interscience)
6. Carroll, B. (editor), (1969). *Physical Methods in Macromolecular Chemistry,* Vol. 1, (New York: Marcel Dekker)
7. Slade, P. E. and Jenkins, L. T. (editors) (1966). *Techniques and Methods of Polymer Evaluation,* Vol. 1. *Thermal Analysis,* (London: Arnold)
8. Stevens, M. T. (editor) (1969). *Techniques and Methods of Polymer Evaluation,* Vol. 3. *Characterisation and Analysis of Polymers by Gas Chromatography.* (New York: Marcel Dekker)
9. Segal, C. L. (editor) (1967). *High Temperature Polymers,* (New York: Marcel Dekker)
10. Frazer, A. H. (1968). *High Temperature Resistant Polymers,* (New York: Wiley)
11. Korshak, V. V. and Vinogradova, S. V. (1968). *Uspekhi Khimii,* **11,** 2024
12. Geddes, W. C. (1967). *Rubber Chem. Technol.,* **40,** 177
13. Palma, G. and Carenza, M. (1970). *J. Appl. Polymer Sci.,* **14,** 1737
14. Geddes, W. C. (1967). *European Polymer J.,* **3,** 267, 733, 747
15. McNeill, I. C. and Neil, D. (1968). *Makromol. Chem.,* **117,** 265
16. McNeill, I. C. and Neil, D. (1970). *European Polymer J.,* **6,** 143, 569
17. Takahashi, T., Yasukawa, T. and Murakami, K. (1969). *Angew. Makromol. Chem.,* **9,** 182
18. Guyot, A., Bert, M., Michel, A. and Spitz, R. (1970). *J. Polymer Sci.* **A**-1, **8,** 1596
19. Bamford, C. H. and Fenton, D. F. (1969). *Polymer,* **10,** 63
20. Hay, J. N. (1970). *J. Polymer Sci. A*-1, **8,** 1201
21. Afonskii, V. K., Berlin, A. A. and Yanovskii, D. M. (1966). *Vysokomol. soyed.,* **8,** 699
22. Neiman, M. V., Papko, R. A. and Pudov, V. S. (1968). *Vysokomol. soyed.,* **A10,** 841
23. Troitskaya, L. S., Myakov, V. N., Troitskii, B. B. and Razuvayev, G. A. (1967). *Vysokomol. soyed.,* **A 9,** 2119
24. Varma, I. K. and Grover, S. S. (1969). *Angew. Makromol. Chem.,* **7,** 29
25. Bengough, W. I. and Grant, G. F. (1968). *European Polymer J.,* **4,** 521
26. van der Van, S. and de Wit, W. F. (1969). *Angew. Makromol. Chem.,* **8,** 143
27. Braun, D. and Bender, R. F. (1969). *European Polymer J. Suppl.,* 269
28. Chytry, V., Obereigner, B. and Lim, D. (1969). *European Polymer J. Suppl.,* 379
29. Caraculacu, A. A. (1966). *J. Polymer Sci. A-1,* **4,** 1829, 1839
30. Braun, D. and Weiss, F. (1970). *Angew. Makromol. Chem.,* **13,** 55
31. Gupta, V. P. and St. Pierre, L. E. (1970). *J. Polymer Sci. A-1,* **8,** 37
32. Murayama, N. (1966). *J. Polymer Sci. B,* **4,** 115
33. Braun, D., Thallmaier, M. and Hepp, D. (1968). *Angew. Makromol. Chem.,* **2,** 71
34. Braun, D. and Thallmaier, M. (1966). *Makromol. Chem.,* **99,** 59
35. Thallmaier, M. and Braun, D. (1967). *Makromol. Chem.,* **108,** 241
36. Matsumoto, T., Mune, I. and Watatina, S. (1969). *J. Polymer Sci. A-1,* **7,** 1609
37. Nakagawa, T. and Okawara, M. (1968). *J. Polymer Sci. A-1,* **6,** 1795
38. Minsker, K. S., Malinskaya, V. P. and Panasenko, A. A. (1970). *Vysokomol. soyed.,* **A-12,** 1151
39. Lundberg, L. L., Jayaraman, A., Rohn, C. L. and Maines, R. G. (1967). *Polymer Preprints,* **8,** 547
40. Kelen, T., Balint, G., Galambos, G. and Tudos, F. (1969). *European Polymer J.,* **5,** 597, 617, 629
41. Kelen, T., Galambos, G., Tudos, F. and Balint, G. (1970). *European Polymer J.,* **6,** 127
42. Loan, L. D. (1970). *Polymer Preprints,* **11,** 224
43. Minsker, K. S., Krats, E. O. and Pakhomova, I. K. (1970). *Vysokomol. soyed.,* **A-12,** 483
44. Varma, I. K. and Grover, S. S. (1970). *J. Appl. Polymer Sci.,* **14,** 2965
45. Cox, W. C., Crawford, D. J. and Peill, P. L. D. (1970). *J. Appl. Polymer Sci.,* **14,** 611
46. Kwei, K. P. S. (1968). *Polymer Preprints,* **9,** 691
47. Kwei, K. P. S. (1969). *J. Polymer Sci. A-1,* **7,** 237, 1075
48. Tikhomirov, L. A. and Buben, N. Y. (1966). *Vysokomol. soyed.,* **8,** 1881
49. Salovey, R., Luongo, J. P. and Yager, W. A. (1969). *Macromolecules,* **2,** 198
50. Salovey, R., Albarino, L. V. and Luongo, J. P. (1970). *Macromolecules,* **3,** 314
51. Salovey, R. and Luongo, J. P. (1970). *J. Polymer Sci. A-1,* **8,** 209
52. Salovey, R. and Bair, H. E. (1970). *Polymer Preprints,* **11,** 230
53. Flory, P. J. (1939). *J. Amer. Chem. Soc.,* **61,** 1518
54. Millan, J. and Smets, G. (1969). *Makromol. Chem.,* **121,** 275

55. Tepelekian, M., Tho, P. Q. and Guyot, A. (1969). *European Polymer J.*, **5,** 795
56. Grant, D. H. (1970). *Polymer*, **11,** 581
57. Aseyeva, R. M., Berlin, A. A., Kasatochkin, V. I. and Smutkina, Z. S. (1966). *Vysokomol. soyed.*, **8,** 2171
58. Burnett, G. M., Haldon, R. A. and Hay, J. N. (1968). *European Polymer J.*, **4,** 83
59. Haldon, R. A. and Hay, J. N. (1968). *J. Polymer Sci. A-1*, **6,** 951
60. Dolozel, A., Pegoraro, M. and Beati, E. (1970). *European Polymer J.*, **6,** 1411
61. Tsuge, S., Okumoto, T. and Takeuchi, T. (1969). *Macromolecules*, **2,** 177
62. Grassie, N. and Grant, E. M. (1967). *J. Polymer Sci. C.*, **16,** 591
63. Chukhadzhyan, G. A., Kalaidzhyan, A. Y. and Petrosyan, V. A. (1970). *Vysokomol. soyed.*, **A-12,** 171
64. Braun, D. and Sayed, I. A. A. (1969). *Angew. Makromol. Chem.*, **6,** 136
65. Grassie, N. and Grant, E. M. (1966). *European Polymer J.*, **2,** 255
66. Ito, H., Tsuge, S., Okumoto, T. and Takeuchi, T. (1970). *Makromol. Chem.*, **138,** 111
67. McNeill, I. C. and McGuchan, R. (1967). *European Polymer J.*, **3,** 511
68. Haldon, R. A. and Hay, J. N. (1967). *J. Polymer Sci. A-1*, **5,** 2297
69. Aslanyan, K. A., Bagdasaryan, R. V. and Kafadarova, Y. A. (1970). *Vysokomol. soyed.*, **A-12,** 434
70. Servott, A. and Desreux, V. (1968). *J. Polymer Sci. C*, **22,** 367
71. David, C., Borsu, M. and Geuskens, G. (1970). *European Polymer J.*, **6,** 959
72. Yonetani, K. and Graessley, W. W. (1968). *Polymer Preprints*, **9,** 229
73. Yonetani, K. and Graessley, W. W. (1970). *Polymer*, **11,** 222
74. Laren, V. A., Markova, Z. A., Yakovenko, V. I. and Bakh, N. A. (1967). *Vysokomol. soyed.*, **A-9,** 1221
75. Davies, R. F. B. and Reynolds, G. E. J. (1968). *J. Appl. Polymer Sci.*, **12,** 47
76. Tsuchiya, Y. and Sumi, K. (1969). *J. Polymer Sci. A-1*, **7,** 3151
77. Matsubara, T. and Imoto, T. (1968). *Makromol. Chem.*, **117,** 215
78. Duncalf, B. and Dunn, A. S. (1967). *J. Polymer Sci., C*, **16,** 1167
79. Dulog, L., Kern, R. and Kern, W. (1968). *Makromol. Chem.*, **120,** 123
80. Grassie, N. McLaren, I. F. and McNeill, I. C. (1970). *European Polymer J.*, **6,** 679
81. Grassie, N., McLaren, I. F. and McNeill, I. C. (1970). *European Polymer J.*, **6,** 865
82. Friedlander, H. N., Peebles, L. H., Brandrup, J. and Kirby, J. R. (1968). *Macromolecules*, **1,** 79
83. Braun, D. and El Sayed, I. A. A. (1969). *Angew. Makromol. Chem.*, **6,** 136
84. Grassie, N. and Hay, J. N. (1962). *J. Polymer Sci.*, **56,** 189
85. Watt, W. (1970). *Proc. Roy. Soc.*, **A319,** 5
86. Turner, W. M. and Johnson, F. C. (1969). *J. Appl. Polymer Sci.*, **13,** 2073
87. Bailey, J. E. and Clarke, A. J. (1970). *Chem. Brit.*, **6,** 484
88. Grassie, N. and McGuchan, R. (1970). *European Polymer J.*, **6,** 1277
89. Gillham, J. K. and Schwenker, R. F. (1966). *Appl. Polymer Symp.*, **2,** 59
90. Thompson, E. V. (1966). *J. Polymer Sci. B*, **4,** 361
91. Reich, L. (1968). *Macromolecular Reviews*, **3,** 49
92. Hay, J. N. (1968). *J. Polymer, Sci., A-1*, **6,** 2127
93. Noh, I. and Yu, H. (1966). *J. Polymer Sci. B*, **4,** 721
94. Ulbricht, J. and Makschin, W. (1969). *European Polymer J. Suppl.*, 389
95. Kubasova, N. A., Shishkina, M. V., Zaliznaya, N. F. and Geiderikh, M. A. (1968). *Vysokomol. soyed.*, **A10,** 1324
96. Kudryavtsev, G. I., Odnoralova, V. N. and Shablygin, M. V. (1966). *Vysokomol. soyed.*, **8,** 821
97. Fock, J. (1968). *J. Polymer Sci., A-1*, **6,** 963, 969
98. Glagoleva, Y. A. and Regel, V. R. (1970). *Vysokomol. soyed., A12*, 948
99. Knight, G. J. (1967). *J. Polymer Sci., B*, **5,** 855
100. Falb, R. D. and Berry, D. A. (1966). *Polymer Preprints*, **7,** 495
101. Wegner, J. and Patat, F. (1970). *J. Polymer Sci. C*, **31,** 121
102. Wall, L. A., Straus, S., Flynn, J. H. and McIntyre, D. (1966). *J. Phys. Chem.*, **70,** 53
103. Wall, L. A., Straus, S. and Fetters, L. J. (1969). *Polymer Preprints*, **10,** 1472
104. Cameron, G. G. and MacCallum, J. R. (1967). *J. Macromol. Sci., C*, **1,** 327
105. Cameron, G. G. and Kerr, G. P. (1968). *European Polymer J.*, **4,** 709
106. Cameron, G. G. and Kerr, G. P. (1970). *European Polymer J.*, **6,** 423
107. Richards, D. H. and Salter, D. A. (1967). *Polymer*, **8,** 139

108. McNeill, I. C. and Haider, S. I. (1967). *European Polymer J.*, **3,** 551
109. McNeill, I. C. and Makhdumi, T. M. (1967). *European Polymer J.*, **3,** 637
110. Richards, D. H. and Salter, D. A. (1967). *Polymer,* **8,** 153
111. Cameron, G. G. (1967). *Makromol. Chem.*, **100,** 255
112. Tsvetkov, N. S., Markovskaya, R. F. and Lukyanets, V. H. (1969). *European Polymer J. Suppl.,* 489
113. Cameron, G. G. and McWalter, I. T. (1970). *European Polymer J.*, **6,** 1601
114. Jellinek, H. H. G. and Lipovac, S. N. (1970). *Macromolecules,* **3,** 231
115. Beachell, H. C. and Smiley, L. H. (1967). *J. Polymer Sci., A-1,* **5,** 1635
116. Grassie, N. and Weir, N. A. (1965). *J. Appl. Polymer Sci.*, **9,** 963, 975, 987, 999
117. Price, T. R. and Fox, R. B. (1966). *J. Polymer Sci., B,* **4,** 771
118. Hendrickson, J. G. (1967). *J. Appl. Polymer Sci.*, **11,** 1419
119. Kells, D. I. C., Koike, M. and Guillet, J. E. (1968). *J. Polymer Sci., A-1,* **6,** 595
120. Bi, L. K. and Kamiya, Y. (1969). *J. Polymer Sci., A-1,* **7,** 1131
121. Selivanov, P. I., Maksimov, V. L. and Kirillova, E. I. (1969). *Vysokomol. soyed.*, **A11,** 482
122. Jellinek, H. H. G. and Flajsman, F. (1969). *J. Polymer Sci., A-1,* **7,** 1153
123. Jellinek, H. H. G. and Kryman, F. J. (1969). *J. Appl. Polymer Sci.*, **13,** 2504
124. Jellinek, H. H. G., Flajsman, F. and Kryman, F. J. (1969). *J. Appl. Polymer Sci.*, **13,** 107
125. Harrington, R. E. (1966). *J. Polymer Sci., A-1,* **4,** 489
126. Moore, D. E. and Ports, A. G. (1968). *Polymer,* **9,** 52
127. Ariswa, A. and Porter, R. S. (1970). *J. Appl. Polymer Sci.*, **14,** 879
128. Eckert, R. E., Maykrantz, T. R. and Salloum, R. J. (1968). *J. Polymer Sci., B,* **6,** 213
129. Amelin, A. V., Glagoleva, Y. A., Pozdnyakov, O. F. and Regel, V. R. (1969). *Vysokomol. soyed.,* **A11,** 1926
130. Tino, J., Capla, M. and Szocs, F. (1970). *European Polymer J.*, **6,** 397
131. Grassie, N. and Bain, D. R. (1970). *J. Polymer Sci., A-1,* **8,** 2653
132. Grassie, N. and Bain, D. R. (1970). *J. Polymer Sci., A-1,* **8,** 2665
133. Grassie, N. and Bain, D. R. (1970). *J. Polymer Sci., A-1,* **8,** 2679
134. Grassie, N. and Farish, E. (1967). *European Polymer J.*, **3,** 305
135. Cameron, G. G. and Kerr, G. P. (1969). *J. Polymer Sci., A-1,* **7,** 3067
136. Grassie, N., McNeill, I. C. and Cooke, I. (1968). *J. Appl. Polymer Sci.*, **12,** 831
137. Richards, D. H. and Salter, D. A. (1967). *Polymer,* **8,** 127
138. Vukovic, R. and Gnjatovic, V. (1970). *J. Polymer Sci., A-1,* **8,** 139
139. Turkova, L. D. and Belen'kii, B. G. (1970). *Vysokomol. soyed.*, **A12,** 467
140. Allport, D. C. (1967). *Polymer,* **8,** 492
141. Jellinek, H. H. G. and Kachi, H. (1968). *J. Polymer Sci., C,* **23,** 97
142. Jellinek, H. H. G. and Luk, M. D. (1969). *European Polymer J. Suppl.*, 149
143. Still, R. H. and Keattch, C. J. (1966). *J. Appl. Polymer Sci.*, **10,** 193
144. Still, R. H. and Jones, P. B. (1969). *J. Appl. Polymer Sci.*, **13,** 2033
145. Still, R. H., Jones, P. B. and Mansell, A. L. (1969). *J. Appl. Polymer Sci.*, **13,** 401, 1555
146. Geuskens, G., Borsu, M., Hellinckx, E. and David, C. (1970). *Makromol. Chem.*, **135,** 235
147. David, C., Verhasselt, A. and Geuskens, G. (1967). *J. Polymer Sci. C,* **16,** 2181
148. David, C., Demarteau, W. and Geuskens, G. (1967). *Polymer,* **8,** 497
149. David, C., Demarteau, W., Derom, F. and Geuskens, G. (1970). *Polymer,* **11,** 61
150. David, C., Demarteau, W. and Geuskens, G. (1969). *Polymer,* **10,** 21
151. Jellinek, H. H. G. and Wang, L. C. (1968). *Polymer Preprints,* **9,** 442
152. Jellinek, H. H. G. and Fox, S. Y. (1967). *Makromol. Chem.*, **104,** 18
153. Angeles, R., Aldridge, M. H., Freeman, D. H. and Wall, L. A. (1969). *J. Polymer Sci., B,* **7,** 609
154. Tsuchiya, L. and Sumi, K. (1968). *J. Polymer Sci., A-1,* **6,** 415
155. Cicchetti, O. (1970). *Advan. Polymer Sci.*, **7,** 70
156. Winslow, F. H., Matreyek, W. and Trozzolo, A. M. (1969). *Polymer Preprints,* **10,** 1271
157. Hartley, H. G. and Guillet, J. E. (1968). *Macromolecules,* **1,** 165
158. Heacock, J. F., Mallory, F. B. and Gay, F. P. (1968). *J. Polymer Sci., A-1,* **6,** 2921
159. Chan, M. G. and Hawkins, W. L. (1968). *Polymer Preprints,* **9,** 1638
160. Shilov, Y. B. and Denisov, Y. T. (1969). *Vysokomol. soyed.*, **A11,** 1812
161. Ivanchenko, P. A., Kharitonov, V. V. and Denisov, Y. T. (1969). *Vysokomol. soyed.,* **A11,** 1622
162. Chien, J. C. W. (1968). *J. Polymer Sci., A*-1, **6,** 375
163. Meltzer, T. H. and Morgano, T. J. (1967). *Polymer Preprints,* **8,** 558

164. Bohm, G. G. A., Currie, J. A. and Dole, M. (1968). *Polymer Preprints*, **9**, 303
165. Silverman, J. and Nielsen, S. O. (1968). *Polymer Preprints*, **9**, 296
166. Bohm, G. G. A. (1967). *J. Polymer Sci., A*-2, **5**, 639
167. Tsuchiya, Y. and Sumi, K. (1969). *J. Polymer Sci., A*-1, **7**, 1599
168. Magrupov, M. A. and Gafurov, I. (1969). *Vysokomol. soyed.*, **A11**, 2798
169. Bailey, W. J. and Statz, R. J. (1970). *Polymer Preprints*, **11**, 244
170. Boss, C. R. and Chien, J. C. W. (1966). *J. Polymer Sci., A*-1, **4**, 1543
171. Chien, J. C. W. (1968). *J. Polymer Sci., A*-1, **6**, 381
172. Chien, J. C. W. and Boss, C. R. (1967). *J. Polymer Sci., A*-1, **5**, 3091
173. Reich, L. and Stivala, S. S. (1969). *J. Appl. Polymer Sci.*, **13**, 17
174. Adams, J. H. (1970). *J. Polymer Sci., A*-1, **8**, 1077
175. Adams, J. H. and Goodrich, J. E. (1970). *J. Polymer Sci., A*-1, **8**, 1269
176. Adams, J. H. (1970). *J. Polymer Sci., A*-1, **8**, 1279
177. Chien, J. C. W. (1968). *J. Polymer Sci., A*-1, **6**, 393
178. Reich, L. and Stivala, S. S. (1967). *Makromol. Chem.*, **103**, 74
179. Abu-isa, I. (1970). *J. Polymer Sci., A*-1, **8**, 961
180. Bawn, C. E. H. and Chaudhri, F. A. (1968). *Polymer*, **9**, 113, 123
181. Reich, L. and Stivala, S. S. (1969). *J. Appl. Polymer Sci.*, **13**, 23
182. Chien, J. C. W. and Boss, C. R. (1967). *J. Polymer Sci., A*-1, **5**, 1683
183. Gromov, B. A. and Shlyatnikov, Y. A. (1967). *Vysokomol. soyed.*, **A9**, 2637
184. Mikhailov, N. V. and Pleshakov, M. G. (1970). *Vysokomol. soyed.*, **A12**, 1491
185. Kelleher, P. G. (1966). *J. Appl. Polymer Sci.*, **10**, 843
186. Stivala, S. S. and Jadrnicek, B. R. (1970). *Polymer Preprints*, **11**, 724
187. Carlsson, D. J., Kato, Y. and Wiles, D. M. (1968). *Macromolecules*, **1**, 459
188. Kato, Y., Carlsson, D. J. and Wiles, D. M. (1969). *J. Appl. Polymer Sci.*, **13**, 1447
189. Carlsson, D. J. and Wiles, D. M. (1969). *Macromolecules*, **2**, 587
190. Carlsson, D. J. and Wiles, D. M. (1969). *Macromolecules*, **2**, 597
191. Carlsson, D. J. and Wiles, D. M. (1970). *Polymer Preprints*, **11**, 760
192. Carlsson, D. J. and Wiles, D. M. (1970). *J. Polymer Sci., B*, **8**, 419
193. Kujirai, C., Hashiya, S., Furuno, F. and Terada, N. (1968). *J. Polymer Sci., A*-1, **6**, 589
194. Richters, T. (1970). *Macromolecules*, **3**, 262
195. Stivala, S. S., Yo, G. and Reich, L. (1969). *J. Appl. Polymer Sci.*, **13**, 1289
196. Ayscough, P. B. and Munary, S. (1966). *J. Polymer Sci., B*, **4**, 503
197. Iwasaki, M., Ichikawa, T. and Toriyama, K. (1967). *J. Polymer Sci., B*, **5**, 423
198. Veselovskii, R. A., Leshchenko, S. S. and Karpov, V. L. (1968). *Vysokomol. soyed.*, **A10**, 760
199. Veselovskii, R. A., Leshchenko, S. S. and Karpov, V. L. (1966). *Vysokomol. soyed.*, **8**, 744
200. Baranwal, K. (1968). *J. Appl. Polymer Sci.*, **12**, 1459
201. Baranwal, K. and Jacobs, H. L. (1969). *J. Appl. Polymer Sci.*, **13**, 797
202. Tsuchiya, Y. and Sumi, K. (1969). *J. Polymer Sci., A*-1, **7**, 813
203. McGuchan, R. and McNeill, I. C. (1968). *European Polymer J.*, **4**, 115
204. Lukhovitskii, V. I., Tsingister, V. A., Sharadina, N. A. and Karpov, V. L. (1966). *Vysokomol. soyed.*, **8**, 1932
205. Porter, R. S., Cantow, M. J. R. and Johnson, J. F. (1967). *J. Polymer Sci., C*, **16**, 1
206. Porter, R. S., Cantow, M. J. R. and Johnson, J. F. (1967). *Polymer*, **8**, 87
207. Ram, A. and Kadim, A. (1970). *J. Appl. Polymer Sci.*, **14**, 2145
208. Porter, R. S., Cantow, M. J. R. and Johnson, J. F. (1967). *J. Appl. Polymer Sci.*, **11**, 335
209. Chandra, S., Roy-Chowdhury, P. and Biswas, A. B. (1966). *J. Appl. Polymer Sci.*, **10**, 1089
210. Reginato, L. (1970). *Makromol. Chem.*, **132**, 113, 125
211. Reference 1, p. 189
212. Hackathorn, M. J. and Brock, M. J. (1970). *J. Polymer Sci., B*, **8**, 617
213. Kavum, S. M., Podkolzina, M. M. and Tarasova, Z. N. (1968). *Vysokomol. soyed.*, **A10**, 2584
214. Golub, M. A. (1969). *Macromolecules*, **2**, 550
215. Rabek, J. F. (1967). *J. Polymer Sci., C*, **16**, 949
216. Evans, D., Morgan, J. T., Sheldon, R. and Stapleton, G. B. (1969). *J. Polymer Sci., A*-2, **7**, 725
217. Bevilacqua, E. M. (1965). *J. Appl. Polymer Sci.*, **9**, 267

218. Bevilacqua, E. M. (1966). *J. Appl. Polymer Sci.*, **10,** 1295, 1287
219. Bevilacqua, E. M. (1968). *J. Polymer Sci., C,* **24,** 285
220. Shimada, J. and Kabuki, K. (1968). *J. Appl. Polymer Sci.*, **12,** 655, 671
221. Berlin, A. A., Mirotvortsev, I. I., Firsov, A. P. and Lyakhin, V. Y. (1969). *Vysokomol. soyed.*, **A11,** 1734
222. Lee, L. H., Stacy, C. L. and Engel, R. G. (1966). *J. Appl. Polymer Sci.*, **10,** 1699, 1717
223. Harmon, D. J. and Jacobs, H. L. (1966). *J. Appl. Polymer Sci.*, **10,** 253
224. Jellinek, H. H. G. and Flajsman, F. (1970). *J. Polymer Sci., A*-1, **8,** 711
225. Gent, A. N. and Hirakawa, A. M. (1967). *J. Polymer Sci., A*-2, **5,** 157
226. Florin, R. E., Parker, M. S. and Wall, L. A. (1966). *J. Res. Natl. Bur. Stand. A,* **70,** 115
227. Fock, J. (1968). *J. Polymer Sci., B,* **6,** 127
228. Florin, R. E. and Wall, L. A. (1968). *Polymer Preprints,* **9,** 1633
229. Florin, R. E. and Wall, L. A. (1970). *Macromolecules,* **3,** 560
230. Hedvig, P. (1969). *J. Polymer Sci., A*-1, **7,** 1145
231. Wall, L. A., Straus, S. and Florin, R. E. (1966). *J. Polymer Sci., A*-1, **4,** 349
232. Cotter, J. L., Knight, G. J. and Wright, W. W. (1968). *J. Polymer Sci., B,* **6,** 763
233. Barlow, A., Lehrle, R. S., Robb, J. C. and Sutherland, D. (1967). *Polymer,* **8,** 523, 537
234. Anufriev, G. S., Pozdnyakov, O. F. and Regel, V. R. (1966). *Vysokomol. soyed.*, **8,** 834
235. McNeill, I. C. (1968). *European Polymer J.*, **4,** 21
236. Cameron, G. G. and Kerr, G. P. (1968). *Makromol. Chem.*, **115,** 268
237. Jellinek, H. H. G. and Luh, M. D. (1968). *Makromol. Chem.*, **115,** 89
238. Jellinek, H. H. G. and Kachi, H. (1968). *J. Polymer Sci., C,* **23,** 87
239. Bagby, G., Lehrle, R. S. and Robb, J. C. (1969). *Polymer,* **10,** 683
240. McNeill, I. C. (1970). *European Polymer J.*, **6,** 373
241. Grassie, N. and Grant, D. H. (1960). *Polymer,* **1,** 445
242. Johnston, N. W. and Harwood, H. J. (1968). *Polymer Preprints,* **9,** 36
243. Fox, R. B. and Price, T. R. (1967). *J. Appl. Polymer Sci.*, **11,** 2373
244. David, C., Fuld, D., Geuskens, G. and Charlesby, A. (1969). *European Polymer J.*, **5,** 641
245. David, C., Fuld, D. and Geuskens, G. (1970). *Makromol. Chem.*, **139,** 269
246. Gardner, D. G. and Henry, G. A. (1967). *Polymer Lett.*, **5,** 101
247. Putyagin, P. Y. (1967). *Vysokomol. soyed.*, **A9,** 136
248. Lazar, M. and Szocs, F. (1967). *J. Polymer Sci., C,* **16,** 461
249. Semen, J. and Lando, J. B. (1969). *Macromolecules,* **2,** 570
250. Semen, J. and Lando, J. B. (1969). *Polymer Preprints,* **10,** 1281
251. Baines, F. C. and Bevington, J. C. (1968). *J. Polymer Sci., A*-1, **6,** 2433
252. Bevington, J. C., Brinson, R. and Hunt, B. J. (1970). *Makromol. Chem.*, **134,** 327
253. Cameron, G. G. and Kane, D. R. (1967). *Makromol. Chem.*, **109,** 194
254. Cameron, G. G. and Kane, D. R. (1968). *Makromol. Chem.*, **113,** 75
255. Cameron, G. G., Davie, F. and Kane, D. R. (1970). *Makromol. Chem.*, **135,** 137
256. Cameron, G. G. and Kane, D. R. (1968). *Polymer,* **9,** 461
257. McGaugh, M. C. and Kottle, S. (1967). *J. Polymer Sci., B,* **5,** 817
258. Eisenberg, A., Yokoyama, T. and Sambalido, E. (1969). *J. Polymer Sci., A*-1, **7,** 1717
259. McGaugh, M. C. and Kottle, S. (1968). *J. Polymer Sci., A*-1, **6,** 1243
260. Monahan, A. R. (1966). *J. Polymer Sci., A*-1, **4,** 2381
261. Monahan, A. R. (1967). *J. Polymer Sci., A*-1, **5,** 2333
262. Leonard, F., Kulkarni, R. K., Brandes, G., Nelson, J. and Cameron, J. J. (1966). *J. Appl. Polymer Sci.*, **10,** 259
263. Grassie, N. and Farish, E. (1967). *European Polymer J.*, **3,** 619
264. Grassie, N. and Farish, E. (1967). *European Polymer J.*, **3,** 627
265. Grassie, N. and Torrance, B. J. D. (1968). *J. Polymer Sci., A*-1, **6,** 3303, 3315
266. McNeill, I. C. (1968). *European Polymer J.*, **4,** 21
267. Grassie, N., Torrance, B. J. D. and Colford, J. B. (1969). *J. Polymer Sci., A*-1, **7,** 1425
268. Dudina, L. A., Karmilova, L. V., Tryapitsyna, E. N. and Enikolopyan, N. S. (1967). *J. Polymer Sci. C,* **16,** 2277
269. Vickers, W. H. J. (1967). *European Polymer J.*, **3,** 199
270. Grassie, N. and Roche, R. S. (1968). *Makromol. Chem.*, **112,** 16
271. Pennewiss, H., Jaacks, V. and Kern, W. (1967). *Makromol. Chem.*, **100,** 271
272. Blyumenfel'd, A. B., Kotrelev, M. V. and Kovarskaya, B. M. (1970). *Vysokomol. soyed.*, **A12,** 81

273. Blyumenfel'd, A. B., Neiman, M. B. and Kovarskaya, B. M. (1966). *Vysokomol, soyed.*, **8,** 1990
274. Nikitina, L. A., Sukhov, V. A., Baturina, A. A. and Lukovnikov, A. F. (1969). *Vysokomol. soyed.*, **A11,** 2150
275. Gur'yanova, V. V., Neiman, M. B., Kovarskaya, B. M., Miller, V. B. and Maksimova, G. V. (1967). *Vysokomol. soyed.*, **A9,** 2165
276. Sukhov, V. A., Nikitina, L. A., Baturina, A. A., Lukovnikov, A. F. and Yenikolopyan, N. S. (1969). *Vysokomol. soyed.*, **A11,** 808
277. Grassie, N. and Roche, R. S. (1968). *Makromol. Chem.*, **112,** 34
278. Blyumenfel'd, A. B. and Kovarskaya, B. M. (1970). *Vysokomol. soyed.*, **A12,** 633
279. Goglev, R. S. and Neiman, M. B. (1967). *Vysokomol. soyed.*, **A9,** 2083
280. Blyumenfel'd, A. B., Neiman, M. B. and Kovarskaya, B. M. (1967). *Vysokomol. soyed.*, **A9,** 1587
281. Price, C. C., Spector, R. and Tumolo, A. L. (1967). *J. Polymer Sci., A*-1, **5,** 407
282. Minoura, Y., Kasuya, T., Kawanura, S. and Nakano, A. (1967). *J. Polymer Sci., A*-2, **5,** 125
283. Thomas, T. H. and Kendrick, T. C. (1969). *J. Polymer Sci., A*-2, **7,** 537
284. Rode, V. V., Verkhotin, M. A. and Rafikov, S. R. (1969). *European Polymer J. Suppl.*, 401
285. Rode, V. V., Verkhotin, M. A. and Rafikov, S. R. (1969). *Vysokomol. soyed.*, **A11,** 1529
286. Aleksandrova, Y. A., Nikitina, T. S. and Pravednikov, A. N. (1968). *Vysokomol. soyed.*, **A10,** 1078
287. Andrianov, K. A., Papkov, V. S., Slonimskii, G. L., Zhdanov, A. A. and Yakushkina, S. Y. (1969). *Vysokomol. soyed.*, **A11,** 2030
288. Novikov, S. N., Kagan, Y. G. and Pravednikov, A. N. (1966). *Vysokomol. soyed.*, **8,** 1015
289. Verkhotin, M. A., Andrianov, K. A., Yermakova, M. N., Rafikov, S. R. and Rode, V. V. (1966). *Vysokomol. soyed.*, **8,** 2139
290. Nikitina, T. S., Khodzhemirova, L. K., Aleksandrova, Y. and Pravednikov, A. N. (1968). *Vysokomol. soyed.*, **A10,** 2783
291. Verkhotin, M. A., Andrianov, K. A., Zhdanov, A. A., Kurasheva, N. A., Rafikov, S. R. and Rode, V. V. (1966). *Vysokomol. soyed.*, **8,** 1226
292. Guyot, A., Cuidard, R. and Bartholin, M. (1969). *J. Polymer Sci., C*, **22,** 785
293. Sidnev, A. I., Vishnevskii, F. N., Moiseyev, A. F., Zubkov, I. A. and Pravednikov, A. N. (1970). *Vysokomol. soyed.*, **A12,** 355, 362
294. Dkachuk, B. V., Kolotyrkin, V. M. and Kirei, G. G. (1968). *Vysokomol. soyed.*, **A10,** 585
295. Siegel, S., Champetier, R. J. and Calloway, A. R. (1966). *J. Polymer Sci., A*-1, **4,** 2107
296. Shaw, M. T. and Rodriguez, F. (1967). *J. Appl. Polymer Sci.*, **11,** 991
297. Nametkin, N. S., Nechitailo, N. A., Durgar'yan, S. G. and Khotimskii, V. S. (1966). *Vysokomol. soyed.*, **8,** 888
298. Chatterjee, P. K. (1968). *J. Appl. Polymer Sci.*, **12,** 1859
299. Patel, K. S., Patel, K. C. and Patel, R. D. (1970). *Makromol. Chem.*, **132,** 7
300. Lipska, A. E. and Wodley, F. A. (1969). *J. Appl. Polymer Sci.*, **13,** 851
301. Kozmina, O. P., Dubyaga, V. P., Belyakov, V. K. and Zaichukova, N. A. (1969). *European Polymer J. Suppl.*, 447
302. Chamberlin, T. A. and Kochanny, G. L. (1969). *Macromolecules*, **2,** 88
303. Koz'mina, O. P., Syutkin, V. N., Slavetskaya, B. A. and Danilov, S. N. (1966). *Vysokomol. soyed.*, **8,** 1196
304. Buxbaum, L. H. (1967). *Polymer Preprints*, **8,** 552
305. Zhuravleva, I. V. and Rode, V. V. (1968). *Vysokomol. soyed.*, **A10,** 569
306. Davis, A. and Golden, J. H. (1968). *European Polymer J.*, **4,** 581
307. Ehlers, G. F. L., Fisch, K. R. and Powell, W. R. (1969). *J. Polymer Sci., A*-1, **7,** 2969
308. Marcotte, F. B., Campbell, D., Cleaveland, J. A. and Turner, D. T. (1967). *J. Polymer Sci., A*-1, **5,** 481
309. Campbell, D., Monteith, L. K. and Turner, D. T. (1968). *Polymer Preprints*, **9,** 250
310. Campbell, D., Monteith, L. K. and Turner, D. T. (1970). *J. Polymer Sci., A*-1, **8,** 2703
311. Campbell, D. and Turner, D. T. (1968). *J. Polymer Sci., B*, **6,** 1
312. Burow, S. D. (1966). *J. Polymer Sci., A*-1, **4,** 613
313. Campbell, D. and Turner, D. T. (1967). *J. Polymer Sci., B*, **5,** 471
314. Oprea, C. V., Neguleanu, C. and Simionescu, C. (1970). *European Polymer J.*, **6,** 181
315. Muinov, T. M., Mavlyanov, A. M. and Marupov, R. (1970). *Vysokomol. soyed.*, **A12,** 1724

316. Heuvel, H. M. and Lind, K. C. J. B. (1970). *J. Polymer Sci., A-2,* **8,** 401
317. Krasnov, Y. P., Savinov, V. M., Sokolov, L. B., Logunova, V. I., Belyakov, V. K. and Polyakova, T. A. (1966). *Vysokomol. soyed.,* **8,** 380
318. Bruck, S. D. (1966). *Polymer Preprints,* **7,** 520
319. Bruck, S. D. (1966). *Polymer,* **7,** 231
320. Davis, A. and Golden, J. H. (1969). *J. Macromol. Sci.,* **3,** 49
321. Sidnev, A. I., Pravednikov, A. N. and Kovarskaya, B. M. (1968). *Vysokomol. soyed.,* **A10,** 1187
322. Turska, E. and Sinierska-Kapuscuiska, M. (1969). *European Polymer J. Suppl.,* 431
323. Berlin, A. A., Levantovskaya, I. I., Kovarskaya, B. M., Dralyuk, G. V. and Raginskaya, L. M. (1968). *Vysokomol. soyed.,* **A10,** 1103
324. Levantovskaya, I. I., Dralyuk, G. V., Pshenitsyna, V. P., Smirnova, O. V., Yefimovich, T. M. and Kovarskaya, B. M. (1968). *Vysokomol. soyed.,* **A10,** 1633
325. Berlin, A. A., Yarkina, V. V. and Firsov, A. P. (1968). *Vysokomol. soyed.,* **A10,** 1913
326. Shulman, G. P. and Lochte, H. W. (1966). *J. Appl. Polymer Sci.,* **10,** 619
327. Berlin, A. A., Yarkina, V. V. and Firsov, A. P. (1968). *Vysokomol. soyed.,* **A10,** 2157
328. Alaminov, H. and Andonova, N. (1970). *J. Appl. Polymer Sci.,* **14,** 1083
329. Learmonth, G. S. and Osborn, P. (1968). *J. Appl. Polymer Sci.,* **12,** 1815
330. Grassie, N. and Meldrum, I. G. (1968). *European Polymer J.,* **4,** 571
331. Kalashnik, A. T., Kardash, I. Y., Shpitonova, T. S. and Pravednikov, A. N. (1966). *Vysokomol. soyed.,* **8,** 526
332. Jellinek, H. H. G. and Lipovac, S. N. (1970). *J. Polymer Sci., A-1,* **8,** 2517
333. Ehlers, G. F. L., Fisch, K. R. and Powell, W. R. (1969). *J. Polymer Sci., A-1,* **7,** 2931
334. Christopher, N. S. J., Cotter, J. L., Knight, G. J. and Wright, W. W. (1968). *J. Appl. Polymer Sci.,* **12,** 863
335. Kelleher, P. G. and Gesner, B. D. (1968). *Polymer Preprints,* **9,** 469
336. Kelleher, P. G. and Gesner, B. D. (1969). *J. Appl. Polymer Sci.,* **13,** 9
337. Davis, A. (1969). *Makromol. Chem.,* **128,** 242
338. Davis, A., Gleaves, M. H., Golden, J. H. and Huglin, M. B. (1969). *Makromol. Chem.,* **129,** 63
339. Gesner, B. D. and Kelleher, P. G. (1968). *J. Appl. Polymer Sci.,* **12,** 1199
340. Brown, J. R. and O'Donnell, J. H. (1970). *Macromolecules,* **3,** 265
341. Anderson, H. C. (1966). *Polymer,* **7,** 193
342. Keenan, M. A. and Smith, D. A. (1967). *J. Appl. Polymer Sci.,* **11,** 1009
343. Patterson-Jones, J. C. and Smith, D. A. (1968). *J. Appl. Polymer Sci.,* **12,** 1601

8
Diffusion in Polymers

V. STANNETT, H. B. HOPFENBERG
and J. H. PETROPOULOS
North Carolina State University

8.1 INTRODUCTION

It is intended in this chapter to review the important developments in the general field of diffusion of small molecules in high polymers that have taken place during the past 5 years. Although the primary emphasis will be on fundamental aspects, some discussion will also be directed towards practical applications. The phrase 'diffusion in polymers' has been broadened to include some discussion on the related topics of solution and permeation in high polymers. The 5-year period to be covered in this review is particularly appropriate since the developments up to that time have been extremely well presented in an authorative text, *Diffusion in Polymers*, which appeared in 1968[1].

A great deal of research has been conducted in the area of transport of small molecules in polymers. Like much polymer research the motivating force has been largely the considerable industrial interest in related applications. In particular, studies of the use of polymer membranes for various separation procedures for gases and for water purifications have been greatly accelerated in the past 5 years. The use of membranes for the reverse osmosis desalination of sea water and for other water purifications by reverse osmosis or ultrafiltration has been studied in great detail. Other applications include biomedical engineering devices; especially for artificial lung and kidney designs. Older applications under continuous study are for the design of better gas and vapour barriers for flexible packaging and for better protective coatings. In addition to these direct applications diffusion in polymers is important in many other areas of industrial importance including stress cracking, corrosion, electrical problems and in the dyeing of textile fibres.

There are many ways of arranging a detailed discussion on diffusion in polymers. The authors have found that basic research can be directed conveniently into these subject areas: gases, water vapour and organic vapours. This division is based upon the observed differences in behaviour; gases normally behave ideally in that their solubility in polymers obeys Henry's law and their diffusion constants are independent of concentration. The diffusion of gases therefore presents a good method for studying details of the structure and morphology of the polymer itself. Water vapour frequently behaves anomalously and the observed diffusion coefficients can increase or decrease or be independent of concentration depending on the nature of the polymer. Finally, organic penetrants behave anomalously in various ways depending on the particular penetrant–polymer system and on whether the polymer is above or below the glass temperature. The whole field of anomalous transport behaviour of organic penetrants in glassy polymers has been essentially opened up and developed during this period. Complications such as stress cracking and solvent crazing accompanying organic penetrant transport have been studied thoroughly during the past 5 years. The results of these studies will be discussed in detail in this chapter.

8.2 THEORETICAL DEVELOPMENTS

Progress in the study of diffusion in polymer–penetrant systems goes hand in hand with advances in diffusion theory. The main recent developments in this field concern the study of 'anomalous' or 'non-Fickian' diffusion.

Non-Fickian diffusion due to accompanying molecular relaxations of the polymer, induced by the penetrant near the glass transition temperature, has been studied earlier and reviewed in considerable detail[1,2]. This work was based very largely on an analysis of sorption kinetics and has continued in the period under review[3,4]. In this connection an elaboration[4] of the method of Kishimoto and Matsumoto[5] for the determination of the diffusion coefficient D of a 'pseudo-Fickian' system, should be mentioned, i.e. a non-Fickian system which nevertheless satisfies some of the criteria for Fickian systems. In this method, the thickness of the membrane l is varied and the value of D corresponding to $l \to \infty$ obtained. The D values deduced in this way for the methyl ethyl ketone–atactic polystyrene system were found to agree with the corresponding ones obtained from the permeation steady-state flux q_S.

Recent interest, however, has centred mainly on the permeation time-lag. In non-Fickian systems, the observed time-lag, L^a, generally differs from that, L^a_S, calculated from the appropriate Fickian expression[6]. An expression has also been given[7] for the discrepancy or 'non-Fickian time-lag increment' $L^a_T = L^a - L^a_S$. This includes the time dependence induced by molecular relaxations on both the diffusion coefficient and the boundary concentrations C_0, C_l, at the upstream ($x = 0$) and downstream ($x = l$) faces of the membrane respectively; i.e., $D = D(C,t), C_0 = C_0(t), C_l = C_l(t)$. No use has been made, so far of this rather complicated expression. Instead of this, Kishimoto and Kitahara[8], in their study of the polyacrylamide–water system, used a well-known simplified model[9] involving only the latter effect, which yields for $C_l = C_l(\infty)$:

$$L^a_T = \beta^{-1}[C_0(\infty) - C_0(0)]$$

where $\beta(= \text{const.})$ is the relaxation frequency. They were able to obtain concordant results from their results of sorption kinetics and time-lags, both of which were obtained as a function of concentration by varying the boundary concentrations from one experiment to the next according to the 'differential' or 'interval' method (in which both C_0, C_l are varied keeping $C_0 - C_l \sim \text{const.}$).

Petropoulos and Roussis[10] were able to obtain a reasonably simple and easy-to-use expression for L^a_T without simplifying assumptions. They used the activity of the penetrant (a) as the dependent variable related to concentration through the solubility coefficient $S = C/a$, which was defined as a quasi-thermodynamic quantity $S(a,t)$ varying between the instantaneous initial value $S(a,0)$ (corresponding to no relaxation in the polymer other than what can occur instantaneously) and the final value $S(a,\infty)$ (corresponding to full relaxation). The latter value is, of course, the true thermodynamic quantity. The boundary concentrations are now given by $C_0 = Sa_0$, $C_l = Sa_l$, where a_0, a_l are constants. Starting then from the diffusion equation with chemical potential gradient of penetrant as the driving force[11] (D_T, P_T = the thermodynamic[10,11] diffusion coefficient and permeability respectively), namely

$$\frac{\partial C}{\partial t} = \frac{\partial}{\partial x}\left(D_T S \frac{\partial a}{\partial x}\right) = \frac{\partial}{\partial x}\left(P_T \frac{\partial a}{\partial x}\right)$$

they obtained the final expression

$$L^a_T = \int_0^\infty \mathrm{d}t \int_{a_l}^{a_0} [P_T(a,\infty) - P_T(a,t)]\mathrm{d}a \Big/ \int_{a_l}^{a_0} P_T(a,\infty)\mathrm{d}a \qquad (8.1)$$

Using this equation, together with appropriate equations for the time-dependence of S and D_T, it was shown[10] that these quantities contribute to L_T^a nearly additively. The behaviour of L_T^a as a function of the boundary concentrations was also examined thoroughly using both 'interval' and 'integral' methods. (In the latter, the upstream boundary concentration is varied, keeping C_l = const. usually zero.) It is, however, considerably simpler to write an expression for $P_T(a, t)$. By analogy with S and D_T, this is given[12] as

$$P_T(a,t) = P_T(a,\infty) - [P_T(a,\infty) - P_T(a,0)] \exp[-\beta(a)t]$$

The relaxation frequency depends on a and this dependence becomes strong near the glass transition region. The main features of the $L_T^a(a_0, a_l)$ function are either a continuous decrease (interval method) or passage through a maximum (integral method). As the dependence of β on a becomes stronger, $L_T^a(a_0, a_l)$ exhibits a sharper decline (interval method) or its maximum shifts to lower activity. In the latter case, the plot of log L_T^a *v.* a_0 is very nearly linear in the region beyond the maximum. This feature is, in fact, shown by the data of Kishimoto[13] on the polyvinyl acetate–methanol system. The temperature behaviour of these data was also reproduced by the theoretical equation[12]. The functions $P_T(a, \infty)$, $P_T(a, 0)$ and $\beta(a)$ were exponential.

This work provides further strong support for the simple model developed by Crank and others[9,14]. Another important conclusion is that, since L_T^a depends only on P_T whereas sorption kinetics is also influenced by S, the information obtained from these two types of experiment is complementary[15]. This fact is not so important in polymer–good swelling agent systems where both S and P_T tend to increase with time. Consequently, simplified models involving S (or D_T) only (as used in Reference 4, for example) can interpret experimental results adequately in a qualitative or semi-quantitative manner. The situation is rather different when the penetrant is a poor swelling agent for the polymer, as is the case with water in hydrophobic polymers. Here, Petropoulos and Roussis[12,15] deduced from experimental data of Stannett *et al.*[16,17], that $L_T^a < 0$, indicating (see equation (8.1)) a tendency of P_T to decrease with time. On the other hand, the sorption *v.* *t* curves[17] still exhibit, though only barely so, the S-shape characteristic of good swelling agents. Calculation of sorption curves with various time dependences of S and P_T has shown[15] these time-lag and sorption kinetic results to be compatible. Accordingly, a tentative interpretation of this behaviour, based again on the molecular relaxation model, has been offered[12,15]. According to this, since the final (relaxed) state is one of lower free energy (at constant a), molecular relaxation should always lead to an increase of S with time. The effect on D_T however, will depend on the type of penetrant. Thus, if the polymer acts as a good solvent for the penetrant, the latter will tend to be molecularly dispersed in both initial (unrelaxed) and final (relaxed) molecular configurations of the polymer; hence P_T will tend to be an increasing function of time ($L_T^a > 0$). If, however, the polymer acts as a poor solvent for the penetrant, the thermodynamically most favourable state is one allowing for a high degree of penetrant clustering. Accordingly, the favoured manner of rearrangement of the macromolecules is one leading to the partial concentration of free volume into cavities large enough to permit such clustering. Since clusters contribute little to transport, this relaxation process may lead to a net decrease of P_T with time ($L_T^a < 0$). The remarkably small dependence of L^a on a_0 (with

$a_l = 0$), which is usually observed[16,17] is due, according to these views[15] to a compensating increase in both L_S^a and L_T^a. Whether this interpretation turns out, upon further experimental evidence, to be correct or not, the above arguments are also consistent with the failure of increasing free volume to lead (in accordance with current theories for good swelling agents[1,2]) to an increase in D in hydrophobic–polymer–water systems[1], as well as with the behaviour of moderately hydrophobic polymers (cf. $L_T^a = 0$ for water in cellulose acetate[18] with $L_T^a < 0$ in ethyl cellulose).

High concentrations of good swelling agents in a glassy polymer induce swelling stresses which can affect transport considerably. In what is known as Case II diffusion, the rate of sorption in a semi-infinite medium becomes proportional to t instead of $t^{\frac{1}{2}}$, the sorbate forming a sharp diffusion front which moves at constant velocity. This suggests a simple convection process and detailed theoretical treatments[19,20] have been developed covering this and intermediate cases. Starting with the general equation

$$\frac{\partial C}{\partial t} = \frac{\partial}{\partial x}\left(D\frac{\partial C}{\partial x} - Bs_0 C\right)$$

where B is the mobility of penetrant and s_0 the proportionality factor between partial stress and total uptake of penetrant[20], a final equation of the form

$$y = k_1 t + k_2 t^{\frac{1}{2}}$$

was obtained[19,20]. Here y may represent either total uptake or distance of penetration and k_1, k_2 are constants. This equation fits most experimental data well[19,20].

Non-Fickian diffusion anomalies can also result from a concurrent reaction between the penetrant and reactive groups in the polymer. Only the simplest systems, involving irreversible reaction, immobile product, constant D and reaction constant k and integral reaction order a, b, have been considered so far. The relevant equations are

$$\frac{\partial C_A}{\partial t} + \frac{\partial C_B}{\partial t} = D_A \frac{\partial^2 C_A}{\partial x^2}\,; \qquad \frac{\partial C_B}{\partial t} = -KC_A^a C_B^b$$

where subscripts A and B refer to penetrant and polymer reactive groups respectively (with C_A^0, C_B^0 denoting the constant boundary concentration of A and initial value of C_B respectively). Interest has centred principally on the reaction and the way it is influenced by diffusion rather than the other way about. Odiam and Kruse[21] have treated the important case of graft copolymerisation, where $K = K_p\, R_i/K_t$, $a = 1$, $b = 0$, (K_p, K_t = propagation and termination constants respectively, R_i = rate of initiation), and analytical solutions exist. They obtained expressions for total graft v. t and the distribution of graft across the membrane, under the usual sorption boundary conditions, and also deduced the condition for 'homogeneous grafting' as opposed to diffusion-controlled, non-uniform, or 'surface' grafting as

$$\alpha = (l/2)(K_p/K_t)(R_i/D_A) \leqslant 0.1$$

As far as real graft-copolymerisation systems are concerned, the theoretical model is obviously oversimplified. Nevertheless it was found[21] to describe accurately the effect of varying the diffusion control of the reaction (due to changing R_i or l) in the polyethylene–styrene–acrylonitrile system.

Petropoulos *et al.*[15,22], in a more general treatment found, for the same boundary conditions, that

$$K_E = l^2K(C_A^0 + C_B^0)(C_A^0)^{a-1}(C_B^0)^{b-1}/4D_A$$

provides a good measure of the extent of diffusion control, provided that the reaction is not too strongly diffusion-controlled (say $K_E < 50$). They also propose methods for determining K, a, b under these conditions.

Evolution of the heat of sorption can also give rise to non-Fickian anomalies. A detailed treatment for the case of relatively fast heat conduction (so that the temperature of the membrane is very nearly uniform) has been developed[23]. The variation in temperature causes S and D_T to vary with time in a manner depending on the rate of generation (proportional to the rate of sorption) and rate of loss of the heat to the surroundings of the membrane. The model is a simple one (Fick's and Henry's laws are assumed), but it has been applied successfully to the ethyl cellulose–water system over not too large vapour pressure intervals.

Another important type of non-Fickian diffusion system, which has attracted considerable interest, involves non-homogeneous membranes. Here S and D_T vary with position, particularly along the diffusion coordinate, as a result of a corresponding variation in some property of the membrane (such as density, crystallinity, etc.). Such inhomogeneity may be introduced during preparation of the membrane either spuriously or deliberately. There has been considerable interest lately in the latter possibility[24], because of the well-known[25] directional (along x or $-x$) dependence of the permeation flux, when the variation of S and D_T across the membrane is asymmetric and a, x in $S(a,x)$ and/or $D_T(a,x)$ are not separable. Sternberg and Rogers[26] used a polyethylene membrane grafted with a linearly varying (across the membrane) amount of polyvinyl acetate. They were able to explain the directional dependence of methanol flux in terms of the variation in the solubility coefficient from the relation $S = v_1S_1 + v_2S_2$, where v_1, v_2 are the volume fractions and S_1, S_2 the solubility coefficients of the two polymers. Their treatment was based on an earlier, more general, one of Frisch[27]. Work along these lines has also been carried out by Petropoulos and Roussis[28,29] and by Ash and Barrer[30]. The latter authors have also considered cylindrical and spherical geometries, but have, so far, only examined cases of separable a, x.

The theoretical study of non-steady-state diffusion of this type has been concerned almost exclusively with the time-lag. Earlier work[31] is limited to laminated membranes with constant D in each component lamination. In the period under review, this work was extended to hollow cylindrical laminates and applied to gas diffusion through tubing of nylon 11 surrounded by a sheath of polyethylene[32]. Petropoulos and Roussis[28] treated the case of continuous variation of S and D_T across the membrane. They then generalised previous treatments of laminated media by allowing for such variation of S and D_T within each component lamination[33]. Frisch and Prager[34] derived a general expression for the time-lag in a non-homogeneous medium and applied this also to a laminated membrane, with particular emphasis on the relation between the scale of the inhomogeneity and the discrepancy between observed and Fickian time-lags. The same question has been considered by Ash *et al.*[35], and with greater generality, by Petropoulos and Roussis[33].

The latter authors[10, 28, 29] carried out a thorough study of the properties of non-Fickian time-lag increments and were able to develop time-lag analysis into a useful general tool for the study of non-Fickian diffusion systems. Their method consists in measuring the amount of penetrant entering or leaving the membrane at the upstream and downstream boundaries, $Q(0,t)$ and $Q(l,t)$ respectively, and obtaining a time-lag from each of these quantities as illustrated in Figure 8.1. In the usual permeation experiment, the membrane

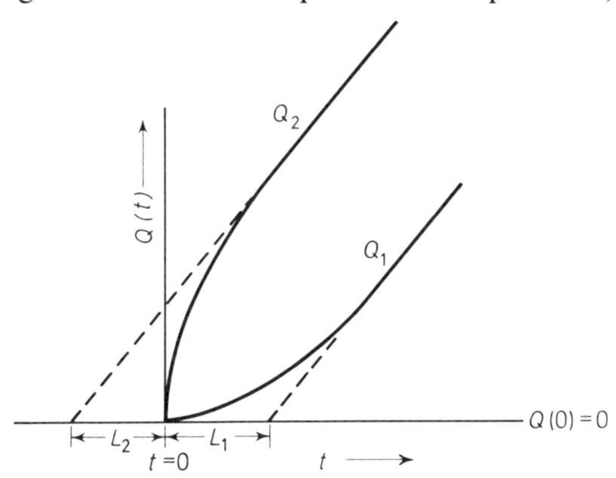

Figure 8.1 Schematic representation of absorption or desorption permeation experiment: (a) Absorption, $a(X,0) = a(l,t) = a_l$; $Q_1 = A(l,t)$, $Q_2 = Q(0,t)$; $L_1 = L^a(1)$, $L_2 = L^a(0)$. (b) Desorption, $a(X,0) = a(0,t) = a_0$; $Q_1 = Q(0,t)$, $Q_2 = Q(l,t)$; $L_1 = L^d(0)$, $L_2 = L^d(l)$

is brought to equilibrium at $a(x,0) = a_l$ and at $t = 0$ the activity of penetrant in the upstream reservoir is increased to a_0. This experiment yields the 'absorption' upstream and downstream time-lags, $L^a(0)$, $L^a(l)$ respectively, the latter being the usual time-lag L^a. In a 'desorption' permeation experiment, on the other hand, the membrane is equilibrated with the reservoirs at $a(x,0) = a_0$ and at $t = 0$, the activity of penetrant in the downstream reservoir is reduced to a_l. One then obtains the 'desorption' upstream and downstream time-lags, $L^d(0)$ and $L^d(l)$ respectively. These four time-lags, whose measurement can be repeated with the direction of flow reversed (the new time-lags being distinguished by an asterisk) thus bringing the total to eight, can be made to yield simpler forms. Thus, it is convenient to base the analysis on the following quantities: (i) the usual time lag L^a; (ii) the 'absorption upstream–downstream (algebraic) time-lag difference' $\Delta L^a = L^a - L^a(0)$; (iii) the 'absorption–desorption (algebraic) time-lag difference' at $x = l$, $\delta L = \delta L(l) = L^a - L^d(l)$; and (iv) the double difference (again in an algebraic sense), $\delta \Delta L = \Delta L^a - \Delta L^d$. The Fickian expressions for these differences are:

$$\Delta L^a_{\mathrm{S}} = q_{\mathrm{S}}^{-1} \int_0^l [C(x,\infty) - C_l]\mathrm{d}x,$$

$$\delta \Delta L_{\mathrm{S}} = 2\delta L_{\mathrm{S}} = l(C_0 - C_l)/q_{\mathrm{S}},$$

where $C(x,\infty)$ represents the steady-state concentration distribution of penetrant in the membrane.

Some properties of the non-Fickian time-lag increments of the above time-lag quantities (denoted by subscripts T and n according to whether they refer to time- or X-dependent non-Fickian systems respectively) were found to be very characteristic and are summarised in Table 8.1. It is thus possible to distinguish between the two most important types of non-Fickian diffusion system and, in addition, to obtain quite detailed information of a qualitative and quantitative nature about S and D_T. Table 8.1 shows that in the case of $S = S(a,X)$, $D_T = D_T(a,X)$, there is a possibility (e.g. concurrent immobilising reversible reaction or diffusion into sinks[10]) that P_T may be independent of time, thus yielding normal time-lags. The non-Fickian nature of such systems is revealed only in transient diffusion measurements such as sorption kinetics (which, as already noted, depend on S as well). In the case of $S = S(a,X)$, $D_T = D_T(a,X)$, one can readily distinguish between symmetrical and non-symmetrical (about the mid-plane of the membrane) variation of S and D_T; also between cases of separable and non-separable a, X in $S(a,X)$, $D_T(a,X)$, even in the absence of flow-reversal properties. A study[29] of these time-lag increments as a function of a_0, a_l has shown that considerable information about the functions $S(a,X)$, $D_T(a, X)$ can be obtained in this way.

Another important point[28] is that, if in a particular system both kinds of non-Fickian behaviour co-exist in a form no more complicated than $S(a,t,X) = S(a,t,0)\phi_1(X), D_T(a,t,X) = D_T(a,t,0)\phi_2(X)[\phi_1(0) = \phi_2(0) = 1]$,then they contribute to all non-Fickian time-lag increments additively. Thus:

$$\delta L = \delta L_S + \delta L_T + \delta L_n,$$

but, since $\Delta L_T^a = 0$ (See Table 8.1),

$$\Delta L^a = \Delta L_S^a + \Delta L_n^a$$

It follows that, by making use of the relation $\delta L_n = -\delta L_n^*$, for example, ΔL_n^a, δL_n and δL_T can be determined (and studied) separately. In the general case of $S = S(a,t,X)$, $D_T(a,t,X)$, however, only the latter equation is valid, thus permitting study of ΔL_n^a only. Another relation from Table 8.1, which survives in this case is $\delta\Delta L = \delta\Delta L_S$. This is particularly useful for determining[36] the 'mean' or 'effective' solubility coefficient of the system, defined by

$$S(a,\infty) = l^{-1} \int_0^l S(a,X,t \to \infty) \, \mathrm{d}X,$$

without resorting to equilibrium sorption studies. The relevant equation (where a_l is zero or else $\tilde{S}(a_l)$ is known from a previous experiment) is

$$\delta\Delta L = l[\tilde{S}(a,\infty)a_0 - \tilde{S}(a_l,\infty)a_l]/a_s.$$

The quantities δL and ΔL^a are also useful in this respect for certain non-Fickian systems[36]. For Fickian systems, δL appears to be the most convenient quantity to use for this purpose[10, 36, 37].

The method of time-lag analysis described above has, so far, been applied only to compacted graphite membranes with nitrogen as penetrant[38]. The

Table 8.1 Properties of non-Fickian time-lag increments and other quantities for $a_o \neq a_l$. $S(a,X)$ $D_T(a,X)$ are, at all values of a, either symmetrical or unsymmetrical about $X=l/2$. The inequalities given above may become equalities in isolated instances[28, 29]; the equalities have general validity. Flow reversal is without effect for $S(a,t)$, $D_T(a,t)$ and *sym* $S(a,X)$, $D_T(a,X)$.
(From Petropoulos and Roussis[29], by courtesy of the American Institute of Physics)

S(a,t), D_T(a,t)		S(a,X), D_T(a,X)					
		independent of a		a,x *separable*		a,x *nonseparable*	
$P_T = P_T(a,t)$	$P_T = P_T(a)$	*unsym*	*sym*	*unsym*	*sym*	*unsym*	*sym*
$L_T^a \neq 0$	$L_T^a = 0$	$L_n^a \neq 0$	$L_n^a \neq 0$	$L_n^a \neq 0$	$L_n^a \neq 0$	$L_n^a \neq 0$	$L_n^a \neq 0$
$\Delta L_T^a = 0$	$\Delta L_T^a = 0$	$\Delta L_n^a \neq 0$	$\Delta L_n^a = 0$	$\Delta L_n^a \neq 0$	$\Delta L_n^a \neq 0$	$\Delta L_n^a \neq 0$	$\Delta L_n^a \neq 0$
$\delta L_T \neq 0$	$\delta L_T = 0$	$\delta L_n \neq 0$	$\delta L_n = 0$	$\delta L_n \neq 0$	$\delta L_n = 0$	$\delta L_n \neq 0$	$\delta L_n \neq 0$
$\delta \Delta L_T = 0$	$\delta \Delta L_T = 0$	$\delta \Delta L_n = 0$	$\delta \Delta L_n = 0$	$\delta \Delta L_n = 0$	$\delta \Delta L_n = 0$	$\delta \Delta L_n = 0$	$\delta \Delta L_n = 0$
		$\delta L_n = -\Delta L_n^a$		$\delta L_n/\delta L_0 = \text{const.}$		$\delta L_n/\delta L_s \neq \textit{const.}$	$\delta L_n/\delta L_s \neq \textit{const.}$
		$L_n^a = L_n^{a*}$		$L_n^a \neq L_n^{a*}$		$L_n^a \neq L_n^{a*}$	
		$\Delta L_n^a = -\Delta L_n^{a*}$		$\Delta L_n^a \neq -\Delta L_n^{a*}$		$\Delta L_n^a \neq -\Delta L_n^{a*}$	
		$\delta L_n = -\delta L_n^*$		$\delta L_n = -\delta L_n^*$		$\delta L_n \neq -\delta L_n^*$	
		$q_s = q_s^*$		$q_s = q_s^*$		$q_s \neq q_s^*$	
		$\delta \Delta L = \delta \Delta L^*$		$\delta \Delta L = \delta \Delta L^*$		$\delta \Delta L \neq \delta \Delta L^*$	

results showed the presence of X dependence in S and D_T. Within the experimental precision, no evidence for time-dependence was found.

A number of theoretical developments concerning Fickian diffusion have been reported. Some of these deal with the analysis of sorption *v.* time curves to obtain the diffusion coefficient as a function of concentration. Fels and Huang[39] use the expression for $D_T(C)$ given by the free volume theory[1] and the sorption isotherm given by the Flory–Huggins theory. They then solve the resulting equation numerically to obtain the desorption *v.* time curve and determine the free volume parameters by repeating the calculation systematically, until the best fit with a particular experimental curve is obtained. The method has been applied to the benzene–polyethylene system. Duda and Vrentas[40] solved the diffusion equation (including the effects of phase volume change and volume change on mixing) to obtain analytical expressions representing approximately the sorption kinetic process. Their method permits deduction of D as a function of C from a single experiment, though the use of a pair of absorption–, desorption–time curves is recommended to improve accuracy. Barrie and Machin[41] have been concerned with the case of D varying inversely with C as in water–hydrophobic polymer systems. They examined the concentration dependence of the time-lag and the relative usefulness for these systems of various standard methods[1, 42] of deducing $D(C)$ from sorption kinetic measurements.

Considerable attention has also been given to systems where the sorbate is divided into mobile (with constant diffusion coefficient D_m) and immobile molecular populations, obeying Henry and Langmuir sorption isotherms respectively. This model was used[43] for gas–glassy polymer systems, where part of the sorbate appears to be 'trapped' in holes in the polymer matrix. A method of variable boundary concentration sorption–time data fitted by numerically computed curves, was employed to determine D_m. By defining an appropriate dimensionless time-scale, little computation was found to be necessary for this purpose. Weisz *et al.*[44, 218], have considered in detail the use in standard (constant C_0) sorption–time experiments of this dimensionless time-scale, which is $t = D_m k_m C_m^0 / l^2 C_0$ (k_m, C_m^0 = Henry law constant and boundary concentration respectively for mobile species; C_0 = total boundary concentration). The treatment of Reference 43 involves the assumptions of constant D_m, immobility of 'trapped' species and fast equilibrium between this and mobile species. Paul[45] pointed out the need of checking the latter two assumptions and suggested measurement of the time-lag as a more sensitive test. This method still has disadvantages, however. Fortunately, it has been shown[46] by appropriate formulation of the diffusion equation, that standard permeation and sorption methods will allow such testing much more simply and efficiently. At the same time, an estimate of the diffusion coefficient of 'trapped' molecules (if appreciable) is obtainable. Sorption kinetic experiments of the type employed in Reference 43 have also been studied theoretically by Pilchowski *et al.*[47].

Permeation experiments with variable boundary concentrations, particularly on the downstream side of the membrane, have also been examined in some detail. Two methods have been proposed for deducing S and D_T (when both are constant) from the $Q(l,t)$ curve. The first[48, 49] is applicable when $n = V_m S / V_l$ (V_m, V_l = volume of membrane and downstream reservoir respectively) is small enough for $Q(l,t)$ to possess a substantial very nearly linear portion. First estimates of D and S are obtained, as in the standard time-lag method, and improved by iteration, i.e. by estimating n and calculating correction factors (which are functions of n) for D and S[48]. Jenkins *et al.*[49] give $n = 0.15$ as the upper limit for the applicability of this method. This is certainly too high. Their claim of a 4% systematic error in the time-lag, as usually measured in practice is also unacceptable. These authors have in mind a technique of measuring L^a which is impractical, involving as it does a single point on $Q(l,t)$ (situated at $t = 3L$ in the case of low n) and drawing of a tangent at that point. The same criticism applies to the method they propose for higher values of n. In this case, the quantity $f = 1-C(l,t)/C_0$ can attain values which are an appreciable fraction of unity and the method of choice is that of Kubin and Spacek[50] in which a linear plot of $\log f$ *v.* t is obtained at sufficiently large t. By solving the relevant transcendental equations, D and S are then obtained from the slope and intercept of this plot. This second method has been generalised[51] to the case of variable boundary concentrations on both sides of the membrane. Here $f = [C(0,t)-C(l,t)]/C(0,0)$. The case of a laminated membrane of the type ABA has also been worked out[52], on the assumption that a steady-state concentration distribution is established in A instantaneously. The above methods have been applied to the diffusion of oxygen and KCl through hydrophilic gels (based on glycol–methacrylic acid[50, 51]). A detailed experimental investigation of the effect of the size, and adsorption on the walls, of the downstream reservoir on permeation measurements has been carried out for water vapour–polymer systems by Yasuda and Stannett[53]. These authors devised means of avoiding or minimising the latter effect (whereas Stern and Britton[54] attempted to correct for it). Finally, Siegel and Coughlin[55] examined the error induced in L^a by experimental error in q_S. Their treatment, however, is too simplified to be fully satisfactory.

Permeation measurements in which the flux $q(l,t) = \mathrm{d}Q(l,t)/\mathrm{d}t$ is measured directly have been carried out by Pasternak *et al.*[56]. Their method of obtaining D (D,S = constant) consists in fitting the theoretical plot of $q(l,t)/q_S$ *v.* $4\Delta t/l^2$ by scaling the experimental one. Evnochides and Henley[57] have determined S and D (both constant) in the system polyethylene–ethane by 'alternating sorption' measurements, in which the boundary concentration was varied sinusoidally. Theoretical analysis shows that measurement of the amplitude of the total amount sorbed of penetrant and the phase angle between this and the variation of external pressure will yield both D and S in one experiment. The results of this method were found to be in reasonably good agreement with those of standard methods[57], but no particular advantage has been claimed for this method so far. However, Paul[58] found a cyclic operation (this time of a square wave form) to be advantageous from the point of view of separation of a binary gas mixture, though it is worse than steady-state operation from the point of view of productivity. It may be worth mentioning here that the idea of periodic variation of boundary concentration is not new[59].

Some of the work done on numerical methods of solution of the diffusion equation may be of interest here. In particular, the explicit method of solution proposed by Barakat and Clark[60] is worth mentioning. This is unconditionally

ally stable and is claimed to yield somewhat better accuracy than other methods. Rick *et al.*[61] have studied the relative merits of analogue and digital computation (using the semi-implicit Crank–Nicolson method in the latter case) for systems with D depending exponentially on concentration with or without simultaneous time dependence. Finally, Petropoulos and Roussis[29] paid particular attention to numerical methods of computing time-lags for the case of $S(a,X)$, $D_T(a,X)$ with non-separable a,X, where analytical expressions are not available. They investigated the accuracy of their results by comparison of the numerical and analytical values for the corresponding cases of separable a,X.

8.3 METHODS OF MEASUREMENT

Almost all the measurements of the diffusion constants for gases in polymers reported have been obtained using the high-vacuum time-lag technique developed in detail by Barrer[62]. Such measurements give also the permeability constants and therefore the solubility coefficients. These can also be measured directly by various volumetric techniques. Good agreement between the values obtained by both methods was found many years ago by Van Amerongen[63]. A more recent method for measuring gas and vapour solubilities by a pressure–volume method has been described by Williams and Peterlin[64]. With many vapours the solubilities and the diffusivities can be measured gravimetrically using a quartz helix or electronic microbalance.

In recent years a number of new methods have been developed which are isobaric in nature. These avoid the problems of good seals and of maintaining high-vacuum systems for long periods of time. In principle, one side of the polymer film is exposed to, or flushed with, a known pressure of the gas under investigation and the other is swept by an inert carrier gas such as helium. The concentration of the gas in the carrier stream can then be measured by a number of techniques. Since at normal pressures the permeability and diffusion constants are independent of pressure, normal atmospheric pressures can be maintained on both sides of the membrane. A number of detecting devices have been used. The hot-wire thermal-conductivity probe was used by Pasternak and McNulty[65] and by Ziegel *et al.*[66], and by Riemschneider and Riedel[67]. Caskey[68] also used gas-chromatographic techniques but with a radiation-source helium detector. In an extension of the use of thermal-conductivity changes to measure gas-transport rates, Yasuda and Rosengren[69] described in considerable detail a new method based on the use of very small thermistors. These are small enough to use very small receiving volumes. Most isobaric methods are only used with a flow technique where direct time-lags cannot be measured and early time- or half-time methods for estimating the diffusion constants need to be used. The method of Yasuda and Rosengren[69] can be used with either a fixed volume time-lag, or as a flow technique. The use of twin cells with a reference film eliminates errors due to loss of occluded gases or vapours in the polymer sample.

It is difficult to estimate at this time how far the newer isobaric techniques will replace the high-vacuum time-lag method. A rather misleading point often cited in favour of techniques which measure transport by flowing two gas streams on either side of the membrane to be tested is that of rapidity of measurement. In all cases the film must be thoroughly degassed, however, before reliable measurements can be made which should be obviously much faster under high-vacuum conditions. For instance, under vacuum conditions there are no boundary conditions at the film surface to overcome. Once the film is properly degassed and the film thickness and all the remaining variables are normalised, each technique should require identical time for the actual measurement itself. The overall time employing the vacuum method at the worst should be equal to or less than the time required by atmospheric methods.

Other points generally given little or no attention in comparative discussions of techniques are the often long calibration procedures required for various gases and the problem of gas mixing associated with several of the atmospheric methods of determining transport parameters, such as conductivity methods. Although these experimental procedures are generally carried out prior to the introduction of the membrane it seems only fair to include them in the overall experimental time. The vacuum method, on the other hand, employs direct sensors of pressure or concentration and thus such calibration procedures and the mixing of gases are rather simple.

Nitrogen, oxygen and carbon dioxide are the most often tested atmospheric gases and these can be easily measured by most of the available techniques. Measurement of many vapours and liquids, however, can usually be carried out using a high-vacuum technique, but are often so difficult to measure by atmospheric methods as to be not worthwhile. Thus, the flexibility of this method almost dictates its continued use for most scientific investigations. Perhaps, for more routine studies, isobaric methods, particularly those using the thermal conductivity method[65, 66, 69] for example, will be much favoured. For many fundamental investigations, however, it is believed that the high-vacuum time-lag technique will continue to be preferred in many cases.

Permeability constants for water and organic vapours have been measured by a number of other ingenious devices using conductivity[70, 71], dielectric constant[72] and phosphorus pentoxide electrolytic cell devices[73].

A number of other special methods have also been described. Eichhorn[74] studied the transport of water vapour through polyethylene tubes by filling them with molten sodium metal and measuring the sodium hydroxide produced by electrical resistance measurements[75]. Jones[75] used the quenching by oxygen of a phosphorescent additive, dispersed in a polymer, to measure oxygen diffusion rates.

Finally, the use of radioisotope tagging to measure diffusion rates[76] has continued with ^{35}S [77] for example and ^{14}C [17] labelling.

8.4 TRANSPORT OF GASES

8.4.1 Introduction

Experimental work concerned with the transport of gases in high polymers has been extensive during recent years. No particular area, however, has been studied in great detail. New data has been obtained both with polymers already studied and with newer polymers such as the fluorocarbons. The effects of grafting and of cross-linking and of changes in morphology have also been investigated.

Because of its importance in a number of biological and biomedical applications, the transfer of dissolved oxygen through polymer membranes has also been studied in some detail.

8.4.2 New data

A considerable amount of new data concerning the diffusion of gases in polymers has been reported during the past 5 years. Unfortunately, most of the studies reported have been rather isolated in nature and with systems chosen for various particular reasons. A complete review of the permeability data through 1965 has been presented by Lebovits[79]. *Polyvinyl chloride* has been studied in considerable detail. Barrer *et al.*[80] investigated the effect of plasticisers on the diffusion, solution and permeation of neon and hydrogen above and below the glass transition temperatures. Plasticising was found to reduce the solubility, and the selectivity of neon and hydrogen in the membranes. Some evidence of a change in activation energy at the glass temperature was presented. Plasticising reduced this temperature considerably. Tikhomirov *et al.*[81] studied the diffusion, solution and permeation of a number of gases in plasticised polyvinyl chloride. A change in the slope of the activation energy plots were found with krypton but not with smaller-diameter gases. Finally, a study of the effect of modifying the polyvinyl chloride by chemically reacting with *N*-methyl dithiocarbonate was reported[82]. The membranes were found to be radiation-stabilised to a considerable degree whereas only minor changes in the gas transport behaviour was noted.

Polyethylene has continued to be studied under various conditions and modifications. Stern *et al.*[83, 84] have studied the permeation of a number of gases at pressures up to 60 atmospheres and temperatures between −10 °C and 60 °C. The principle of corresponding states was used to correlate the solubility of gases in polyethylene over a wide range of temperatures and to predict the pressure at which pressure-dependence of the permeability constant begins. The use of polyethylene for separating gas mixtures was also discussed in some detail. Studies of the effects of pressure and temperature on the solubility and permeability of a number of gases and vapours in polyethylene have also been reported by Li and Long[85], and Li[86] also extended their work to gas and vapour mixtures. The paper by Li[86] and earlier papers by Li and Henley[87] and Casper and Henley[88] also included some interesting data on polypropylene and polytetrafluorethylene. The best selectivity was found at low pressures and the best permeabilities at high pressures. The solubility and transport of various gases in both polyethylene and nylon were determined by Ash, Barrer and Palmer[89]. Good correlation between the solubilities and the Lennard-Jones interaction parameters were found with both polymers.

Shterenzon *et al.*[90] have measured the diffusion and sorption of hydrogen chloride in polyethylene, both in aqueous acid and gaseous forms. Gaseous hydrogen chloride behaved 'ideally' in that the solubilities obeyed Henry's law and the diffusion constants were independent of concentration. With the aqueous acid the behaviour was more complex due to the presence of hydrates in the solution.

The effect of γ-radiation on the transport of nitrogen and methane in polyethylene has been studied further by Kanitz *et al.*[91]. In agreement with earlier work[92], the diffusivities and permeabilities of both gases were found to decrease progressively with dose when the polymer had been irradiated in air or in *vacuo*, the latter being the most pronounced. The solubilities, on the other hand, increased considerably with the air-irradiated samples but were only slightly affected when the polyethylene was irradiated *in vacuo* until very high doses were reached. This was probably due to a decrease in crystallinity at the high doses and to oxidation with the air-irradiated samples. The separation factors decreased with increasing radiation dose.

The permeation of gases through the fluorocarbon polymers has received considerable attention in recent years. Polytetrafluorethylene[14] and a tetrafluorethylene–hexafluorpropylene copolymer[15] have both been studied by Pasternak *et al.*[93,94]. Nitrogen, oxygen and carbon dioxide were found to permeate and diffuse at rates similar to those in low-density polyethylene at 25 °C. The activation energies for diffusion were, however, consistently about 2 kcal mol^{-1} less for polytetrafluorethylene. This was ascribed[94] to the diffusion in polytetrafluorethylene being at least in part along pore and grain boundaries. This seems reasonable as the films were prepared by a sintering process. The permeation of nitric acid from aqueous solution through a number of fluoropolymers was studied by Russian workers[95]. Huang and Kanitz[96] have investigated the effect of γ-radiation on the gas permeation in both polytetrafluorethylene and in tetrafluorethylene–hexafluorpropylene copolymer. The diffusion and permeability constants were found to decrease with increasing radiation dose whereas the solubilities were not greatly affected. As with polyethylene[91], the separation factors were increased by radiation but at the expense of lower permeation rates.

Finally, a study of the transport of hydrogen and deuterium was reported by Ziegel *et al.*[97] in polyvinyl fluoride. Deuterium was found to diffuse at a lower rate and with a slightly higher activation energy than hydrogen as predicted by the theoretical considerations of Frisch and Rogers[98]. Polyvinyl fluoride is a remarkably good gas and vapour barrier and deserves further attention from both the practical and theoretical points of view.

Gas transport in various *graft and block polymers* has been studied by a number of workers. The progressive grafting of styrene to polyethylene was found by Huang and Kanitz[99] to first decrease and then increase the gas permeabilities. The decreases were ascribed to a decrease in the free volume of the films due to a 'filling' effect of the grafted side chains. With further increases in grafting it was hypothesised that the crystallinity was disrupted leading to the observed increase in gas permeability. Somewhat similar conclusions were drawn from the results of a study by Williams and Stannett[100] on grafting to polyoxymethylene. The work with polyethylene was extended by Huang and Kanitz[101] to include acrylonitrile grafting and both styrene and acrylonitrile to polytetrafluorethylene and a tetrafluorethylene–

hexafluorpropylene copolymer. In all cases the diffusivities and permeabilities decreased with grafting and this was again attributed to a decrease in the free volume of the polymers. The separation factors were improved in some cases but at the expense of reduced gas permeability. A somewhat different study was reported by Wellons *et al.*[18]. The films were cast from purified graft copolymers of cellulose acetate and styrene rather than prepared by grafting onto pre-existing polymer films as in the previous papers discussed above. Most of the study was with water vapour but some gas-transport measurements were made; and, in general, the diffusivities and permeabilities were found to be intermediate between those of the two homopolymers. However, the actual values were found, in the case of the graft copolymers, to be highly dependent on the conditions of film formation. This constitutes a problem of considerable interest in itself and well worthy of further study. The water-vapour studies showed clear evidence of domain structure in the graft polymer films. A recent study[102] on gas transport in a number of block copolymers also gave evidence of domain structure especially with the larger molecular-size gas-molecules and a mathematical model was proposed for this type of behaviour. An important industrial development in recent years has been the development of partly grafted transparent polymers with a high acrylonitrile or methacrylonitrile content[103]. These polymers are transparent, tough and with a remarkably low permeability to oxygen and other gases. They show great promise for the manufacture of plastic bottles and containers as well as for high barrier films.

An intensive study of the effect of *cross-linking* on the permeation, diffusion and solution of a number of gases in copolymers of ethyl acrylate and tetraethyleneglycol dimethacrylate was published by Barrer *et al.*[104]. The permeabilities and diffusivities of all the gases studied decreased with increasing degrees of cross-linking whereas the solubilities were only slightly affected. The gas separation factors increased with cross-linking but at the expense of reduced permeability. The results were in good agreement with the radiation-induced cross-linking studies of Huang *et al.*[91, 96]. The transport of the noble gases in polymethyl acrylate was studied in detail by Burgess *et al.*[105]. The results were contrasted with the isomeric polymer, polyvinyl acetate. The polymethyl acrylate was found to have consistently lower activation energies for diffusion and the results were interpreted in terms of lower cohesive forces but tighter packing of the acrylate polymers. The solubility and diffusivity of methane in polyisobutylene was studied by Lundberg and Rogers[106] at high temperatures and pressures. Deviations from Henry's law were found and interpreted in terms of a Langmuir-type site-filling mechanism. The diffusivities and solubilities of gases in a number of molten polymers were re-calculated and included and discussed with the polyisobutylene data. Data on the diffusion and solution of a number of gases in various other heat-softened or molten polymers were also reported by Durrill and Griskey[107, 108]. In general, Henry's law was obeyed and both the solubilities and diffusivities followed the Arrhenius relationship. A new technique for such measurements was briefly discussed by Morrison[109]. Finally, the sorption and diffusions of carbon dioxide in a number of epoxy, phenolic, alkyd and other coating compositions was studied by Slabaugh and Kennedy[110].

8.4.3 Effect of polymer morphology

Earlier work on the effect of crystallinity on gas transport in polymers was almost exclusively concerned with polyethylene terephthalate and polyethylene[111]. During the past 5 years, however, this work has been extended to include polypropylene. Russian workers[112,113,114], for example, have studied the effects of varying quenching and annealing conditions on gas diffusion and permeation in polyethylene[112] and on polypropylene[112,113,114]. The thermal history was found to have a marked effect on the diffusion, on permeation, and also on the effect of radiation on the diffusion of gases[112]. The permeability of gases in polypropylene was found to be independent of spherulite size up to about 80 Å diameter, then to gradually increase. Finally, with very large spherulites there was found to be a considerable increase due to the development of microcracks in the structure[114].

A much more extensive study with polypropylene was reported by Vieth and Wuerth[115]. Solubilities, diffusivities and permeabilities of helium, argon and carbon tetrafluoride were determined in a large number of samples of different thermal histories and a sample of atactic polypropylene. All the quenched unannealed samples were found to be of similar amorphous contents and solubilities. Samples annealed above 90 °C, however, showed changes in the amorphous contents with annealing conditions. The solubility coefficients were found to increase with increasing amorphous content to a certain amount dependent on the gas and then levelled off and became independent of the amorphous content rather than continuing to increase up to the value found with purely amorphous atactic polypropylene. This was attributed to the existence of two types of crystals leading, however, to similar amorphous contents but different densities. This behaviour is quite different from that found with polyethylene where the solubilities were found to be well correlated with the amorphous contents.

The diffusion constants were found to decrease to a minimum and then to actually increase with decreasing amorphous content, again in marked contrast to polyethylene where they were always found to decrease with increasing crystallinity. The activation energies were essentially independent of the crystallinity and thermal history. This indicated that the increase in the diffusion constants was due to decreased impedance by the crystallites rather than the restricted mobility of the chain segments. This is consistent with the results of Kosovova and Reitlinger[114] showing the increase in the permeability with increasing spherulite size above a certain size.

8.4.4 Transport of dissolved oxygen in membranes

The rate of *dissolved oxygen* transfer through membranes is an important aspect of biological systems and the actual oxygen transfer rate through various membranes including organic polymer film has been studied using a variety of techniques. However, the study of dissolved oxygen transfer rates through polymeric membranes in comparison to the gaseous oxygen permeability has not been pursued until relatively recent years. Yasuda and Stone[116] have approached this problem from the viewpoint that the essential step of oxygen transport through organic polymer film is dissolved oxygen (in the polymer) and the partial pressure of oxygen across a membrane is the driving force. Yasuda[117] has later expanded the consideration to all possible cases in which two different phases face each side of a membrane. It was shown that the gas permeability constant of a homogeneous polymer measured under the conventionally-used gas-phase experiment provides the ideal true membrane permeability of oxygen regardless of the nature of the liquid which faces the membrane if the liquids do not change the polymer properties.

The oxygen transfer rate through a membrane is dependent upon the partial pressure gradient of oxygen across the membrane but not on the difference of concentration of oxygen in liquids. Therefore, the use of a liquid which has high solubility does not affect the oxygen transfer rate in the membrane if the partial pressure of oxygen is the same.

Although the true (ideal) dissolved oxygen permeability can be given by the gas-phase experiment, the actual dissolved oxygen transfer-rate through a polymer membrane can be considerably less if the dissolved oxygen permeability is measured, e.g. the rate of increase of oxygen dissolved in water separated by a polymer membrane such as silicone rubber from oxygen-saturated water or gaseous oxygen is considerably smaller than the calculated value from the oxygen permeability of the polymer. This discrepancy is caused by the boundary layer resistance (or the concentration polarisation at the polymer–liquid interface).

Hwang, Fang and Kammermeyer[118] have recently studied the extent of boundary layer resistance using silicone rubber of different thickness and have shown that over 50% of the total resistance of dissolved oxygen transfer through a thin silicone rubber is due to the boundary resistance.

Aiba and Hwang[119] have recently investigated the effects of electrode response and the liquid-layer resistance with membrane-covered electrodes. The intimate contact of cathode surface to the membrane was found to be essential for an accurate measurement when a platinum cathode was used; however, with a silver cathode this factor was insignificant and the first-order response was always obtained even when the membrane was separated from the cathode surface by an electrolyte solution.

The magnitude of boundary layer resistance may depend on the definition of true membrane permeability, i.e. the true membrane permeability may be either the gas permeability of the polymer obtained by gas phase experiment, based on the thermodynamic consideration[117], or the empirical values obtained by extrapolating data with varying membrane thickness[118], or with varying stirring rate of the liquid which contacts membrane surface[119].

In any case, the contribution of the boundary layer resistance is small if the transfer rate in the polymer membrane is low. Therefore, dissolved oxygen transfer rates through either a polymer of low oxygen permeability or a thick membrane of relatively high oxygen permeability are nearly equal to the transmission rates which can be calculated from the characteristic oxygen permeability of polymers. However, the boundary layer resistance becomes more and more significant as the overall transport resistance of a membrane becomes smaller. The dissolved oxygen permeability of silicone rubber which has the highest gas permeabilities among existing polymers has been found to be considerably smaller than the corresponding gas permeability[117,118]. Consequently, the actual dissolved-oxygen transfer rate through a polymer membrane such as is the case in a blood oxygenator depends more on engineering design to minimise the boundary layer resistance than the absolute value of oxygen permeability of the polymer if the latter is high enough to be in the vicinity of that of silicone rubber.

The above discussions are limited to homogeneous polymer membranes which are not affected by the liquid, and special caution is needed for the proper understanding of the situation if porous membranes are used in place of homogeneous polymer membranes, since the gas permeability coefficient of porous membranes measured by gas-phase experiment merely represents the gas-effusion rate and does not correspond to the gas-permeability constants discussed above. Yasuda and Lamaze[120] have recently studied this subject and demonstrated the significance of liquid penetration into pores to overall dissolved oxygen permeability of porous membranes. It should be noticed that the dissolved oxygen permeabilities of many porous membranes are lower than those of homogeneous membranes of silicone rubber in spite of extremely high gas-effusion rates of those membranes. This may mislead one into expecting high dissolved oxygen permeability.

If a membrane has high affinity for the liquid, such as in the case of hydrophilic polymer membranes or water-swollen membranes, the dissolved-oxygen permeability (true membrane permeability) can be estimated only by the direct measurement of dissolved oxygen transport. In these cases, the diffusivity and particularly the solubility of oxygen in the solvent play the decisive role in the dissolved oxygen permeability of such membranes.

Imbibition of solvent increases the characteristic permeability of a polymer due to the plasticising effect of the solvent; however, the actual transport rate of dissolved gas may or may not be increased by the swelling depending on the characteristic permeability (diffusivity and solubility) of the solvent relative to that of dry polymer. For instance, if the liquid is water the dissolved oxygen permeability of water-swollen or hydrophilic membranes increases from that of dry polymer; however, the dissolved oxygen permeabilities of such membranes are generally much smaller than that of silicone rubber due to the very small solubility of oxygen in water and rather low oxygen-permeability of hydrophilic polymers in the dry state. This situation is somewhat analogous to porous membranes which allow the penetration of liquid water into the pores[120].

8.5 TRANSPORT OF ORGANIC PENETRANTS

8.5.1 Introduction

Fundamental studies of small organic molecule transport in polymers free from cracks, pinholes and other flaws have been actively continued during the past 5 years[121]. These investigations have included a wide variety of polymers and penetrants over a broad range of experimental conditions.

Many investigators were concerned with a specific transport feature which they implicitly suggested was characteristic of the given polymer–penetrant pair. These classes of behaviour include:

Concentration-independent Fickian diffusion[62]
Concentration-dependent Fickian diffusion[122–125]
Time-dependent diffusion anomalies[126–128]
Case II transport[20, 129, 130, 132, 133]
Solvent-crazing–stress-cracking[128, 131, 132, 134–137]

All the other modes of transport to be considered are diffusive whereby net transport is a consequence of a gradient of chemical potential of the diffusing species across the membrane. A third limiting case of transport involves polymer relaxation as the rate-determining step rather than viscous flow or pure Fickian diffusion. When the relaxations are entirely controlling the transport kinetics, so-called Case II transport is observed. The superposition of relaxation-controlled transport and either of the other limiting transport mechanisms leads to time-dependent anomalies.

The transport of small molecules in solid polymers is defined as Fickian when the transport satisfies one of Fick's equations for diffusion:

$$\frac{\partial c}{\partial t} = D\frac{\partial^2 c}{\partial x^2}$$

$$\frac{\partial c}{\partial t} = \frac{\partial}{\partial x}\left[D(c)\frac{\partial c}{\partial x}\right]$$

The following boundary conditions are conveniently met for Fickian sorption into a film:

$$c(x,0) = 0$$
$$c(0,t) = c(l,t) = C_0$$

where D is the diffusion coefficient, c the penetrant concentration, C_0 the equilibrium concentration, t and x the independent variables time and distance and l the film thickness. Due to the concentration-dependence of the diffusivities for most organic vapour–polymer systems, the second equation normally applies for the diffusion of organic penetrants above T_g [121]. Normal or Fickian sorption is characterised by the following features:

(a) A linear relationship exists between the initial weight gain of the sample undergoing sorption and square root of time.

(b) A smooth and continuous concentration profile exists through the film.

Below T_g the transport process often becomes anomalous. Frequently, the diffusivity not only depends on concentration but on time and position as well.

The transport features observed for many organic penetrants in glassy polymers, e.g. normal hydrocarbons in polystyrene, are qualitatively quite similar to those observed in other, rather diverse, polymer–penetrant systems[20, 133]. The similarities in qualitative behaviour suggest that the diverse behavioural features noted in glassy polymers such as polystyrene probably occur for most amorphous systems if a sufficient range of temperature and activity (traversing the glass-transition range) is encompassed by experimental conditions. The patterns of behaviour are quite diverse and complicated. The relationship between the various transport features are more easily assimilated, however, by examining the various regions of the temperature–activity plane presented in Figure 8.2.

The following features are readily apparent:

(a) Time-dependent anomalies and Case II sorption including solvent crazing are confined to relatively high penetrant activities and temperatures

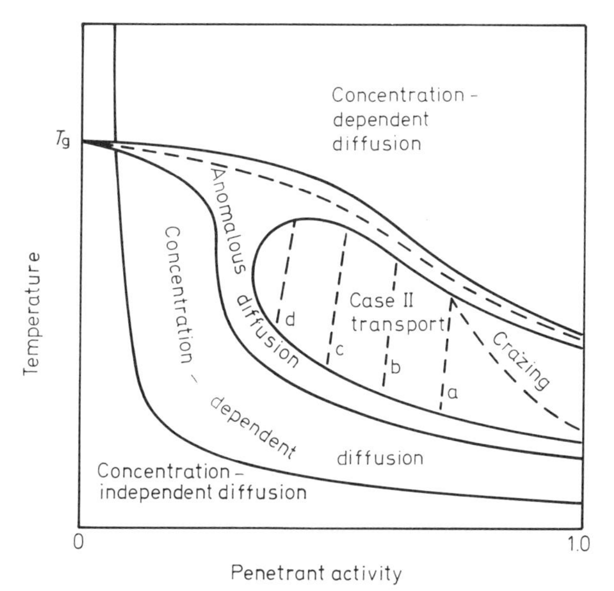

Figure 8.2 Transport behaviour in the various regions of the temperature–penetrant activity plane. Lines a, b, c, d are of constant activation energy. The effective glass - transition temperature is represented by the dashed line extending from T_g through the anomalous diffusion region.
(From Hopfenberg and Frisch[141], by courtesy of Interscience)

in the vicinity of and below the effective T_g of the system. The effective T_g of the system is represented by the dashed line extending through the anomalous diffusion region.

(b) The region of Case II sorption (relaxation-controlled transport) is separated from the Fickian diffusion region by a region where both mechanisms are operative giving rise to diffusional anomalies.

(c) The activation energy characterising Case II sorption decreases as penetrant activity is reduced.

(d) Concentration-independent diffusion is only apparent at very low temperatures and/or activities.

(e) Concentration-dependent diffusion occurs above the glass temperature with the more highly-sorbed organic penetrants or at high activities.

It should now be readily apparent that diverse transport behaviour occurs for a specific polymer–penetrant pair if the investigator is willing to extend his analysis to a broad range of temperature and penetrant activity. One must specify a limited temperature–activity regime when referring to the likelihood of observing a given transport feature. All the limiting and intermediate transport features can most likely be observed experimentally if a sufficiently broad temperature activity plane is explored. For some systems, insufficient penetrant levels may be achieved at equilibrium to generate an osmotic stress large enough to bias relaxations controlling Case II transport and/or solvent crazing. For these special systems, the crazing region and possibly the Case II transport regime would disappear from the temperature–activity plane. It is proposed to discuss specific features of the transport of small organic molecules in high polymers under the broad headings of Fickian diffusion and of anomalous, relaxation-controlled diffusion.

8.5.2 Transport in rubbery polymers

The sorption, diffusion and permeation of liquid and/or vapour-phase organic penetrants in and through rubbery synthetic high polymers has been focal to the overall topic of diffusion in polymers for the past 100 years. Recent work in this special area has involved the extension of already developed analytical and experimental techniques to new and diverse polymer–penetrant systems. In addition, this broad experimental effort has been supported by novel and intriguing theoretical and analytical considerations. These considerations inevitably complement the experimental program in that new measurement techniques and more explicit interpretation of data have been made available. This review separates the basically experimental contributions from those primarily concerned with theoretical aspects of the problem.

8.5.2.1 Experimental contributions

A great deal of data has been presented relating to organic penetrant transport in polyolefins. The work deals with single-component sorption and permeation as well as binary mixture separation by polyolefins. In the latter case, rather involved treatments of the polyolefin membrane including solvent, thermal, mechanical, and radiation pre-treatments have been studied in an attempt to improve the flux and/or separate the relative permeabilities of a permeant pair.

A number of studies[142–151] involved the measurement of the sorption kinetics and equilibria of a variety of organic penetrants in one or more polyolefins. Kochan *et al.*[142] studied the sorption of chlorinated hydrocarbons including chloroform, carbon tetrachloride and tetrachloroethylene in polyethylene films. Kinetic analysis of these sorption data revealed rather low activation energies for diffusion for these systems over the temperature range 20–60 °C. Averev *et al.*[143] conducted a related study involving benzene sorption in polypropylene fractions; the fractions were obtained by extraction of polypropylene by a variety of organic liquids at their boiling point. Sorption kinetics and equilibria were correlated with film density; the densities were controlled by the original extraction procedure.

The effect of structure formation, such as spherulitic development in polypropylene was studied by monitoring sorption behaviour over a wide range of temperatures.

Bogayevska, Gotovskya and Kargin[144] used this technique to study structure formation in polypropylene over the temperature range 100–230 °C which embraces a region above the melting point of the polymer.

The diffusion of substituted benzophenones, which are used commercially as ultraviolet absorbers in polymers, was studied in stereoblock polypropylene as well as high and low density polyethylene[145]. The studies were carried out in the temperature range 20–90 °C and activation energies for diffusion are reported. Jackson *et al.*[146] used a radio-tracer method based on ^{14}C for measuring the diffusion coefficient of an additive in a polymer. Rather large and complex penetrants (additives) were studied in polyethylene, polypropylene, and poly-4-methylpent-1-ene. They reported that although Arrhenius behaviour was observed in all cases studied, the activation energy for diffusion was a function of polymer structure alone and independent of the size and shape of diffusing molecule over the range of penetrant geometries encountered.

Laine and Osburn[147] reported a 'sorption technique' to monitor permeation whereby polyethylene film was formed into a pouch, filled with silica gel, sealed, and then suspended in a saturated atmosphere composed of one of 19 organic penetrants. The steady-state permeability was calculated from the linear weight gain with time which was observed.

Michaels, Vieth, Hoffman and Alcalay[148] studied the effect of solvent annealing of polypropylene on the resultant morphology and transport properties. Changes in spherulitic texture resulting from these high-temperature solvent treatments were correlated with the large increase in film permeability to aliphatic and aromatic hydrocarbons which accompanied decreased selectivity toward penetrants. Siegel and Coughlin[149] extended the concept of solvent-annealing of polyethylene by preceding the solvent annealing by high-energy irradiation. The authors suggest sufficient improvement in membrane properties to merit commercial interest in this membrane-treatment process. Twenty-five combinations of binary liquid mixtures permeating low density polyethylene were studied by Huang and Lin[150]. Several interesting conclusions relating to temperature and concentration effects were noted; permeation increases, resulting in binary permeation rates higher than either pure component comprising the permeating mixture, were observed.

The effect of temperature *per se* on the permeation of organic liquids through polyethylene was studied by Ghosh and Rawat[151]. They observed that although permeability increased exponentially with temperature, permselectivity decreased with increasing pressure at low downstream pressures but increased with temperature at high permeate pressures.

Transport in polyamide membranes was characterised by Shikusawa *et al.*[152] and Vasenin and Chernova[153]. Shikusawa studied a variety of nonionic penetrants diffusing from water into nylon 6 to model a dyeing process while Vasenin studied the sorption of a wide variety of organic vapours in polyamide as well as polyisobutylene and polypropylene membranes.

Garrett and Park[154] measured the diffusion coefficients of benzene in a

46% vinyl acetate–54% methyl acrylate copolymer and interpreted their data in the light of free volume theory. Their equilibrium data suggest that sorption in this system is characterised by a Flory–Huggins interaction parameter χ of 0.36, much closer to the value for polyvinyl acetate than for polymethylacrylate.

Munari *et al.*[155] grafted styrene to polytetrafluoroethylene by a pre-irradiation technique. The resultant membrane is cation selective with mechanical properties quite similar to the PTFE matrix.

The opposite of the often-observed phenomenon of increasing diffusion coefficient with increasing penetrant concentration was reported by Barrie[156] for the permeation and sorption of methanol in polydimethylsiloxane. The observed decrease in the diffusion coefficient with increasing concentration has been explained by penetrant clustering which becomes more predominant as concentration is increased in this system. Hopfenberg, Schneider and Votta[157] report somewhat similar behaviour for ethanol transport in a rubbery polyurethane. They suggest that the sigmoidal increase of diffusion coefficient with increasing concentration is related to a competition between penetrant clustering and the often-observed polymer plasticisation. The sorption of dioxane by polyurethanes with different degrees of cross-linking was studied by Lipatov *et al.*[158].

Huang and Jarvis[159] used the concepts of plasticisation and clustering to explain their data describing the separation of a variety of alcohol–water mixtures by polyvinyl alcohol membranes. They also observed that although permeation rate increased with increasing temperature the separation factor or selectivity, describing the relative transport rates of alcohol with respect to water, decreased with increasing temperature.

Clearly, the emphasis of the experimental programmes in this area under study has shifted from rubber-like polymers to the investigation of transport in the thermoplastic polymers already discussed. Rather detailed studies of organic penetrant transport in a series of rubbers was presented, however, by Chen and Ferry[78] as well as Alexopoulos *et al.*[160]. Chen and Ferry used a tracer technique for characterising diffusion and concomitant relaxation in natural rubber, styrene–butadiene rubber, polyisobutylene, polydimethylsiloxane, and twelve polybutadienes with varying *cis*-, *trans*- and vinyl components. Alexopolous *et al.*[160] studied two effects explicitly: the effect of *cis*-, *trans*-, isomerisation in polyisoprene on the permeation of n-butane and the permeation of propane through a variety a substituted alkyl and halogenated alkyl polysiloxanes. Sorption and diffusion of butane in the isoprenes was only slightly affected by *cis–trans* isomerisation. Conversely, the replacement of a methyl group by a bulky phenyl group in the polysiloxane series markedly depressed both the permeability and the diffusion coefficient of propane.

*8.5.2.2 Theoretical contributions**

Paul and Ebra-Lima[161] have combined classical thermodynamics, diffusion theory and careful experimentation with an elegant extension of reverse

*See also Section 8.2

osmosis behaviour to organic permeants. They studied the pressure-induced transport of 12 organic liquids through a highly swollen rubbery membrane and explained their results in terms of a concise, albeit general, model describing pressure-induced transport of liquids across polymeric membranes. Since these highly swollen membranes can yield very high fluxes at moderate pressure, these pressure-driven organic separations may be of commercial interest for performing certain separations.

The free volume theory for diffusion has been used as the basis of a theoretical model which interprets the effect of mixture composition on liquid separation by polymer membranes[162]. The model developed by Fels and Huang takes into account the effect of one component in the mixture on the diffusion of the other component. They present data which is in reasonable agreement with their model.

Rogers has continued his pioneering work in the area of concentration-dependent diffusion in collaboration with Sternberg[26, 163]. They consider the effect of membrane asymmetry on directional flow in membranes with a gradient of composition which exhibit concentration-dependent diffusion. Hypothetical membranes with varying structures and solubility properties have been considered; the resultant equations have been applied to predict the magnitude of penetrant fluxes as well as the differences in membrane fluxes effected by the directional or asymmetric nature of the barrier. Peterlin and Williams[164] have very recently extended the theory of vectorised membranes to a consideration of the effect of applied pressure.

Petropolous and Roussis[12, 15] have presented a quantitative treatment for the diffusion time-lag in amorphous polymer–penetrant systems exhibiting anomalous diffusion behaviour. Their treatment is based upon the molecular relaxation model, used for the interpretation of anomalous features of sorption kinetics. For the system methanol–polyvinyl acetate, a consistent interpretation of time-lag and sorption kinetics was achieved.

8.5.3 Anomalous diffusion and solvent crazing

8.5.3.1 General

In the Introduction it was emphasised that the mode of transport depended markedly on temperature, solvent activity and other system parameters. Below the glass transition temperature the transport process often becomes 'anomalous'. Frequently the diffusivity not only depends upon concentration, but upon time and position as well and is characterised by the following features:

(a) At temperatures well below T_g, a linear relationship exists between the initial weight gain of a polymeric film undergoing sorption and time. (In contrast, Fickian sorption leads to a linear relationship between the initial weight gain and the square root of time.)

(b) A sharp boundary separates an inner glassy core of essentially zero-penetrant concentration from an outer swollen, rubbery shell of uniform concentration (not a sufficient criterion for non-Fickian diffusion since sharp advancing boundaries have been observed for Fickian diffusion with a strongly concentration-dependent diffusivity).

(c) The boundary advances at a constant velocity.

This model, which Alfrey developed for a simple limiting case of anomalous diffusion, fits several of the anomalous features of the transport process below T_g observed by King and others. King[138] reported a linear relationship between the initial weight gain and time in his study of the sorption of alcohol vapours in wool and keratin. His explanation for the linear advance was a concentration-dependent diffusion coefficient which led to a build-up of a steep front which then moved through the medium. Other authors have reported anomalous behaviour in their studies of organic vapour transport in glassy polymeric systems, but have offered no simple explanation for the behaviour. Crank[139] did develop a model to describe the behaviour with a time-dependent diffusion coefficient and/or a non-constant boundary condition, $C_0(t)$; but in each case he had to include several additional parameters to satisfactorily describe the behaviour.

Since Case II transport is a relaxation-controlled transport process, the parameters affecting relaxation, such as polymer orientation, molecular weight, molecular-weight distribution, temperature, penetrant activity, and penetrant physicochemical properties, appear to be the important parameters to study in an investigation of the transport of organic molecules in glassy polymers. The effects of temperature, penetrant activity and penetrant structure on the sorption kinetics and equilibria of hydrocarbons in bi-axially oriented and cast-annealed polystyrene have been presented by Hopfenberg, Holley and Stannett[132,178]. The effects of molecular weight and subtle orientation effects were presented by Bray and Baird *et al.*[140,179]. A survey and generalisation of the various transport features observed for alkanes in polystyrene was presented by Hopfenberg and Frisch[141].

The effects of polmer molecular weight, molecular weight distribution and orientation on the rate of relaxation-controlled sorption of n-pentane by glassy polystyrene were studied by Baird *et al.*[179]. The sorption of n-pentane follows Case II kinetics but for films which sorb slowly the sorption rate increases at relatively long times until sorption is sharply terminated. This rate increase may be explained by the development of dispersed micro-voids within the unrelated film core. Overshoot of the equilibrium n-pentane content occurs in sorption experiments in which accelerated sorption is pronounced.

The sorption rate is independent of polymer molecular weight and molecular weight distribution *per se* over a broad range of these parameters. Essentially identical vapour sorption kinetics were observed for well-annealed polystyrene films of different molecular weights and distributions. Conversely, for vapour sorption by uniaxially oriented films and for liquid sorption by partially annealed films, high molecular weight film (1 880 000) exhibits greater sorption rates than low molecular weight film (c. 200 000). These differences in rate are not due to molecular weight differences *per se*, but are a consequence of the dissimilar response of free volume and strain development for films of different molecular weight prepared with a given time–temperature–strain history.

Crazing of carefully annealed polystyrene films occurs during desorption of n-pentane from partially saturated films. The depth of craze penetration reflects the point of advance of the discontinuous Case II sorption boundary.

The stresses exerted on the polymer may be a combination of external applied stress, osmotic swelling stresses, or internal residual pre-orientation stresses.

Organic polymeric glasses display a variety of responses to stress. Three distinct forms of behaviour may be identified in response to a tensile stress: (a) an apparent brittle response which leads to failure after an elongation of a few per cent, (b) a cold drawing which involves extensive deformation of the sample, and (c) crazing. This last mode refers to the formation of localised zones at stresses below the fracture stress identified as containing both voids and polymer orientated in the stress direction and may be common to all crack-initiation processes[165]. Like fracture itself, the development of crazes is a time-dependent phenomenon and is accelerated by the presence of a plasticising environment.

Environmental stress cracking has been defined by Holley[166] as 'the cracking or crazing of a polymer specimen under stress (external and/or internal) in the presence of a second component without which the polymer would not fail under the given stress.' Solvent- and environmental stress-cracking are differentiated by their modes of initiation. Initiation of solvent stress-crazing involves a liquid which is at least a partial solvent at elevated temperatures and is absorbed, perhaps with mild swelling, at ordinary temperatures. Crazing will occur when the combined internal and external stresses exerted on the polymer result in a local strain which is greater than the material can withstand.

Thus crazing may be induced under a variety of test conditions. Kambour[167] has observed crazing preceding fracture in eight glassy polymers in air. Most studies on environmental stress-cracking have used a glassy polymer subjected to an external stress while in the presence of an organic environment. Rudd[135] examined polystyrene in butanol under a tensile stress of 1000 p.s.i. Bernier and Kambour[134] studied stress crazing and cracking of polyphenylene oxide bars under strains from 0.1% to 1.5% exposed to environments of 28 different organic vapours. They also studied crazing of equilibrated films. As described previously, Michaels, Bixler and Hopfenberg[131] have observed solvent crazing where the swelling stresses were of such a magnitude that crazing resulted without application of an external stress. Mercury-cast annealed polystyrene films supposedly free of internal stress were also shown to craze in organic liquids[131].

8.5.3.2 The nature of the craze

Kambour has focused his research on the nature or morphology of microformations in polymers and has discussed a distinction between what he terms crazes compared with larger scale cracks. Like normal cracks, crazes reflect light and are planar. In contrast to earlier concepts, however, crazes are understood today not to be true cracks but rather thin plate-like regions containing polymer material which is interconnected with the normal polymer surrounding the craze[168]. No holes or discontinuities of an optical size are observed in the craze.

Craze formation takes place by a plastic deformation in the stress direction[169]. The degree of deformation residing in any craze depends on its history. Freshly formed but unstressed crazes have been observed to undergo a 60 % plastic elongation[170]. The elongation process appears to occur without macroscopic reduction in cross-section. This results in creation of a substantial void content. Kambour[171] concludes that the void content of the craze is large but relatively invariant from one polymer to another and independent of crazing environment.

Dry crazes have been known to sustain surprisingly high stresses for long times. The strength of the craze is relatively low when the crazing agent is present, but upon drying the craze strength can increase markedly and in the case of polycarbonate, at least, can sometimes exceed the normal yield stress of the polymer[165].

X-ray scattering and electron microscopy studies of craze material have confirmed that the void content appears to consist of an open-celled foam, the holes and polymer elements of which appear to be approximately 200 Å in diameter for PPO crazes grown in ethanol. Michaels, Bixler and Hopfenberg[131] observed a hole size of about 3 micrometers (3000 Å) in diameter in polystyrene crazed in n-heptane in the absence of isothermal stress. The holes in the craze are interconnected on the evidence of the ease of exchange of liquids in the craze[168]. The internal specific surface area of a freshly formed craze has been roughly calculated to be 100–200 m^2/cm^3 of craze as determined by x-ray scattering[165].

Kambour has observed two kinds of craze structure[172]: an unfibrillated kind in the peripheral region and an orientated fibrillar kind constituting the interior portions of the craze. He suggests that craze development might sometimes take place in two steps: first a nucleation and growth of spheroidal voids followed by an elongation of elements in the stress direction which begins to occur only when the crazing region is somewhat removed from the craze–polymer interface. He also shows a craze which split during formation where the irregular tearing process resembles 'that kind easily observed on the macroscopic scale with soft tissue paper'. Kambour suggests that substantial amounts of plastic deformation occur locally during the final act of craze breakdown.

8.5.3.3 Crack propagation

During studies of crack propagation, Berry[173] drew attention to the interference colours exhibited by fracture surfaces in polymethyl methacrylate and suggested that they arose from a highly orientated thin layer polymer material. The orientation process was suggested to occur at the crack tip and the orientated region then broken to produce a thin layer on each surface (Figure 8.2). Recently[169], Kambour has measured the refractive index of the surface layer and found it identical with that of the craze matter. He concluded that the surface layer was quantitatively similar to the craze, and that crack propagation involves the formation and then breaking of a craze. He observed this phenomenon in crack propagation at all temperatures up to T_g. When a crack tip is examined with a microscope, a configuration of the crack tip and the unbroken craze preceding it may be determined. The craze is initiated far ahead of the crack tip; the crack itself has a parabolic configuration. Kambour found that the craze extended 25 μm beyond the crack tip for PMMA and 550 μm beyond for polystyrene under the cracking conditions he studied.

8.5.3.4 Theories on environmental craze formation

The exact role of organic agents in the enhancement of the mechanism of environmental stress-crazing and cracking is not known. There do exist, however, a number of theories which attempt to rationalise the crazing phenomenon in the light of processes known or suspected to take place in a stressed glassy polymer in contact with a liquid.

Nielsen[174] discussed the concept of stress crazing in the light of his 'domain' theory. A 'domain' consists of imperfections or inhomogeneities in a polymer sample which are revealed by crazing. Robertsen[175] has suggested that glassy polymers consist of randomly orientated regions inside which an imperfect alignment of chains exists, that the average length of such a 'domain' is of the order of 100 Å and that the elemental step in craze growth may be the opening up of a void in a 'domain'.

The role of an organic crazing agent is further complicated by the differential expansion of the sample due to swelling. Rosen[176] explains that in order to reach swelling or solubility equilibrium, an organic glass must frequently undergo a large expansion to accommodate the vapour or liquid. Since this pronounced swelling is at first restrained by the glassy film, the rate of solvent sorbed is often largely controlled by the stress relaxation of this macromolecular diffusion medium rather than being controlled by simple diffusion. As the film begins to absorb vapour or liquid at its surface, the films initial resistance to the localised osmotic swelling creates severe compressions acting along the planes of the surface. This initial swelling occurs perpendicular to the surface with an increase in the thickness of the film. This condition is obtained even at the expense of introducing long-range macromolecular orientation near the surfaces during this, partly irrecoverable, tensile strain. Thus the vapour or liquid sorption must initially be retarded in awaiting these viscous adjustments. As the film absorbs more vapour, it undergoes a belated, osmotically-forced area extension which aggravates the biaxial tensions acting in the plane of the film at the central core. This internal tension can be severe enough to cause internal fracture or crazing. During this abrupt area expansion, the film absorbs solvent at a very high rate and a tensile-strain-enhanced diffusion takes place. After the glass reaches swelling equilibrium, an abrupt removal of vapour or liquid from the atmosphere initiates an overtly rapid desorption of solvent from the film. This initial vapour desorption creates a severe tension at the films surfaces. Even biaxial, tensile stress crazing can result at the surface from this severe condition[176].

One theory that attempts to rationalise solvent crazing holds that the organic agent acts as a plasticiser. Originally it was suspected that solvent crazing occurred when the combination of stress and plasticiser lowered the polymer T_g to ambient temperature. Because of our present knowledge of the structure and mechanical properties of crazes and because of current concepts about flow in the glassy state under stress, a modern casting of this hypothesis would be that limited plasticisation lowers T_g to a degree, and application of sufficient stress promotes a liquid-like flow of the glass in the stress direction. Lowering T_g depends directly on plasticiser concentration. Bernier and Kambour[134] attempted to measure the lowered T_g of a PPO film containing several organic agents and compare this with the critical strain necessary to induce crazing. Solvents with a high equilibrium volume solubility acted as cracking agents rather than crazing agents. They concluded that most small organic molecules act as plasticisers in reducing resistance to craze formation in PPO.

In the same articles, Bernier and Kambour found that even liquids of negligible solubility in the polymer increased the crazing tendency. They suggested that these liquids act by reducing the surface energy of the holes in the craze making the craze formation easier. This theory was first proposed by analogy with certain effects in inorganic systems and now exists in the form stated by Bernier and Kambour. Hopfenberg *et al.*[140,178] hold that solvent crazing or environmental stress-cracking is a simple extension of relaxation-controlled, or Case II, transport. If the summation of stresses applied at the boundary between swollen and unswollen polymer is sufficient to cause polymer failure over and above the rate-determining relaxation, then crazing accompanies the Case II sorption process.

They noted a sharp boundary between the crazed outer surfaces and the unaffected central core advancing with a constant velocity through a film. A study of the temperature-dependence of this process revealed an apparent activation energy of approximately 60 kcal mol^{-1}, which is in the range of activation energies for stress relaxation of polystyrene and well above that for a normal diffusion process. More importantly, this activation energy is sufficiently high to reflect primary bond breakage and free-radical formation as a consequence of the advancing penetrant front. They concluded that the rate-controlling step of this transport process is the osmotically-induced relaxation of the polymer at the boundary between the crazed outer layers and the unaffected central core. Diffusion to the boundary is rapid and does not affect the observed transport kinetics. If the stresses exerted are large enough to bias a relaxation but not polymer failure then transport of penetrant is limited by polymeric relaxations and the sample remains craze free.

In summary, it should be re-emphasised that because of the complex nature of stress crazing, drastic changes in behaviour may be observed depending upon subtle changes in (a) test conditions, e.g. rate of strain, type of strain, temperature, activity of liquids and vapours present, and physicochemical properties of penetrants (b) sample history, e.g. pre-orientation, annealing procedures and (c) secondary structural variables, e.g. molecular weight, molecular weight distribution, branching, cross-linking, and crystallinity[178,179].

8.5.4 Drawing and its influence on diffusion

Orientation effects on transport properties in polymeric materials have been known for a considerable span of time. Perhaps textile chemists and colourists were the first to recognise that dye diffusion was substantially altered by the degree of orientation of the synthetic fibre being dyed. In the early days of synthetic fibres, when draw ratio and extrusion conditions were not held rigorously constant, dyed fabrics were often streaked as a result of the non-uniformity in draw ratio during fibre spinning. An even more quantitative observation made by the earlier dye chemists was the lack of dyeability of certain fibres that had been cold drawn to impart higher tensile strength by orientating the chains in the amorphous regions. They noted that dyeing times or dye concentrations had to be increased in order to obtain the same shade as obtained with the unorientated fibre, both which are indicative of a lower rate of diffusion in the highly drawn fibre.

These earlier observations have in recent years been examined more carefully by several authors. For instance, Davies and Taylor[180] found that diffusion constants decreased considerably and the activation energies increased greatly with increasing draw ratio for the sorption of dyes in nylon 6,6. Takagi and Hittori[181] in experiments with nylon 6 fibres, found there was an increase in the values of the diffusion constants followed by a considerable decrease, with increasing draw ratios, accompanied with a corresponding decrease and increase in the activation energy. In a subsequent paper, Takagi[182] described work with dye diffusion parallel and perpendicular to the fibre axis. It was shown that in the experiments perpendicular to the axis results similar to those previously obtained were found. In the parallel direction, however, the diffusion constants decreased, and the activation energies increased steadily, with increasing draw ratio.

Although dye diffusion is drastically altered by uniaxial orientation owing to the large molecules involved, studies of simple molecules have also shown marked reductions in transport properties accompanying the drawing process. Bixler and Michaels[183], for example, in their study using small organic molecules found a substantial decrease in the permeation rate at higher draw ratios (500 %) in the case of stretched polyolefins. Later, Michaels *et al.*[184] found similar trends with this system using atmospheric gases.

The full extent of the reduction in solvent sorption was first dramatically demonstrated by wide line n.m.r. measurements by Peterlin and Olf[185] with the sorption of tetrachloroethylene in linear polyethylene.

In further studies and using very well-defined drawing conditions Peterlin *et al.*[186] investigated the influence of draw ratio on the transport properties of organic vapours in polyethylene. These workers found that the solubility decreased by tenfold at high draw ratios (900 %) accompanied by an even more drastic reduction in diffusion rate – indeed, more than 150 times lower than the undrawn material. They also noted that considerable relaxation, however, took place at high penetrant activities and/or at high temperatures. These workers concluded that cold drawing reduces the sorption sites without changing their energy content, but drastically cuts down diffusion and increases the activation energy. A smaller part of the increase of the latter is a consequence of the lower enthalpy of the amorphous material and a larger part is probably due to the increased distance between sorption sites. In extensions of this study, Williams and Peterlin[187] examined more closely the exact influence of draw ratio on the transport properties of methylene chloride in polyethylene. For draw ratios (τ) between $\tau = 6$ and 25 the diffusion constant dropped drastically between $\tau = 8$ and 9 and then remained nearly constant, dropping only slightly up to $\tau = 25$. Also, the exponential dependence of diffusion constants on concentration of sorbent increased abruptly in the same draw interval and then remained constant at the higher draw ratios.

Similar effects were also found by Williams and Peterlin[64] in the case of water diffusion into polyethylene. They explained their results[186] based on a composite model: a low-permeability fibre structure embedded in a high-permeability spherulitic matrix. As the draw ratio is increased, the initial spherulitic film is gradually transformed into the fibre structure with the transformation being completed in this case between $\tau = 8$ and 9. During subsequent drawing to $\tau = 25$ the mutual arrangement of microfibrils, the basic elements of the fibre structure, changes by longitudinal sliding. However, their transport properties remain nearly constant. The diffusion constant drops a little as a consequence of the increased fraction of tie molecules which reduces the number of unperturbed sorption sites.

Although cold drawing reduces the diffusion rates as a result of the orientation of tie molecules in the amorphous regions, this type of behaviour is not necessarily general for uniaxial deformation. For example, Williams and Peterlin[188] have shown that drawing in the presence of a solvent can result in large increases in the diffusion and permeability rates in contrast to the behaviour observed for cold drawing. This was shown clearly, however, to be due to the development of a porous structure in the polymer.

Most of the work to date on cold drawing implies that orientation of chains in the amorphous regions results in a substantial decrease in the overall sorption and diffusion. This alignment, however, increases the tensile strength so substantially that little or no consideration is given to detrimental aspects of slow rates of diffusion as found in dyeing. In fact, these slow rates of diffusion can themselves offer many advantageous properties to the substrate, e.g. stain resistance.

8.5.5 Diffusion of dyestuffs

Approaches to the diffusion of dyestuffs are becoming more consistent with the ideas and developments in the physical sciences. The limitations of simple theoretical models are becoming more generally recognised. The rise in importance of the synthetic fibres is leading to a greater recognition of the utility of the existing theories and models for the diffusion of organic penetrants in polymers.

An analysis of mass transfer in dyeing systems, on the basis of non-equilibrium thermodynamics, emphasises the complexity of dyeing systems and the need for independent thermodynamic data[189]. The transport equations of non-equilibrium thermodynamics reduce to the simple form of Fick's laws only under especially simple circumstances. In general, it is the gradients of the electrochemical potentials which appear as the 'driving forces' in the diffusion equations[189]. The transport equations alone do not suffice for the description or solution of a problem. They must be supplemented by appropriate initial and boundary conditions. Here we enter a region of greater uncertainty. A dimensional analysis of the transport equations emphasises the importance of the conventional dimensionless groups of mass transport theory[190]. The conditions at the boundary between a solid polymer and a flowing dye solution can be related to the numerical values of certain dimensionless groups[190, 191], which determine the extent to which the sorption kinetics conforms to the simple model of sorption controlled by the diffusion of dye into the solid from a constant surface concentration of dye[191].

Time-dependent surface concentrations of dye have been observed during dyeing[192], but it is not clear whether convective diffusion effects are involved or whether time-dependent sorption or dye aggregation processes in the solid are responsible. Since dyeing processes may not conform to the simple model of diffusion-controlled sorption, all diffusion coefficients determined from integral sorption rate curves are inherently suspect.

Microspectrophotometric and rolled-film studies of dye distributions in solid polymers during dyeing confirm earlier observations that some dyeing systems do conform, within experimental error, to the simple sorption model. Provided that the solid has been stabilised in the dyebath medium before dyeing, the diffusion of non-ionic dyes and some organic penetrants in polymer films conforms to Fick's equations, with a constant diffusion coefficient[193, 194, 195].

In contrast, the diffusion of ionic dyes often shows a pronounced dependence on dye concentration and electrolyte concentration[194–203]. This concentration-dependence makes it difficult to use the more general experimental diffusion equations which take account of interactions between the different diffusion flows. Nevertheless, the diffusion of dye mixtures has been studied in this way and the cross-diffusion coefficients are stated to be far from negligible[204, 205].

The origins of the concentration-dependence of the dye diffusion coefficient are a matter for controversy. Since there is no independent thermodynamic data, the use of the simple activity gradient and chemical potential gradient models is without a firm physical basis, even though these models give a surprisingly useful account of experimental data[194–203]. An explanation based on the Nernst–Planck equations appears more attractive as a simple first approximation[195, 196, 206], but it must be recognised that activity coefficient variations may still play a role. A resolution of this difficulty must await more information on the thermodynamic properties of these systems, and a more extensive application of isotopic-labelling and other techniques for the determination of the 'self-diffusion' coefficients in these systems[207, 208].

So much for the problems associated with the transport equations and boundary conditions. A significant new development is the growing interest in the relationships between fibre structure and dye diffusion. Two apparently opposing viewpoints are emerging[209–215].

On the one hand, the observed correlations between dye diffusion and certain mechanical or dynamic mechanical properties of fibres have stimulated interest in the so-called 'free volume' theories of diffusion[1]. These theories regard the accessible regions of fibres as homogeneously fluid-like in character, albeit of high 'viscosity'. No permanent pore structure is envisaged, except insofar as the crystalline or ordered regions form boundaries for the accessible regions, and the dye diffusion is regarded as determined by the polymer segment mobility. This two-phase model has recently been abandoned by Prevorsek and the existence of a third phase of intermediate order in certain fibres has been postulated[216]. Of course, all fibres have complex structures, and all the useful models are grossly over-simplified.

An opposite approach has been advocated which extends earlier theories in which the dye diffusion is believed to take place in solvent-filled pores in the fibre, with simultaneous sorption on the pore walls[217–220]. This model is attractive in its simplicity and leads to calculated 'true' diffusion coefficents for the 'free' dye in the pores of the fibre, which, for aqueous systems, do not differ by more than one or two orders of magnitude from normal aqueous diffusion coefficients[217–221].

However, the corrections for porosity and tortuosity in this model do not appear to conform to the requirements of mathematical analyses of diffusion in porous media[222] and seem to be essentially empirical. The assumption of internal sorption equilibrium cannot be universally justified and, especially in ionic systems, it is by no means obvious that the external, macroscopic sorption isotherm for the system will give a correct picture of any local, internal sorptive interactions during diffusion in the fibre. The claim that this model has general validity, and that it leads to 'true' diffusion coefficients does not seem, to the reviewers at least, to be entirely tenable. However, this model is a useful contribution to the mathematics of diffusion in systems with non-linear sorption isotherms[217] and may correspond quite well to the real situation in some highly-swollen fibre systems. The structural and other effects which may influence dye diffusion in polymers have also been reviewed in recent years[223–226].

8.6 TRANSPORT OF WATER VAPOUR

Water presents a number of special difficulties in measuring the diffusion constants in polymers both with the time-lag method and from sorption studies. This is due to its high hydrogen-bonding capacity and cohesive energy leading to adsorption on the surfaces of the measuring equipment and to excessive heat evolution on sorption. These difficulties have been thoroughly discussed by Yasuda and Stannett[53, 227] and by Barrie and Machin[228] and methods of overcoming these experimental problems presented.

Water transport in high polymers follows four types of behaviour: (a) Henry's solution law is obeyed and the diffusion constants are independent of concentration, (b) the solubility coefficients increase with increasing vapour pressure but the diffusion constants are independent of the concentration, (c) the solubility coefficients increase with vapour pressure and the diffusion constants increase with concentration, and (d) the solubility coefficient increases with vapour pressure and the measured diffusion constants decrease with concentration. This last behaviour appears to be confined to water and methanol and has been generally attributed to the clustering of the penetrant in the polymer. Curiously, the diffusion constants measured by the time-lag technique are often quite constant and no satisfactory explanation of this has so far been presented. Barrie and Machin[228, 229] have shown that the time-lag is less sensitive, however, to concentration effects. Petropoulos and Roussis[12, 15] have presented an interesting explanation based on the concept that the polymeric relaxation times are comparable with that of diffusion in these systems. They showed that this could lead to the proportion of clustered molecules increasing with time which would lead, in turn, to the measured time-lag being smaller than the true time-lag. (See also Section 8.2.)

Much of the research concerned with the diffusion of water in high polymers during recent years has been concerned with the clustering phenomenon. Extensions of the original Zimm and Lundberg[230] theory have been presented by Lundberg[231] and by Orofino *et al.*[232]. Clustering has been invoked to explain the diffusion behaviour of water in a number of different polymers including the polyurethanes[233–235], the silicone rubbers[236], polyvinyl chloride[81, 237] and cellulose nitrate, and a number of copolymers of hydroxyethyl methacrylate[235]. Vieth *et al.*[235] have extended the concept of clustering to reverse osmosis studies and found good agreement with results obtained with sorption measurements. It was shown in the cases of polyoxymethylene[237] and cellulose acetate[18, 237] that the time-lag diffusion constant was independent of the concentration but that the values obtained from the quotient of the permeability constants and the equilibrium solubility coefficients measured by sorption were identical to the time-lag values. The sorption isotherms of both polymers were, however, of the type where the solubility coefficients increase with increasing vapour pressure. Apparently little or no clustering was associated with isotherms and this was reflected in the lower degree of curvature compared with polymers where extensive clustering was found to occur by parallel diffusion measurements. In an interesting paper by Rosenbaum and Cotton[238], the steady-state distribution of water in cellulose acetate was determined using a multilayer stack of films and the results reported at a number of pressure gradients. A linear concentration gradient through the film was shown, again indicating the ideal nature and lack of clustering of water in this polymer. The results obtained with these and other polymers were also correlated with the Flory–Huggins interaction parameter[237]. The more formal relationships between the Zimm–Lundberg and Flory–Huggins treatments of sorption were discussed by Orofino *et al.*[232].

Two other papers[239–240] reported data on the transport of water in polyoxymethylene and its copolymers. The solubility of water in this polymer was found to increase with increasing temperature. Data was also presented on the effects of the thermal history of the film on the solubilities and diffusivities of water. Both were found to increase steadily with decreasing density of the polymer. In a paper by Prosser[241], a method for determining simultaneously the amount of adsorbed and absorbed water was described and used with polyethylene terephthalete films. Less than 1.5 % of the sorbed water was found to be on the surfaces of the film. The mechanism of water absorption in a series of different polyamides has been reported by Puffr *et al.*[242, 243]. Rate and equilibrium studies lead to a sorption mechanism in which the water is bound by hydrogen bonding to the amide groups together with more loosely bound water.

There has been comparatively little activity in the field of water diffusion in polymers which show increasing diffusivities with concentration. There has been great interest, however, in reverse-osmosis studies with polymers of this general class. In addition, there has been considerable research into water transport in highly swollen hydrophilic polymers. Both these areas were motivated by interest in polymeric membranes for water purification and in various biological and biomedical applications.

Two interesting papers were concerned with the diffusion and solution of water in polyvinyl alcohol. Spencer and Ibrahim[244] measured the rates of sorption into single-layer films and also the concentration gradient driving a steady-state permeability experiment by using a stack of single films and analysing each individually. The method used was similar to that reported previously by Gillespie and Williams[245]. Excellent agreement between these measurements and those calculated for the analytical expression of Long[246] using the single-film sorption measurements was found.

Takizawa *et al.*[247] studied the effect of crystallinity on water-vapour sorption into polyvinyl alcohol. The small initial Langmuir-type sorption was found to be absent from the sorption isotherms of the crystalline material. The isotherms themselves were analysed in terms of both B. E. T. and Flory–Huggins theories.

Sorption studies of water in modified and cross-linked cellulose films were reported by Newns[248]. The results were interpreted in terms of the coupled diffusion and relaxation mechanisms often discussed with diffusion into glassy polymers. Water flow through swollen cellulose films was discussed by Volgin *et al.*[249, 250] and found to be largely capillary flow. A study of water transport in polypropylene glycol monoacrylate hydrogel showed a mixture of diffusion and viscous flow mechanisms depending on the water content and thermal history of the hydrogel[251].

Acknowledgements

We would like to thank Professor R. McGregor of the School of Textiles, North Carolina State University for his help with the section on dyestuff diffusion and Dr. H. Yasuda and Dr. J. L. Williams of the Camille Dreyfus Laboratory for helpful suggestions and comments.

References

1. Crank, J. and Park, G. S. (1968). *Diffusion in Polymers* (N. Y. Academic Press)
2. Fujita, H. (1961). *Fortschr, Hochpolym. Forsch.*, **3,** 1
3. Odani, H., Kida, S. and Tanuira, M. (1966). *Bull. Chem. Soc. Japan*, **39,** 2378
4. Odani, H. (1967). *J. Appl. Polymer. Sci.*, A2, **5,** 1189
5. Kishimoto, A. and Matsumoto, K. (1964). *J. Polymer Sci. A*, **2,** 679
6. Crank, J. and Park, G. S., Chap. 1 in Ref. 1
7. Frisch, H. L. (1962). *J. Chem. Phys.*, **37,** 2408
8. Kishimoto, A. and Kitahara, T. (1967). *J. Polymer Sci., A1*, **5,** 2147
9. Long, F. A. and Richman, D. (1960). *J. Amer. Chem. Soc.*, **82,** 513
10. Petropoulos, J. H. and Roussis, P. P. (1967). *J. Chem. Phys.* **47,** 1491
11. Rogers, C. E. (1965). *Physics and Chemistry of the Organic Solid State*, Vol. 2, D. Fox, M. M. Labes and A. Weissberger, Eds) (N. Y. John Wiley),
12. Petropoulos, J. H. and Roussis, P. P. (1969). *J. Polymer Sci. C*, **22,** 917

13. Kishimoto, A. (1964). *J. Polymer Sci. A*, **2**, 1421
14. Crank, J. (1953). *J. Polymer Sci.*, **11**, 151
15. Petropoulos, J. H., and Roussis, P. P. (1969). in *Organic Solid State Chemistry*, ch. 9 (B. Adler Editor) (London: Gordon and Breach); *Mol. Cryst. Liq. Cryst.*, **9**, 343
16. Stannett, V. and Williams, J. L. (1965). *J. Polymer Sci. C*, **10**, 45
17. Wellons, J. D. and Stannett, V. (1966). *J. Polymer Sci. A1*, **4**, 593
18. Wellons, J. D., Williams, J. L. and Stannett, V. (1967). *J. Polymer Sci. A1*, **5**, 1341
19. Peterlin, A. (1965). *J. Polymer Sci. B*, **3**, 1083; *idem. Makromol. Chem.*, **124**, 136 (1969)
20. Frisch, H. L., Wang, T. T. and Kwei, T. K. (1969). *J. Polymer Sci. A2*, **7**, 879; *idem.*, ibid., **7**, 2019
21. Odiam, G. and Kruse, R. L. (1969). *J. Polymer Sci. C*, **22**, 691
22. Petropoulos, J. H. (1970). *Kinetic Analysis of Simple Heterogeneous Reactions with particular reference to Reactive Dyeing Systems*, Paper read at the "First Perkin Discussion", Ashbridge, England
23. Armstrong, A. A. and Stannett, V. (1966). *Makromol. Chem.*, **90**, 145; Armstrong, A. A., Wellons, J. D. and Stannett, V. (1966). ibid., **95**, 78
24. Rogers, C. E. (1965). *J. Polymer Sci. C*, **10**, 93; Stannett, V., Williams, J. L., Gosnell, A. B. and Gervasi, J. A. (1968). *J. Polymer. Sci. B*, **6**, 185; Rogers, C. E., Sternberg, S. and Salovey, R. (1968). *J. Polymer Sci. A1* **(6)** 1409
25. Hartley, G. S. (1948). *Disc. Faraday Soc.*, **3**, 233; Rogers, C. E., Stannett, V. and Sewarc, M. (1957). *Ind. Eng. Chem.*, **49**, 1933
26. Sternberg, S. and Rogers, C. E. (1968). *J. Appl. Polymer Sci.*, **12**, 1017
27. Frisch, H. L. (1964). *J. Polymer Sci. A*, **2**, 1115
28. Petropoulos, J. H. and Roussis, P. P. (1967). *J. Chem. Phys.*, **47**, 1496
29. Petropoulos, J. H. and Roussis, P. P. (1969). *J. Chem. Phys.*, **50**, 3951
30. Ash, R. and Barrer, R. M. (1971). *J. Phys. D*, **4**, 888
31. Barrer, R. M., Chap. 6 in Ref. 2.
32. Ash, R., Barrer, R. M. and Palmer, D. G. (1969). *Trans. Faraday Soc.*, **65**, 121
33. Petropoulos, J. H. and Roussis, P. P. (1969). *J. Chem. Phys.*, **51**, 1332
34. Frisch, H. L. and Prager, S. (1971). *J. Chem. Phys.*, **54**, 1451
35. Ash, R., Baker, R. W. and Barrer, R. M. (1967). *Proc. Roy. Soc. A*, **299**, 434
36. Petropoulos, J. H. and Roussis, P. P. (1970). *J. Polymer Sci. A2*, **8**, 1411
37. Paul, D. R. (1969). *J. Polymer Sci. A1*, **7**, 2031
38. Roussis, P. P. (1970). *Ph.D. Thesis* (Athens University)
39. Fels, M. and Haung, R. Y. (1970). *J. Appl. Polymer Sci.*, **14**, 523
40. Duda, J. L. and Vrentas, J. S. (1971). *Amer. Inst. Chem. Eng. J.*, **17**, 466
41. Barrie, J. A. and Machin, D. (1967). *J. Polymer Sci. A2*, **5**, 1300; *idem., Trans. Faraday Soc.* (in press)
42. Crank, J. (1956). *Mathematics of Diffusion* (Oxford University Press)
43. Vieth, W. R. and Sladek, K. J. (1965). *J. Colloid Sci.*, **20**, 1014
44. Weisz, P. B. (1967). *Trans. Faraday Soc.*, **63**, 1801
45. Paul, D. R. (1969). *J. Polymer Sci. A2*, **7**, 1811
46. Petropoulos, J. H. (1970). *J. Polymer Sci. A2*, **8**, 1797
47. Pilchowski, K., Danes, F. and Wolf, F. (1969). *Kolloid Z. Z. Polymer*, **230**, 328; *idem., Ber. Bunsenges. Phys. Chem.*, **73**, 99 (1969)
48. Paul, D. R. and DiBenedetto, A. T. (1965). *J. Polymer Sci. C*, **10**, 17
49. Jenkins, C. L., Nelson, P. M. and Spirer, L. (1970). *Trans. Faraday Soc.*, **66**, 1391
50. Kubin, M. and Spacek, P. (1965). *Collect. Czech. Chem. Commun.*, **30**, 3294
51. Spacek, P. and Kubin, M. (1967). *J. Polymer Sci. C*, **16**, 705
52. Kubin, M. and Spacek, P. (1967). *Collect. Czech. Chem. Commun.*, **32**, 2733
53. Yasuda, H. and Stannett, V. (1969). *J. Macromol. Sci. B*, **3**, 589
54. Stern and Britton, *J. Polymer Sci. A2* (in press)
55. Siegel, R. D. and Coughlin, R. W. (1970). *J. Polymer Sci. A2*, **14**, 3145
56. Pasternak, R. A., Schimscheimer, J. F. and Heller, J. (1970). *J. Polymer Sci. A2*, **8**, 467
57. Evnochides, S. K. and Henley, E. J. (1970). *J. Polymer Sci. A2*, **8**, 1987
58. Paul, D. R. (1971). *Ind. Eng. Chem. Process Des. Develop*, **10**, 375
59. Stewart, C. R., Lubinski, A. and Blenkarn, K. A. (1961). *J. Petroleum Technol.* **13**, 383
60. Barakat, H. Z. and Clark, J. A. (1966). *J. Heat Transfer*, **88**, 421
61. Rick, R. F., McAvoy, T. J. and Chappeleon, D. C. (1968). *J. Polymer Sci. A2*, **6**, 1863
62. Barrer, R. M. (1948). *J. Polymer. Sci.*, **3**, 549

63. Van Amerongen, G. J. (1946). *J. Appl. Physics,* **17,** 972
64. Williams, J. L. and Peterlin, A. (1968). *Makromol. Chem.,* **122,** 215
65. Pasternak, R. A. and McNulty, J. A. (1970). *Mod. Packaging*
66. Ziegel, K. D., Frensdorff, H. K. and Blair, D. E. (1969). *J. Polymer Sci., A2,* **7,** 809
67. Riemschneider, R. and Riedel, E. (1968). *Z. Naturforsch.,* **1,** 116
68. Caskey, T. L. (1967). *Modern Plastics,* **4,** 148
69. Yasuda, H. and Rosengren, K. (1970). *J. Appl. Polymer Sci.,* **11,** 2839
70. Riemschneider, R. and Riedel, E. (1966). *Kunstoffe,* **56,** 355
71. Sewall, P. A. and Skirrow, G. (1970). *Polymer,* **1,** 2
72. Riemschneider, R. and Riedel, E. (1969). *Kunstoffe,* **59,** 169
73. Rust, G. and Herreo, F. (1969). *Material Prufung,* **11,** 166
74. Eichhorn, R. M. (1970). *Polymer Eng. Sci.,* **10,** 32
75. Jones, P. F. (1968). *J. Polymer Sci. B, Polymer Lett.,* **6,** 487
76. Reference 1, pages 30–33
77. Mozisek, M. (1970). *Polymer Eng. Sci.,* **6,** 383
78. Chen, S. P. and Ferry, J. D. (1968). *Macromolecules,* **1,** 270
79. Lebovits, A. (1966). *Modern Plastics,* **43,** 139
80. Barrer, R. M., Mallinder, R. and Wong, P. S. L. (1967). *Polymer,* **8,** 321
81. Tikhomirov, B. P., Williams, J. L., Hopfenberg, H. B. and Stannett, V. (1968). *Makromol. Chem.,* **118,** 177
82. Nakagawa, T., Hopfenberg, H. B. and Stannett, V. (1971). *J. Appl. Polymer Sci.,* **15,** 231
83. Stern, S. A., Mullhaupt, J. T. and Garris, P. J. (1969). *Amer. Inst. Chem. Eng. J.,* **15,** 64
84. Stern, S. A., Fang, S. M. and Jobbins, R. M. (1971). *J. Macromolec. Sci.,* **B5,** 41
85. Li, N. N. and Long, R. B. (1969). *Amer. Inst. Chem. Eng. J.,* **15,** 73
86. Li, N. N. (1969). *Ind. Eng. Chem. (Product Res. and Develop.)* **8,** 281
87. Li, N. N. and Henley, E. J. (1964). *Amer. Inst. Chem. Eng. I.,* **10,** 666
88. Casper, V. G. and Henley, E. J. (1966). *J. Polymer Sci.,* **B4,** 417
89. Ash, R., Barrer, R. M. and Palmer, D. G. (1970). *Polymer,* **11,** 421
90. Shterenzon, A. L., Reitlinger, S. A. and Toping, L. P. (1969). *Vysokomol. Soyed.,* **11** A, 887
91. Kanitz, P. J. F. and Huang, R. Y. M. (1970). *J. Appl. Polymer Sci.,* **14,** 2739
92. Reference 1, page 62
93. Pasternak, R. A., Christensen, M. V. and Heller, J. (1970). *Macromolecules,* **3,** 366
94. Pasternack, R. A., Burns, G. L. and Heller, J. (1971). *Macromolecules,* **4,** 470
95. Gillinskaya, N. S., Reitlinger, S. A., Galil-Ogly, F. A. and Novikov, A. S. (1969). *Vysokomol. Soed.,* **B11,** 215
96. Huang, R. Y. M. and Kanitz, P. J. F. (1971). *J. Macromol. Sci.,* **B5,** 71
97. Ziegel, K. D., Frensdorff and Blair, D. E. (1969). *J. Polymer Sci.,* **A2.7,** 809
98. Frisch, H. L. and Rogers, C. E. (1964). *J. Chem. Phys.,* **40,** 2293
99. Huang, R. Y. M. and Kanitz, P. J. F. (1969). *J. Appl. Polymer Sci.,* **13,** 669
100. Williams, J. L. and Stannett, V. (1970). *J. Appl. Polymer Sci.,* **14,** 1949
101. Huang, R. Y. M. and Kanitz, P. J. F. (1971). *J. Appl. Polymer Sci.,* **15,** 67
102. Ziegel, K. D. (1971). *J. Macromol. Sci.,* **B5,** 11
103. Hughes, E. C., Idol, J. D., Duke, J. T. and Wick, L. M. (1969). *J. Appl. Polymer Sci.,* **13,** 2567
104. Barrer, R. M., Barrie, J. A. and Wong, P. S. L. (1968). *Polymer,* **9,** 609
105. Burgess, W. H., Hopfenberg, H. B. and Stannett, V. (1971). *J. Macromol. Sci.,* **B5,** 23
106. Lundberg, J. L. and Rogers, C. E. (1969). *J. Polymer Sci.* **A2.7,** 947
107. Durrill, P. L. and Griskey, R. G. (1966)' *Amer. Inst. Chem. Eng. J.,* **12,** 1147
108. Durrill, P. L. and Griskey, R. G. (1969). *Amer. Inst. Chem. Eng. J.,z***15,** 106
109. Morrison, M. E. (1967). *J. Appl. Polymer Sci.,* **11,** 2588
110. Slabaugh, W. H. and Kennedy, G. H. (1967). *J. Appl. Polymer Sci.,* **11,** 179
111. Barrer, R. M. (1968). Reference 1, Chapter 6
112. Savin, A. G., Shaposhnikova, T. K., Karpov, V. I., Sogolova, T. I. and Kargiu, V. A. (1968). *Vysokomol. Soyed.,* **A10**(7)1584
113. Savin, A. G., Karpov, V. I., Shaposhnikova, T. K. and Sogolova, T. I. (1967). *Vysokomol. Soyed.,* **B9**(7)496
114. Kosovova, Z. P. and Reitlinger, S. A. (1967). *Vysokomól. Soyed.,* **A9**(2)415
115. Vieth, W. R. and Wuerth, W. F. (1969). *J. Appl. Polymer Sci.,* **13,** 684
116. Yasuda,, H. and Stone, W. (1966). *J. Polymer Sci.,* **A1,** 4, 1314

117. Yasuda, H. (1967). *J. Polymer Sci.*, **A1**,5, 2952
118. Hwang, S. T., Fang, T. E. S. and Kammermeyer, K. (1971). *J. Macromol. Sci.*, **B5**, 1
119. Aiba, S. and Hwang, S. Y. (1969). *Chem. Eng. Sci.*, **24**, 1149
120. Yasuda, H. and Lamaze, C. E., *J. Appl. Polymer Sci.* (in press)
121. Reference 1, Chapters 3 and 5.
122. Aitken, A. and Barrer, R. M. (1955). *Trans. Faraday Soc.*, **51**, 116
123. Prager, S. and Long, F. A. (1951). *J. Amer. Chem. Soc.*, **73**, 4072
124. Kokes, R. J. and Long, F. A. (1953). *J. Amer. Chem. Soc.*, **75**, 6142
125. Fujita, H., Kishimoto, A., and Matsumoto, K. (1960). *Trans. Faraday Soc.*, **56**, 424
126. Park, G. S. (1951). *Trans. Faraday Soc.*, **48**, 11
127. Mandelkern, L. and Long, F. A. (1951). *J. Polymer Sci.*, **6**, 457
128. Hayes, M. J. and Park, G. S. (1955). *Trans. Faraday Soc.*, **51**, 1134
129. Alfrey, T., Gurnee, E. F. and Lloyd, W. O. (1966). *J. Polymer Sci.*, **C12**, 249
130. Alfrey, T. (1965). *Chem. Eng. News*, **43**, No. 41, 64
131. Michaels, A. S., Bixler, H. J. and Hopfenberg, H. B. (1968). *J. Appl. Polymer Sci.*, **12**, 991
132. Hopfenberg, H. B., Holley, R. H. and Stannett, V. (1969). *Polymer Eng. Sci.*, **9**, 242
133. Kwei, T. W. and Zupko, H. M. (1969). *J. Polymer Sci.*,**A2 7**, 867
134. Bernier, G. A. and Kambour, R. P. (1968). *Macromolecules*, **1**, 393
135. Rudd, J. F. (1963). *J. Polymer Sci.*, **B1**, 1
136. Ceresa, R. J. (1962). *Block and Graft Copolymers*, **97**, (London: Butterworths)
137. Hopfenberg, H. B. (1964). *Ph.D. Thesis*, M.I.T., Dept. of Chemical Eng., Cambridge, Mass.
138. King, G. (1945). *Trans. Faraday Soc.*, **41**, 325
139. Crank, J. (1953). *J. Polymer Sci.*, **11**, 151
140. Bray, J. and Hopfenberg, H. B. (1969). *J. Polymer Sci.*, **B7**, 679
141. Hopfenberg, H. B. and Frisch, H. L. (1969). *J. Polymer Sci.*, **B7**, 405
142. Kochan, A. A., Chernyavskii, G. V. and Shrubovich, V. A. (1967). *Vysokomol. Soyed.*, **9**, B, 40
143. Averev, M. P., Kostina, T. F. and Klimenkov, V. S. (1967). *Vysokomol. Soyed.*, **9**, B, 333
144. Bogayevska, T. A., Gotovskya, T. V. and Kargin, V. A. (1968). *Vysokomol. Soyed.*, **10**, A, 1357
145. Cicchetti, O., Dubini, M., Parrini, P., Vilario, G. P. and Bua, E. (1968). *Europ. Polymer J.*, **4**, 419
146. Jackson, R. A., Oldland, S. R. D. and Pajackowdki, A. (1968). *J. Appl. Polymer Sci.*, **12**, 1297
147. Laine, R. and Osburn, J. O. (1971). *J. Appl. Polymer Sci.*, **15**, 327
148. Michaels, A. S., Vieth, W. R., Hoffman, A. S. and Alcalay, H. F. (1969). *J. Appl. Polymer Sci.*, **13**, 577–598
149. Siegel, R. O. and Coughlin, R. W. (1971). *J. Appl. Polymer Sci.*, **14**, 2431
150. Huang, R. Y. M. and Lin, V. J. C. (1968). *J. Appl. Polymer Sci.*, **12**, 2615
151. Ghosh, S. K. and Rawat, B. S. (1966). *Indian J. Technol.*, **5**, 101
152. Shikusawa, T. and Iijima, T. (1970). *J. Appl. Polymer Sci.*, **14**, 1553
153. Vasenin, R. M. and Chernova, I. V. (1966). *Vysokomol. Soyed.*, **8**, 2006
154. Garrett, T. A. and Park, G. S. (1967). *J. Polymer Sci.*, **C16**, 601
155. Munari, S., Vigo, F., Telado, G. and Rossi, C. (1967). *J. Appl. Polymer Sci.*, **11**, 1563
156. Barrie, J. A. (1966). *J. Polymer Sci. A*, **4**, 3081–3088
157. Hopfenberg, H. B., Schneider, N. S. and Votta, F. (1969). *J. Macromol. Sci. B*, **4**, 751
158. Lipatov, Y. S., Sergeeva, L. M. and Kovalenko, G. F. (1968). *Vysokomol. Soyed.*, **10**, *B*, **3**, 205
159. Huang, R. Y. M. and Jarvis, N. R. (1970). *J. Appl. Polymer Sci.*, **14**, 2341
160. Alexopolous, J. B., Barrie, J. A., Tye, J. C. and Fredrickson, M., (1968). *Polymer*, **9**, 56
161. Paul, D. R. and Ebra-Lima, O. M. (1970). *J. Appl. Polymer Sci.*, **14**, 2201
162. Fels, M. and Huang, R. Y. M. (1971). *J. Macromol. Sci.*, **B5**, 89
163. Rogers, C. E. and Sternberg, S. (1971). *J. Macromol. Sci.*, **B5**, 189
164. Peterlin, A. and Williams, J. L. (1971). *J. Appl. Polymer Sci.*, **15**, 1493
165. Kambour, R. P. (1968). *Appl. Polymer Symposia*, **7**, 215
166. Holley, R. H. (1969). *Ph.D. Thesis*, North Carolina State University
167. Kambour, R. P. (1966). *J. Polymer Sci. A2*, **4**(1)17

168. Kambour, R. P. (1968). *Polymer Eng. Sci.*, **8**(4)281
169. Kambour, R. P. (1969). *J. Polymer Sci. A2*, **7**, 1393
170. Kambour, R. P. (1964). *Polymer*, **5**, 107
171. Kambour, R. P. (1964). *J. Polymer Sci. A*, **2**, 4159
172. Kambour, R. P. (1968). *J. Polymer Sci. A2*, **7**, 1393
173. Berry, J. P. (1961). *J. Polymer Sci.*, **50**, 107, 313
174. Nielsen, L. E. (1959). *J. Polymer Sci.*, **1**(24)27
175. Robertsen, R. E. (1965). *J. Phys. Chem.*, **69**, 1575
176. Rosen, B. J. (1961). *Polymer Sci.*, **49**, 177
177. Newman, S. (1968). *Appl. Polymer Symp.*, **7**, 161
178. Hopfenberg, H. B., Holley, R. H. and Stannett, V. (1970). *Polymer Eng. Sci.*, **10**, 376
179. Baird, B. R., Hopfenberg, H. B. and Stannett, V. (1971). *Polymer Eng. Sci.*, **11**, 274
180. Davies, G. and Taylor, H. (1965). *Text. Res. J.*, **35**, 405
181. Takagi, Y. and Huttori, H. (1965). *J. Appl. Polymer Sci.*, **9**, 2167
182. Takagi, Y. (1965). *J. Appl. Polymer Sci.*, **9**, 3887
183. Bixler, H. and Michaels, A. S. (1964). Paper presented at 53rd Natl. Meeting AIChE, Pittsburg, Pa.
184. Michaels, A. S., Vieth, W. R. and Bixler, H. (1964). *J. Appl. Polymer Sci.*, **8**, 2735
185. Peterlin, A. and Olf, H. G. (1966). *J. Polymer Sci., A2*, **4**, 587
186. Peterlin, A., Williams, J. L. and Stannett, V. (1967). *J. Polymer Sci., A2*, **5**, 957
187. Williams, J. L. and Peterlin, A. (1971). *J. Polymer Sci., A2*, **9**, 1483
188. Williams, J. L. and Peterlin, A. (1970). *Makromol. Chem.*, **135**, 41
189. Miliceire, B. and McGregor, R. (1966). *Helv. Chim. Acta*, **49**, 1302, 1319
190. Miliceire, B. and McGregor, R. (1966). *Helv. Chim. Acta*, **49**, 2098
191. Peters, R. H., McGregor, R. and Varol, K. (1970). *J. Soc. Dyers. Col.*, **86**, 437, 442
192. Blacker, J. G. and Patterson, D. (1969). *J. Soc. Dyers Col.*, **85**, 598
193. Peters, R. H., McGregor, R. and Romachandoan, C. R. (1968). *J. Soc. Dyers Col.*, **84**, 19
194. Hossain, T. M. A., Iijima, T., Morita, Z. and Maeda, H. (1969). *J. Appl. Polymer Sci.*, **13**, 541
195. Sand, H. (1967). *Kolloid, Z. Z. Polymere*, **218** (2), 124; also *Ber. Bunsenges, Physik. Chem.*, **69**, 333 (1965)
196. Hopper, M. E., McGregor, R. and Peters, R. H. (1970). *J. Soc. Dyers Col.*, **86**, 117
197. Brody, H. (1965). *Text. Res. J.*, **35**, 844
198. Shimizu, T., Ohya, S., Ito, K. and Kumeda, H. (1967). *Sen-i-Yakkaishi*, **23**, 602
199. Morita, Z., Iijima, T., Yoshida, T. and Sekido, M. (1967). *Kogyo Kagaku Zasshi*, **70**, 180
200. Sekido, M., Iijima, T. and Takahashi, A. (1965). *Kogyo Kagaku Zasshi*, **68**, 524
201. Bell, J. P., Carter, W. C. and Felty, D. C. (1967). *Text. Res. J.*, **37**, 512
202. Morizane, H., Suda, Y. and Shirota, T. (1971). *Sen-i-Yakkaishi*, **27**, 113
203. Takazawa, H., Katayama, A. and Knroki, N. (1971). *Sen-i-Yakkaishi*, **27**, 96
204. Sekido, M. and Morita, A. (1963). *Bull. Chem. Soc. Jap.*, **36**, 1601
205. Morita, Z. (1965). *Bull. Tokyo Inst. Tech.*, **67**, 541.65
206. Doremus, R. H. (1966). *Polymer Lett.* **4**, 755
207. Chantrey, G. and Rattee, I. D. (1969). *J. Soc. Dyers Col.*, **85**, 618
208. Medley, J. A. (1965). *Proc. 3rd Int. Wool Textile Res. Conf.*, 1
209. Rosenbaum, S. (1965). *J. Polymer Sci. A*, **3**, 1949
210. Sprague, B. S. (1967). *J. Polymer Sci. C*, **20**, 159
211. Bell, J. P. (1968). *J. Appl. Polymer Sci.*, **12**, 627
212. Bell, J. P. and Murayama, T. (1968). *J. Appl. Polymer Sci.*, **12**, 1795
213. Dumbleton, J. H., Bell, J. P. and Murayama, T. (1968). *J. Appl. Polymer Sci.*, **12**, 2491
214. Fukuda, T. and Omori, S. (1971). *Sen-i-Yakkaishi*, **27**, 83
215. Kawai, S., Kano, F., Igarashi, Y. and Nakayasu, H. (1971). *Sen-i-Yakkaishi*, **27**, 138
216. Prevorsek, D. C. (1971). Paper delivered at A.A.T.C.C. Golden Jubilee Conference, Boston, Mass.
217. Weisz, P. B. (1967). *Trans. Faraday Soc.*, **63**, 1801
218. Weisz, P. B. and Zollinger, H. (1967). *Trans. Faraday Soc.*, **63**, 1807
219. Weisz, P. B. and Zollinger, H. (1967). *Trans. Faraday Soc.*, **63**, 1815
220. Weisz, P. B. and Zollinger, H. (1967). *Melliand. Textilber.*, 70
221. Afinogenova, L. V., Blinicheva, I. B. and Moryganov, P. V. (1970). *Tech. Textile Industry USSR*, No. 5, 82
222. Barrer, R. M., Reference 1, Chapter 6

223. Peters, R. H. and McGregor, R. (1968). *J. Soc. Dyers Col.,* **84,** 267
224. Holme, I. (1969). *Review of Progress in Coloration and Related Topics,* **1,** 31
225. Jones, F. (1969). *Reviews of Progress in Coloration and Related Topics,* **1,** 15
226. Marshall, J. (1971). *Review of Textile Progress* 196607. p. 311, Textile Institute & Society of Dyers & Colorists
227. Yasuda, H. and Stannett, V. (1962). *J. Polymer Sci.* **57,** 907
228. Barrie, J. A. and Machin, D. (1967). *J. Polymer Sci. A* **2 5,** 1300
229. Barrie, J. A. and Machin, D. (1968). *J. Appl. Polymer Sci.* **12,** 2633
230. Zimm, B. H. and Lundberg, J. L. (1956). *J. Phys. Chem.,* **60,** 425
231. Lundberg, J. L. (1969). *J. Macromol. Sci. B*3 **(4),** 693
232. Orofino, T. A., Hopfenberg, H. B. and Stannett, V. (1969). *J. Macromol. Sci. B*3 **(4),** 777
233. Schneider, N. S., Dusablon, L. V., Spano, L. A., Hopfenberg, H. B. and Votta, F. (1968). *J. Appl. Polymer Sci.,* **12,** 527
234. Schneider, N. S., Dusablon, L. V., Snell, W. E. and Prosser, R. A. (1969). *J. Macromol. Sci., B3* **(4),** 623
235. Vieth, W. R., Douglas, A. S. and Block, R. (1969). *J. Macromol. Sci., B*3 **(4),** 737
236. Barrie, J. A. and Machin, D. (1969). *J. Macromol. Sci., B*3 **(4),** 645
237. Williams, J. L., Hopfenberg, H. B. and Stannett, V. (1969). *J. Macromol. Sci., B*3 **(4),** 711
238. Rosenbaum, S. and Cotton, D. (1969). *J. Polymer Sci. A1,* **(7),** 101
239. Hardy, G. F. (1967). *J. Polymer Sci., A2* **(5)** 671
240. Braden, M. (1968). *J. Polymer Sci., A1* **(6),** 1227
241. Prosser, R. A. (1970). *J. Appl. Polymer Sci.,* **14,** 989
242. Puffr, R. and Sebenda, J. (1967). *J. Polymer Sci.,* **C16,** 79
243. Puffr, R. (1968). *Kolloid Z. Z. Polymere,* **222,** 130
244. Spencer, H. G. and Ibrahim, I. M. (1968). *J. Polymer Sci., A2* **(6),** 2067
245. Gillespie, T. and Williams, B. M. (1966). *J. Polymer Sci., A1* **(4),** 933
246. Long, R. B. (1965). *Ind. Eng. Chem. (Fundamentals)* **4,** 445
247. Takizawa, A., Negishi, T. and Minoura, Y. (1968). *J. Polymer Sci., A1* **(6),** 475
248. Newns, A. C. (1968). *Kolloid Z. Z. Polymere,* **218,** 355
249. Volgin, V. D., Dytnerskit and Planovskii, A. (1968). *Kolloid Zhurnal,* **30(3),** 342
250. Volgin, V. D., Dytnerskit and Planovskii, A. (1969). *Zh. Prik. Khim.,* **42,** 449
251. Refojo, M. F. (1967). *J. Appl. Polymer Sci.,* **11,** 407